Globalization and Sustainability - Ecological, Social and Cultural Perspectives

Edited by Levente Hufnagel

Published in London, United Kingdom

Globalization and Sustainability - Ecological, Social and Cultural Perspectives
http://dx.doi.org/10.5772/intechopen.1001808
Edited by Levente Hufnagel

Contributors
Adrielle B.O. Sardinha, Aidi Ahmi, Akungah O'Nyangeri, Angel Ernesto Jimenez Bernardino, Basak Ozoral, Carlos A. Leonetti, Damir Purković, Emad F. Saleh, Epaminondas J. Messias, Ewa Jadwiga Lipińska, Faiza El Mejjad, Ferenc Mics, Ganchimeg Gombodorj, Gaudencio Lucas Bravo, Hammad Zahid Muharram, John Habwe, José Daniel Padilla De la Rosa, Khalid Aada, Khululekani Ntakana, Károly Pető, Laoucine Amzil, Levente Hufnagel, Lorelei Hanson, Luis Fernando Sánchez Supelano, Lya Sañudo-Guerra, M.J. van Hulten, Magdy M. Mostafa, Malaika Hassan, Maria Reyes Hernández-Castilla, Michael McAllum, Nikolai Mihailov, Noer Fauzi Rachman, Noha Essam Khamis, Octavio Luis Pineda, Retno Hanggarani Ninin, Rodolfo Martinez Gutierrez, Sikunder Ali, Sonia Martins Felix, Teresa J. Kennedy, Vesselina Valkanova, Wafa Al-Maamari, Yamina El Kirat El Allame, Zaja Omboga

First published in London, United Kingdom, 2024 by IntechOpen
IntechOpen is the global imprint of INTECHOPEN LIMITED, registered in England and Wales, registration number: 11086078, 167-169 Great Portland Street, London, W1W 5PF, United Kingdom

British Library Cataloguing-in-Publication Data
A catalogue record for this book is available from the British Library

Additional hard and PDF copies can be obtained from orders@intechopen.com

Globalization and Sustainability - Ecological, Social and Cultural Perspectives
Edited by Levente Hufnagel
p. cm.

This title is part of the Sustainable Development Book Series, Volume 24
Topic: Sustainable Economy and Fair Society
Series Editor: Usha Iyer-Raniga
Topic Editor: Antonella Petrillo

Print ISBN 978-0-85466-698-0
Online ISBN 978-0-85466-697-3
eBook (PDF) ISBN 978-0-85466-699-7
ISSN 2753-6580

For EU product safety concerns:
IN TECH d.o.o., Prolaz Marije Krucifikse Kozulić 3, 51000 Rijeka, Croatia,
info@intechopen.com or visit our website at intechopen.com.

IntechOpen Book Series

Sustainable Development

Volume 24

Aims and Scope of the Series

Transforming our World: the 2030 Agenda for Sustainable Development endorsed by United Nations and 193 Member States, came into effect on Jan 1, 2016, to guide decision making and actions to the year 2030 and beyond. Central to this Agenda are 17 Goals, 169 associated targets and over 230 indicators that are reviewed annually. The vision envisaged in the implementation of the SDGs is centered on the five Ps: People, Planet, Prosperity, Peace and Partnership. This call for renewed focused efforts ensure we have a safe and healthy planet for current and future generations.

This Series focuses on covering research and applied research involving the five Ps through the following topics:

1. Sustainable Economy and Fair Society that relates to SDG 1 on No Poverty, SDG 2 on Zero Hunger, SDG 8 on Decent Work and Economic Growth, SDG 10 on Reduced Inequalities, SDG 12 on Responsible Consumption and Production, and SDG 17 Partnership for the Goals

2. Health and Wellbeing focusing on SDG 3 on Good Health and Wellbeing and SDG 6 on Clean Water and Sanitation

3. Inclusivity and Social Equality involving SDG 4 on Quality Education, SDG 5 on Gender Equality, and SDG 16 on Peace, Justice and Strong Institutions

4. Climate Change and Environmental Sustainability comprising SDG 13 on Climate Action, SDG 14 on Life Below Water, and SDG 15 on Life on Land

5. Urban Planning and Environmental Management embracing SDG 7 on Affordable Clean Energy, SDG 9 on Industry, Innovation and Infrastructure, and SDG 11 on Sustainable Cities and Communities.

The series also seeks to support the use of cross cutting SDGs, as many of the goals listed above, targets and indicators are all interconnected to impact our lives and the decisions we make on a daily basis, making them impossible to tie to a single topic.

Meet the Series Editor

Usha Iyer-Raniga is a professor in the School of Property and Construction Management at RMIT University. Usha co-leads the One Planet Network's Sustainable Buildings and Construction Programme (SBC), a United Nations 10 Year Framework of Programmes on Sustainable Consumption and Production (UN 10FYP SCP) aligned with Sustainable Development Goal 12. The work also directly impacts SDG 11 on Sustainable Cities and Communities. She completed her undergraduate degree as an architect before obtaining her Masters degree from Canada and her Doctorate in Australia. Usha has been a keynote speaker as well as an invited speaker at national and international conferences, seminars and workshops. Her teaching experience includes teaching in Asian countries. She has advised Austrade, APEC, national, state and local governments. She serves as a reviewer and a member of the scientific committee for national and international refereed journals and refereed conferences. She is on the editorial board for refereed journals and has worked on Special Issues. Usha has served and continues to serve on the Boards of several not-for-profit organisations and she has also served as panel judge for a number of awards including the Premiers Sustainability Award in Victoria and the International Green Gown Awards. Usha has published over 100 publications, including research and consulting reports. Her publications cover a wide range of scientific and technical research publications that include edited books, book chapters, refereed journals, refereed conference papers and reports for local, state and federal government clients. She has also produced podcasts for various organisations and participated in media interviews. She has received state, national and international funding worth over USD $25 million. Usha has been awarded the Quarterly Franklin Membership by London Journals Press (UK). Her biography has been included in the Marquis Who's Who in the World® 2018, 2016 (33rd Edition), along with approximately 55,000 of the most accomplished men and women from around the world, including luminaries as U.N. Secretary-General Ban Ki-moon. In 2017, Usha was awarded the Marquis Who's Who Lifetime Achiever Award.

Meet the Volume Editor

Dr. Levente Hufnagel is a full professor and the head of the Research Institute of Multidisciplinary Ecotheology of John Wesley Theological College (Budapest, Hungary) working on ecology, biogeography, ecological research methodology, ecotheology, and sustainability with over 20 years of experience at leading Hungarian academic institutions teaching Ph.D., MSc, and BSc students from various social and cultural backgrounds, having more than 260 scientific publications (in both aquatic and terrestrial ecological aspects of plants, animals, and microbes, at the community as well as population level) and more than 1,500 independent citations. He has been a participant in several large ecological research and development projects, having significant experience in multidisciplinary cooperations, with more than 200 coauthors in different publications. As the supervisor of several Ph.D., BSc, and MSc theses, and as editor-in-chief of an international scientific journal indexed by Web of Science and Scopus, he has wide supervising and editing experience. Dr. Hufnagel graduated from Eötvös Lorand University with a master's degree in ecology and evolutionary biology and a Ph.D. in hydrobiology; he also graduated from Szent István University with a Ph.D. in agricultural science and with other degrees from Corvinus University of Budapest and Adventist Theological College.

Contents

Preface

The 1987 report of the United Nations World Commission on Environment and Development entitled "Our Common Future" defines sustainable development as "meeting the needs of the present without compromising the ability of future generations to meet their own needs". A sustainable society is a targeted, but not yet realized, global form of mankind that ensures long-term (even for centuries or millennia) the survival of the human race, the long-term preservation of environmental and social living conditions, the protection of the human environment (climate, atmosphere, water, soil, wildlife), the functioning of the biosphere and the protection of the biodiversity, the functioning of economy, the reduction of social tensions (unequal opportunities, famine, deep poverty, crime, riots, terrorism, aggression, wars), scientific and technological progress and the rise of humanity, and the preservation and development of our natural and cultural heritage. There are both macro-level (legislation, political will, consensus, social support) and micro-level (affecting the daily functioning of individuals, families, businesses, and small communities) conditions for creating a sustainable society.

Humankind is an important but only a small part of the biosphere, the human race is just one of the many millions of species living on our planet. The existence of our species and the functionality and sustainability of its society, economy, and culture depend on the ecosystem services provided by the rest of the biosphere. The ecological processes that determine the living conditions of humanity (climate system, ocean currents, and biogeochemical cycles) are fundamentally global, just as everyday economic processes are also global today (just think of the manufacturers of objects in our homes or the producers of our food), the science that brings us new knowledge now operates on a global scale, as are commerce, tourism, as well as the Internet, which conveys everyday information.

The global ecological crisis is basically due to the overpopulation of the human race, the overexpansion of artificial land use types (e.g., residential, industrial and transport zones, and agricultural areas), and, consequently, the shrinking of natural habitats. However, only these natural ecosystems are capable of regulating the climate of our planet, maintaining its biodiversity, and providing ecosystem services for our society (e.g., breathable air, clean water, and regulated climate).

The crisis complex has hit humanity when it is divided into about 195 independent states without common global legislation, common executive power, common justice, common reliable information provision, and effective global coordination of its activities, that is, meaningful cooperation.

In the chapters of this book, we can find a lot of interesting information about these questions.

Our book titled *Globalization and Sustainability – Ecological, Social and Cultural Perspectives* gives a useful report about this important applied scientific area.

Levente Hufnagel
Research Institute of Multidisciplinary Ecotheology,
John Wesley Theological College,
Budapest, Hungary

Chapter 1

Introductory Chapter: Globalization – Phenomenon, Process, and Challenge – With the Chance of a Sustainable Society

Levente Hufnagel and Ferenc Mics

1. Introduction

The ecological processes that determine the living conditions of humanity (climate system, ocean currents, and biogeochemical cycles) are fundamentally global, just as everyday economic processes are also global today (just think of the manufacturers of objects in our homes or the producers of our food), the science that brings us new knowledge now operates on a global scale, commerce, and tourism, as well as the Internet, which conveys everyday information. Many large multinational companies and NGOs also integrate their activities on a global level. Today's ecological crisis, which involves mass species extinction and threatens the survival of modern human civilization, can only be interpreted on a global scale. The global ecological crisis is basically due to the overpopulation of the human race, the over-expansion of artificial land use types (e.g., residential, industrial and transport zones, and agricultural areas), and, consequently, the shrinking of natural habitats. However, only these natural ecosystems are capable of regulating the climate of our planet, maintaining its biodiversity, and providing ecosystem services for our society (e.g., breathable air, clean water, and regulated climate). Overpopulation, of course, also causes social problems, such as deteriorating public health conditions, epidemics, increasing aggression, supply disruptions, shock waves of refugees, increasing environmental pollution, wasteful energy and water management, excessive use of chemicals, dangerous industries. Today, we do not have an effective common strategy for dealing with the global ecological crisis; the set of principles and objectives aimed at dealing with the problem can be marked with the summary slogan "sustainable society."

The crisis complex hit humanity when it was divided into about 195 independent states; without common global legislation, common executive power, common justice, common reliable information provision, and effective global coordination of its activities, Azaz meaningful cooperation.

It is clear that decisions related to the coordination of our activities should be made at the level where the common good and public interest can be most effectively represented against private interests or narrow group interests. Today, the democratic political power of a small nation-state is clearly illusory compared to the economic power of a multinational corporation with a global network. The situation is further worsened by the fact that among the 195 nation-states, there are also

IntechOpen

many where the democratic nature of political power is not fully ensured, that is, it itself partly serves private or group interests instead of the public good (e.g., due to corruption, authoritarian governments, or due to limited or possibly manipulated access to information).

The "nation" is a cultural community that basically lives in its language (also in its traditions, folk art, shared memories of its past). Nowadays, however, the actual activity, communication, and interest relations between people are no longer of an ethnic nature, and in many cases, they cannot even be restricted to geographical boundaries. The geographical boundaries of the 195 historically formed political formations do not show a clear connection with the spoken languages and the ethnic distribution, but they cannot be meaningfully covered by the actual activity, communication, and interest relations pattern. This is therefore an outdated network, an outdated system, the modernization and redesign of which is in our common interest and one of the most important prerequisites for the road to sustainability. In addition to creating an institutional system for effective global coordination, it is very important to create the self-government of real professional and interest protection communities, and within the framework of geographically-based self-government, the only actual entity that really exists: the strengthening of the self-government of the settlement community.

Globalization as a phenomenon means the increasing interconnection and interdependence of countries and their economies, cultures, and societies. It is a multifaceted and continuous process that leads to greater integration and interaction between people, regardless of where they live.

The most important aspects of globalization are:

- Economic globalization: The expansion of international trade, investment and financial flows. It includes the movement of goods, services, capital, labor, technology, knowledge, and practices across national borders.

- Cultural Globalization: The sharing of cultural elements such as language, science, arts, religions, beliefs, fashion, and entertainment among people worldwide. This also means the mixing and sharing of everyday lifestyles and traditional practices.

- Technological Globalization: The rapid spread and development of technology, especially in communication and transportation, which facilitates global connectivity and information exchange.

- Political globalization: The growing interconnectedness of political institutions on a global scale. Not only the creation of international organizations and international agreements, but also the voluntary association and merger of states, confederations, and federations (as well as the deepening and expansion of existing continental (e.g., EU) and network (e.g., BRICS or the Commonwealth) integrations) are also included.

- Environmental globalization: Recognizing that environmental problems are fundamentally global (we cannot say that "the ship will not sink in our cabin," "the Earth will not be overpopulated in our village, or the global climate system will not change in our city"), and effective solutions require well-coordinated global cooperation.

While globalization has brought many benefits, such as increased economic opportunities, access to different cultures, freer movement of people, increased adaptability, and more effective scientific and technological development, at the same time, legal and social inequality, lack of equal opportunities, loss of cultural values, also raised concerns about nature conservation, epidemics, and environmental challenges. The impact and perception of globalization varies from region to region and community to community, and there are ongoing debates about the consequences of globalization and possible reforms to address the disadvantages.

The assessment of globalization is controversial: while some associate it with trade, freedom, and development as advantages, others believe that globalization threatens local cultures, local social cohesion, previously stable local economies, historically established social relations, and its disadvantages are emphasized [1]; in the following, without claiming to be exhaustive, we will review some positive and negative aspects, illuminating the "two sides of the coin."

2. Advantages, positive effects, and opportunities

The term globalization was originally mentioned by an American scientist T. Levitt in 1983 in an article published in the magazine "Harvard Business Review." The author saw this as a process of unification of the various regional product markets produced by large transnational corporations [2].

The growth and diversification of various products and services played a decisive role in the development of the changes that occurred in the nineteenth century, and the twentieth century was also influenced by developments in the field of info-communication technology and transportation. Globalization facilitates international trade and investment, leading to accelerated economic growth. Countries can specialize in the production of goods and services where they have a comparative advantage, which promotes efficiency and productivity [3].

Culture is also vital in the process of globalization [4]. The positive impact of globalism on national culture is due to the fact that it contributes to its development, while recognizing in its achievements the characteristics corresponding to national cultures [5]. Tourism, as an activity inherently involving cultural exchange, can increase awareness of cultural differences by enhancing intercultural communication, while fostering cultural hybridization, which enables the coexistence and integration of different cultures [6].

Globalization is one of the most important channels through which the penetration of digital technology affects innovation, but it is not the only one. It boosts technology and knowledge transfer, increases the adoption of digital technology, and strengthens globalization and development [7].

In the context of globalization, politics takes place above the state through political integration systems such as the European Union and through intergovernmental organizations such as the International Monetary Fund, the World Bank, the United Nations, and the World Trade Organization. When nation-states sign treaties with international organizations, it is an agreement that they will operate according to norms and standards that all signatories will abide by. An example of this is the establishment of norms for war crimes, which are supervised by the International Criminal Court in Brussels [8].

One notable positive environmental impact of globalization is that it has provided a mechanism for governments and other international actors to cooperate and address pressing environmental issues. Undoubtedly, globalization has created an opportunity to discuss and solve environmental issues on the international stage, thereby enabling the use of international relations to promote environment-oriented international cooperation. This cooperation resulted in the creation of contracts, more specific policies, and standards or benchmarks [9].

3. Disadvantages, negative effects, and challenges

Globalization can bring disproportionate growth both between and within nations. Although globalization increases economic growth for many countries, the growth is not equal – richer countries often benefit more than developing countries. Within countries, globalization is often associated with an increase in immigration, which, in addition to obvious advantages, also brings temporary disadvantages. From a macroeconomic point of view, immigration increases the gross domestic product (GDP), which means an economic boom for the host nation. However, immigration can reduce GDP per capita in the short term [10].

The widespread use of English has a beneficial effect on the creation of economic opportunities for people in the third world, but at the same time, it threatens the indigenous languages. Local cultures may erode, and ancient customs may disappear due to technological changes caused by globalization [11].

The digital divide is the phenomenon where the more affluent have better access to technology than the poor. The inequality in the distribution of technology provides a competitive advantage to richer developed countries. Although technological globalization has allowed technologies to spread more rapidly around the world, the digital divide still exists between and within the nations of the world [12]. The introduction of new technologies has been good for the production of cheaper goods, but many worry that it will make entire industries redundant, leaving people unemployed and disillusioned [13].

In theory, globalization can be an opportunity to spread values and practices such as environmental protection and labor rights throughout the world. In practice, this propagation is slow and imperfect. For example, instead of exporting the labor standards that a company must comply with in Western countries, it may follow lower standards in another third-world country where labor is not protected [14].

Globalization has led to a "race to the bottom" situation in which companies actively seek countries with the weakest labor and environmental protections and the lowest wages. While globalization has increased the flow of goods, services, and capital, there are still plenty of tax havens, meaning that much of the value added by globalization is not captured and redistributed by governments [15].

One of the primary consequences of globalization is that it opens up new markets for businesses to sell goods and resources, raw materials, and components. Both of these facts mean that finished products now travel farther than ever, possibly halfway around the world. In the past, products were more likely to be produced, sold, and consumed locally. This increased transportation of goods can affect the environment in many ways. The further the product travels, the more fuel is used, and the more greenhouse gas emissions are produced.

Transport – especially overland – requires good infrastructure such as roads and bridges. Building such infrastructure can lead to problems such as habitat loss and environmental pollution. Most of the goods are also transported by ship. The more ships there are at sea, the greater the risk that oil spills or leaks, as well as noise pollution, will damage the fragile marine environment [16].

4. Conclusion

A much more effective global economic, political, and cultural integration of humanity than at present is, on the one hand, essential for the survival of our species and civilization, but on the other hand, it does not solve the problems of the transition to a sustainable society by itself. The global ecological and social crisis can only be managed in the framework of a globally coordinated, unified legal system that creates equal opportunities, knowledge-based, democratic, environmentally conscious, and solidary integration.

Author details

Levente Hufnagel* and Ferenc Mics
Research Institute of Multidisciplinary Ecotheology, John Wesley Theological College, Budapest, Hungary

*Address all correspondence to: leventehufnagel@gmail.com

References

[1] Shayo M. Social identity and economic policy. Annual Review of Economics. 2020;**12**:355-389

[2] Levitt T. The globalization of markets. Harvard Business Reviews. 1983;**61**:92-102

[3] Surugiu M-R, Surugiu C. International trade, globalization and economic interdependence between European countries: Implications for businesses and marketing framework. Procedia Economics and Finance. 2015;**32**:131-138

[4] Sofronov B. The economic impact on global tourism. Annals of Spiru Haret University, Economic Series, Universitatea Spiru Haret. 2017;**17**(2):127-139

[5] Ergashev I, Farxodjonova N. Integration of National Culture in the process of globalization. Journal of Critical Reviews. 2020;7(2):477-479

[6] Pieterse JN. Global multiculture, flexible acculuration. In: Archer K, Bosman MM, Amen MM, Schmidt E, editors. Cultures of Globalization: Coherence, Hybridity, Contestation. Abingdon-on-Thames, Oxfordshire: Routledge; 2013

[7] Skare M, Soriano DR. How globalization is changing digital technology adoption: An international perspective. Journal of Innovation & Knowledge. 2021;**6**(4):222-233

[8] Perlas N. Shaping Globalization: Civil Society, Cultural Power and Threefolding. Forest Row, E. Sussex: Temple Lodge Publishing; 2019

[9] Doytch N, Uctum M. Globalization and the environmental Spillovers of sectoral FDI. Economic Systems. 2012;**40**(4):582-594

[10] Wade EH. Global growth, inequality, and poverty: The globalization argument and the 'political' science of economics. In: Ravenhill J, editor. Global Political Economy. Kettering, Northamptonshire: Oxford University Press; 2017

[11] Boussebaa M. From cultural differences to cultural globalization: Towards a new research agenda in cross-cultural management studies. Critical Perspectives on International Business. 2020;**17**(3):381-398

[12] Bhanu Murthy KV, Kalsie A, Shankar R. Digital economy in a global perspective: Is there a digital divide? Transnational Corporations Review. 2021;**13**(1):1-15

[13] Korinek A, Stiglitz JE. Artificial intelligence, globalization, and strategies for economic development. In: NBER Working Paper No. 28453. Cambridge: National Bureau of Economic Research; 2021

[14] Jaumotte F, Buitron CO. Inequality and labor market institutions. In: IMF Staff Discussion Notes 2015/014. Washington D.C.: International Monetary Fund; 2015

[15] Evertsson N. Corporate tax avoidance: A crime of globalization. Crime, Law and Social Change. 2016;**66**(2):199-216

[16] Le HP, Ozturk I. The impacts of globalization, financial development, government expenditures, and institutional quality on CO_2 emissions in the presence of environmental Kuznets curve. Environmental Science and Pollution Research. 2020;**27**:22680-22697

Chapter 2

Advancing Ocean Sustainability for Climate-Resilient Change

Teresa J. Kennedy

Abstract

The United Nations (UN) Decade of Ocean Science for Sustainable Development spans from 2021 to 2030 as a global call to action for marine scientists and Early Career Ocean Professionals (ECOPs) to collaboratively create, implement, and communicate science-based solutions to the critical challenges faced by our shared Ocean. The Decade integrates science, policy, and international engagement of a broad base of civil society stakeholders through interdisciplinary research challenges aimed at establishing sustainable marine practices to confront issues related to climate change. This chapter presents a synopsis of the UN-led initiatives that culminated in the launch of the Ocean Decade and describes the global vision articulated in the UN 2030 Agenda for Sustainable Development, focusing on Sustainable Development Goal 14, *Life Below Water*. The critical need for multidisciplinary Ocean research and innovation, collaborative sharing of findings grounded in data, and recommendations for global mitigation and policy development are discussed. The Global Ocean Ship-based Hydrographic Investigations Program (GO-SHIP) is recognized as an exemplary Ocean Decade endorsed action, offering collaborative research experiences and opportunities to contribute to data-driven recommendations influencing policy development on an international scale. Enhancing Ocean literacy through the dissemination of precise, data-driven information is essential for advancing Ocean sustainability for climate-resilient change and safe-guarding the well-being of our Ocean.

Keywords: climate change research, Early Career Ocean Professional (ECOP), GO-SHIP Evolve, Ocean Decade, Ocean literacy, sustainable development goals (SDGs), sustainable marine practices

1. Introduction

International cooperation is paramount in addressing global challenges such as climate change and its impact on our Ocean. As a response to this need, the United Nations (UN), established in 1945 in the aftermath of World War II, plays a crucial role as a global organization uniting 193 countries to collaboratively address challenges affecting our well-being and the health of our planet [1]. Policies are agreed upon during the annual UN General Assembly (UNGA), which convenes each year in September at the United Nations Secretariat Building in New York City to engage in concurrent sessions until all global issues on their agenda have been addressed. Given the critical nature of the topics discussed, sessions typically extend for a 12-month period and conclude shortly before the commencement of the subsequent General Assembly the following year. Diplomatic dialog during high-level meetings and summits at each UNGA result in resolutions (formal statements that carry significant political and moral influence, prompting international norms and practices) and declarations (reports outlining the collective views of all 193 member states). Additionally, the UNGA organizes specific "conventions" to negotiate transparent and inclusive frameworks for international agreements and treaties that promote cooperative multilateral diplomacy, global problem-solving, international legal frameworks, standardization of practices, and mechanisms for monitoring and enforcing compliance of internationally agreed upon initiatives.

Thirteen years following the establishment of the United Nations, concerns related to the world's Ocean gained prominence and became a priority. The first formal UN Conventions on the Law of the Sea (UNCLOS) occurred in 1958 in Geneva, Switzerland, and focused on four separate themes which established formal maritime regulations defining the rights and responsibilities of the UN member states. **Table 1** provides specific actions established at each of the four initial conventions [2].

The Geneva UNCLOS essentially outlined the first global framework for Ocean usage. However, while it touched on certain environmental safeguards, such as sanitation, it granted member states the power to decide how to implement strategies and enforce policies. Consequently, the consistency and effectiveness of the planned environmental protections differed from one country to another.

Twenty-four years later, in 1982, the UN member states agreed upon a more concerted global effort aimed at addressing the implementation of the Geneva UNCLOS. This international treaty provided the legal framework for use, and importantly, the protection of the world's Ocean. It is a widely accepted global agenda for Ocean governance and is often referred to as one of the most comprehensive and significant instruments of international law. Nonetheless, it took an additional 12 years to secure approval in 1994 from all member states to uniformly adopt the updated UNCLOS. Still, no formal reporting on the global state of the Ocean had occurred to that point, and furthermore, none was planned for in the foreseeable future.

Convention theme	Focus and outcome
Convention 1: Territorial Sea and the Contiguous Zone (CTS)	Established the concept of the territorial seas (the belt of coastal waters extending 12 nautical miles from a coastal state), and the contiguous zone (waters extending an additional 12 miles beyond the territorial sea); gave states jurisdiction to enforce laws related to customs, fiscal matters, immigration, and sanitary regulations.
Convention 2: The High Seas (CHS)	Outlined the freedoms of the high seas for all states; established principles of navigation, fishing, and scientific research.
Convention 3: Fishing and Conservation of the Living Resources of the High Seas (CFCLR)	Established international rules for the conservation and management of marine living resources in the high seas.
Convention 4: Continental Shelf (CCS)	Established principles for the delimitation of the continental shelf between adjacent or opposite states.

Table 1.
Geneva 1959 UN conventions on the law of the sea (UNCLOS).

Another nine years passed before the UNGA successfully established an agreed upon process for regular reporting and assessment of marine environments on a global scale through UNGA Resolution 57/141, *Oceans and the Law of the Sea* in 2003. The UNCLOS spent the next two years formalizing plans to conduct an *Assessment of Assessments* (AOA) and assembled an international Group of Experts in 2005 to oversee the four-year process. The final report on the AOA was released in 2009, inspiring UNGA Resolution 64/72 to request a global assessment of the state of the world's Ocean, specifically focusing on the relationships between the Ocean and humans, covering ecological, economic, and social dimensions [3].

In 2010, the First World Ocean Assessment (FWOA) was released. It successfully drew global attention to the concerning condition of the Ocean and emphasized the urgent need for improved management of coastal and marine environments to safeguard crucial ecosystem services [4]. Results of the FWOA report were shared globally at the UNGA in September 2015, adding the Ocean to a longer list of areas

Year	United Nations and General Assembly Actions	Time span for actions to occur	Total years
1945	Establishment of the United Nations		
1958	Geneva UN Conventions on the Law of the Sea (UNCLOS), first Global Framework for Ocean usage	13 years after UN established	
1982	United Nations Convention on the Law of the Sea (UNCLOS) Treaty proposed protection of the Ocean	24 years after the Geneva UNCLOS	24
1994	Adoption of the UNCLOS Treaty	12 years for agreed upon adoption	36
2003 (23 December)	UNGA Resolution 57/141, *Oceans and the Law of the Sea,* established regular process for global reporting and assessment of the world's marine environment	9 years to establish need for on-going reporting process	45
2005	UNGA document A/60/91 formalized plans to conduct an Assessment of Assessments (AOA); convened Group of Experts	2 years to plan the Assessment of Assessment	47
2009 (1 September) (4 December)	AOA Report presented by the Group of Experts UNGA Resolution 64/72, *Sustainable fisheries...,* approves First Global Integrated Marine Assessment	4 years to complete AOA and approve FWOA	51
2010	First Global Integrated Marine Assessment (First World Ocean Assessment, FWOA)	1 year to create FWOA	52
2015 (September 1) (September 25) (December 23)	FWOA assessment and summary released; UNGA Resolution 70/1, *Transforming our World: The 2030 Agenda for Sustainable Development;* Sustainable Development Goals (SDGs) defined UN Resolution 70/235, *Oceans and the Law of the Sea,* approved the FWOA assessment and summary	5 years to conduct assessment, release global report, and establish Agenda 2030 and the SDGs	57

Table 2.
United Nations General Assembly (UNGA) actions leading to the 2030 agenda.

of concern regarding the maintenance of a globally sustainable planet. The adoption of UNGA Resolution 70/1, *Transforming our World: The 2030 Agenda for Sustainable Development*, was the first collaborative global initiative to create a foundational plan aimed at cultivating health, peace, and prosperity for both humanity and the planet, spanning the present and future [5]. **Table 2** provides an overview of the historic initiatives undertaken by the UN, beginning in 1958 with the Geneva UNCLOS and culminating in the formulation of the *2030 Agenda for Sustainable Development* in 2015. This table documents the chronological progression leading to global awareness of Ocean health, highlighting the pivotal year of 2003 when the groundwork for consistent monitoring of worldwide marine environments was established and the creation of Agenda 2030 and the Sustainable Development Goals (SDGs) in 2015.

Finally, 57 years after the Geneva UNCLOS, the UN successfully drew attention to the critically urgent need of governments around the world to work together to protect our shared Ocean. Many member states began calling for urgent action to safeguard the health of our Ocean, and widespread dissemination of reports documented an alarming picture [6].

The 2030 Agenda stresses the need for increased Ocean research and data collection, better dissemination and communication of findings, and improved implementation of marine management policies and strategies worldwide. A fundamental aim of the 2030 Agenda is to halt global environmental degradation, which naturally encompasses efforts to address concerns related to the well-being of our Ocean.

Secretary-General Antonio Guterres of Portugal, the Chief Administrative Officer of the UN, explained that although the Ocean connects us all, we have taken it for granted to the point that today, "we face an 'Ocean Emergency' and the tide must be turned ... our failure to care for the Ocean will have ripple effects across the entire 2030 Agenda" ([7], paras. 6–7).

2. Addressing the Ocean Emergency

Through the UNGA adoption of the 2030 Agenda, member states acknowledged the health of the world's Ocean as one of the most pressing developmental challenges we face today. They further emphasized its significance, along with the legal frameworks governing global activities related to promoting sustainable and equitable use of the Ocean and its resources through Resolution 72/73, *Oceans and the Law of the Sea* [8]. The 2030 Agenda and Resolution 72/73 would not have occurred without the collaborative agreement of governments around the world to share information about the state of marine environments globally. "The politically controlled process and scientifically compiled FWOA were successful in establishing a politically accepted set of credible Ocean facts" ([9], p. 11). In December 2015, UN Resolution 70/235, *Oceans and the law of the sea,* approved the FWOA assessment and related conclusions.

It is important to note that international input into the resolution referred to different regions of this vast body of water as separate Oceans (e.g., the Atlantic Ocean, the Pacific Ocean, the Indian Ocean, etc.), however the concept of one interconnected global Ocean gained prominence immediately following Resolution 72/73, reflecting the growing awareness of the unity and interdependence of Earth's oceanic systems, encompassing everything from currents and marine life to geological features and climate interactions. The emphasis on the singular "Ocean" highlights the holistic approach needed to successfully address global environmental issues, bringing member states together to work collaboratively to address the current Ocean emergency.

2.1 UN Sustainable Development Goals

In support of the 2030 Agenda for Sustainable Development, a global set of Sustainable Development Goals (SDGs), along with accompanying targets and indicators, were established to mobilize international efforts between governments, businesses, civil society, and individuals ([5], p. 13). The SDGs are a cohesive set of interdependent objectives aligned with the 2030 Agenda covering "global challenges that are crucial for the survival of humanity" ([10], p. 6). The SDGs address a range of global challenges, including social, economic, and environmental issues [11]. Each of the 17 goals (see **Figure 1**) addresses different aspects of sustainable development and encourages governments around the world to set national targets and incorporate their goals into national planning processes, policies, and strategies. The interconnectedness of the goals recognizes that achieving one goal often involves addressing multiple aspects of sustainable development, therefore uniting efforts between the goals [11, 12].

Figure 1.
UN sustainable development goals. Image source: https://sdgs.un.org/goals.

2.2 Life Below Water

Numerous factors contribute to the poor state of our marine environment. SDG 14 challenges the international community to acknowledge the transboundary nature of marine resources and threats to Ocean sustainability, and to collectively work toward solutions to shared problems through innovative research and governance ([13], p. 5). *Life Below Water* calls on all countries and stakeholders to conserve and sustainably use the Ocean and marine resources through ten targets [14]. These include:

14.1 Prevent and reduce marine pollution.

14.2 Protect and restore marine and coastal ecosystems.

14.3 Minimize and address the impacts of Ocean acidification.

14.4 Promote sustainable fishing.

14.5 Conserve coastal and marine areas.

14.6 End subsidies contributing to overfishing and unregulated fishing.

14.7 Increase the economic benefits from sustainable use of marine resources.

14.8 Increase scientific knowledge, research capacity, and technology for Ocean health.

14.9 Support small scale fishers.

14.10 Enhance the conservation and sustainable use of the Ocean and its resources as reflected in the UN Convention on Law of the Sea.

While SDG 14 focuses on *"Life Below Water"* and aims to conserve and sustainably use marine resources for sustainable development, it is directly interconnected

Figure 2.
The central role of Ocean health to global sustainable development. Image source: Institute for Advanced Sustainability Studies ([13], p. 9).

with the other SDGs since actions taken to achieve this goal generally result in either positive or negative impacts on the others. For example, a healthy Ocean and healthy marine ecosystems contribute to the livelihoods of millions of people, especially those in poverty who rely on fisheries and coastal resources for sustenance and resources (e.g., SDG 1: No poverty). **Figure 2** ([13], p. 9) and **Figure 3** [15] highlight connections between SDG 14 and the other SDGs, emphasizing the central role of Ocean health to attaining sustainable development globally. While **Figure 3** specifically focuses on the key interactions between three of the SDGs (SDG 2: Zero Hunger; SDG 3: Good Health and Wellbeing; SDG 7: Affordable and Clean Energy) to provide an example entry point, **Table 3** provides example connections between each of the SDGs and SDG 14, *Life Below Water* [16–20].

A literature review conducted by Barakat et al. [21] revealed strong interlinkages between the SDGs (pp. 23–24), suggesting the importance of wide dissemination of scientific research findings to reinforce the five "Ps of Sustainable Development"—people, planet, prosperity, peace, and partnership (p. 7). The health of the Ocean is crucial for advancing global efforts toward sustainable development, and therefore actions must

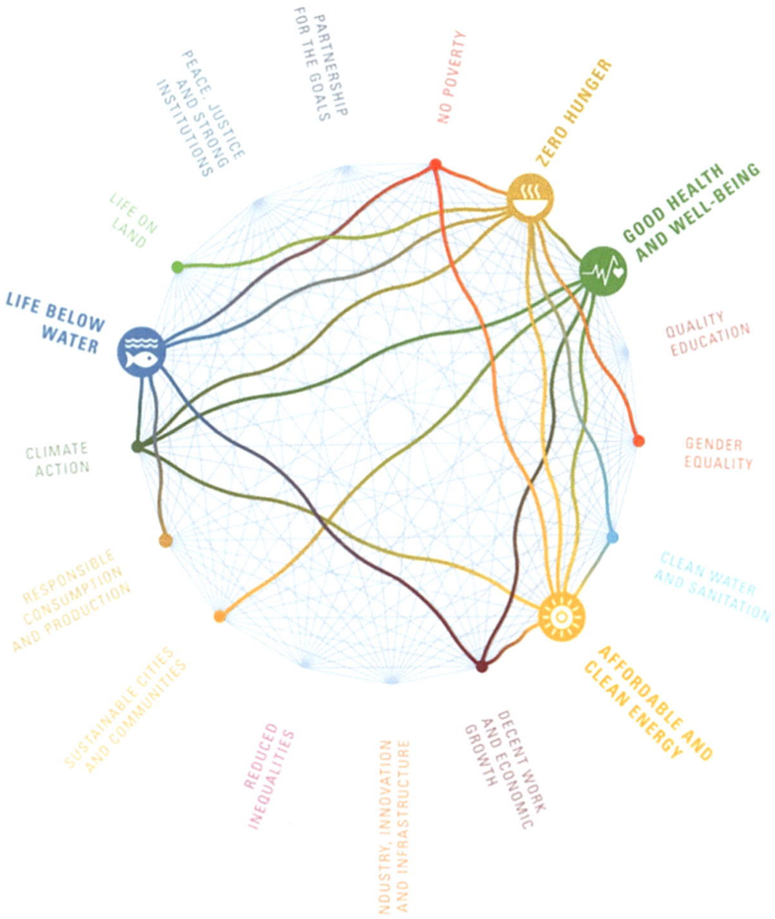

Figure 3.
SDG interactions analyzed using a seven-point rating scale and based on reviews of scientific findings and expert knowledge. Image source: (ICSU, 2017; [15]).

Sustainable development goal	Example connection to SDGs 14
1 End poverty in all its forms everywhere	Livelihood of more than three billion people depend on coastal and marine ecosystems.
2 End hunger, achieve food security and improved nutrition and promote sustainable agriculture	"Blue foods" are the primary source of protein for around three billion people; sustainable marine resource management ensures food security.
3 Ensure healthy lives and promote well-being for all at all ages	Microorganisms in the Ocean produce half of the oxygen we breathe; the Ocean regulates freshwater cycles providing drinking water; blue foods provide protein to billions; mental well-being.
4 Ensure inclusive and equitable quality education and promote lifelong learning opportunities for all	Ocean literacy fosters life-long emotion connections to the environment and prepares the "blue workforce."
5 Achieve gender equality and empower all women and girls	Gender equity in the blue workforce boosts global GDP and contributes to the blue economy.
6 Ensure availability and sustainable management of water and sanitation for all	Sea-level rise results in saltwater intrusion into freshwater aquifers and water sources; warming Ocean water disrupts freshwater cycles; Ocean plastic flows from around 1600 rivers globally.
7 Ensure access to affordable, reliable, sustainable, and modern energy for all	Growing markets for offshore wind, tidal stream, and wave energy.
8 Promote sustained, inclusive, and sustainable economic growth, full and productive employment, and decent work for all	Aquaculture and fishing, energy, shipping, tourism; and jobs in conservation, marine research and science, communications, engineering, maritime law.
9 Build resilient infrastructure, promote inclusive and sustainable industrialization and foster innovation	90% of globally traded goods travel by sea; shipping contributes to global greenhouse gas emissions; "green" infrastructures protect people, homes, buildings, power networks from storm damage and sea level rise.
10 Reduce inequality within and among countries	Small island developing states are the most vulnerable to sea-level rise, extreme weather, marine heatwaves, and Ocean acidification.
11 Make cities and human settlements inclusive, safe, resilient, and sustainable	Coastal communities are vulnerable to sea-level rise and extreme weather; serving as hubs for innovative climate solutions such as blue-green infrastructures (protecting/creating wetlands, oyster beds, barrier islands, etc.).
12 Ensure sustainable consumption and production patterns	Sustainable use and management are essential to reduce undesirable disturbances to marine ecosystems such as litter, plastic waste, chemical contamination, high demand for blue foods.
13 Take urgent action to combat climate change and its impacts	Ocean regulates Earth's climate; warming waters and melting glaciers result in sea-level rise and Ocean acidification; blue carbon projects fight climate change and biodiversity loss.
15 Protect, restore, and promote sustainable use of terrestrial ecosystems, sustainably manage forests, combat desertification, and halt and reverse land degradation and halt biodiversity loss	Ocean provides 75% of the rainwater and 50% of the oxygen needed for life on land; helps to protect terrestrial ecosystems from excess heat.

Sustainable development goal	Example connection to SDGs 14
16 Promote peaceful and inclusive societies for sustainable development, provide access to justice for all and build effective, accountable, and inclusive institutions at all levels	UN-led high-level meetings, summits, and conventions; resolutions and declarations such as the UNCLOS resulting in global actions, such as the March 2022 resolution to end plastic consumption and governance of marine resources.
17 Strengthen the means of implementation and revitalize the global partnership for sustainable development	Funding partnerships support a sustainable Ocean economy and global Ocean health (e.g. partnerships with Indigenous communities).

Table 3.
Connections between SDG 14, Life Below Water *and other SDGs.*

be approached through a holistic lens that implements ongoing impact assessments to monitor and adapt strategies in an informed manner. It is critically important that scientific research efforts consider the broader context and potential "ripple effects" that may result from well-intended efforts to avoid unintended consequences.

In 2019, the preliminary draft of the *Technical Guidance on Ocean Accounting for Sustainable Development* was created in response to country demands for a set of guidelines and methodologies to integrate the value of the Ocean into their national accounting systems [22].

This international effort aimed to develop sustainable development practices and inform policy decisions worldwide. After undergoing international review and revision, the official document was circulated throughout the UN.

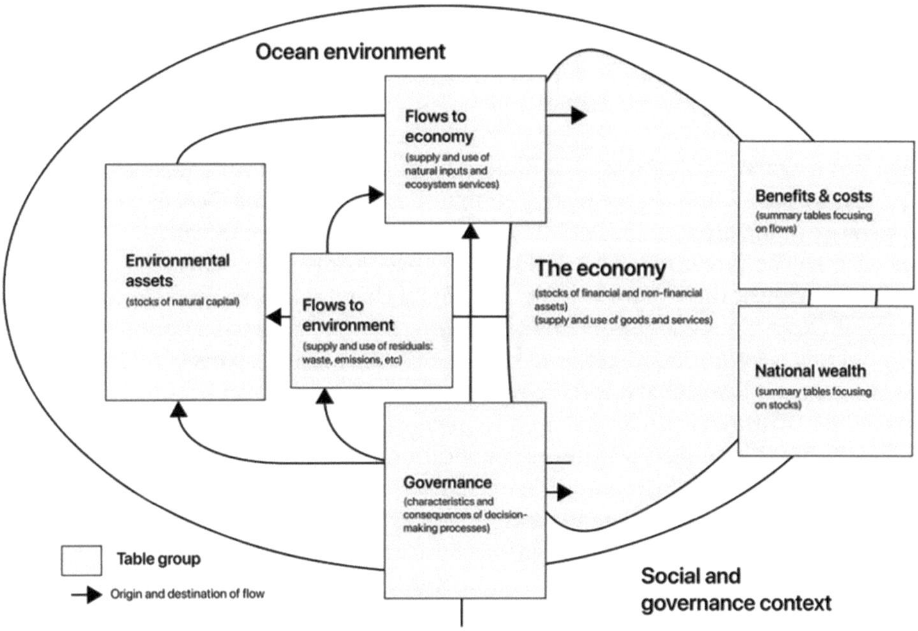

Figure 4.
The Ocean Accounts Framework ([23], p. 28).

The Ocean Accounts Framework, depicted in **Figure 4** includes both a social and governmental context, showing the interrelationships between the two ([23], p. 28, [24]). These include involving Ocean assets (natural capital); flows to economy (Ocean services, including goods); flows to environment (residuals including ecosystem impacts); the Ocean economy and the economy; governance; combined presentation (recording a "report card" of summary information, physical quantities, monetary value, and/or qualitative status) and indicators concerning the flows of benefits and costs (the latter broadly defined as maintenance and restorations costs, disservices and externalities) between the Ocean environment and the economy; national wealth; and stock of Ocean wealth (environmental assets and economic and financial assets).

The information recorded on **Figures 2–4** and on **Table 3** is intended to support broad decision-making when compiled on a regular basis including strategic development planning of the Ocean or "blue" economy; management of Ocean space of protected areas, marine spatial planning and integrated coastal zone management; finance and investment related to the Ocean; and Ocean analyses, monitoring, and assessment [23, 25].

3. The UN Decade of Ocean Science for Sustainable Development

The vast expanse of the marine environment, Earth's largest domain, plays a crucial role in stabilizing climate and sustaining life on our planet, with far-reaching implications for human well-being. In 2015, the UN shared the preliminary findings of the *First Global Integrated Marine Assessment* (First World Ocean Assessment, FWOA), which revealed significant degradation in much of the Ocean, noting losses in the structure, function, and benefits from marine ecosystem services. The assessment addressed ten main themes: (1) climate change and related changes in the atmosphere; (2) overexploitation of marine living resources; (3) food security and food safety; (4) changes in the patterns of biodiversity; (5) pressures from increased uses of Ocean space; (6) threats from increased pollution; (7) effects of anthropogenic (human) impacts; (8) inequalities in the distribution of benefits from the Ocean; (9) the importance of coherent management of human impacts on the Ocean; and (10) delays in implementing measures and solutions to address known problems ([26], pp. 7–10).

The findings of the FWOA led to the 2017 UN General Assembly declaration of the *Decade of Ocean Science for Sustainable Development* as an essential component of the 2030 Agenda. This initiative, referred to as the Ocean Decade, spans from 2021 to 2030, and aims to mobilize global efforts to reverse the decline in Ocean health and unite stakeholders worldwide under a shared framework. It calls for robust scientific research to effectively assist countries to improve conditions for sustainable development of the Ocean. The Ocean Decade is unique in comparison to other global initiatives focusing on Ocean science since it covers all disciplines within the field ([27], p. 3).

The UN designated the United Nations Educational, Scientific, and Cultural Organization (UNESCO) to facilitate implementation activities and ensure that the Ocean Decade benefits from international scientific expertise, partnerships, and coordination capabilities. UNESCO's Intergovernmental Oceanographic Commission (IOC) plays a key role in coordinating the Decade's preparatory processes, working collaboratively with the global Ocean community to plan activities and research in Ocean science and technology throughout the Decade. The IOC coordinates research programs, services, and capacity-building for Ocean observations, hazard mitigation, tsunami warnings, and marine spatial planning, among other critically important initiatives, with guidance from the Ocean Decade Advisory Board, an international group of experts and leaders in Ocean science, policy, and management.

The Ocean Decade launched on 1 January 2021, after four years of strategic international planning facilitated by the IOC and guided by the vision statement *The science we need for the Ocean we want* [28]. This tagline summarizes the fundamental idea that scientific knowledge and comprehension of the Ocean are critical to attaining the envisioned state of the Ocean, emphasizing the role of science to inform policy, action, and sustainable development initiatives to protect the Ocean's health and resources for current and future generations.

The Decade's mission, to "catalyze transformative Ocean science solutions for sustainable development, connecting people and [the] Ocean" accentuates the importance of ensuring the well-being of both humanity and marine environments through impactful Ocean research and information dissemination. Educating both civil society and government officials to change policy and actions based on scientific knowledge is at the core of the Ocean Decade.

3.1 Ocean Decade Outcomes

The 'Ocean we want' for a sustainable future at the end of the Decade is represented by the following seven overarching Decade outcomes ([29], pp. 4–5):

1. A clean Ocean where sources of pollution are identified, reduced, or removed.
2. A healthy and resilient Ocean where marine ecosystems are understood and managed.
3. A productive Ocean supporting sustainable food supply and a sustainable Ocean economy.
4. A predicted Ocean where society understands and can respond to changing Ocean conditions.
5. A safe Ocean where life and livelihoods are protected from Ocean-related hazards.
6. An accessible Ocean with open and equitable access to data, information, and technology and innovation.
7. An inspiring and engaging Ocean where society understands and values the Ocean in relation to human well-being and sustainable development.

The success of the Ocean Decade relies on global involvement of scientists to engage in groundbreaking research, make recommendations for innovative solutions, and collaboratively address pressing challenges for the sustainable well-being of our Ocean and our planet. At its core, the involvement of youth and Early Career Ocean Professionals (ECOP) is imperative, as the dynamic perspectives they contribute will impact the future of Ocean conservation and sustainable practices. Integration of generational perspectives merges innovative thinking with wisdom and experience, fostering effective succession planning and enabling a more comprehensive examination of long-term implications and sustainability at broader levels encompassing all civil society.

The Ocean Decade serves as a catalyst for collective commitments to promote Ocean sustainability now and into the future. It empowers scientists to gather climate data, leverage technological advancements, foster international collaborations, raise awareness of critical issues, and educate the public, increasing Ocean literacy on a worldwide scale. These activities enable the design and implementation of effective mechanisms to deepen our understanding of the Ocean, and consequently, develop sustainable solutions for Ocean-based development. By disseminating scientific breakthroughs globally through the Ocean Decade network, scientists have the ability to shape policies and empower businesses, governments, and civil society to adopt sustainable practices in their everyday activities.

3.2 Ocean Decade Action Framework

The Ocean Decade facilitates strategic planning and involvement in scientific research, with a focus on climate as the central component. This collaborative approach includes experts in coastal zone management, marine spatial planning, aquaculture and fishery management, disaster risk reduction, adaptation mitigation and climate services, as well as governance to create policies that ensure global peace and security. The IOC and its partners played a significant role in shaping SDG 14, *Life Below Water* [11], which focuses on conserving and sustainably using Ocean, sea, and marine resources for sustainable development. The Ocean Decade Action Framework consists of three objectives and ten challenges aimed at inspiring action from the scientific community [30].

Decade Objectives

1. Identify required knowledge for sustainable development.
2. Generate comprehensive knowledge and understanding of the Ocean.
3. Increase the use of Ocean knowledge.

Decade Challenges

1. Understand and beat marine pollution.
2. Protect and restore ecosystems and biodiversity.
3. Sustainably feed the global population.
4. Develop a sustainable and equitable Ocean economy.
5. Unlock Ocean-based solutions to climate change.
6. Increase community resilience to Ocean hazards.
7. Expand the Global Ocean Observing System (GOOS).
8. Create a digital representation of the Ocean.
9. Develop skills, knowledge, and technology for all.
10. Restore society's relationship with the Ocean.

The global challenges were identified through a consultive process facilitated by the UNESCO IOC, and involve global stakeholders including scientists, policymakers, non-governmental organizations (NGOs), and the private sector. The objectives of the global challenges contribute significantly to the UN Sustainable Development Goals (SDGs). The 17 SDGs, with their natural intersection with one another, are deeply interrelated with the Decade Challenges (see **Figure 5**).

3.3 Decade Actions

The Decade Challenges serve as 'calls to action' to the international scientific community to develop initiatives, projects, and activities that improve scientific understanding of the Ocean, promote sustainable use of Ocean resources, and enhance international cooperation and collaboration [24]. These actions will continue to be the focus over the next six years to meet the Decade's challenges, and their results will

Figure 5.
Decade challenges ([31], p. 10).

be communicated widely through strategic and multi-faceted approaches via the UN General Assembly and its Member States.

The Ocean Decade seeks to generate the scientific knowledge, infrastructures, and partnerships necessary to inform policies through global and region planning meetings. To achieve this, all actors need to come to the table. This includes scientists, academia, policy makers and managers, civil society and non-governmental organizations, businesses and private sectors, donors, and foundations, and of course, our youth. To move from "the Ocean we have" right now to the "Ocean we want" in 2030, a multistep implementation process involving three non-linear, overlapping steps has been established by the IOC.

- First, identify the knowledge that is required for sustainable development.
- Second, gather data, information, and knowledge for the development of a comprehensive understanding of the Ocean, its components, and its interactions.
- Third, use the generated knowledge and understanding of the Ocean to deploy solutions for sustainable development.

The Ocean Decade call to action inspires marine scientists and oceanographers studying all aspects of the Ocean to set specific goals as an effort to work together to contribute to a global understanding of marine environments and anthropogenic impacts on the Ocean, and to develop strategies for sustainable Ocean management. Audrey Azoulay, UNESCO Director-General, proclaimed, "How can we succeed in protecting the Ocean when we know so little about it? Only 20% of the seabed is mapped. We need to go further and mobilize the international community so that at least 80% of the seabed is mapped by 2030" [32].

In response to this challenge, UNESCO made a grand announcement at the One Ocean Summit in 2022 stating that through the collaborative actions of the Ocean Decade, at least 80% of the seabed will be mapped by 2030, compared to the 20% that has currently been recorded. The Ocean Decade places a high priority on developing a collaborative and international Ocean data management system, given that substantial parts of the global Ocean remain inadequately monitored.

The High-Level Panel for a Sustainable Ocean Economy endorsed this vision and further stressed the importance of collaboration in the Ocean Decade to share Ocean-related data through an open and equitable digital ecosystem [33]. As explained by Dr. Richard Spinrad, Under Secretary of Commerce for Ocean and Atmosphere and NOAA Administrator, "the *New Blue Economy* is a knowledge-based economy, looking to the sea not just for extraction of material goods, but for data and information to address societal challenges and inspire their solutions" ([34], para. 2).

4. Ocean Decade Call to Action

Vladimir Ryabinin, Assistant Director General (ADG) of UNESCO in charge of the IOC, proposed, "By 2030, we will put in place a global monitoring tool, which will report annually on the progress of the mapping and identify remaining gaps. This global seabed map will be one of the legacies of the UN Decade of the Ocean" [35]. According to the 2022 State of the Ocean Report, "The current Ocean observing system embraces around 10,000 Ocean observing platforms, with some 84 countries contributing to the system. ... The observing system provides essential data and products to weather, climate and Ocean forecasters, maritime commerce, fisheries and coastal communities. However, there are major gaps in coverage and the system is not currently able to provide data where it is most needed, namely in areas with high biodiversity and intense human pressures" ([32], pp. 10–11). Organizations, governments, and individuals around the world are responding to the Ocean Decade call to action for Ocean conservation and sustainable management of the Ocean. This involves developing an "all of society approach" in a manner that strengthens the science-policy-society interface; working within a framework to not only engage in data collection, but also produce recommendations and engage in the implementation of potential solutions to challenges uncovered through the research process. An example of one such Ocean Decade research initiative follows.

4.1 Spotlight on an exemplar Ocean Decade initiative

The Global Ocean Ship-based Hydrographic Investigations Program (GO-SHIP) is an international initiative investigating long-term (decadal) changes in Ocean heat content, carbon uptake, acidification, circulation, and other parameters throughout the global Ocean [36]. Fifteen nations are represented by the GO-SHIP scientific committee, with nineteen nations contributing ship time. GO-SHIP unites researchers specializing in physical oceanography, the carbon cycle, marine biogeochemistry, and ecosystems, along with other stakeholders involved in hydrographic data gathering. Additionally, GO-SHIP research cruises provide training and capacity building opportunities for ECOPs, including undergraduate and graduate students in oceanography. As a result, the program contributes to the next generation of marine scientists and broadens the availability of real-time data that ensures fit-for-purpose provision of calibration data through collaborations with partner networks and the global oceanographic community. It also focuses on the development of new data products driven by stakeholder needs and promotes innovation considering data-based solutions.

GO-SHIP is a part of the Global Climate Observing System (GCOS) sponsored by the World Meteorological Organization (WMO), the Intergovernmental Oceanographic Commission of the United Nations Education, Scientific and Cultural Organization (IOC-UNESCO), the United Nations Environment Programme (UN Environment), and the International Science Council (ISC). GO-SHIP contributes to the Global Climate Observing System (GCOS), which conducts periodic evaluations of the state of worldwide climate observations in the atmosphere, on land, and in the Ocean, and provides recommendations for enhancement of research activities. Expert panels within GCOS are responsible for upholding the definitions of "Essential Climate Variables" (ECVs), crucial for systematically monitoring the Earth's evolving climate ([37], para. 2). The observations endorsed by GCOS play a vital role in addressing climate research challenges and form the foundation for climate services, as well as measures related to adaptation and mitigation. In addition, GO-SHIP is a major contributor to the World Climate Research Program's (WCRP) Climate Variability and Predictability Experiment (CLIVAR) and the International Ocean Carbon Coordination Project.

As an endorsed initiative of the United Nations Ocean Decade, *GO-SHIP Evolve*, aims to enhance the collection of biological and ecosystem data, revealing trends and variability in crucial indicators of Ocean health [38]. Of particular importance, this initiative conducts comprehensive water column measurements along designated hydrographic transects across the world's Ocean with the aim of investigating temporal changes in physical and hydrographic conditions. Information gathered during GO-SHIP research cruises plays a vital role in monitoring deep Ocean parameters, specifically those below a depth of 2 km. This segment of the Ocean, constituting more than half of the total Ocean volume, remains largely inaccessible to researchers except through ship-based hydrography.

As a result, the program contributes to a universally coordinated network of continuous hydrographic sections, supporting the broader global Ocean/climate observing system through coast-to-coast or coast-to-ice research transects. **Figure 6** shows the GO-SHIP reference sections consisting of 55 core lines strategically positioned to track long-term trends and variations in oceanic conditions. Although not all transect lines are currently occupied, they represent key pathways for comprehensive oceanic monitoring, serving as important reference points for future research endeavors and facilitating a deeper understanding of the complex interactions within the global Ocean system. These core lines enable scientists to assess changes in temperature, salinity, oxygen levels, and other critical parameters over time, helping to identify emerging patterns and trends that inform climate models

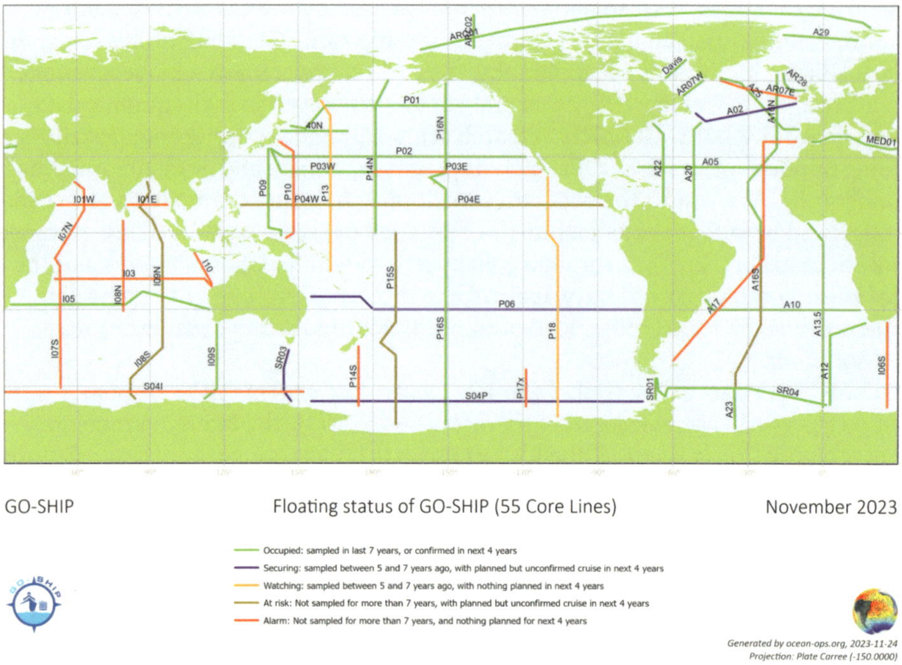

Figure 6.
GO-SHIP reference sections https://www.go-ship.org/RefSecs/goship_ref_secs.html.

and support informed decision-making regarding Ocean conservation and management strategies.

The GO-SHIP program provides invaluable insights into the profound changes occurring in our Oceans. By meticulously collecting and analyzing data from some of the most remote and inaccessible regions of the Ocean, GO-SHIP contributes crucial information to our understanding of how climate change is reshaping marine ecosystems. These data not only enhance our ability to predict future climate scenarios but also inform policymakers and stakeholders worldwide, guiding the development of effective strategies for mitigating the impacts of climate change and preserving the health of our Ocean- for generations to come.

Addressing Ocean Decade Challenge 7 (expand the global Ocean observing system) and Challenge 8 (create a digital representation of the Ocean), GO-SHIP contributes to global monitoring, mapping, and identification of observational gaps throughout the global Ocean. GO-SHIP facilitates multinational voyages and shared facilities, fosters training opportunities, and accelerates the utilization of knowledge and multinational capacity, encouraging contributions from a wider array of countries [39]. A description of a GO-SHIP research cruise recently completed follows [40].

4.2 GO-SHIP A13.5: 2024

GO-SHIP A13.5 is a decadal hydrographic long-line research cruise occupying the 0E meridional line in the South Atlantic Ocean. Conducted aboard the R/V Marcus G. Langseth, a 235-foot (71.6 m), 3834-gross ton research vessel operated by Columbia University's Lamont-Doherty Earth Observatory (LDEO) on behalf of the NSF, the 2024 mission was led by scientists from NOAA's Pacific Marine Environmental

Laboratory (PMEL) and Earth and Space Research (ESR). Data gathered along the A13.5 follows cruises along the same transect in 1983/84, 1995, and 2010. Analyses and comparisons to past findings will provide insight into how this area in the south-eastern Atlantic is changing due to anthropogenic carbon emissions. **Figures** 7 and 8 provide geographic views of the cruise track.

The recent re-occupation of the A13.5 transect (Ghana to Bouvet Island, February-March 2024) contributed to a universally coordinated network of continuous hydrographic sections, thereby enhancing the broader global Ocean/climate observing system. The western margin and southern limit of the South Atlantic Ocean have been significantly impacted by anthropogenic carbon (C_{ant}) uptake, as shown by a C_{ant} column inventory change in the region that was higher than the global mean storage rate [41]. Predictions have warned that aragonite undersaturation will occur in the next two decades along the eastern margin if anthropogenic changes maintain the present trend and have recommended more extensive monitoring in the eastern margin of the South Atlantic to identify the extent to which ecosystems health, marine life, and societies will be affected in the future.

The 2024 decadal investigations of the A13.5 transect were overdue as this line was delayed due to COVID-19. Since CTD (conductivity, temperature, and depth) instrumentation is the primary tool used for determining the physical properties of sea water [42, 43], these data are needed to continue monitoring this area. The CTD instrumentation detects how the conductivity and temperature of each water sample taken have changed relative to depth since 2010. Additionally, 2024 data will provide information about changes in the physical, chemical, and biological properties of the water column since 1983/1984, and further document changes in Ocean heat content, carbon uptake, acidification, circulation, and other parameters of interest in the southeastern Atlantic Ocean. Global CO_2 and climate variability programs can use these data to complement their research on an international scale.

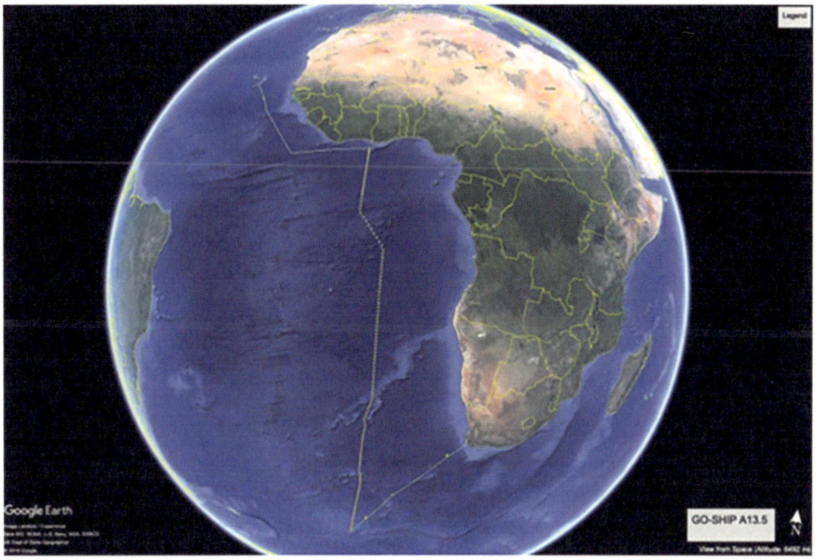

Figure 7.
Google Earth depiction of the 2024 GO-SHIP A13.5 cruise track.

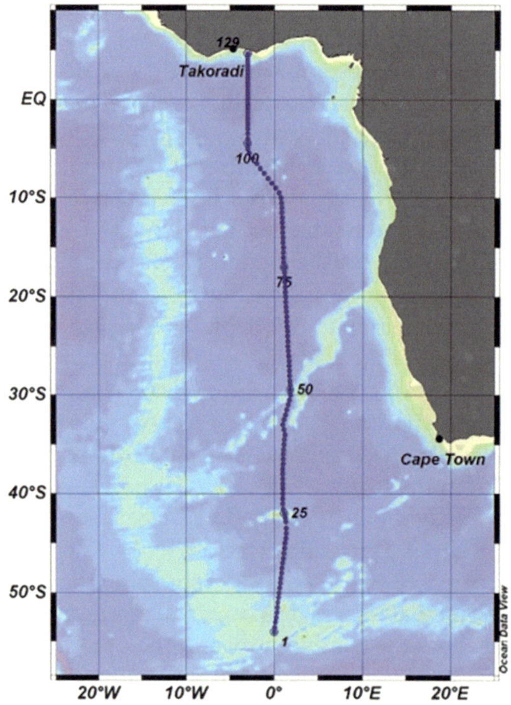

Figure 8.
GO-SHIP 2024 A13.5 Transect https://usgoship.ucsd.edu/2024/03/23/weekly-reports-from-a13-5/.

The 2024 A13.5 science team represented 14 different organizations, and included undergraduate and graduate-level oceanography students, resulting in members of the next generation of Ocean scientists becoming trained and engaged in this decadal initiative, ultimately growing the ECOP network. Additionally, to further expand education and outreach efforts to K–12 teachers and their students, scientists onboard the ship partnered with the Adopt-a-Float program to inspire and educate students about global Ocean biogeochemistry and climate change, creating a powerful opportunity for students of all ages to learn about the research underway by naming and tracking BGC (biogeochemical) floats [44]. All data collected have been made freely available by the Global Ocean Biogeochemistry Array (GO-BGC) Project funded by the National Science Foundation, Division of Ocean Sciences (NSF OCE 1946578) and by the Argo Program, which is part of the Global Ocean Observing System) and the national programs that contribute to it (http://www.argo.ucsd.edu).

5. Conclusion

More than 80% of the Ocean remains unexplored, yet human well-being is dependent on the activity beneath the waves [45]. Considering the Ocean's role in sustaining us through nourishment, climate regulation, and the production of over half of the world's oxygen, alongside its capacity to absorb 50 times more carbon dioxide than our atmosphere, it is clear why the United Nations (UN) prioritized the Ocean in its 2015 mandate, *Transforming our World: The 2030 Agenda for Sustainable Development.*

Reversing the current decline in Ocean health and educating diverse stakeholders, including all sectors of civil society, is a clear priority of the 2030 Agenda, implemented as a response to high level reports and actions issued by the United Nations General Assembly (UNGA) for more than 65 years (from 1958 to 2024). These initiatives included groundbreaking resolutions, declarations, and conventions such as the *Global Framework for Ocean Governance* established by the UN Convention of the Law of the Sea in 1982, the 2003 UNGA *Resolution 57/141* which initiated the first process for global reporting and assessment activities of marine environments worldwide, along with the 2010 *First World Ocean Assessment* that gathered data for a period of nearly five years, resulting in the *Global Ocean Science Report 2020: Charting Capacity for Ocean Sustainability* which called for a "comprehensive and holistic digital representation of the Ocean, including a dynamic Ocean map, allowing the matching of data to needs by multiple stakeholders" ([46], p. 225).

Importantly, the 2030 Agenda led to the UN proclaimed Decade of Ocean Science for Sustainable Development (2021–2030), providing a convening framework for scientists, stakeholders, diverse segments of civil society, and governments at all levels to collaboratively build awareness about the severe and alarming state of our Ocean and take action to mitigate further destruction of our global marine ecosystems. The Ocean Decade, managed by the Intergovernmental Oceanographic Commission (IOC) of the United Nations Educational, Scientific, and Cultural Organization (UNESCO), prominently emphasizes the interdisciplinary nature of the 17 UN Sustainable Development Goals through Goal 14, which focuses on *Life Below Water* [47]. Through the development of scientific knowledge leading to a more comprehensive understanding of the Ocean, Goal 14 aims to deeply embed Ocean literacy into all aspects of society through the sharing of scientific evidence and collaborative management strategies. This is particularly important since the Ocean is the primary source of a substantial portion of the global economy, which is crucial in industries such as tourism, fisheries, international shipping, and transportation.

The underlying connections that link Ocean sustainability with sustainable blue economies, creates a set of shared societal goals to achieving a cleaner, more data transparent, and safe Ocean [28]. The primary goal of the Ocean Decade is to develop data-driven solutions that address the ongoing challenges and issues impacting our Ocean, and consequently inform and contribute to the development of policies, technologies, and strategies that support the growth of the New Blue Economy while ensuring responsible use and conservation of Ocean resources for the future. The transformative potential of engagement within an Ocean Decade action prepares ECOPs for future multidisciplinary research activities, provides valuable experiences communicating data-based finding on a global scale, and holds great potential to change behaviors within all sectors of civil society through broad-scale communications, dissemination of valuable research findings, and enhanced Ocean literacy worldwide.

This chapter highlighted GO-SHIP Evolve, one example of a multitude of Ocean Decade research initiatives. GO-SHIP contributes to the "multistakeholder, multicomponent digital Ocean ecosystem" described in the 2020 *Global Ocean Science Report* ([46], p. 225) through its focus on global monitoring, mapping, and identification of observational gaps throughout the global Ocean. Its current scope is concentrating on emerging possibilities for the fourth decade of GO-SHIP surveys in novel scientific domains due to evolving scientific questions, technology, and research needed to monitor and protect the world's Ocean.

The time is now to engage in research contributing to the overarching Ocean Decade goal of achieving sustainable Ocean management in support of the New Blue Economy by 2030; contributing to the "*Science We Need for the Ocean We Want*." We cannot wait another 65 years to begin seeking solutions to resolve the critically poor condition of our world's Ocean. Governments and international organizations around the world rely on research-based data to develop policies and regulations that can protect the Ocean. Join the Ocean Decade today. Marine scientists must help civil society to better understand marine ecosystems, including their biodiversity, dynamics, and interdependencies, as understanding the significance of these data are essential to making informed decisions about conservation and sustainable development of our Ocean, mitigation of anthropogenic threats, and raising awareness of the importance of Ocean conservation [48].

For more information about the Ocean Decade, see: https://oceandecade.org/sign-up/.

For information about the GO-SHIP research activities, see: https://usgoship.ucsd.edu/.

Author details

Teresa J. Kennedy
The University of Texas at Tyler, USA

*Address all correspondence to: tkennedy@uttyler.edu

References

[1] National Archives. Charter of the United Nations. 1945. Available from: https://www.archives.gov/milestone-documents/united-nations-charter#transcript

[2] United Nations. 1958 Geneva conventions on the law of the sea. Audiovisual Library of International Law. 1958. pp. 1-5. Available from: https://legal.un.org/avl/ha/gclos/gclos.html

[3] United Nations General Assembly. Sustainable Fisheries Resolution 64/72. 2009. Available from: https://www.gc.noaa.gov/documents/un-64-72-N0946615.pdf

[4] Evans K, Zielinski T, Chiba S, Garcia-Soto C, Ojaveer H, Park C, et al. Transferring complex scientific knowledge to useable products for society: The role of the global integrated ocean assessment. Frontiers in Environmental Science. 2021;**9**:1-9. Available from: https://www.frontiersin.org/journals/environmental-science/articles/10.3389/fenvs.2021.626532/full

[5] United Nations General Assembly. Transforming our World: The 2030 Agenda for Sustainable Development. Resolution 70/1. 2015. Available from: https://documents.un.org/doc/undoc/gen/n15/291/89/pdf/n1529189.pdf?OpenElement

[6] United Nations. The Second World Ocean Assessment. Division for Ocean Affairs and the Law of the Sea. 2021. Available from: https://www.un.org/regularprocess/woa2

[7] United Nations. Guterres Outlines Four Recommendations to Help Us All 'Save Our Ocean.' Department of Economic and Social Affairs. 2022a. Available from: https://www.un.org/tr/desa/guterres-outlines-four-recommendations-help-us-all-save-our-ocean

[8] United Nations General Assembly. Oceans and the Law of the Sea. Resolution 72/73. 2017. Available from: https://documents-dds-ny.un.org/doc/UNDOC/GEN/N17/421/90/PDF/N1742190.pdf?OpenElement

[9] Fawkes KW, Cummins V. Beneath the surface of the first world ocean assessment: An investigation into the global process' support for sustainable development. Frontiers in Marine Science. 2019;7(612):1-14. Available from: https://www.frontiersin.org/articles/10.3389/fmars.2019.00612/full#B68

[10] UNESCO. Education for Sustainable Development Goals: Learning Objectives. 2017. Available from: https://unesdoc.unesco.org/ark:/48223/pf0000247444?posInSet=5&queryId=cd7b8f91-6b28-4edf-a2b4-2d3963e8da2c

[11] UNESCO. Sustainable Development Goals. 2024. Available from: https://www.unesco.org/en/sdgs

[12] International Science Council. Executive Summary: SDGs and Interactions. 2017. Available from: https://council.science/wp-content/uploads/2017/05/SDGs-interactions-executive-summary.pdf

[13] Unger S, Müller A, Rochette J, Schmidt S, Shackeroff J, Wright G. Achieving the Sustainable Development Goals for the Oceans. Institute for Advanced Sustainability Studies; 2017. Available from: https://sdgs.un.org/sites/default/files/2022-03/archieving_the_sdgs_for_oceans.pdf

[14] United Nations. 14 Conserve and Sustainably Use the Oceans, Seas, and Marine Resources. Available from: https://www.un.org/sustainabledevelopment/oceans/

[15] IISD. Focus on Interactions to Make the SDGs a Success: Our Key Messages for the Ocean Conference. SDG Knowledge Hub; 2017. Available from: https://sdg.iisd.org/commentary/guest-articles/focus-on-interactions-to-make-the-sdgs-a-success-our-key-messages-for-the-ocean-conference/

[16] Le Blanc D, Freire C, Vierros M. Mapping the Linkages Between Oceans and Other Sustainable Development Goals: A Preliminary Exploration. 2017. Available from: https://www.un.org/ru/desa/mapping-linkages-between-oceans-and-other-sustainable-development-goals

[17] Ntona M, Morgera E. Connecting SDGs 14 with other sustainable development goals through marine spatial planning. Marine Policy. 2018;**93**:214-222. Available from: https://www.sciencedirect.com/science/article/pii/S0308597X17300520?via%3Dihub

[18] United Nations. High-Level Political Forum on Sustainable Development. 2022b. Available from: https://hlpf.un.org/sites/default/files/2022-06/BN%20HLPF%202022%20SDG%2014.pdf

[19] UN DESA. Expert Group Meeting on SDGs 14 (Life below Water) and its Interlinkages with Other SDGs: Concept Note. 2022. Available from: https://sdgs.un.org/events/expert-group-meeting-sdg-14-life-below-water-and-its-interlinkages-other-sdgs-46167

[20] World Ocean Initiative. The ocean flows through all 17 of the UN sustainable development goals. The Economist. 2023. Available from: https://impact.economist.com/ocean/ocean-sustainable-development-goals/

[21] Barakat B, Bengtsson S, Muttarak R, Kebede EB. Education for people and planet: Creating sustainable futures for all. International Institute for Applied Systems Analysis. 2016. pp. 23-24. Available from: https://unesdoc.unesco.org/ark:/48223/pf0000245580

[22] Global Ocean Accounts Partnership (GOAP) The GOAP Secretariat is hosted by the Sustainable Development Reform Hub at the University of New South. Wales, Sydney, Australia; 2019. Available from: https://www.oceanaccounts.org/about-the-global-ocean-accounts-partnership/#:~:text=The%20GOAP%20Secretariat%20is%20hosted,India%20Indo%2DPacific%20Oceans%20Initiative

[23] Global Ocean Accounts Partnership. Technical Guidance on Ocean Accounting for Sustainable Development. 2021. Available from: https://oceanaccounts.atlassian.net/wiki/spaces/DTGOOA/overview?ref=oceanaccounts.org

[24] Loureiro TG, Milligan B, Gacutan J, Adewumi IJ, Findlay K. Ocean accounts as an approach to foster, monitor, and report progress towards sustainable development in a changing ocean—The systems and flows model. Marine Policy. 2023;**154**:1-9. Available from: https://www.sciencedirect.com/science/article/pii/S0308597X23001951

[25] Global Ocean Accounts Partnership. Technical Guidance on Ocean Accounting. GOAP secretariat. 2022. Available from: https://www.oceanaccounts.org/technical-guidance-on-ocean-accounting-2/

[26] United Nations. The First Global Integrated Marine Assessment. Division

for Ocean Affairs and the Law of the Sea. Available from: https://www.un.org/regularprocess/content/first-world-ocean-assessment; 2016 https://www.un.org/Depts/los/global_reporting/WOA_RPROC/WOACompilation.pdf

[27] Guan S, Qu F, Qiau F. United Nations decade of ocean science for sustainable development (2021-2030): From innovation of ocean science to science-based ocean governance. Frontiers in Marine Science. 2023;**9**:1-8. Available from: https://www.frontiersin.org/articles/10.3389/fmars.2022.1091598/full

[28] Intergovernmental Oceanographic Commission. The Science We Need for the Ocean We Want. UNESCO; 2020b. Available from: https://unesdoc.unesco.org/ark:/48223/pf0000265198

[29] Ryabinin V, Barberie J, Haugan P, Kullenberg G, Smith N, McLean C, et al. The UN decade of ocean science for sustainable development. Frontiers in Marine Science. 2019;**6**(470):1-10. Available from: https://www.researchgate.net/publication/334806579_The_UN_Decade_of_Ocean_Science_for_Sustainable_Development

[30] Intergovernmental Oceanographic Commission. Implementation Plan Summary. UNESCO. 2021b. Available from: https://unesdoc.unesco.org/ark:/48223/pf0000376780

[31] Intergovernmental Oceanographic Commission. Co-Designing the Science We Need for the Ocean We Want. Paris, France: UNESCO; 2021a. Available from: https://www.ioc.unesco.org/en

[32] UNESCO. State of the Ocean Report 2022. International Oceanographic Commission. 2022b. Available from: https://unesdoc.unesco.org/ark:/48223/pf0000381921

[33] Leape J, Abbott M, Sakaguchi H. Technology, Data and New Models for Sustainably Managing Ocean Resources. World Resources Institute; 2020. Available from: https://oceanpanel.org/publication/technology-data-and-new-models-for-sustainably-managing-ocean-resources/

[34] NOAA. New Blue Economy. 2023. Available from: https://www.noaa.gov/blue-economy

[35] UNESCO. One Ocean Summit: UNESCO Pledges to Have at Least 80% of the Seabed Mapped by 2030. 2022a. Available from: https://www.unesco.org/en/articles/one-ocean-summit-unesco-pledges-have-least-80-seabed-mapped-2030 . (Updated April 20, 2023)

[36] GO-SHIP. GO-SHIP Repeat Hydrography Manual: A Collection of Expert Reports and Guidelines. IOCCP Report No. 14, ICPO Publication Series No. 134. 2019. Available from: https://www.go-ship.org/HydroMan.html

[37] World Meteorological Organization. Global Climate Observing System (GCOS). 2023. Available from: https://wmo.int/activities/global-climate-observing-system-gcos

[38] UNESCO. GO-SHIP Evolve. GO-SHIP Program Office. 2021a. Available from: https://oceandecade.org/actions/go-ship-evolve/

[39] GO-SHIP. GO-SHIP: Towards a Sustained Global Survey of the Ocean Interior. Available from: https://www.go-ship.org/

[40] GO-SHIP A13.5. Decadal re-occupation of the U.S. GO-SHIP meridional hydrographic line known as the A13.5 (0°E in the Atlantic Ocean). GO-BGC, Monterey Bay Aquarium Research Institute. 2024. Available from:

https://www.go-bgc.org/expedition-logs/south-atlantic-2024

[41] Piñango A, Kerr R, Orselli IBM, Carvalho ACO, Azar E, Karstensen J, et al. Ocean acidification and long-term changes in the carbonate system properties of the South Atlantic Ocean. Global Biogeochemical Cycles. 2022;**36**(9):1-20. Available from: https://agupubs.onlinelibrary.wiley.com/doi/full/10.1029/2021GB007196

[42] NOAA. How to use CTD data. Gulf of Mexico Expedition 2012. 2012. Available from: https://oceanexplorer.noaa.gov/okeanos/explorations/ex1202/background/edu/media/ex1202_ctd.pdf

[43] Woods Hole Oceanographic Institution. Conductivity, Temperature, Depth (CTD) Sensors. 2023. Available from: https://www.whoi.edu/what-we-do/explore/instruments/instruments-sensors-samplers/conductivity-temperature-depth-ctd-sensors

[44] GO-BGC Global Ocean Biogeochemistry Array. Adopt-a-Float. Available from: https://www.go-bgc.org/outreach/adopt-a-float

[45] Duarte CM, Agusti S, Barbier E, Britten GL, Castilla JC, Gattuso JP, et al. Rebuilding marine life. Nature. 2020;**580**:39-51. Available from: https://www.nature.com/articles/s41586-020-2146-7

[46] Intergovernmental Oceanographic Commission. Global Ocean Science Report 2020—Charting Capacity for Ocean Sustainability. UNESCO. 2020a. Available from: https://unesdoc.unesco.org/ark:/48223/pf0000375147

[47] The Global Goals. 14 Life Below Water. Available from: https://www.globalgoals.org/goals/14-life-below-water/

[48] Clausen A, Kennedy TJ, Dufourneaud O, Castillo MP, Deane S, Durand J, et al. Ocean decade manual for non-governmental organizations (NGOs). Intergovernmental Oceanographic Commission of the United National Education, Scientific and Cultural Organization (UNESCO). 2024

Chapter 3

The Function of the Distribution of ICMS Tax Collection in Brazil (Ecological ICMS) as a Potentially Beneficial Measure for Reducing Greenhouse Gas Emissions

Adrielle B.O. Sardinha, Epaminondas J. Messias and Carlos A. Leonetti

Abstract

The global effort to combat climate change hinges on the effectiveness of local actions. Consequently, Brazil, being a crucial carbon sink, confronts a multifaceted challenge: the imperative to reduce emissions originating predominantly from wildfires and deforestation across various municipalities within the framework of Fiscal Federalism. Therefore, the Ecological ICMS (or Green ICMS) involves using sustainability criteria for the distribution of this tax, in Portuguese Imposto sobre Circulação de Mercadorias e Serviços (ICMS). This book chapter conducts a study of the extra-fiscal function of the Ecological ICMS as a potentially benefit measure to reduce greenhouse gas emissions (GHG). We will use a bibliographic review and documentary sources. We hope to discover trends that show correlations of this fiscal instrument with qualitative implementation of low-emission areas.

Keywords: Brazil, ecological ICMS, extra-fiscal measures, environmental taxes, environmental protection

1. Introduction

In the global pursuit of addressing climate change, the effectiveness of local actions becomes paramount. This holds particularly true for Brazil, a crucial carbon sink facing a complex challenge in reducing emissions stemming mainly from wildfires and deforestation across diverse municipalities within the framework of Fiscal Federalism. The Ecological ICMS, also known as Green ICMS, is a type of extra-fiscal measure that plays a pivotal role by incorporating sustainability criteria into the distribution of this tax (Imposto sobre Circulação de Mercadorias e Serviços—ICMS in Portuguese).

This system, referred to as ICMS Verde, operates by the State reallocating percentages of tax collections in accordance with specific socio-environmental requirements. Essentially, it incentivizes municipalities to meet certain environmental goals by offering financial rewards based on their performance in areas such as conservation, sustainable development, and biodiversity protection. Through this mechanism, funds are directed toward initiatives that promote environmental sustainability and community well-being, fostering a more balanced and eco-friendly approach to economic development at the local level.

On a national scale, initiatives geared toward reducing greenhouse gas emissions are poised to make significant contributions to the Global Stocktake. This is an instrument of The Paris Agreement, a paramount international regulation, that binds participating states to progressively submit nationally determined contributions with escalating ambition. This commitment aligns with the overarching framework of the United Nations Framework Convention on Climate Change.

This book chapter delves into an exploration of the extra-fiscal function of the Ecological ICMS, examining its potential as a beneficial measure in reducing greenhouse gas emissions (GHG). This study examines the potential impact of a particular instrument on the distribution of ICMS tax collection, aiming to assess its effectiveness as a measure for mitigating greenhouse gas emissions. The study relies on bibliographic and documentary reviews to inform its analysis and findings. The research is structured into three main sections, apart from this introductory section and concluding remarks. Section 2 comprises three distinct topics: firstly, an exploration of the theoretical underpinnings of greenhouse gas emissions in the Brazilian context; secondly, a succinct discussion on the ecological ICMS tax; and lastly, an endeavor to establish a connection between these two domains.

2. The paths of development in research

2.1 Brazilian greenhouse gas emissions

Climate change stands as one of the most pressing and far-reaching challenges of our time, casting a profound impact on ecosystems, weather patterns, and human societies worldwide. Characterized by a discernible alteration in Earth's climate, driven largely by human activities such as the burning of fossil fuels and deforestation, climate change poses unprecedented threats to the environment and biodiversity. Rising temperatures, extreme weather events, and shifts in precipitation patterns are becoming increasingly apparent, underscoring the urgent need for a concerted global effort to mitigate the causes and adapt to the consequences of this complex and interconnected phenomenon.

As the world grapples with the multifaceted implications of climate change, there is a critical call for collaborative solutions, sustainable practices, and informed policies to safeguard the planet for future generations. Brazil, renowned for its breathtaking natural diversity, grapples with significant challenges regarding greenhouse gas (GHG) emissions. While the country's economy thrives in sectors such as agriculture and livestock, the foremost source of these emissions is unequivocally linked to deforestation, particularly in the vast expanse of the Amazon and Cerrado biomes [1].

National Greenhouse Gas Inventories play a crucial role in global efforts to address climate change, and the United Nations (UN) is at the forefront of coordinating such endeavors. The UN, through its specialized body, the Intergovernmental Panel on Climate Change (IPCC), provides guidance and methodologies for countries to develop comprehensive inventories of their GHG emissions [2].

These inventories serve as valuable tools for understanding the sources and trends of emissions, aiding in the formulation of effective climate policies. Each country, in accordance with UN guidelines, prepares its National Greenhouse Gas Inventory, detailing the emissions from various sectors such as energy, agriculture, and industrial processes. These inventories are submitted periodically to the UN Framework Convention on Climate Change (UNFCCC) as part of the reporting obligations. The data collected allows for comparisons across nations, facilitating a better understanding of global emission patterns and helping to identify areas where collective efforts to achieve containment of greenhouse gas emissions.

Based on **Figure 1**, it is possible to observe that the agriculture and forestry and other land use sectors together represent the main sources of greenhouse gas emissions in Brazil. As suggested by the Guidelines of IPCC for National Greenhouse Gas Inventories: Enteric Fermentation, some of the subcategories included Land Converted to Cropland, Grassland Remaining Grassland, Biomass Burning, Urea Application, etc.

In this sense, two of these subcategories place Brazil in 5th place in the ranking of largest emitters, deforestation, and wildfires. At the top of the list are China, the United States, the European Union, and India, due to the industry and energy sectors.

If compared to data from The Global Carbon Project (GCP), which integrates knowledge of greenhouse gases for human activities and the Earth system, it is possible to observe that Brazil has become a significant emitter of GHGs due to the increase in deforestation [4].

This trajectory deprives this country of a very important source of mitigation, as the tropical forest is a large carbon sink. Brazil possesses significant potential for mitigating greenhouse gas emissions through a variety of strategies.

Some of these include expanding renewable energy sources like wind, solar, and hydropower to diminish emissions from the energy sector; implementing sustainable agricultural practices such as precision farming and improved fertilizer management to reduce emissions from agricultural activities; conserving existing forests and restoring degraded lands to enhance carbon sequestration and mitigate emissions from land use change and forestry; and improving waste management techniques like recycling, composting, and landfill gas capture to decrease emissions from the waste sector.

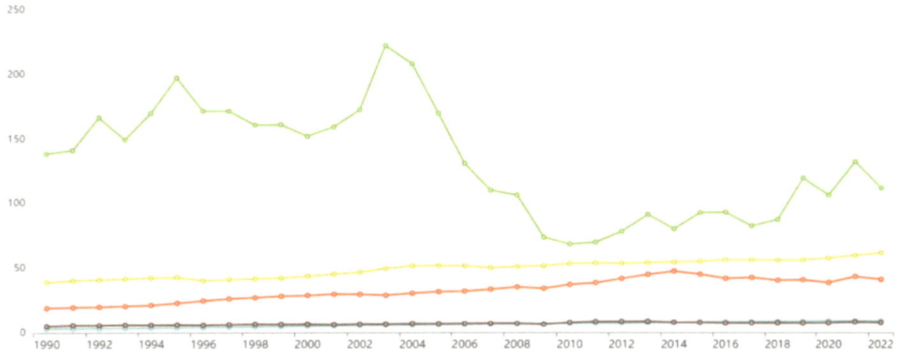

Figure 1.
Brazilian emissions in CO2e (t) by sectors according to GWP-AR5. Yellow—Agriculture; green—Forestry and other land use; red—Energy; Brown—Industrial processes and product use; blue—Waste. Source: Adapted from greenhouse gas emissions estimation system (SEEG) of the climate observatory [3].

Instead of valuing the form of mitigation, what has been happening is the reduction of Amazon's capacity to act as a carbon sink [5]. This situation is very worrying, as the easiest way for Brazil to achieve its mitigation goal is to maintain its forests.

As the international community intensifies its focus on mitigating climate change, accurate and transparent GHG inventories are indispensable for tracking progress, assessing the effectiveness of climate policies, and fostering collaboration between countries. The UN's commitment to facilitating the development and exchange of these inventories underscores its role as a pivotal force in the collective global response to the challenges posed by greenhouse gas emissions.

Brazil has adopted various policies and initiatives to combat deforestation, including the implementation of stricter environmental laws, enhanced surveillance, and the promotion of sustainable practices in agriculture. Therefore, the primary source of greenhouse gas emissions in Brazil is intrinsically linked to human activities of land use and land cover changes. As a result, it alters surface-atmosphere interactions, changing the timing and magnitude of energy fluxes, which impacts the partitioning of available energy, and consequently, the climate and water balance [6].

Between 1999 and 2018, Brazil witnessed a total of 16,141,383 fire foci, with the Cerrado and Amazon biomes being the most impacted, constituting 41.56% and 38.34% of the total, respectively. Carbon emissions resulting from these fires totaled 8089.17 Tg between 1999 and 2018, with the Amazon contributing 60.71% and the Cerrado 32.04%. Despite an increase in fire foci in the Cerrado from 2010 onwards, the Amazon remained the primary source of carbon emissions. Future projections indicate a rise in carbon emissions, in Ref. [1].

2.2 Considerations on the ecological ICMS

Environmental protection necessitates an environmental consciousness grounded in ethical values, transcending the commodification of the environment [7]. Within this framework, the Brazilian Constitution upholds the environment as an inherent right for all citizens, necessitating collaborative, multidisciplinary endeavors to secure its preservation. Article 225 of the 1988 Federal Constitution explicitly affirms the entitlement of every individual in Brazil to an environment that is ecologically balanced, underscoring the joint responsibilities of both the government and the collective to safeguard and preserve it for the benefit of present and future generations [8].

Consequently, environmental protection and the fight against pollution have become a shared responsibility among the Union, States, the Federal District, and Municipalities, article 23, VI, in Ref. [8]. Examining this constitutional provision prompts a question: How can municipalities effectively fulfill this constitutional mandate and protect the environment. Considering their financial constraints, municipalities, generally weaker compared to the Union and States, face budgetary limitations, except for major metropolises such as São Paulo, Rio de Janeiro, Belo Horizonte, Salvador, and Recife.

Despite being the weakest entity in the Brazilian federal system in financial and budgetary terms, municipalities play a significant role by opting for greater participation in the distribution of the amount collected by the State through the ICMS in exchange for embracing the Ecological ICMS.

The ICMS, a key state tax, holds significant importance in shaping the financial framework of local governments in Brazil. Nevertheless, its conventional distribution approach tends to benefit economically advanced municipalities, creating difficulties for areas characterized by environmental conservation efforts. To counter this imbalance, the notion of Green ICMS surfaced, seeking to rectify disparities and promote eco-friendly practices.

By adhering to the Ecological ICMS, municipalities commit to establishing and maintaining conservation units, public parks with green areas, protection of water supply catchment areas, environmental sanitation, selective waste collection, preservation of historical and cultural heritage, indigenous reserves, and other potential actions aimed at environmental protection. These actions, as counterparties, benefit society, extend beyond municipal boundaries, and ultimately contribute to the welfare of the entire community. The Ecological ICMS transcends the scope of Tax Law, traditionally concerned with the establishment, supervision, collection, and revenue of taxes; it serves as a significant mechanism within Financial Law, enabling the implementation of environmental policies through extra-fiscal means.

The criteria for distributing the ICMS are outlined in Article 158, Section IV, and the sole paragraph, II, of the Brazilian Constitution. Municipalities are entitled to 25% of the total ICMS collected, with 1/4 (one-fourth) of this amount being transferred according to State Law. They were supported by the 1988 Federal Constitution, under Section IV, sole paragraph, II, allowing for up to one-fourth, in accordance with state law, in Ref. [8]. This provision enables the implementation of public policies, allowing federated states to influence environmental preservation behaviors in municipalities through the Ecological ICMS.

The state of Paraná, in the southern region of Brazil, was the first to institute the Ecological ICMS in 1991. This initiative aimed to compensate municipalities for land use restrictions in areas protected by conservation units, where economic activities may be restricted or prohibited to ensure environmental preservation [6]. The experience in Paraná yielded positive results, with an increase in the number of municipalities benefiting from the Green ICMS and a rise in the volume of financial resources allocated. Additionally, there was an observed increase in municipal expenses related to environmental programmatic functions, in Ref. [9].

Several other states in the federation have also implemented the Ecological ICMS, with distribution criteria tailored to their realities and interests. Beyond environmental factors, other considerations may include the number of children enrolled in elementary education, dropout rates, infant mortality rates, the availability of hospital beds, etc. [10].

Moreover, as per the author's findings, among the 27 states comprising the Brazilian Federation, 17 have formally adopted the Ecological ICMS, (see **Table 1**) while the remaining 10 are yet to enact dedicated legislation on the subject. However, it is crucial to note that not the entire collected revenue is allocated toward sustainability initiatives. For instance, in São Paulo, a mere 0.5% of the total collected funds is directed toward specific actions, and in Minas Gerais, this figure stands at 1.1%. This distribution pattern suggests that the extra-fiscal function is diffused across various aspects of public administration.

Furthermore, municipalities within the 17 states that have implemented the Ecological ICMS, upon participation, not only operationalize the mandate prescribed in Article 23, VI of the Constitution but also contribute substantively to Brazil's adherence to commitments outlined in international treaties and agreements, particularly those with organizations such as the UN. This collective effort is oriented toward advancing sustainability objectives and promoting environmental preservation.

	States of the federation	Tax legislation	%
1	Acre	1.530/04	5,00
2	Amapá	322/96	1,40
3	Cerará	14.023/07	2,00
4	Goiás	complementary law 90/11	5,00
5	Minas Gerais	18.030/09	1,35
6	Mato Grosso	complementary law73/00	7,00
7	Mato Grosso do Sul	4.216/12	5,00
8	Pará	7.638/12	8,00
9	Pernambuco	11.899/00	3,00
10	PernanbucoB	9.600/11	5,00
11	Paraná	9.491/90	5,00
12	Piauí	6.581/14	5,00
13	Rio de Janeiro	5.100/07	2,5
14	Rio Grande do Sul	13.188/09	7,00
15	Rondônia	complementary law 147/96	5,00
16	São Paulo	17.348/21	2,00
17	Tocantins	2.959/15	13,00

Source: Elaborated by the authors.

Table 1.
Brazilian states with legislation on ecological ICMS and percentage allocated to municipalities.

2.3 Unraveling the connection between greenhouse gas reduction and the green ICMS

At the national level, initiatives aimed at reducing greenhouse gas emissions are poised to make a substantive contribution to the Global Stocktake. Among the paramount international regulations is the Paris Agreement, wherein each participating State commits to submit its nationally determined contributions progressively and with escalating ambition. This commitment is integral to the overarching framework of the United Nations Framework Convention on Climate Change [11].

Brazil has adopted several initiatives to reduce greenhouse gas emissions (GHGs), reflecting the country's commitment to mitigating climate change. One of the main strategies is the reduction of deforestation in the Amazon, considered one of the country's largest sources of carbon emissions. Brazil has implemented monitoring and control policies for deforestation, such as the Real-Time Deforestation Detection System (Deter), Terra Brasilis Project, and MAPBiomas Alerta project.

The Terra Brasilis Project is an initiative by the National Institute for Space Research (INPE) aimed at monitoring and combating deforestation in the Brazilian Amazon. Utilizing remote sensing technologies, particularly high-resolution satellite imagery and advanced data processing algorithms, the project provides precise and up-to-date information on deforestation activities in near real time.

The MAPBiomas Alerta project is an innovative initiative focused on monitoring land use changes and deforestation across Brazil. Developed by a consortium of research institutions, including the Amazon Environmental Research Institute (IPAM) and the Federal University of Minas Gerais (UFMG), the project utilizes cutting-edge satellite imagery analysis and machine learning algorithms to detect and alert stakeholders to changes in land cover in near real time.

By providing timely and accurate information on deforestation, forest degradation, and land use conversion, MAPBiomas Alerta and The Terra Brasilis Project empowers government agencies, environmental organizations, and local communities to take proactive measures to combat illegal logging, land clearing, and other destructive activities. Through its open-access platform and collaborative approach, the project facilitates data-driven decision-making and supports efforts to conserve Brazil's precious natural resources and biodiversity.

The Action Plan for Prevention and Control of Deforestation in Legal Amazon (PPCDAm) and the Plan for Prevention and Combating Deforestation and Fires in the Cerrado (PPCerrado) are initiatives implemented in Brazil aimed at addressing deforestation and wildfires in specific regions. PPCDAm focuses on the Legal Amazon region, offering strategies for monitoring, law enforcement, and sustainable land management to reduce deforestation rates and promote conservation efforts in this critical biome. On the other hand, PPCerrado targets the Cerrado biome, aiming to prevent and combat deforestation and wildfires through strengthened enforcement of environmental laws, promotion of sustainable agricultural practices, and community engagement.

The discontinuation of the PPCDAm and PPCerrado projects represents a concerning interruption in efforts to combat deforestation and wildfires in two of Brazil's most important and vulnerable regions: the Legal Amazon and the Cerrado biome. These projects have played a crucial role in reducing environmental degradation, protecting biodiversity, and promoting sustainable development in these areas. However, the lack of continuity of these initiatives may lead to an increase in deforestation rates, loss of natural habitats, and a rise in greenhouse gas emissions, exacerbating climate change and harming local communities that depend on natural resources for their livelihoods.

Furthermore, Brazil has invested in renewable energy, primarily in hydroelectric, solar, and wind power generation. The country's energy matrix is one of the cleanest in the world, with a significant portion of its electricity coming from renewable sources. Brazil also has incentive programs for the production and consumption of biofuels, such as sugarcane ethanol, which replaces part of fossil fuels and reduces greenhouse gas emissions in the transportation sector.

REDD+ (Reducing Emissions from Deforestation and Forest Degradation, as well as conservation, sustainable forest management, and enhancement of forest carbon stocks) is a global initiative aimed at combating climate change by incentivizing deforestation reduction and forest conservation. The program seeks not only to mitigate carbon emissions resulting from forest loss but also to promote biodiversity conservation, protect the livelihoods of local communities, and contribute to sustainable development. REDD+ provides financial incentives to developing countries that implement effective forest conservation policies and practices, providing a framework for financing projects that combat deforestation and promote sustainable management of natural resources.

Another significant initiative is the Low Carbon Agriculture Program (ABC), which promotes sustainable agricultural practices such as no-till farming, integrated crop-livestock-forest systems, and land restoration. These practices not only reduce carbon emissions but also contribute to climate change adaptation, increase the resilience of agricultural systems, and promote food security.

Brazil also actively participates in international agreements on climate change, such as the Paris Agreement, in which the country committed to reducing its emissions by 37% by 2025, with a perspective of reducing by 43% by 2030, compared to 2005 levels. To achieve these goals, the country has sought to promote sustainable development policies, incentivize the adoption of clean technologies, and strengthen international cooperation in combating climate change.

In summary, Brazil has implemented a series of initiatives to reduce greenhouse gas emissions, addressing key areas such as deforestation, energy, agriculture, and transportation. These efforts are crucial to contribute to global efforts to combat climate change and promote more sustainable and resilient development in the country. Furthermore, the lack of ongoing commitment to environmental conservation undermines Brazil's credibility on the international stage and hampers global efforts to protect the environment.

Currently, one of the most significant initiatives is Brazil's Nationally Determined Contribution (NDC) under the Paris Agreement, which outlines the country's commitment to reducing greenhouse gas emissions and addressing climate change. Brazil has pledged to decrease its emissions by 37% below 2005 levels by 2025, with an additional indicative target of 43% by 2030. The NDC underscores Brazil's acknowledgment of its role as a major emitter and its dedication to transitioning toward a low-carbon economy while concurrently fostering sustainable development.

Consequently, Nationally Determined Contributions are influenced by self-determination, the steps taken to mitigate greenhouse gases appear to be more of a political decision-making process rather than a purely technical necessity driven by the imperative to curb global temperatures [12].

Therefore, "the Green ICMS policy has been operational in several Brazilian states for several years, serving as an economic off-budget mechanism designed to fulfill constitutional objectives related to environmental preservation, the advancement of tax justice, and the encouragement of voluntary initiatives by municipalities to augment their revenue sources" [13]. A successful example of extra-fiscal measures policy has been observed.

The complete legitimacy of utilizing extra-fiscal mechanisms is achieved as stipulated in the Federal Constitution, when the aim is to champion fundamental rights, particularly in environmental protection, while simultaneously reconciling national economic development with the responsibility to preserve the environment and advance socio-environmental justice [14].

Extra-fiscality, serving as an instrument of fiscal (tax) policy, not only caters to fundraising requirements but also emerges as a tool to fortify fundamental socio-environmental rights and foster socio-environmental justice. Extra-fiscality aligns with the concept of social responsibility, striving not only to finance the State but also to enhance social well-being through public policies focused on sustainable social development, in Ref. [13].

Despite the existence of an old norm provision, it can still be regarded as an incipient instrument in Brazilian law due to its limited usage. Currently, two major categories of tax solutions are employed to attain an ecologically balanced environment: the imposition of environmental taxes, relying on conduct standards and cost internalization, and the establishment of incentives for sustainable production, utilizing the "incentive-awards" model to prompt ecologically responsible decisions in both industry and its products.

The statement suggests that the Green ICMS can be easily introduced as an instrument based on the existing system with only a change in the State's law. It indicates that its impact is limited to the revenue transferred by the State to cities for compensating environmental efforts. The underlying logic is straightforward, proposing that economic development, achieved primarily through land use, compensates those unable to utilize it or protect practical sustainable use. The Green ICMS is presented as a financial option for municipalities struggling to generate sufficient sales tax to ensure a high income, in Ref. [13].

The Green ICMS model, already implemented in Brazilian states, anticipates diverse distribution criteria, with conservation units being the most prevalent. Within the spectrum of conservation units, various levels of protection are established. The Integral Protection Conservation Unit strives to preserve nature by permitting only indirect utilization of its natural resources, except in cases specified by law. Conversely, the Sustainable Use Conservation Unit endeavors to harmonize nature conservation with the sustainable exploitation of a portion of its natural resources. Additionally, there exists an interplay between policies concerning solid waste management, biodiversity conservation, and the safeguarding of indigenous and quilombola territories.

Indigenous and quilombola areas stand out in mitigating climate change due to sustainable land management, preservation of forest cover, and active resistance against predatory activities such as deforestation and mining. Their practices of subsistence agriculture and biodiversity conservation contribute to carbon emissions reduction, while their traditional knowledge promotes sustainability and climate resilience.

The implementation of the Ecological ICMS (Tax on Circulation of Goods and Services) faces several challenges, predominantly rooted in administrative complexities, inadequate monitoring systems, and insufficient financial resources.

The lack of continuity in updating data produced by the Brazilian government stands out, for example, the last National GHG Inventory was published in 2016.

Another weakness is the fact that policies to combat and prevent deforestation in the Cerrado and Amazon biomes are no longer legally valid between 2019 and 2023, with deforestation reaching the mark of 13 thousand km^2 in 2021, which has not occurred since 2006, moving the country away from the goals established in international agreements [15].

The Cerrado, as a hot spot for biodiversity and the source of nine river basins, receives little legal protection to combat deforestation. One of the main difficulties lies in the lack of standardized criteria to determine the distribution of funds between municipalities, leading to disparities and potential disputes between local governments. The reason for this can be pointed out to be the fact that there are 5568 Brazilian municipalities, plus two districts (Fernando de Noronha and Distrito Federal). Therefore, the individualities of each location must be taken into account, making the process of standardizing ICMS Verde implementation criteria difficult.

Additionally, there are challenges related to accurately measuring and verifying environmental indicators, such as biodiversity conservation and ecosystem services, which are crucial in determining each municipality's eligibility for funding.

Furthermore, the effectiveness of the Ecological ICMS depends on the availability of adequate financial resources, which may be limited due to competing budgetary priorities and economic constraints. These restrictions make command and control action by inspection bodies unfeasible.

Overcoming these challenges requires robust governance structures, better data collection mechanisms, and greater cooperation between government agencies, between federal entities, and local stakeholders to ensure the equitable and transparent implementation of the Ecological ICMS. Ultimately, concerted efforts are required to overcome these obstacles and effectively implement the Ecological ICMS, ensuring the conservation of Brazil's natural resources for future generations.

3. Conclusions

The Ecological ICMS holds significant importance as a tool for environmental preservation and as an incentive for municipalities that adhere to it. In summary, the recognition of this importance is mainly attributed to the following factors:

It has proven to be an excellent instrument for encouraging municipalities to establish or advocate for the creation of new conservation areas and improve the quality of existing protected areas. This represents a significant increase in their revenues due to a greater share in the distribution of the collected amount, as outlined in Article 158, IV, § unique, II, of the Brazilian Constitution.

The Ecological ICMS provides an opportunity for the Federal State to influence the process of sustainable development in municipalities, allowing for the compensation of environmentally desirable activities. This makes the revenue sharing with the ICMS a crucial instrument for public policy and the operationalization of a set of innovative principles to enhance Brazilian environmental management, particularly that of the provider-receiver.

Extra-fiscality through the Ecological ICMS, besides being an important source of revenue for many municipalities, serves as a tool to promote sustainable development, nature conservation, and biodiversity protection. This is manifested by: (a) an increase in the number and extent of conservation units and other specially protected areas (quantitative dimension); (b) the regularization, planning, implementation, and pursuit of sustainability of conservation units (qualitative dimension).

The implementation of the Ecological ICMS (Tax on Circulation of Goods and Services) encounters numerous challenges, mainly due to administrative complexities, inadequate monitoring systems, and insufficient financial resources. A critical issue lies in the irregular updates of data provided by the Brazilian government.

Another significant obstacle is the absence of standardized criteria for distributing funds among Brazil's vast number of municipalities, leading to disparities and potential disputes among local governments.

Furthermore, accurately measuring and verifying environmental indicators such as biodiversity conservation and ecosystem services pose significant challenges in determining each municipality's eligibility for funding. Furthermore, the effectiveness of the Ecological ICMS hinges on the availability of sufficient financial resources, which are often constrained due to competing budget priorities and economic limitations, rendering enforcement actions by inspection bodies impractical.

Addressing these challenges requires robust governance structures, improved data collection mechanisms, and enhanced cooperation among government agencies, federal entities, and local stakeholders to ensure equitable and transparent Ecological ICMS implementation. It is imperative to prioritize regular updates of environmental data, strengthen legal enforcement measures against deforestation, and develop standardized criteria for fund distribution among municipalities. Furthermore, investing in innovative technologies and capacity-building initiatives can enhance monitoring systems and facilitate accurate measurement of environmental indicators and instruments that already exist.

Author details

Adrielle B.O. Sardinha*, Epaminondas J. Messias and Carlos A. Leonetti
Federal University of Santa Catarina, Florianópolis, Brazil

*Address all correspondence to: adrielle.sardinha@mail.uft.edu.br

References

[1] da Silva Junior CA. Fire in Brazilian biomes in the face of emissions and international agreement. In: Anais do XX Simpósio Brasileiro de Sensoriamento Remoto. Florianópolis, SC, Brasil: São José dos Campos: INPE; 2023. Available from: https://www.proceedings.science/sbsr-2023/trabalhos/fire-in-brazilian-biomes-in-the-face-of-emissions-and-international-agreement?lang=pt-br

[2] IPCC. 2019 Refinement to the 2006 IPCC Guidelines for National Greenhouse Gas Inventories. Available from: https://www.ipcc.ch/report/2019-refinement-to-the-2006-ipcc-guidelines-for-national-greenhouse-gas-inventories/

[3] SEEG (Greenhouse Gas Emissions Estimation System) is the First Nongovernmental Initiative in the World to Calculate Annual Emissions from All Sectors of the Economy. Available from: https://plataforma.seeg.eco.br/?_gl=1*1j4r5pt*_ga*NzQ3NzQxMDk3LjE3MDcyMjYwNjE.*_ga_XZWSWEJDWQ*MTcwNzU4ODUxNy40LjAuMTcwNzU4ODUxNy4wLjAuMA

[4] Global Carbon Atlas. Available from: https://www.globalcarbonatlas.org/emissions/land-use/

[5] Gatti LV et al. Amazonia as a carbon source linked to deforestation and climate change. Nature. 2021;**595**(7867):388-393. DOI: 10.1038/s41586-021-03629-6

[6] Caballero CB, Ruhoff A, Biggs T. Land use and land cover changes and their impacts on surface-atmosphere interactions in Brazil: A systematic review. Science of the Total Environment. 2022;**808**:1-7. DOI: 10.1016/j.scitotenv.2021.152134

[7] Montero CEP. Tributação Ambiental: Reflexões Sobre a introdução da variável Ambiental no Sistema tributário. São Paulo: Saraiva; 2014

[8] Brasil [Constituição (1988)]. Constituição da República Federativa Do Brasil de 1988. Brasília, DF: Presidência da República; 1988. Available from: http://www.planalto.gov.br/ccivil_03/constituicao/constituicao.htm [Accessed: December 8, 2023]

[9] Loureiro W. Contribuição Do ICMS Ecológico à Conservação da Biodiversidade no Estado Do Paraná. Tese de Doutorado. Curitiba: (Curso de Pós-Graduação em Engenharia Florestal)—Setor de Ciências Agrárias, UFPR. Available from: https://www.acervodigital.ufpr.br/bitstream/handle/1884/25377/T%20-20LOUREIRO?sequence=1; 2002 [Accessed: November 20. 2017]

[10] Lucik MR. Instrumentos tributários Para a sustentabilidade: Uma análise comparativa da destinação do ICMS pelos estados segundo critérios ambientais. In: CAA De; CL de Andrade; RR Lodi (Org.). Tributação e Sustentabilidade Ambiental. Rio de Janeiro: FGV; 2015

[11] Derani C, Oliveira AB. Intrumentalization of the normative bottom-up paradigm of the Paris agreement by the rulebook. Veredas do Direito, Belo Horizonte. 2021;**18**(40):331-358. DOI: 10.18623/rvd.v18i40.1943

[12] Bodansky D, Brunnée J, Rajamani L. International Climate Change Law. Oxford: Oxford University Press; 2017

[13] Tupiassu L et al. Green ICMS: Brazil's tax revenue distribution based

on environmental criteria. In: Hymel M, Kreiser L, Milne JE, Ashiabor H, editors. Innovation Addressing Climate Change Challenges Market-Based Perspectives. Cheltenham, UK: Edward Elgar Publishing; 2018. DOI: 10.4337/9781788973366.00023

[14] Caliendo P et al. Tributação e sustentabilidade ambiental: A extrafiscalidade como instrumento de proteção do meio ambiente. Revista de Direito Ambiental. 2014;**76**:471 | Out / 2014 | Dez / 2014 DTR\2014\20451

[15] Brazil. Desmatamento na Amazônia Tem Queda histórica de 66% Em Julho. 2023 [Internet]. Available from: https://www.gov.br/pt-br/noticias/meio-ambiente-e-clima/2023/08/desmatamento-na-amazonia-tem-queda-historica-de-66-em-julho

Chapter 4

Environmental Taxation and Well-Being in the Context of Globalisation and Sustainability

M.J. van Hulten

Abstract

Environmental taxation exemplifies and emphasises the global dynamics of sustainability. Combining environmental taxation and well-being can contribute to broad aims of sustainability, such as the survival of humankind, protection of the biosphere and its biodiversity, social solidarity and justice and the preservation and development of our natural and cultural heritage. As a result of environmental market failures, hardly any market price reflects the true overall cost. Environmental taxation can provide suitable counters to such market failures. Global dynamics of sustainability clearly show in issues such as marine litter, air pollution, and climate change. Recognising the reality of 195 different states, environmental taxes would often be most effective if true international cooperation would be achieved. Unilateral environmental taxes may trigger relocation or dumping/burning abroad. The choice for, and design of, environmental taxes affects the distribution of environmental burdens and benefits, now and in the future. Critical scrutiny is needed, for example, if effects of environmental taxes are particularly felt in certain regions or parts of the population, or if such taxes unfairly affect competition. In this chapter, and in relation to environmental taxes and the global dynamics of sustainability, practices and possible strategies are assessed, and recommendations are made.

Keywords: environmental taxes, taxation, well-being, globalisation, sustainability, sustainable development goals, economy, lifestyle

1. Introduction

Environmental taxation provides key examples of, and attention points regarding, the global dynamics of sustainability. Namely, and in the context of sustainability and of the Sustainable Development Goals (SDGs), environmental taxation can be seen to start from the proposition that an environment adequate for health and well-being is essential for all human beings including future generations and that sustainability requires views of human needs and well-being that also incorporate variables such as health, clean air and water, and the protection of natural beauty [1, 2]. Taxes can be used to stimulate behaviour where it is beneficial to the environment (one may think of incentivising less-polluting automobiles) or to discourage behaviour where such behaviour is detrimental to the environment (for example, specific tax levies on polluting consumption of fuels or plastics) [3]. More in general, environmental taxes can be implemented for instance as charges to cover costs, as instruments to incentivise behaviour, and/or as revenue-raising tools for spending on environmental causes [4]. There is therefore a clear potential role of environmental taxation in supporting sustainability transitions [5].

In a world which is currently divided into 195 states, the classic argument is for state intervention to focus on the externalities of behaviour which is harmful for the environment. However, and especially with the current global forms of organisations and enterprises, it is not always easy to make the polluter pay [6]. Here, the global dynamics of sustainability clearly show. Namely, and notwithstanding that many environmental issues are local, there are also many environmental aspects that have a global dimension to them [7]. Consider, for instance, the cross-border problems caused by marine litter, air pollution, and climate change. Simply put, what is consumed and produced in one country can generate waste elsewhere [8]. Natural processes are to a large extent global, since the impacts of matters such as climate change, overpopulation, and contamination of oceans, rivers, and atmosphere are not contained within states' borders. In this regard, the possibility of negative as well as positive environmental spillover effects implies that the impacts of actions aimed at the environment oftentimes carry over to also affect well-being abroad [7]. A clear example is that environmental taxes can have cross-border effects on business competition, which may affect the environment via relocation of production and/or via a shift of environmental damage elsewhere [4]. Considering the global aspects of environmental effects (for instance, climate change due to carbon dioxide emissions), such relocation can even result in overall environmental losses in case polluting activities are relocated to a low/non-taxing states that score worse on environmental efficiency [4]. Also in view of the before mentioned, a global distribution of environmental burdens and benefits has been argued in the literature [9].

Moreover, and again in the sphere of sustainability, environmental impacts also carry over into the future. Also here, various dynamics of sustainability are involved. As example: climate change more heavily affects future generations, while emission reductions more heavily affect current and near-future generations [10]. Environmental capacities which are weakening now may reduce nature's possibilities to sustain the well-being of future generations [8]. Environmental challenges therefore clearly call not only for a short-term but also for a long-term vision [8].

Taking a step back, the current and practical reality is that of the existence of different states that – by and large – each implement and apply their own systems and rules of (environmental) taxation. At the same time, whenever states take on an environmental focus, such requires significant considerations by states in determining who or what are in scope of their environmental taxes. Likewise, there are choices to be made regarding the distribution of environmental benefits and the distributional impacts of environmental taxes, whereby the appreciation of a certain policy instrument also depends on the weight accorded to such distributional aspects [11]. This may directly or indirectly affect dimensions of sustainability and of the SDGs. More specifically, environmental taxes might detract from obtaining distributional objectives, especially insofar lower-income households, relatively speaking, consume more energy-intensive goods [12]. In terms of environmental action, such as environmental taxes, the choice for a specific measure can thus affect whether or not, and if so to what extent, redistribution from rich to poor occurs, for example, through targeting international emission reduction at richer regions [10, 13].

There are examples of environmental taxation that were designed specifically with the international aspects of environmental issues in mind. One example is the European Union (EU)'s Carbon Border Adjustment Mechanism (CBAM), intended to put a fair price on carbon emitted during the production of carbon-intensive goods that are entering the EU and to thus encourage cleaner industrial production in countries outside the EU [14]. Such measures, at the same time, may raise concerns, for instance, regarding unfair welfare losses in the form of overburdening of developing countries [15], which may adversely affect other dimensions of sustainability and of the SDGs.

The before-mentioned illustrates the global dynamics of sustainability involved in environmental taxation. In this chapter, and in relation to well-being aspects of environmental taxes, the above-introduced questions, practices and possible strategies are further addressed, and recommendations are made.

1.1 Materials and methods

The research was performed through desk studies, most notably literature review and also case law, legislation and policy research. Materials used and referenced are predominantly in the English language. The topic is addressed from multiple perspectives: from the perspective of (tax) law and also, to some extent, from perspectives such as economics, social sciences, politics, and psychology.

The search for the literature was initiated through applying keyword searches – for example, environmental taxes/taxation, well-being, welfare, globalisation, sustainability, sustainable development goals, effectiveness, efficiency and combinations thereof - in various online databases, including the online databases of Tilburg University, Westlaw International and HeinOnline. Snowball search and citation search methods were also applied.

Selection of environmental taxation examples involved non-probability sampling, primarily purposive sampling (i.e., selection of examples based on the expertise and opinion of the author) and also convenience sampling (i.e., selection of easily accessible examples). Examples were sought to allow illustration with a broad potential relevance and with plausible or at least intuitive connections with well-being and/or the global dynamics of sustainability. As examples are primarily used for illustrative purposes, the research output is not expected to be invalidated by possible bias in the example selection process.

Recommendations regarding design and implementation of environmental taxes are often drawn from research focusing on welfare and/or environmental assessments, which research is expected to be relevant, although not all-encompassing and therefore second-best, for drawing inferences regarding well-being and the global dynamics of sustainability.

2. Environmental taxation and well-being in the context of globalisation and sustainability

2.1 Well-being aims in environmental taxes

The focus in this chapter is on environmental taxes and well-being in the context of globalisation and sustainability. A few words here on the terminology used. The environment is taken to be the natural world and all that is related to it. The term 'environmental taxes' is employed here in a broad sense to refer to taxes with a specific base that has a clear negative effect on the environment, or which seek to tax certain activities, goods or services so that environmental costs may be included in the respective price and/or so that producers and consumers are oriented towards activities that better respect the environment [16]. The well-being concept is used here to describe that which is ultimately, or non-instrumentally, good for a person while leaving room for the realistic possibility of different concepts of well-being and divergent views on what constitutes or contributes to well-being. Besides human well-being, concern could furthermore be had for the well-being of (other) animals or other species (such as plants) [17], or for the well-being of the environment as such [8]. It should be noted that views on the (desired) scoping of environmental taxes, in the context of well-being and sustainability, depend to a large extent on ecological, social and cultural perspectives, which perspectives may in turn differ depending on, amongst other things, time and place.

What seems rather uncontroversial is the proposition that environmental conditions can affect well-being in many ways, for instance, the joy of living in an environment rich with nature or the impact on health of clean water, pollution levels, noise or natural disasters [18]. Access to green spaces is one of the significant factors (acting independently) to predict mental well-being, whereby dissatisfaction with green spaces is a predictor of poor mental well-being [19]. Areas that face a rapid loss of wild nature often face relatively high poverty and dense human settlement and are also more often characterised by social conflict [17]. The Organisation for Economic Co-operation and Development (OECD) concluded that our well-being relies on resources and environmental services that natural assets provide and that, in order to ensure that progress in living standards does not grind to a halt, new ways of producing and consuming would have to be found, and (how to measure) progress would have to be redefined to include amongst other things contributions to well-being [7]. While information may still sometimes be patchy, the consequences of the great erosion of wild nature for human well-being are likely on the whole to be both negative and substantial, which gives reason for due consideration. There is a need, in this regard, for models that describe the linkages between nature, human well-being and the drivers of ecosystem change, even if those may initially be crude and inaccurate, as well as a considerable need for local field studies quantifying the impacts on human well-being of natural changes [17]. Ideal approaches (such as studying how climate change affects human welfare through health and nature in each area around the world) are, however, expensive and time consuming, as a result of which a choice is often made for monetised estimates using economic values [20]. Expressed in monetary value, and as an example, the environmental damage caused by emissions to air, earth, and water in 2015 in the Netherlands was calculated to translate into a loss of more than 31 billion euro [21]. In any case, environmental taxation is one of the ways which can be employed to try and better reflect (the consequences of) such environmental damage into market prices [21].

As mentioned in the introduction, environmental taxes can be seen to be targeted towards environmental protection (reducing harm done to the environment) [22] or towards environmental promotion (improving environmental conditions). A recommendation that is often found in the literature is that environmental taxes are targeted at the pollutant or polluting behaviour, and aims other than environmental may best be addressed using other or complementary policy instruments [23]. A risk is that attempts to address, via taxation, both environmental and other aims (for example distributional concerns) may result in the tax not achieving either of those aims (or not to satisfaction) [23].

Environmental aims and well-being do not always fully correlate, especially in the short run. For instance, environmentally beneficial innovations and green overhauls can in the short-term result in well-being losses in the forms of, for example, obsolescence of certain current employee skills and of physical capital, unemployment, transport to new industries and new locations, (re)training and equality costs in terms of windfall gains and losses imposed on particular societal groups [24]. Notwithstanding, in the long run and in the absence of proper environmental policy measures, predicted future losses of wildlife are very likely to considerably erode well-being [17]. An example in the field of environmental taxation is that the phasing out of non-environmentally friendly tax exemptions can have negative consequences for some countries or population groups in the short term, while such phasing out may be expected to have overall positive economic, environmental and well-being impacts in the long term [7].

Arguably, an important role that well-being can play in matters of the environment, and in matters of environmental taxation, is to ensure that public policy consideration goes beyond economic (welfare) aspects [25]. Sustainability issues require states to consider environmental measures and effects not in isolation (for instance only with regard to income, or only with regard to health) but in integration, taking into account how different government policies affect each other. In that sense, the combination of environmental taxation and well-being can contribute to pursuing broader aims of sustainability, such as the survival of humankind, the protection of the biosphere and its biodiversity, social solidarity and justice and the preservation and development of our natural and cultural heritage in the long term.

Most often, positive well-being effects of environmental 'gains' (e.g. reduced pollution or improved natural conditions) attained through environmental taxes will not be limited to one specific group, as such environmental gains can be expected to spread through positive externalities to positively affect the well-being of many. Poorer regions and persons may be more exposed to environmental damages (e.g. fine-particles pollution, traffic noises or desertification of land) [26], and at the same time, high-income households' well-being can be affected by 'luxuries' such as pristine wilderness or high-quality public parks [24]. The before-mentioned again illustrates the global dynamics of sustainability, which warrant careful consideration when dealing with environmental taxation.

2.2 State intervention in the environment

Given the current lack of an authorised global organisation or effective cooperation system (undisputedly) representing the common interests of humankind efficiently, it is typically held that states should intervene in the environment, considering associated and otherwise unaddressed environmental market failures.

For instance, air pollution, noise and parks and green spaces all have important (negative or positive) externalities, since related decisions made and actions undertaken by individuals affect other individuals as well [27]. As mentioned, public intervention can be designed to increase positive externalities and decrease negative externalities, for instance, via environmental taxation [25].

Benefits of nature are often substantial and at the same time also generally non-marketed and public (e.g. fresh air) [17]. As a result, decisions and activities that carry negative consequences for the environment may make narrow economic sense, in that such decisions and activities – for instance, habitat conversion, or overexploitation - can generate private gains for the firms involved [17]. There is no, or almost no, market incentive for firms and households to take into account environmental damages caused, as environmental damages often come with little to no (direct) cost to the polluter [23]. However, those private gains are typically more than offset by accompanying societal losses, so that overall, the activities are detrimental and can from that perspective be considered undesirable [17].

Next to negative externalities, environmental actions also encounter market failures regarding positive externalities, namely in those situations where environmental benefits are enjoyed by others than those investing in the environmental actions, which may result in underinvestment [16].

Furthermore, there are environmental market failures caused by coordination failures such as lacks in effective and efficient cooperation on an international level and by asymmetric information (for instance, resulting in uncertainty regarding risks and returns of projects, which uncertainty tends to be high for environmental investments that typically are long term) [16].

As a result of environmental market failures such as those discussed before, hardly any market price reflects the true overall cost (including environmental costs and benefits) [8]. These market failures are a main cause of environmental decline and require adequate intervention that is often best achieved through collective – and international – action [28]. For many environmental concerns and especially in open economies, the possibilities for effective national policy intervention through environmental taxes are limited if international coordination between states is not feasible [28]. Environmental taxes would in many situations be most effective and lead to the highest welfare and well-being gains if effective international cooperation would be achieved (for instance, in the form of global carbon emission taxation) [27, 29]. Such a global way of international and local cooperation in environmental taxation, for instance, through an authorised global organisation or effective cooperation system which would represent the common interests of humankind efficiently, is, however, currently not the status quo. As mentioned in the introduction, there are examples of environmental taxation designed with the international aspects of environmental issues in mind, such as the EU's CBAM. However, for now, such measures seem second-best, for instance, when considered from the perspective of international inclusivity versus policing/coercive mechanics.

2.3 Environmental taxes compared to alternatives

When considering environmental taxation as a potential strategy in the context of sustainability, assessments should be made in comparison to other public policy instruments [12]. Also, in this regard, the current tendency – or practical reality – is to focus on state, rather than global, intervention.

Environmental taxes can be employed as a form of price- and market-based policy instruments, that is, policy instruments that aim to achieve market prices that better reflect all costs involved [21]. Tradable emission rights or subsidies for pollution reduction are examples of other market-based policy instruments which can contribute to correcting market failures as regards environmental prices [21].

A benefit of – stable – environmental taxes in comparison to tradable emission rights or other market-based instruments is that environmental taxes can provide an ongoing, strong incentive to keep on innovating, whereas in the case of tradable emission rights, adoption of new technology may simply reduce the prices of the tradable rights/permits to a level below the optimal tax rate – and thus weaken incentives for innovation in the future [28]. In addition, it is sometimes argued that environmental taxes have a benefit over tradable emission rights in the form of revenue collection and the subsequent possibility to recycle such revenues into reduction of other, economically distortionary taxes [11]. However, it can be questioned whether such a benefit actually exists, as environmental taxes come with their own distortionary effects [30], and considering that successful environmental taxes may generate reduced budgetary income in the future. The choice for one or the other policy therefore seems more properly directed to considerations other than revenue, such as effectiveness, ease of administration, enforceability, compliance or political viability [31]. When environmental aims are pursued through tax incentives, costs associated with those incentives may not always be proportionate to the environmental gains achieved. As an example, tax incentives that were introduced in the Netherlands in 2007 to stimulate fuel-efficient and electric driving for cars resulted in relatively low carbon dioxide emissions related to newly sold automobiles and in a relatively high percentage of electric cars in use [32]. These results came at the cost, however, of increased complexity, reduced tax revenue, high expenses and increased fraud [32].

In comparison to environmental taxes, subsidies can allow for better targeting and can alleviate free-rider problems that may occur if taxes are broadly set. Targeted subsidies, however, bring complexity and possibly related higher administration and compliance costs [22]. Also, where subsidies and other environmental policy tools typically have to consider long-term horizons in order to see effects, environmental taxes can work over shorter periods of time as well [4]. Moreover, subsidies effectively do not make the polluter pay for environmental damage but rather, and effectively, have all taxpayers (who may already be suffering negative externalities from such environmental damage) pay in order to induce pollution reduction, which can go against perceptions of fairness in society [11].

Governments can also use a whole range of non-market environmental instruments, for instance, command-and-control regulations – legislation, permits, bans, minimum performance standards, liability rules and pollution quotas – state financing, public-private partnerships, information campaigns or green tech support policies [4].

In general, environmental taxes are quite favourably assessed in comparison to regulatory policy instruments such as standards, quotas or product bans [22]. One of the main advantages of environmental taxes, and of other market-based instruments, such as tradable emission rights, is that market-based instruments provide an efficient way to reduce environmental damage that can avoid deadweight losses [11]. Environmental taxes, if properly set, can induce polluters in a flexible and rather cost-efficient way to optimally reduce pollution [33]. Moreover, environmental taxes can provide a continuing, dynamic incentive for polluters to reduce their pollution (for instance, through green innovation), while regulatory instruments typically only incentivise polluters up to a certain point [28]. Environmental taxes are likely to be particularly valuable when wide-ranging behavioural changes are needed across a large number of activities, in which cases the costs of direct regulation would be (too) large [33]. That said, conventional command-and-control instruments have proven to be very effective in directing change and can be especially useful when environmental problems are complex or when international coordination, needed for many market-based instruments, turns out to be unattainable [28]. Caution should be had to avoid that environmental taxes are seen as a legitimation of polluting behaviour [33] or that environmental taxes generate unintended behaviour that detracts from the environmental goal, such as how energy taxes may accelerate the depletion of fossil fuel if such taxes raise the expected future energy prices [33, 34].

Ultimately, the assessment of environmental taxes in comparison to other policy instruments depends significantly on the specifics of the measures at hand. For instance, emission taxes come with uncertainty and associated monitoring and enforcement costs that can make other instruments preferable, notwithstanding the fact that emission taxes can be effectively targeted to the externality in question [11]. Sometimes, a mixture of different policy instruments is optimal [11].

2.4 Recommendations regarding effectiveness and efficiency of environmental taxes

Next, and taking into account the global dynamics of sustainability, general recommendations are made regarding effective and efficient design and implementation practices and strategies for environmental taxes:

- For reasons as discussed before, it is recommendable to focus on the careful design of environmental taxes that aim for important long-term environmental objectives, rather than focusing on the revenue-raising function of such environmental taxes [28].

- Careful coordination, design and planning with other policy instruments are required to prevent suboptimalities caused by conflict or overlap [28].

- The introduction of complementary policy instruments that stimulate innovation is recommendable if environmental taxes target technological development in areas that tend to build on existing technologies (such as based on fossil fuel) and that therefore face high costs associated to the adoption of new technologies [28].

- .Incentivising effects of environmental taxes can be increased if the environmental taxes result in a gradual and predictable increase of the market prices over long time periods or, alternatively, if existing exemptions or other tax expenditures are gradually reduced [4].

- Environmental taxes typically work best when they are part of a broader policy package with other, complementary instruments which, for instance, stimulate green innovation or raise environmental awareness [35].

- Environmental taxes are most effective if they are directly targeted at the polluter or polluting externality that causes the environmental damage [28].

- Taxes that result in a fixed amount of charge per unit of quantity (e.g. tonnes of emissions) are preferable to taxes that are charged as a percentage of the price of such units or goods, since the latter also cover activities not directly related to the environmental damage (such as management or marketing costs), and since the latter effectively favour relatively cheap activities, which are often the most polluting, and may therefore exacerbate undesired shifting of production abroad [28].

- Environmental taxes would in many situations be most effective and lead to the highest gains in terms of the environment, well-being and sustainability if effective international cooperation would be achieved (for instance in the form of global carbon emission taxation) [29].

- To increase efficiency, environmental taxes are recommended to have a broad scope and be simple, containing as little exceptions as possible [23].

- Administrative and compliance costs can be reduced where environmental taxes can be easily included in already existing tax systems (such as excise taxes, sales taxes or value added taxes), instead of implementation via separate, new tax systems [28].

- It is often argued that optimal environmental taxes should be set below the marginal social benefit (i.e. below the Pigouvian rate that reflects the marginal external damage from the polluting activity) [11]. This takes into account that environmental taxes bring additional costs, for instance, in the form of monitoring costs, distributional effects and increases of labour distortions [11].

- Environmental taxes are recommended to be sufficiently substantial in order to outweigh related administrative and compliance costs and to justify their use in terms of effectiveness [24].

- Environmental taxes need to be sufficiently flexible and dynamic to develop alongside new technologies and unexpected opportunities and with economic or environmental developments [24]. Such flexibility, however, can generate conflict with the certainty required for long-term investments and therefore requires a thorough effectivity and efficiency assessment [24].

- It is in any case recommended to ensure that critical assessments of environmental taxes are performed, both beforehand as well as after implementation, to ensure regular monitoring and evaluation of the effectiveness and efficiency of the measures at hand.

2.5 International legitimacy aspects of environmental taxes

In the context of globalisation, sustainability and well-being, the consideration of environmental taxes also requires due concern for their potential international legitimacy issues [36].

For instance, international and supranational bodies of rules or laws to a certain extent restrict the possibilities for environmental taxes. Environmental taxes can be in conflict with the EU's internal market, for example, or with the World Trade Organisation (WTO) rules [4].

Within the EU, some aspects of environmental taxes are regulated at EU level. Already mentioned was the EU's CBAM. As another example, Council Directive 2003/96/EC of 27 October 2003 sets an EU framework (including minimum levels of taxation and obligatory exemptions) for the taxation of energy products and electricity [37]. Tax laws of EU Member States need to adhere to the requirements set out in this Directive, so as to not infringe on the legality of the taxes imposed [38]. Furthermore, environmental taxes of EU member states also need to adhere to the EU rules on fundamental freedoms and state aid, which seek to foster fair and open competition in the European market [39]. This, for instance, means that a tax on stopovers for tourist purposes by aircrafts used for the private transport of persons cannot be imposed only on operators whose tax domicile is outside the territory of the EU member state concerned [40]. Furthermore, the effective protection and sustainable management of forests and forest areas are environmental issues of a cross-border kind that imply a common responsibility of the EU member states, considering that one forest area can extend across multiple jurisdictions [41]. Thus, an inheritance tax exemption applicable only to sustainable forests located in a certain EU member state, with exemption aims of environmental protection, constituted an unjustified infringement of the European free movement of capital [41].

Environmental taxes, applied by WTO members, also need to conform to the requirements following the WTO agreements. The non-discrimination requirements following from these agreements, such as requirements of national and most-favoured-nation treatment, can affect environmental taxes, notwithstanding that exceptions, carve-outs, lack of use in practice and/or political stalemates may mitigate their impact [42].

Other restrictions may, for instance, be imposed through international treaties (such as the European Convention on Human Rights, and the International Covenant on Civil and Political Rights), (other) trade agreements (such as the North American Free Trade Agreement) or double tax conventions or may concern more political, non-binding restrictions via soft-law instruments (such as the European's Code of Conduct for business taxation, or the reviews conducted by the OECD's Forum on Harmful Tax Practices) [43]. States are sometimes held accountable and obliged, under current law, to take environmental action. An example can be found in the judgement of the Court of Appeals of The Hague from October 2018, in which the Netherlands was found to violate its responsibilities towards the protection of civilians within its jurisdiction under Articles 2 (right to life) and 8 (right to respect for private and family life) of the European Convention on Human Rights by not striving for an at least 25 per cent reduction of greenhouse gasses by the end of 2020 [44]. The mentioned judgement of the Court of Appeals of The Hague was upheld by the Dutch Supreme Court in December 2019, in case No. 19/00135 [45].

From the perspective of globalisation and sustainability, environmental taxes can raise equal treatment concerns, for instance, if the effects of such environmental taxes are particularly felt in certain regions or parts of the population or if those taxes lead to discrimination between (otherwise comparable) economic activities [46]. More generally, environmental taxes, at least when viewed in isolation, can conflict with conceptions of ability-to-pay, as environmental taxes are typically based on the contribution to pollution (and may therefore have stronger incentivising effects on lower-income taxpayers) [28]. Environmental taxes can also affect competition in a way that is perceived to be unfair, when environmental taxes are in effect primarily borne by certain groups (for instance, by domestic rather than by foreign enterprises) [42]. As mentioned in the introduction, the EU's CBAM has been criticised for causing unfair welfare losses in the form of overburdening of developing countries [15, 47].

2.6 Environmental taxes and behavioural impacts

Behavioural impacts of environmental taxes can in various ways affect global dynamics of sustainability. If environmental taxes make polluters pay for the pollution they cause, this prevents free-rider behaviour (i.e. behaviour whereby businesses or consumers burden the environment without facing the related costs) and thus can create conditions which increase well-being [28]. However, and as previously mentioned, environmental taxes could have adverse effects, for instance, if taxpayers believe that paying those taxes legitimises polluting behaviour [33]. One view in this regard is that pollution is a 'crime against nature' that should be stigmatised by legal regulations (e.g. bans or punishments) rather than being condoned through tax payment [24].

If environmental taxes come in the form of tax expenditures and subsidise environmentally beneficial actions (a common example is accelerated tax depreciation for environmentally friendly investments), environmental taxes might partially be seen as rewarding polluters who have up to now performed undesirable, environmentally damaging behaviour. In such cases, environmental taxes do not make the polluters pay but rather distribute the tax burden amongst the recipients of pollution to pay for inducing pollution reductions [11].

Environmental taxes can also spark, or curb, tax competition between states. Competition can be beneficial from a sustainability perspective [48]. An example is where environmental gains arise as a result of more optimal allocation or innovation triggered by tax competition [49]. If, however, environmental taxes merely result in competition between states for tax revenues, for instance, in the case of neighbouring countries involved in a race to the bottom for excise taxes on fuel, such competition is likely overall harmful in a beggar-thy-neighbour sense –achieving low or no environmental gains, or even causing environmental losses, while simultaneously resulting in an overall erosion of tax revenues with related detrimental effects on well-being [50]. Or, in the absence of international cooperation, environmental taxes can trigger legal or illegal dumping (or burning) abroad or other forms of behaviour with negative spillover effects to other regions or states [51]. International cooperation, setting minimum taxation that restricts the inefficiencies arising from a tax race to the bottom, while still allowing non-harmful competition, can assist in stimulating desirable behaviour [49]. However, as long as there is a lack of international law forcing international cooperation, there will be incentives for free-rider behaviour, as states can enjoy positive externality spillovers from the environmental actions of other states, without (fully) bearing the associated costs [52]. As previously discussed, environmental taxes also affect the distribution of costs and benefits over different market participants and, in doing so, directly or indirectly affect dimensions of sustainability and of the SDGs next to the environmental dimensions. Win-win situations are rare, especially when viewed in the short-term, as environmental taxes are often intended to generate welfare gains for pollution victims and welfare losses for polluters [28]. Environmental taxes that might be seen as efforts to try and curb tax competition (or even to stimulate a race to the top) – such as the EU's CBAM which puts a price on carbon emitted during the production of carbon-intensive goods that are entering the EU – come with concerns of unfairness from a distributive perspective [53].

In general, public support for environmental taxes seems to be favourable. A Eurobarometer Poll conducted in 1995 concluded that the vast majority of Europeans favourably view (gradually introduced) green taxes as a solution to reduce the environmentally harmful effects of our lifestyles, even if such taxes would cause a slight economic slowdown [54]. It was deduced that almost 90% of Europeans are in favour of application of the polluter-pays principle [54]. However, where it concerns views on the effectiveness of ways to tackle air pollution, European respondents favoured command-and-control and information mechanisms (such as stricter pollution controls on industrial and energy production activities) over, for instance, tax incentives for low-emission products or increasing taxation on air polluting activities [55].

The before-discussed potential regressive effects of environmental taxes can decrease public support for such measures [24]. The possibility that low-income consumers and households pay proportionally more external environmental costs, because they spend proportionally more on most environmentally sensitive goods than do richer groups, calls for serious political attention in order to ensure that there is sufficient public support for environmental taxes [4]. In France in 2018, a proposed increase in fossil fuel tax, intended to fight climate change through encouraging conservation and alternative energy use, was suspended after violent protests [56]. Analysts suggested that the proposed tax increase would fall the hardest on people with stagnating incomes and with limited transportation options outside of automobile usage [56]. As reporters observed, "Rather than spurring the effort to cut fossil fuel use, the misstep now threatens to set it back" [56]. To counterbalance undesired distributive effects, states can complement environmental taxes with other, progressive measures (e.g. via revenue-recycling) [7].

3. Conclusions

In this chapter, and in relation to environmental taxes and the global dynamics of sustainability, practices and possible strategies are assessed, and recommendations are made.

Environmental taxation starts from the proposition that an environment adequate for health and well-being is essential for all human beings, including future generations. It is possible to design tax measures that effectively and efficiently provide incentives for behaviour beneficial to the environment or disincentives for behaviour detrimental to the environment. The combination of environmental taxation and well-being can contribute to pursuing broad aims of sustainability, such as the survival of humankind, the protection of the biosphere and its biodiversity, social solidarity and justice and the preservation and development of our natural and cultural heritage in the long term.

As a result of environmental market failures, hardly any market price reflects the true overall cost including environmental burdens and benefits, and environmental taxation can sometimes provide a suitable instrument to counter such market failures. In the practical reality of a world currently divided into 195 states, the classic argument is for state intervention to focus on the externalities of behaviour which is harmful for the environment. It is, however, not always possible for states to make polluters pay. With the current multinational forms of organisations and enterprises, the global dynamics of sustainability clearly show. Environmental taxes would in many situations be the most effective and lead to the highest welfare and well-being gains if true international cooperation would be achieved. There are many environmental aspects that have a clear global dimension to them, such as cross-border issues and environmental spillover effects caused by marine litter, air pollution and climate change. The choice for and design of a specific environmental tax affects the distribution of environmental burdens and benefits, for instance, through targeting international emission reduction at richer regions to try and prevent overburdening of developing countries. Win-win situations are rare, especially when viewed in the short term. To counterbalance undesired distributive effects, states can complement environmental taxes with other policy measures.

Environmental issues impact intergenerational sustainability dynamics. Whereas emission reductions more heavily affect today's and near-future generations, the current weakening of environmental capacities may reduce nature's possibilities to sustain the well-being of later generations. Focus can be had for the careful design of environmental taxes that aim for important long-term environmental objectives, notwithstanding that environmental taxes can achieve effect over shorter time periods as well.

In the context of globalisation, sustainability and well-being, the consideration of environmental taxes also requires due concern for potential international legitimacy issues. Environmental taxes might come in conflict with the EU's internal market, or with the WTO rules, for instance. Environmental taxes can also raise concerns, for example, if the effects of such taxes are particularly felt in certain regions or parts of the population or if such taxes affect competition in a way that is perceived to be unfair. Tax measures can spark, or curb, competition between states. In the absence of international cooperation, environmental taxes can trigger legal or illegal dumping or burning abroad or other forms of behaviour with negative spillover effects to other regions or states. International cooperation and setting minimum taxation that restricts the inefficiencies arising from a tax race to the bottom, while still allowing non-harmful competition, can assist in stimulating desirable behaviour. As long as there is a lack of effective and inclusive international cooperation, there will be incentives for free-rider behaviour that is detrimental or suboptimal from a global sustainability perspective.

Acknowledgements

Parts of this chapter build on the doctoral thesis research performed by the same author at the Fiscal Institute Tilburg of Tilburg University, published for example in: Van Hulten MJ. Aiming for Well-Being Through Taxation: A Framework of Caution and Restraint for States. Fiscale Monografieën No. 160. Deventer. Kluwer; 2019. ISBN: 9789013156850.

Author details

M.J. van Hulten[1,2]

1 Fiscal Institute of Tilburg University, Tilburg, The Netherlands

2 Lubbers, Boer & Douma, The Hague, The Netherlands

*Address all correspondence to: m.j.vanhulten@tilburguniversity.edu

References

[1] United Nations. Report of the World Commission on Environment and Development: Our Common Future. New York, United Nations. 1987. See, e.g., Chapter 2, points 39 and 76, and Chapter 7, point 109

[2] For some examples of SDGs that are more closely linked to the environment dimension and/or to well-being, reference can be had to SDG 3 (Good Health and Well-Being), SDG 6 (Clean Water and Sanitation), SDG 11 (Sustainable Cities and Communities), SDG 12 (Responsible Consumption and Production), SDG 13 (Climate Action), SDG 14 (Life Below Water), and SDG 15 (Life on Land). United Nations. Transforming Our World: The 2030 Agenda for Sustainable Development. A/RES/70/1. New York, United Nations; 2015

[3] Rodi M. Tax expenditures for environmental purposes. In: International Fiscal Association, editor. Environmental Taxes and Charges. Vol. 18d. London, UK: Kluwer Law International; 1993. ISBN: 9789041100689

[4] EEA. Environmental taxes: Implementation and environmental effectiveness. In: Environmental Issues Series No. 1. Copenhagen: EEA; 1996. ISBN: 0201670006

[5] EEA. The role of (environmental) taxation in support of sustainability transitions. In: Briefing No. 22/2021. Copenhagen: EEA; 2022. DOI: 10.2800/24487

[6] Institute for European Environmental Policy. Green Taxation and Other Economic Instruments: Internalising Environmental Costs to Make the Polluter Pay. Publications Office of the European Union. Luxembourg; 2021. DOI: 10.2779/326501

[7] OECD. Towards Green Growth: A Summary for Policy Makers. Paris: OECD Publishing; 2011. DOI: 10.1787/9789264111318-en

[8] EEA. Well-being and the environment: Building a resource-efficient and circular economy in Europe. In: EEA Signals. Luxembourg: Publications Office of the European Union; 2014. DOI: 10.2800/13662

[9] Caney S. Cosmopolitan justice, responsibility, and global climate change. Leiden Journal of International Law. 2005;**18**:747-775. DOI: 10.1017/s092215605002992

[10] Tol RS. Welfare specifications and optimal control of climate change: An application of fund. Energy Economics. 2002;**24**(4):367-376. DOI: 10.1016/s014098830200010-5

[11] Bovenberg AL, Goulder LH. Environmental taxation and regulation. In: National Bureau of Economic Research Working Paper Series. Working Paper 8458. Cambridge, Massachusetts: National Bureau of Economic Research; 2001

[12] Heine D, Norregaard J, Parry IW. Environmental tax reform: Principles from theory and practice to date. In: IMF Working Paper. WP/12/180. Washington: IMF; 2012

[13] Domguia EN. Taxing for a better life? The impact of environmental taxes on income distribution and inclusive education. Heliyon. 2023;**9**(11):1-24. DOI: 10.1016/j.heliyon.2023.e21443

[14] Van Hulten MJ. The green Deal consultations: Well-being aims and tax related reforms. EC Tax Review.

2020;**6**:263-270. DOI: 10.54648/ecta2020054

[15] Sun X, Mi Z, Cheng L, Coffman DM, Liu Y. The carbon border adjustment mechanism is inefficient in addressing carbon leakage and results in unfair welfare losses. Fundamental Research. 2024;**4**:660-670. DOI: 10.1016/j.fmre.2023/02/026

[16] EC. Communication from the Commission: Guidelines on State Aid for Environmental Protection and Energy 2014-2020. 2014/C 200/01. Luxembourg: Publications Office of the European Union; 2014

[17] Balmford A, Bond W. Trends in the state of nature and their implications for human well-being. Ecology Letters. 2005;**8**:1218-1234. DOI: 10.1111/j.1461-0248-2005.00814.x

[18] Environment (or at least its present and future condition) is for instance indicated as one of the key dimensions of well-being in Stiglitz JE, Sen AK, Fitoussi JP. Mismeasuring our Lives: Why GDP Doesn't Add Up. New York: The New York Press; 2010. ISBN: 9781595585196

[19] Guite HF, Clark C, Ackrill G. The impact of the physical and urban environment on mental well-being. Public Health. 2006;**120**:1123-1124

[20] Tol RS. The economic effects of climate change. The Journal of Economic Perspectives. 2009;**23**(2):29-51. DOI: 10.1016/j.puhe.2006.10.005

[21] Drissen E, Vollebergh HR. Monetaire milieuschade in Nederland: Een verkenning. The Hague: Policy Brief Planbureau voor de Leefomgeving; 2018. p. 3206

[22] Kosonen K, Nicodème G. Taxation papers: The role of fiscal instruments in environmental policy. In: Working Paper No. 19. Luxembourg: Office for Official Publications of the European Communities; 2009. DOI: 10.2778/12205

[23] OECD. Environmental Taxation: A Guide for Policy Makers. Paris: OECD Publishing. 2011

[24] Fullerton D. A framework to compare environmental policies: Association lecture. Southern Economic Journal. 2001;**68**(2):224-248. DOI: 10.1002/j.2325-8012.2001.tb00417.x

[25] Diener E, Lucas RE, Schmimack U, Helliwell JF. Well-Being for Public Policy. New York: Oxford University Press; 2009. ISBN: 9780195334074

[26] Jacobs BB, Van der Ploeg RF. Should Pollution Taxes Be Targeted At Income Redistribution? Tinbergen Institute Discussion Paper. Rotterdam, Amsterdam; 2017. TI 2017-070/VI

[27] Smith S. Taxation and the environment: Theoretical framework. In: International Fiscal Association, editor. Environmental Taxes and Charges. Vol. 18d. London, UK: Kluwer Law International; 1993. p. 8. ISBN: 9789041100689

[28] Vollebergh HR. Environmental Taxes and Green Growth: Exploring Possibilities within Energy and Climate Policy. PBL Netherlands Environmental Assessment Agency. Policy Studies. The Hague: PBL Publishers; 2012. ISBN: 978-94-91506-31-4

[29] Ryszka K, Withagen C. Unilateral climate policies: Incentives and effects. Environmental and Resource Economics. 2016;**63**:471-504. DOI: 10.1007/s10640-014-9867-8

[30] Jacobs BB, De Mooij RA. Pigou meets Mirrlees: On the irrelevance of tax

distortions for the second-best Pigouvian tax. Journal of Environmental Economics and Management. 2015;**71**:90-108. DOI: 10.1016/j.jeem.2015.01.003

[31] Smulders S, Vollebergh HR. Green taxes and administrative costs: The case of carbon taxation. In: Carraro C, Metcalf GE, editors. Behavioral and Distributional Effects of Environmental Policy. Chicago: University of Chicago Press; 2001. pp. 91-130. ISBN: 0-226-09481-2

[32] Staatssecretaris van Financiën. Keuzes voor een beter belastingstelsel. AFP/2014/7802014. pp. 14-15, and the answers provided by the Dutch State Secretary of Finance to questions posed by the Dutch parliament. 2018-2019. Appendix. No. 1335. p. 3

[33] Fullerton D, Leicester A, Smith S. Environmental taxes. In: Working Paper w14197. Cambridge, Massachusetts: National Bureau of Economic Research; 2008

[34] Brink C, Trinks A, Vollebergh HRJ, Zwaneveld P. Abolishing Fossil Fuel Subsidies: A Brain Teaser Rather than a no-Brainer. The Hague: CPB/PBL Publication; 2023

[35] Allcott H, Mullainathan S, Taubinsky D. Energy policy with externalities and internalities. Journal of Public Economics. 2014;**112**:72-88. DOI: 10.1016/j.jpubeco.2014.01.004

[36] Reference can be had, in this regard, to SDG 16 (peace, justice and strong institutions). United Nations. Transforming Our World: The 2030 Agenda for Sustainable Development. A/RES/70/1. New York, United Nations; 2015

[37] For a discussion of Council Directive 2003/96/EC, see Rosenstock M. Environmental taxation within the European Union. Cyprus Economic Policy Review. 2014;**8**(2):113-123

[38] See, for instance, ECJ. C-90/17 (Turbogás — Produtora Energética SA v Autoridade Tributária e Aduaneira), on the Interpretation of Council Directive 2003/96/EC with Respect to a Portuguese Excise Duty on Electricity Consumption. 27 June 2018. ECLI:EU:C:2018:498

[39] For an extensive discussion of the European rules surrounding State aid for environmental protection, reference is made to Quigley C. European State Aid Law and Policy. Third ed. Oxford: Hart Publishing; 2015. ISBN: 9781782254201

[40] ECJ. C-169/08 (Presidente del Consiglio dei Ministri v Regione Sardegna). 17 November 2009. ECLI:EU:C:2009:709

[41] ECJ. Case C-679/17 (Vlaams Gewest v Johannes Huijbrechts). 22 November 2018. ECLI:EU:C:2018:940

[42] Lodin SO. Eco-taxes, neutrality and competition. In: International Fiscal Association, editor. Environmental Taxes and Charges. Vol. 18d. London, UK Boston: Kluwer Law International; 1993. p. 20. ISBN: 9789041100689

[43] Pitrone F. Environmental Taxation: A Legal Perspective. Luiss Guido Carli University; 2013/2014. pp. 1-226

[44] Dutch Court of Appeals The Hague. Case Number 200.178.245/01. 2018

[45] Available from: https://climatecasechart.com/wp-content/uploads/non-us-case-documents/2020/20200113_2015-HAZA-C0900456689_judgment.pdf [Accessed: August 28, 2024]

[46] Oberson X. Some legal issues raised by eco-taxes. In: International Fiscal

Association, editor. Environmental Taxes and Charges. Vol. 18d. Kluwer Law International; 1993. p. 22. ISBN: 9789041100689

[47] Smith ID, Overland I, Szulecki K. The EU's CBAM and its 'significant others': Three perspectives on the political fallout from Europe's unilateral climate policy initiative. Journal of Common Market Studies. 2024;**62**(2):603-618. DOI: 10.1111/jcms.13512

[48] For a critical examination of the dynamics of competition and sustainability, reference is made to Haucap J, Podszun R, Rohner T, Rösner A. Competition and Sustainability; Economic Policy and Options for Reform in Antitrust and Competition Law. Edward Elger Publishing; 2024. ISBN: 9781035325382

[49] Evers M, De Mooij RA, Vollebergh HR. Tax competition under minimum rates: The case for European diesel excises. In: CESifo Working Paper No. 1221. 2004

[50] Lodin SO. Effects on international trade and competition – The Swedish case. In: International Fiscal Association, editor. Environmental Taxes and Charges. Vol. 18d. Kluwer Law International; 1993. p. 49. ISBN: 9789041100689

[51] Bucciol A, Montinari N, Piovesan M. Do not trash the incentive! Monetary incentives and waste sorting. In: Harvard Business School Working Paper 11-093. 2011

[52] Hoel M. Should a carbon tax be differentiated across sectors? Journal of Public Economics. 1996;**59**:18-19. DOI: 10.1016/0047-2727(94)01490-6

[53] Lowe S. The EU's Carbon Border Adjustment Mechanism; How to Make it Work for Developing Countries. Centre for European Reform; 2021

[54] INRA (Europe) – E.C.O. Europeans and the Environment in 1995: Survey Conducted in the Context of the Eurobarometer 43.1 Bis. 1995

[55] TNS opinion & social. Special Eurobarometer 468: Attitudes of European Citizens towards the Environment, Survey Requested by the European Commission, Directorate-General for Environment and co-Ordinated by the Directorate-General for Communication. 2017

[56] Rubin AJ, Sengupta S. 'Yellow Vest' Protests Shake France. Here's the Lesson for Climate Change. The New York Times; 2018. Available from: https://www.nytimes.com/2018/12/06/world/europe/france-fuel-carbon-tax.html [Accessed: July 10, 2024]

Chapter 5

Sustainable Development, Socially Responsible and Ecologically Managed, Increases the World's Ecological Security: Research on Poland and Polish Regional Cities

Ewa Jadwiga Lipińska

Abstract

The essence of the research was to fill the gap in knowledge about ecological safety consistent with social responsibility, sustainable development and environmental management. The conceptualization of research problems allowed the preparation of methodological assumptions for the analysis and assessment of the ecological security of the country, on the example of its regional cities. Theoretical and empirical studies were carried out. The scientific and utilitarian purposes of the research are given. Research thesis: the concept of the state as an ecologically safe organization is based on the concept of cities as an object of knowledge—as opposed to cities as a tool of action. The assumptions were related to environmental aspects. Among other things, the adopted development strategies for regional cities in Poland were analyzed. Models for the development of ecological security have been created. Overall statement: the lack of a single ecologically safe country is sufficient proof of the lack of global ecological security. The author formulated and provided definitions of: city, smart city and ecological security of the state.

Keywords: environmental aspects as a value, security as a property, global threats, smart cities, development strategy problems

1. Introduction

The term security refers to all measures taken to *protect* the area it covers and its resources [1–7]. A state that gives a sense of confidence and a guarantee of behavior as well as a chance to improve safety is therefore a special *value* that an individual may have.

The concept of global, regional, international and national security is based on the integrity of three types of security, which are variable and negotiable tools, but offer a specific value, due to their relationship to the area of the state [1–7]: (1) *ecological security*, which determines the durability and continuity of the process aimed at achieving the desired state in nature and securing the peaceful and healthy existence of all elements of the ecosystem, using various instruments consistent with the principles of internal coexistence of the state and the international community, (2) *internal security*, creating stability and harmony of a given entity (e.g. system, facility or system) and (3) *external security*, determining the absence of threats from other entities and forces of nature.

These three types of security are further defined in 10 areas of security: ecological safety, social safety, cultural safety, physical safety, economic safety, energy safety, identification safety, information technology (IT) safety (i.e. cyber and information and Communications Technology (ICT)), political safety and military safety. They are *values* that confirm a specific *ownership* in various forms. Value and the resulting ownership are characterized by the impossibility of trading security, i.e. transferring value or ownership from one entity to another. Security is inalienable—owed to every individual. Moreover, security can be certified to confirm its value and ownership.

Generally, it can be said that the importance of a given type or field of security is based on its *value* or *ownership* that is revealed in a state of peace or a state of emergency, with specific state resources requiring protection. The states of emergency here include a state of emergency, martial law and a state of natural disaster. However, it cannot be ignored that the scope of restrictions in human rights and freedoms during martial law, a state of emergency or a natural disaster cannot limit the rights and freedoms defined as human dignity, citizenship, protection of life, humane treatment, criminal liability, access to court, personal, conscience, religion, petitions, family and child. It is also unacceptable to limit freedom and human rights solely because of race, gender, language, religion Or lack thereof, social origin, birth or even property. However, the scope of restrictions in human freedoms and rights in a state of natural disaster may limit their freedoms and rights defined as freedom of economic activity, personal freedom, inviolability of the home, freedom of movement and residence within the territory of the state, the right to strike, the right to property, the freedom to work, the right to safe and hygienic working conditions and the right to rest.

The author assumes: the hierarchy of types or areas of security results from other values professed by the state (or another entity operating on the territory of the state), which values determine the actual existence of the state in order to protect all its resources—in a state of peace or in a state of emergency. A state of emergency includes a state of exception, martial law and a state of natural disaster.

The author also assumes: ecological security is a value coherent with other types and fields of security, due to its natural properties. Although natural safety properties are difficult to maintain in their entirety, it is possible to fulfill this obligation [1–3, 5, 6, 8–10].

Research on *the state as an ecologically safe organization* was made by referring to regional development works in Poland. The term *organization* cannot be omitted, because it means [11]: a company, corporation, firm, enterprise, authority or institution, located inside or outside the Community, Or part or combination thereof, whether incorporated or not, public or private, which has its own functions and administration.

The studied community were regional cities in Poland. The statistical significance of the obtained research results was therefore multi-aspect to prove whether the country is or is not an ecologically safe organization. Additionally, the tests were carried out on the sensitivity limit of the newly developed and improved method. Moreover, the research was complicated by difficult epidemic conditions; the research period was 2019–2022. The research method made it possible to establish connections between their theoretical and empirical parts. The differences have helped to establish and explain the meaning of global, international, regional and national security [8, 9, 12, 13].

2. These essences of research

2.1 Achievements

The success of research on the state as an ecologically safe organization is based on the continuity of analyses and assessments of the multi-aspect development of the organization that is the state, because each state pursues many interests, while functioning in the environment of other states. By participating in international relations, the state strives to ensure and maintain security, because it guarantees its survival, ensures comprehensive development and gives a sense of security to its citizens. Socially responsible participation in global sustainable development depends on the political and environmental aspects of all countries, the entire biosphere and humanity. In their context, social and economic aspects serve sustainable innovation changes and public-private partnerships.

In the model study, the author relates *political aspects* to common problems in maintaining the constant readiness of state structures to take up challenges, use opportunities and reduce risks, as well as the ability of state structures to eliminate threats. There are six political threats: (1) weakness of the rule of law and civil society structures, (2) the inability of state institutions to maintain the stability of the constitutional order and protect citizens from discrimination and intimidation by armed groups or criminal organizations, (3) failure to respect human rights and freedoms, (4) terrorist acts and practices, (5) various extremisms and fundamentalisms (e.g. ideological and religious) and (6) lack of international cooperation.

The benefit in the political aspect is ensuring a high level of political security of the state. This benefit can be obtained by implementing nine strategic goals: (1) protection of the constitutional order, (2) ensuring the functioning of state institutions, (3) respecting the rights and freedoms of its citizens, (4) protection of citizens against lawlessness, (5) eliminating terrorist threats, (6) creating appropriate conditions for economic, social and intellectual development, (7) counteracting extremism and fundamentalism, (8) efficient functioning of the crisis management system and (9) maintaining positive relations with other stakeholders of international relations.

In the model study, the author relates *environmental aspects* to the obligations of a democratic state of law, whose territory is the common good of all citizens. These obligations are intended to implement the principles of human rights, social justice and sustainable development [7, 10–12, 14]. This is possible based on environmental management. Such a state, while pursuing a policy ensuring security, fulfills four constitutional tasks, i.e.: (1) treats equally and does not discriminate in political, social or economic life for any reason, (2) indicates areas of health protection, (3) fights epidemic diseases and (4) prevents the negative health effects of environmental degradation.

From the point of view of state security policy, the process of managing the state's territory as a common good is a tool used in a state of peace and a state of emergency. The results of management in a state of peace and in a state of emergency are losses in resources (social, spatial and economic). The scale of losses expressed in *the value* of lost resources is sufficient to constantly search for *other* solutions to state security. The author relates the effectiveness of these other solutions to *the speed* and *continuity* of identifying *positive* and *negative environmental aspects* of the state's development, i.e. the indicator is still discretionary.

According to the author, in global security policy, *the effectiveness* of *other* development solutions should rather be based on an increase in the *obligation* to identify positive environmental aspects in all countries and eliminate negative environmental aspects in these countries. Therefore, it is necessary to create such an "investment climate" that there is a genuine willingness to change. This could be a promising start to the full realization of resilient and competitive change, because it is sustainable in creating real security.

The state's ecological security policy therefore requires considering the *difference* between the public and the privacy of the resources of all organizations in the state, also the difference between the intentional and unintentional actions of these organizations, as well as the difference between the confidentiality, integrity and availability of information about the environmental aspects of the same organizations. This act of differentiation makes it possible to determine the state and quality of the state's security policy [1, 2, 5, 11, 13]. We cannot ignore the fact that the difference-in-differences method is a quasi-experimental approach, i.e. a comparison of changes in results over time between the population covered by the study and the population not participating in the study. Therefore, the value lies in the differences between these environmental aspects of development that determine the condition and quality of the national regional development policy, voivodeship development strategy, city development strategy or commune development strategy [1, 2, 7, 11].

The state's ecological policy requires engineering and technical, urban and architectural, medical and pharmaceutical tasks to ensure a decent *quality of life* for citizens. Therefore, four subsystems are important in security policy: management, executive, operational (i.e. defense and protection) and support (i.e. social, spatial and economic). In this context, the national regional development strategy refers to four objective realities: (1) spatial development plans, government and local government administrative decisions and environmental protection programs, (2) the effectiveness of urban development, (3) institutional development and urbanization processes and (4) diversification of development tasks in order to satisfy the public interest.

From the point of view of the obligations of public authorities, the conceptualization of research required focusing on the analysis and assessment of the coherence of three development concepts: social responsibility, sustainable development and environmental management. The author relates this composition to ecological security, understood as protection of the health and life of citizens and environmental protection, in the context of knowledge about the environmental aspects of the state, **Figure 1**. Environmental management is here a real process revealing the obligation—and even the right—to determine the environmental aspects of the state. The narrowing of development policy is the result of considering only the positive environmental aspects of this development, in the inverted truncated cone illustrating the space in **Figure 1**.

Figure 1.
The idea of coherence of social responsibility and sustainable development, in the context of the law on the protection of human health and life, environmental protection law and the law on environmental management of organizations, in the country. Color codes: Brown—Green transformation of the economy; lavender—A symbol of the fight against cancer; green—Zero emissions; blue—Use of specific precautions or personal protection; yellow—Various types of dangers or their risks.

The idea of coherence of three concepts expressed in **Figure 1**—social responsibility, sustainable development and environmental management—refers to the rights of every human being. This law is understood here as: (1) the maintenance of health and life, by an individual or group, and (2) sticking to a healthy environment. The term *to hold* means being in good condition and quality. However, the term *protection* means that there is a danger that must be defended against. This means *a different* approach to the functioning of the internal body of the state and its external connections, i.e. it is necessary to know where this law *ends*. This idea takes into account *a different* analysis and assessment of development factors (i.e. determinants), because it establishes the obligation to determine *the positive* and *negative environmental aspects* of any development, not only their voluntary recognition [11]. Only in this way can one obtain knowledge—as it seems—where the ecological effectiveness of changes in the development of each organization ends [3–5, 15–17].

Maintaining focus on the ecological effectiveness of development is unpredictable in the diverse socio-economic conditions of the country. It requires optimal demonstration of an ecological development plan or its modification to model and shape safe environments (social, spatial and economic). Therefore, it is advisable to analyze various environments in the context of knowledge about environmental aspects, as indicators for assessing social responsibility and sustainable development, expressed in the development strategies of various organizations in the country.

The research method I adopted helped me to regulate the methodological assumptions of the research and improve research supervision. Additionally, I referred in my research to the knowledge used in them, which concerns legal issues, social and economic issues, urban issues, the nation, the region, international affairs and the World [15–17]. Moreover, the concept of otherness here refers to being, a new life—based on positive environmental aspects and in the context of human prosperity and well-being. This means that strategic development planning, operational development planning and the effects of development activities are predicted on the basis of previously identified positive environmental aspects of the organization, and their negative environmental aspects are excluded.

This concept allowed me to determine how the results of theoretical research will be used to develop and present empirical data.

Specifying the differences between four concepts—ecological security, social responsibility, sustainable development and environmental management—revealed changes that need to be introduced in development strategies (mentioned earlier).

In general, it can be said that the lack of a single ecologically safe country is sufficient evidence of the lack of global ecological security. This statement is based on the concept of sustainable development [18]: the lack of one of the sustainabilities, i.e. social, spatial or economic, constitutes a lack of sustainability at all.

2.2 Objectives, thesis and scope of research

In the social sciences, it is not possible to distinguish subdisciplines according to one criterion [3, 5, 11, 12, 19]. I adopted the concept of *ecological security* as a criterion for theoretical and empirical searches. Theoretical research has been structured and directed to empirically identify differences in socially responsible and sustainable development as it is environmentally managed.

I reflected the issues related to the conditions of the organization of an ecologically safe state in the scientific and utilitarian purposes of the research.

The scientific aim of the research was to theoretically describe the state as an ecologically safe organization and to establish a uniform theoretical framework for the development of the state as an ecologically safe organization, because it is responsible for protecting the life and health of citizens and the protection of the natural environment.

The utilitarian aim of the research was to indicate a new approach to the importance of four concepts of state development policy: *state ecological security*, *state sustainable development*, *state social responsibility* and *state environmental management*. A continuation of the utilitarian goal was to provide comprehensive definitions of *the city*, the definition of *a smart city* and the definition of *the ecological security of the state*, created by the author.

The research thesis was as follows: the concept of the state as an ecologically safe organization is based on the concept of cities as an object of knowledge—as opposed to cities as a tool of action.

I assumed, first, that the development of the state as an ecologically safe organization results from the environmental aspects of the state—its cities and their regions—which aspects determine the impact of development on the territory of the state. Environmental aspects are elements of the state's internal and external security policy, therefore ecological security is both a type and a field of security. The definition of *an environmental aspect* states that it is a component of an organization's activities, products or services that affect or may affect the environment [11].

Second, I assumed that environmental aspects determine the reality to be presented by regional development strategies, city development strategies, municipal development strategies and even development strategies of other organizations. The lack of environmental aspects means utopian development strategies [13, 15, 16].

The overall research concept understood in this way allowed me to organize a coherent search for answers to four questions. Is ecological safety a lasting trend and a recognized international standard relating to the management of the country and its cities? How is the perception of urban development changing around the world, in Europe and in Poland? Was the previously dominant analytical concept of the state sufficient to understand the role of ecological security in contemporary spatial organization? Shouldn't this concept be expanded to include an objective concept?

Looking for answers to these four questions, I conducted research on a deliberately selected group of 18 regional cities in Poland. I assumed that the collective nature of regional cities has features resulting from the advanced social and economic development of these cities. Cities should conduct international cooperation and have widely used digitalization, because only then can such a city be considered smart [20–22]. Additionally, each regional city should consider direct or potential socio-economic benefits from the natural environment, in accordance with the theoretical framework and initiatives supported by the government, regional or local authorities, on the basis of the development policies of these cities. During the research period, these also included obligations resulting from the global COVID-19 pandemic in 2019–2022; it was also a time of research. The essence of the latter context results from its leading role in the world of modern technologies. The author relates the government's responsibility for the sustainable creation of a mobile and digital society to three activities at the level of the city, commune, district and region in the country: (1) increase in the use of digital technologies in education, (2) developing people's digital competences and skills, and (3) data analysis to improve future forecasting.

I took into account the negation of the purpose of the study, i.e. the possibility of missing a certain amount of data and other information in the research material.

In response to the questions (posed above), I took into account the complexity of the political importance of cities and their regions against the background of changes taking place in Europe and around the world. I also took into account the fact that state management is linked in a superior way to city management. Cities, on the other hand, are related to the infrastructure of municipal management, housing management, real estate management and spatial development [10, 22].

Additionally, I analyzed the "National register of the participation of organizations in the Community Eco-Management and Audit Scheme (EMAS)" and my own research questionnaires (surveys) on the topic "Model of an integrated city eco-management and audit system with the goals of sustainable city development and city social responsibility." They were addressed to the presidents of the surveyed regional cities.

I analyzed and assessed 18 development strategies for Polish regional cities, because these strategies determine long-term city development cycles. I compared historical development data with the current state of development of Poland's regional cities. I analyzed the differences in the development of these cities. I assessed variability trends in the development of the same cities. I verified the development challenges, adopted development directions, strategic development goals and operational development goals, and assigned development tasks and expected effects of development activities—which were approved in the development strategies of Polish regional cities. The validity of data in city development strategies referred to 2020 with a perspective to 2023 and 2025 or to 2030 and even to 2050.

I used a multi-criteria comparative and descriptive analysis of the collected research material as an instrument supporting social responsibility and sustainable development in spatial and urban planning.

The complexity of the methodological problem of research is illustrated by *the consolidation model of the state as an ecologically safe organization*, **Figure 2**, which takes into account three types of security: ecological, external and internal. This model also expresses environmental management of relationships within a nation and its economy. These relations are developed in a public-private partnership, managed according to the principles of social responsibility and sustainable development.

Figure 2.
The concept of the consolidation model of the state as an ecologically safe organization.

The consolidation model of the state as an ecologically safe organization creates six concentrically overlapping circles with factors (which are determinants) of development. The system of determinants makes us realize the main role of ecological security in the development of every country. This role is based on various interrelated responsibilities: public authorities, organizations operating within the territory of the state and the country's stakeholders. The structure of concentrically overlapping determinants of the country's development represents a system of inclusive relationships (i.e. without mutual exclusion). Each stakeholder of the inner circle also belongs to the outer (i.e. wider) circle, i.e. to the more inclusive outer circle (but never the other way round), **Figure 2**.

From the point of view of the international concept of *a better world* [23], an important feature of the model of concentrically overlapping circles with determinants of (for each country) development are the positive environmental aspects of the country (see **Figure 1**), because the circles interpenetrate. Negative environmental aspects of any development are excluded, because their presence creates utopian security. It is worth noting that all organizations are at the center, because their decisions and practices are influenced by continuous and interconnected other determinants (**Figure 2**).

The originality of research on *the state as an ecologically safe organization* results from the adopted five research practices:

1. inventory of the elements of the structure of the development strategy of regional cities;

2. assessment of the elements of the structure of the development strategy of regional cities, in the context of the sources (or lack thereof) of sustainable urban development, social responsibility of cities and environmental management of cities, with particular emphasis on the environmental aspects of these cities;

3. demonstrating and justifying the need to introduce activities identifying the environmental aspects of cities and their regions into city development tasks, also in the context of the cross-border impact of negative environmental aspects of cities; and

4. defining the scientific basis for managing the state and its cities to the extent that allows meeting the needs of the current generation and the aspirations of future generations, in the context of only the positive environmental aspects of cities.

DOI: http://dx.doi.org/10.5772/intechopen.1004073

The research concept formulated in this way allowed us to organize knowledge about cities in order to harmonize their development in the country and the development of the state. Additionally, the developed research concept allows for the integration of knowledge from various fields and scientific disciplines. The discussed research concept allows for a different explanation and prediction of the results of the state's development activities.

In general, it can be said that the essence of the values and properties of *ecological security* result directly—i.e. intentionally—from the importance of the state's *social responsibility*, *sustainable development* of the state and *environmental management* of the state, in a state of peace and under the complex burdens of a state of emergency.

The significant contribution of this research to global science is to fill the scientific gap in the field of "social sciences" and the discipline of "safety science" and the discipline of "management and quality science" as well as in the field of "engineering and technical sciences" and the disciplines of "environmental engineering" and "environmental protection." This fulfillment refers to the research results and description of gaps in knowledge about the mechanisms of changes in the strategies of utility development of cities and their regions, in the context of the ecological security of the state.

3. Research results

The basis of the new paradigm of ecological security are three political premises, i.e. external relations of the state, internal relations within the state and factors of state development. Therefore, we can assume the existence of real scopes of competences and skills in managing global ecological security, when the new paradigm is characterized by strong external relations of all countries. It is worth remembering, however, that the new paradigm is characterized only by internal relations and state development factors, because fundamental change occurs only from within the state or states.

From the point of view of the adopted research practice, I distinguished 10 research results on *the state as an ecologically safe organization*, based on the example of Poland and its regional cities, i.e.:

1. I formulated the definition of a city and the definition of a smart city;

2. I demonstrated the impact of social responsibility, sustainable development and environmental management on the increase in safety;

3. I specified the sub-policies for the development of regional cities;

4. I developed catalogs of elements of the structure of the development strategy of these cities;

5. I specified the impact of environmental aspects of cities on the current and future state and quality of the state's ecological security;

6. I identified gaps in meeting the requirements of city management;

7. I proposed changes to documents important for the development of cities;

8. I proposed changes to the spatial databases of central and local government administration, so that development policies determine ecological safety;

9. I requested a state inventory of environmental aspects as an element of the constitutional obligation to ensure safety; and

10. I provided the definition of the state's ecological security, so that science related to urban planning and architecture could plan the development of countries differently.

The first result of my research was the lack of a clear definition of the concept of *a city*, sometimes used interchangeably with the—also essentially undefined—concept of *a smart city*, but no attempt has ever been made to justify such interchangeability. The diversity of the spatial structure of cities is one of the reasons why the definition of a city is not agreed at the international level. However, it is accepted that the concept of *a city* is not always limited to its administrative borders, and therefore the number of its inhabitants does not reflect their actual population [21, 22, 24, 25]. As a result of my research, I created and provided a comprehensive definition of *the city* and it is as follows [26]: *socially responsible and sustainable development of the city as an organization means universally accepted ecological issues, i.e. in the process of city management, in the system of organizing the urban community and in economic relations with interested parties. The city as an organization creates a coherent system of values for the community using the city's natural and technical space. This space is used to achieve real ecological security of the city. In this context, the environmental aspects of the city are continuously identified in its six spatial forms, i.e. production, consumption, power, symbolism, exchange and residence*.

I also created and provided a synthetic definition of *a smart city* [26] and it is *an ecologically safe place for social and economic life. The site is environmentally managed. The city's organizational structure, city planning activities and social responsibilities—in the context of the city's practices, procedures, processes and all resources—serve to develop, implement, achieve, review and maintain that city's environmental policy. The city's ecological policy is the basis of the smart city development policy based on the city's environmental aspects* [26]. This definition is unambiguous for all urban areas because it is not only based on the needs of social space, cultural heritage and city stakeholders, but also on the context of ecological safety policy.

These two definitions present *a different* approach to the issues of development, because it is related to the functions of the supreme authorities, based on the best available techniques, justified by European and even global achievements. Both definitions are clear for urban areas, as they concern spatial needs, i.e. their harmonious development. These definitions exclude ecologically sound development when urban development policies are not consistent with information and communication technology services in the physical structures of cities. Moreover, security relates to cultural heritage.

I related the second research result to the relationship between ecological safety and the degree of coherence of social responsibility, sustainable development and environmental management (of each organization). This relationship is sufficiently justified by international law [e.g. 3–7, 11, 23]. The fact that law is "voluntary" [11]—as a feature of ethical action, constituting a necessary or sufficient condition for being subject to moral assessments—in the context of ecological safety is a legislative defect. The lack of ecological security here means a lack of ethical action and moral judgments. Research on the development potential of Polish regional cities showed that 91 entities, i.e. 0.002% of all entities operating in the territory of Poland, were registered *voluntarily* in the national eco-management and audit system in the Community in 2020. This legislative defect can be removed by the existence of a necessary or sufficient condition: *the organization's consent* to the obligation of eco-management and audit in the organization management process. Consent is the efficient cause of organization. The place of consent may be—as it seems—a specific or random application for financing any activity affecting the socio-natural environment of cities and their regions. As a result, each organization identifies its environmental aspects. The essence of *the consent* is to increase the security of national regional development, **Figure 3**.

The real obligation of public authorities is based on the scenario of socially responsible and sustainable development, which is environmentally managed by all organizations operating in the territory of the state (**Figure 3**). This scenario includes a consolidated report, i.e. informing about the state and quality of health and life of citizens, in connection with the state and quality of the natural environment. In this way, in five development areas, all local government units are responsible for creating and adopting the state's ecological security. These are: (1) studies of conditions and directions of spatial development, (2) local spatial development plans, (3) environmental protection programs, (4) information on environmental aspects and (5) decisions of expert units assessing environmental aspects for the organization's development strategy.

I related the third research result to *the state's ecological security policy*, which should optimize various sub-policies of development of a given city, so that its development success is based on the positive effects of this development. A comparative analysis of 18 development strategies of Polish regional cities showed the existence of 18 development sub-policies, and these are: (1) sub-policy of the environmental protection, (2) sub-policy of the spatial order, (3) sub-policy of the health care, (4) sub-policy of

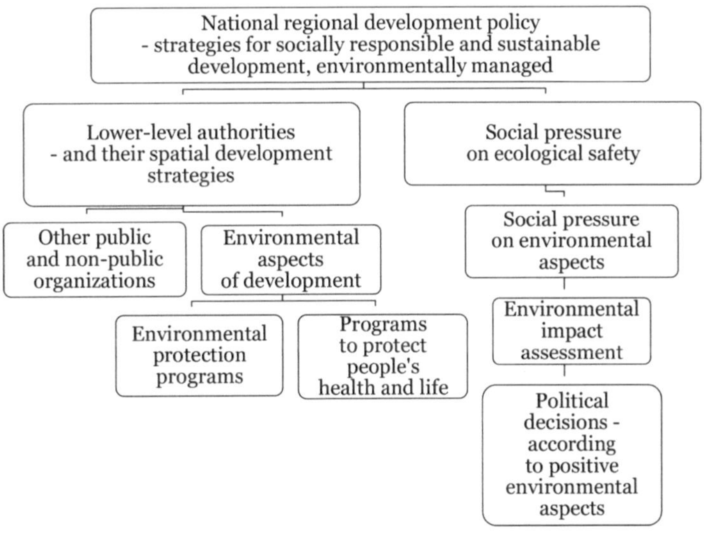

Figure 3.
Scheme of the national regional development policy, taking into account the development policies of lower-level authorities in the country, in the context of socially responsible and sustainable development, environmentally managed.

the employment promotion, (5) sub-policy of the development culture, (6) sub-policy of the development of sports, tourism, recreation and leisure (7) sub-policy of the development of cities and metropolitan areas, (8) sub-policy of the development of rural areas, (9) sub-policy of the development of science and increasing the innovativeness of the economy, (10) sub-policy of the development of pro-social behavior communities and structures of civil society, (11) sub-policy of the human resources development, (12) sub-policy of stimulating the creation of new jobs, (13) sub-policy of creating and modernizing social infrastructure, (14) sub-policy of creating and modernizing technical infrastructure, (15) sub-policy of supporting modernization state institutions, (16) sub-policy of supporting the development of entrepreneurship, (17) sub-policy of supporting economic growth and (18) sub-policy of increasing the competitiveness of the economy. These sub-policies are related to unequal access to resources of strategic importance for the state.

I referred the fourth result of the research to the development of catalogs of elements of the structure of the development strategy of regional cities in Poland. The catalogs contain information about sets of investment services and the limits of the impact of these services on space in the context of a given development sub-policy (i.e. subset). Previously, there were no such catalogs in the development system of regional cities in Poland. As a result of the research, I developed 13 catalogs of elements of the structure of the development strategy of Polish regional cities, containing 1738 services provided to satisfy the public interest. These are: (1) a catalog of urban development challenges; (2) a catalog of the strengths of urban development; (3) a catalog of development opportunities for cities; (4) a catalog of cities' development weaknesses; (5) a catalog of urban development threats; (6) a catalog of urban development directions; (7) a catalog of urban development activities; (8) a catalog of strategic city development goals; (9) a catalog of detailed urban development objectives; (10) a catalog of objectives for the operational development of cities; (11) a catalog of strategic city development tasks; (12) a catalog of tasks for detailed urban development; and (13) a catalog of the results of urban development activities.

Cataloging and presenting these services is based on five activities: (1) analysis and assessment of elements of urban development strategy structures, in order to isolate and collect development services and determine what they concern, (2) dichotomous verification that the services apply to all units and do not contain noticeable errors to determine their usefulness and clarify any ambiguities, (3) determining the differences in services so as not to deform their spatial significance, as a consequence of the fact that the differences are to some extent dependent on the assumptions made, (4) summing up how many of the designated criteria a given service meets and (5) the number of criteria met indicated how problematic the service was from the point of view of development sub-politics. The analysis of the dynamics and directions of development changes in Polish regional cities was carried out both in the entire individual set and in subsets designated by the typological procedure.

The fifth result of my research was focused on creating a catalog of positive and negative environmental aspects in order to assess the expected effects of development activities on their basis. In this context, *positive environmental aspects* determine ecologically safe development policies, referred to as *a smart city*. However, *negative environmental aspects* constitute a lack of ecological security in development policies.

The sixth research result kept this concentration in check, but showed the lack of identified environmental aspects by those approving the development strategies of Polish regional cities. Therefore, there is no specific, non-utopian, prediction of the effects of the development of regional cities. This important public interest in states of peace and states of emergency is also evidence of the lack of effective environmental protection programs in urban areas.

The seventh research result is my proposal for changes in knowledge and development documents consisting in eliminating the shortcomings (mentioned above) in all formal requirements and a constructive approach to spatial data. Additionally, the change is intended to supplement knowledge about the real and sustainable impact of the "investment climate" on the environment, i.e. according to identified environmental aspects.

The eighth research result expresses three premises for institutional and legal changes, i.e.: (1) the emergence of new concepts for managing public space in urban areas, resulting from monitoring the development of cities, (2) changes in the economic space of urban areas as a result of the use of all the resources of these areas and (3) adapting the theory and practice of public space management to international and national urban policy in the field of functional areas and metropolitan areas.

The ninth research result distinguishes three institutional and legal effects of environmental management, i.e.: (1) improving the reliability of the estimation of the consolidation model of development of the country, its regions and cities, in the context of socially responsible and sustainable development, (2) updating the assessment of the state territory in the context of natural capital, social capital and human capital and the ways of using them, (3) redesigning the development activities undertaken—including corrective, restrictive or temporary ones—to increase the importance of security in the function of the state, its regions and cities.

And finally, the tenth research result emphasizes the general cause-and-effect relationship between global security and local security. In this context, I created and presented [26] a definition of *the state's ecological security, which states that it is comprehensive security, i.e. protection of health and life of people and environmental protection, protection of cultural heritage and protection of property, as well as a rigorous and wide-ranging system of advance protection, which involves preparing for and effectively responding to a state of emergency (e.g. martial law, state of emergency or state of natural disaster)*.

This fundamental definition of *ecological security* argues that the concept of state development is based on the concept of city development as an object of knowledge—as opposed to the city as a tool of action. The city and its region are therefore always understood as an object of knowledge. The essence of this knowledge are environmental aspects and respecting them.

In general, it can be said that the developed method of researching the state as an ecologically safe organization creates an opportunity for a constructive approach to improving the assessment of the state and quality of planning for sustainable development of urban areas and making changes for the security of the state.

4. Discourse—a synthesizing approach

The term *ecological security* is included only in one Polish legal act—the Constitution—as the fundamental law of the Republic of Poland. At the same time, neither this legal act nor other legal acts contain any definition of *ecological safety*. Meanwhile, the concept of *ecological security*, according to the Constitution of the Republic of Poland, requires public authorities to take actions to protect citizens and their future generations. This means that you are already experiencing threats to focus on creating barricades against the risk of their unpredictability. This issue refers to the state of peace and states of emergency resulting from insufficiently harmonized socio-ecological, economic-ecological and socio-economic development [26].

The term *ecological safety* may indicate strict, technical tools for receiving applications according to previously adopted assumptions. The mathematical models of urban development used in engineering and management practice are mainly phenomenological. This fact does not ensure ecological security, i.e. one that does not involve significant threats due to the capital-intensive economy.

In terms of the state as an ecologically safe organization, the author believes that there is one appropriate administrative division of the state, i.e.: (1) combining sustainable development with social responsibility and eco-management in the development strategies of cities and their regions, and (2) multi-level approach to cities and their regions, taking into account all levels of government, in the context of ecological security. The author relates the research concept formulated in this way to "the obligation" to identify environmental aspects of cities and their regions in order to strive for an idealistic interpretation of the availability and use of their resources. In the traditional approach to *accessibility*, importance is attributed to the development of cities within their administrative borders [1, 4, 5, 17, 20, 22, 27–29].

Table 1 illustrates general data about 18 *smart regional cities* in Poland. From the point of view of managing the digitization process in Poland, 94% of all cities with more than 200,000 citizens and 84% of all cities with 100,000 to 200,000 citizens have strategic documents covering digitization issues. Only in cities with more than 200,000 citizens do the fastest connections prevail, i.e. above 1 Gbit/s. The essence of regional city offices is to offer services and handle matters online. However, it is significant that one regional city—Wrocław—is registered in the national EMAS register and has an EMAS certificate.

Another approach to development stipulates *the obligation* to identify the environmental aspects of *each organization*. This approach should inspire representatives of all fields of science and scientific disciplines to further research on their essence in the context of the external and internal security of each country. From the point of view of *global ecological security* (**Figure 4**), the discrediting of its understanding means that any discourse can be conducted without knowledge about the environmental aspects of countries. Inspiring representatives of various fields of science and scientific disciplines to further research work is of comparative importance here; especially when development is differentiated by states of peace or states of emergency. The attitude of citizens toward global ecological security policy can therefore be considered in the context of two terms after analyzing the dynamics of negative (i.e. aggressive) dependencies between countries: (1) *prosperity*, which here means good economic conditions as a balance in structural changes, social and economic development; excludes bad economic conditions, i.e. a decline Or stagnation in production, prices, national income, trade in goods and a disturbed general balance in development, and (2) *well-being*, which here refers to the field of psychology and means the subjectively perceived sense of satisfaction with the physical, mental and social state of one's life.

Importantly, the method of further research on the development of *the state as an ecologically safe organization* must take into account environmental aspects—so far "programmatically" omitted in Poland. This method is not based on the question of whether to identify environmental aspects in all organizations operating in the territory of the country—it is based on the obligation to identify environmental aspects of each urban or rural area. The lack of compliance of the assumptions of the urban development strategy with the planned positive effects of urban development refers

No.	City Name	City area in square kilometers	Number of citizens	Unemployment rate in percent	Number of entities of the national economy	Apartments put into use	Dealing with matters online	An EMAS-certified city
1.	Bialystok	102	292,600	5,0	41,852	2131	yes	no
2.	Bydgoszcz	176	354,006	2,2	46,037	2288	yes	no
3.	Gdansk	683	486,345	2,5	95,275	4575	yes	no
4.	Gorzow	86	122,141	2,5	18,717	555	yes	no
5.	Katowice	165	302,397	1,1	54,824	1725	yes	no
6.	Kielce	110	195,942	4,6	31,189	941	yes	no
7.	Krakow	327	766,683	1,9	175,542	8206	yes	no
8.	Lublin	147	342,039	4,3	53,368	2915	yes	no
9.	Lodz	293	696,708	4,4	103,953	3704	yes	no
10.	Olsztyn	88	173,718	1,9	25,672	930	yes	no
11.	Opole	149	118,303	3,0	23,707	585	yes	no
12.	Poznan	262	540,365	1,1	130,003	3874	yes	no
13.	Rzeszow	129	183,901	4,1	35,326	2159	yes	no
14.	Szczecin	301	404,461	3,6	75,304	1942	yes	no
15.	Torun	116	199,469	2,9	28,893	1516	yes	no
16.	Warszawa	517	1,861,975	1,4	548,529	11,620	yes	no
17.	Wroclaw	293	674,132	1,6	147,620	7154	yes	yes
18.	Zielona Gora	278	139,278	2,8	25,040	373	yes	no

Table 1.
Regional cities in Poland—basic data.

Figure 4.
Scheme of pursuing a global ecological security policy by increasing the quality of life, prosperity, well-being and well-being of citizens, in the context of peace—Expressed in the internal and external security of countries—Which is based on socially responsible, sustainable and environmentally managed development.

here to the lack of environmental aspects, and these appear in local conditions; before they are observed in global conditions.

From the point of view of urban policies, standardizing and aggregating information on environmental aspects (generated by diverse investments) —made available and used—is the first stage in establishing cultural conditions. Taking into account the influence of cultural factors on the selection and application of environmental aspects is important when determining potential opportunities for increasing the spatial and architectural culture of cities. Additionally, when taking into account national, regional or local identity, such activities can have a positive impact on global security (**Figure 4**).

From Poland's point of view, the existing spatial data (i.e. location, size and topological relationships between these objects, phenomena or processes), which are used to describe and calculate the resources of cities and even the country, are not fully reliable. These are estimated and approximate data, and some of them may, and even should, change after recognizing the environmental aspects of cities and the country [2–5, 16, 17]. For this reason, estimates of spatial resources are not reliable. Previous efforts to identify urban spatial resources do not provide sufficient information about the negative results of city management. In the case of many spatial resources of cities (and the country), the degree of reliability of individual urban parameters and sociological maps, in the context of the shape and structure of these resources—in order to create their digital model and simulate the exploitation of the same resources—depends on the degree of their actual estimation, and there are several estimation methods.

It cannot be ignored that global governance of countries' development is rightly linked to global urban development policy, but these connections are usually not properly analyzed and presented. In this context, policy is a system of goals and means as well as procedures for achieving them. Measures and procedures result from applicable national law and spatial development. It can therefore be assumed that development planning, as an activity superior to further stages of the management process, should more or less precisely define: (1) directions of development, but only according to positive environmental aspects and (2) protection programs, due to negative environmental aspects.

In general, it can be said that the discussed method of researching the state as an ecologically safe organization makes a significant contribution to better understanding and characterizing the current state and determining the development prospects of each state. Nowadays, countries can and even should have the status of ecologically safe states, important in sensitive states of peace and states of emergency. Examples include: wars (biological, chemical or nuclear) and industrial accidents (in mining, transport and others) and natural crisis events (climate changes, earthquakes, floods or strong storms).

5. Conclusions

The author's intention was to present organized threads that appear in the four questions given in subsection 2.2. "Significant research achievements." The author's answer to these questions is to continue her research from the point of view of three issues: (1) the concept of development of other countries, (2) prospective development of their urban areas and (3) the importance of environmental management in protecting the health and life of citizens and protecting the natural environment. Determining these three issues may be crucial to achieving ecological security in other countries. In this context, the state and its cities are undoubtedly among the important issues of the scientific community. The main task of the science system—any system—was and is the legitimization of the political *status quo*, as well as the scientific justification of all political and social transformations.

Research on the state as an ecologically safe organization showed the mechanisms of development in the local system and revealed the barriers encountered and ways of overcoming them, by Polish regional cities. The developed method for determining the ecological safety of the state and its cities can be used for scientific analyses in the context of urban structures, consolidation of scientific concepts of spatial development and opinions on political decisions, mainly military ones. This method facilitates the selection and location of each investment considered in the context of its real environmental aspects, not only those voluntarily identified.

The basic theory used to catalog and present the elements of the structures of the development strategy of Polish regional cities is based on the assumption that each service is characterized by a certain natural size—an *environmental aspect*, usually *hidden* in the service. This result is based on catalogs with sets of services that are interconnected, therefore some services can be included in at least two sub-policies for the development of regional cities. Importantly, these services include environmental aspects in a non-obvious way—the importance of each of them may increase or decrease—but none of them can disappear. This has very important consequences. It can be said that positive environmental aspects "reward" the progress made in socially responsible and sustainable development. They also allow changes to be financed from "significant" funds. Polish needs in this matter are enormous.

The state as an ecologically safe organization is significant due to the environmental aspects of the state, because they are identified with strengthening development and minimizing, or even eliminating risks or threats of this development [11]. Therefore, such methods and means in the processes of identifying environmental aspects are expected to precisely adjust the investment only to a positive development result—difficult to predict at the development planning stage.

In general, it can be said that the success or failure of the global ecological security policy is determined by those environmental aspects that may hinder or facilitate conclusions about the trends in the development of the geo-political security system [11, 20, 28]. For this reason, the concept of a consolidation *model of an ecologically safe state* should be the starting point for developing new teaching modules in university study programs to complement the theory and practice of managing the state's internal and external security. Future research: it is worth taking into account the responsibilities of the authorities and the resulting environmental aspects in the context of Russia's war with Ukraine and Hamas' war with Israel inside Gaza.

Conflict of interest

The authors declare no conflict of interest.

Author details

Ewa Jadwiga Lipińska
The Faculty of Management, Rzeszow University of Technology, Poland

*Address all correspondence to: e.lipinska@prz.edu.pl

References

[1] Hoijtink M, Muehlenhoff H. Mint: The European Union as a masculine military power: European Union security and Defence policy in 'times of crisis'. Political Studies Review. 2019;**18**(3):362-377. DOI: 10.1177/1478929919884876

[2] Pettersson FJ. Mint: Defensive integration through cooperative re-bordering? How member states use internal border controls in Schengen. Journal of European Public Policy. 2023:1-23. DOI: 10.1080/13501763.2022.2162104

[3] Regulation (EU) 2023/988 of the European Parliament and of the Council of 10 May 2023 on general product safety, amending Regulation (EU) No 1025/2012 of the European Parliament and of the Council and Directive (EU) 2020/1828 of the European Parliament and the Council, and repealing Directive 2001/95/EC of the European Parliament and of the Council and Council Directive 87/357/EEC. Official Journal of the European Union. 2023. L 135/1

[4] Regulation (EU) 2021/1119 of the European Parliament and of the Council of 30 June 2021 establishing the framework for achieving climate neutrality and amending Regulations (EC) No 401/2009 and (EU) 2018/1999 ('European Climate Law'). Official Journal of the European Union. 2021. L 243/1

[5] Communication from the commission to Parliament. The European Council, The Council, The European Economic and Social Committee and the Committee of the Regions 'The European Green Deal'. Brussels; 2019

[6] Decision of the High Representative of the Union for Foreign Affairs and Security Policy of 19 September 2017 on the security rules for the European External Action Service. Official Journal of the European Union. 2018. C 126/01

[7] Directive 2003/4/EC of the European Parliament and of the Council of 28 January 2003 on public access to environmental information and repealing Council Directive 90/313/EEC. Official Journal of the European Union. 2003. L 41/26

[8] Hasan A, Yim D, Khuntia J. Mint: Citizens' adherence to COVID-19 mitigation recommendations by the government: A 3-country comparative evaluation using web-based cross-sectional survey data. Journal of Medical Internet Research. 2020;**22**(8):1-11. DOI: 10.2196/20634

[9] United Nations: Coronavirus Spread Now a Global Emergency Declares World Health Organization. 2020. Available from: https://news.un.org/en/story/2020/01/1056372 [Accessed: February 02, 2023]

[10] Tetrevova L, Jelinkova M. Mint: Municipal social responsibility of statutory cities in the Czech Republic. Sustainability. 2019;**11**(2308):1-19. DOI: 10.3390/su11082308

[11] Regulation (EC) No 1221/2009 of the European Parliament and of the Council of 25 November 2009 on the voluntary participation by organisations in a Community eco-management and audit scheme (EMAS), repealing Regulation (EC) No 761/2001 and Commission Decisions 2001/681/EC and 2006/193/EC. Official Journal of the European Union. 2009. L 342/1

[12] Albuquerque R, Koskinen Y, Zhang C. Mint: Corporate social

responsibility and firm risk: Theory and empirical evidence. Management Science. 2018;**65**(10):4451-4469. DOI: 10.1287/mnsc.2018.3043

[13] Fierke KM. Critical Approaches to International Security. 2nd ed. Cambridge: Polity Press; 2015 248 p. ISBN-10 0745670547; ISBN-13 978-0745670546

[14] Hibbert N. Human rights and social justice. Laws. 2017;**6**(7):1-16. DOI: 10.3390/laws6020007

[15] Lu T, Li C, Zhou W, Liu Y. Mint: Fuzzy assessment of ecological security on the Qinghai–Tibet plateau based on pressure–state–response framework. *Remote Sensing*. 2023;*15*:1293. DOI: 10.3390/rs15051293

[16] Onderco M, Portela C. Mint: External drivers of EU differentiated cooperation: How change in the nuclear nonproliferation regime affects member states alignment. Contemporary Security Policy. 2023;**44**(1):150-175. DOI: 10.1080/13523260.2022.2146336

[17] Rekunenko I, Zhuravka F, Nebaba N, Levkovych O, Chorna S. Mint: Assessment and forecasting of Ukraine's financial security: Choice of alternatives. Problems and Perspectives in Management. 2022;**20**:2. DOI: 10.21511/ppm.20(2).2022.11

[18] Klarin T. Mint: The concept of sustainable development: From its beginning to the contemporary issues. Zagreb International Review of Economics and Business. 2018;**21**(1):67-94. DOI: 10.2478/zireb-2018-0005

[19] Zagonari F. Mint: Four sustainability paradigms for environmental management: A methodological analysis and an empirical study based on 30 Italian industries. Sustainability. 2016;**8**(504):1-34. DOI: 10.3390/su8060504

[20] Ravetz J. Mint: City-region 2020: Integrated planning for a sustainable development. International Journal of Sustainability in Higher Education. 2000;**1**(3):305-308. DOI: 10.1108/ijshe.2000.1.3.305.3

[21] Krivý M. Mint: Towards a critique of cybernetic urbanism: The smart city and the society of control. Planning Theory. 2016;**17**(1):8-30. DOI: 10.1177/1473095216645631

[22] Ringenson T, Höjer M. Smart city planning and environmental aspects? Lessons from seven cities. In: Proceeding of the 4th International Conference on ICT for Sustainability (ICT 4S); 29 August - 1 September 2016; Amsterdam. Vol. 46:1. 2016. pp. 159-166. DOI: 10.2991/ict4s-16.2016.19

[23] Commission of the European Communities. Communication from the Commission. A Sustainable Europe for a Better World: A European Union Strategy for Sustainable Development (Commission's proposal to the Gothenburg European Council). Brussels; 2001. p. 264

[24] Bibri SE. Mint: Backcasting in futures studies: A synthesized scholarly and planning approach to strategic smart sustainable city development. European Journal of Futures Research. 2018;**6**(13):1-27. DOI: 10.1186/s40309-018-0142-z

[25] Caparros-Midwood D, Barr S, Dawson R. Mint: Spatial optimization of future urban development with regards to climate risk and sustainability objectives. Risk Analysis. 2017;**37**(11):2164-2181. DOI: 10.1111/risa.12777

[26] Lipińska EJ. Outline of the Theory of the State as an Ecologically Safe Organization in Empirical Research. 1st ed. Rzeszów: Publishing House of the Rzeszów University of Technology; 2022 527 p. ISBN 978-83-7934-598-4

[27] Fabio Di A. Mint: Positive healthy organizations: Promoting well-being, meaningfulness, and sustainability in organizations. Frontiers in Psychology. 2017;**8**:1-6. DOI: 10.3389/fpsyg.2017.01938

[28] Epstein MJ, Elkington J, Leonard HB. Making Sustainability Work: Best Practices in Managing and Measuring Corporate Social, Environmental, and Economics Impacts (Business). Oakland: Berrett-Koehler; 2008 ISBN-13 978-1576754863

[29] Horton P, Horton BP. Mint: Re-defining sustainability: Living in harmony with life on earth. One Earth. 2019;**1**(1):86-94. DOI: 10.1016/j.oneear.2019.08.019

Chapter 6

A Socially Inclusive and Sustainable Development under Globalization: A Challenge for Latin America Vis-à-Vis Europe

Octavio Luis Pineda

Abstract

This chapter aims to analyze the behavior of a selected group of European and Latin American countries using a set of socioeconomic and environmental parameters from international organizations. Alternative sources are also considered to derive some socioeconomic and ecological implications and externalities resulting from the market liberalization process and globalization within each group. The widespread adoption of neoliberal strategies by most Latin American economies has led, over time, to a pattern of unbalanced, unsustainable, and socially non-inclusive development in these countries. This contrasts with the situation in most developed nations in the European group, which tend to exhibit more balanced, socially inclusive, and sustainable economic development, particularly in the case of Scandinavian countries. Nowadays, these countries stand out as the most successful examples of such strategies within a market-oriented type of economy in today's globalized context. The period analyzed is 1980–2020.

Keywords: unbalanced development, socially non-inclusive, unsustainable development, non-inclusive development, socially inclusive growth

1. Introduction

The economic liberalization concomitant with globalization has become the new world economic paradigm, including not only economic aspects but also social, environmental, political, and cultural issues. This shift has affected nations and societies of the 21st century in varying ways and to different degrees, impacting both developed and peripheral or emerging countries. It has also challenged many traditional paradigms, such as "national sovereignty," "economic stewardship," and the "nation-state," seeking to minimize the presence and intervention of the State in the economic management of both developed and peripheral economies [1].

Through diverse hegemonic pressures and multiple mechanisms, nation-states have been induced to adopt a pro-capitalist reorientation and neoliberal economic strategies that prioritize the global market while often overlooking social inclusion and sustainability. This status quo unfortunately persists, especially in peripheral countries, prompting the need to rethink local growth strategies. These strategies should focus on achieving a healthy balance between growth and social wellbeing, contrary to the approaches imposed by the Washington Consensus, a set of hegemonic policies designed by the International Monetary Fund (IMF), the World Bank, and the US government, as pointed out by Williamson [2]. These policies, often referred to as "straitjackets," were specifically designed for the economies of the Latin American region, with dire implications as highlighted by scholars like Casilda [3] and Rangel [4]. Bermúdez [5] also acknowledges the pernicious role of the Washington Consensus not only in Latin America but elsewhere across the globe.

Under such premise, this work aims to analyze the economic development of two select groups of European and Latin American economies through various economic, social, and environmental indicators stemming from international organizations and other alternative sources. The goal is to derive some implications associated with the market liberalization process and the advent of globalization within each group, as well as the economic strategies adopted by each in the face of economic globalization. The approach taken by most Latin American countries, including Mexico, has ultimately led to the current unequal, unsustainable, and socially exclusive development scheme prevailing throughout Latin America. This contrasts with what is happening among the European economies, particularly in Scandinavia where an ideal prototype or model of balanced, socially inclusive, and sustainable development has been established within the current global context, as highlighted by Oro [6].

1.1 Part I. Overview at the interior of the Latin American region

For purposes of this study, we selected a sample of economies from the most to the least developed in the Latin American region. This includes Brazil, Mexico, Argentina, Venezuela, Chile, and Colombia on one hand, and some of the more lagging economies such as Nicaragua, El Salvador, Costa Rica, Guatemala, and Paraguay on the other. For comparison, we considered Germany, France, and Spain on the European side. And for the Scandinavian region, we focus on Sweden and Norway since we consider them prototypes of developed, balanced economies with some of the highest economic, social, and, environmental standards in the world. These countries are often ranked as exemplars of "social welfare economies" according to ABC.com [7] and serve as models for development that other European and peripheral countries could aspire to in the future. The "Scandinavian model," noted by Robert Erikson [8], has been successfully applied for decades across all Nordic countries, maintaining resilience under the current globalized environment.

It is important to note that we have included Spain within the European group despite its historical lags and ongoing challenges. As pointed out by Marin [9] and O. Luis-Pineda [10], Spain has managed to foster economic growth and maintain social wellbeing, aligning itself with the more advanced economies of Europe, such as Germany and France, as well as the Scandinavian countries, even in the face of trade liberalization and globalization.

Under this premise, the first section of this chapter examines the prevailing situation within the Latin American region. We then proceed to analyze the European and Scandinavian economies in comparison with Latin America through three sets of indicators: economic, social, and environmental. The historical data for this analysis, covering the period 1980–2020[1], is based on statistics from international sources.

1.2 Economic outlook (selected Latin American economies)

The economic parameters considered in this analysis include GDP, GDP per capita, foreign direct investment (FDI) (% GDP), and investment in Research and Development (R&D) (% GDP). These indicators are presented in **Table 1** for the period 1980–2020. To ensure consistency and facilitate international comparison, all monetary figures are expressed in US dollars at current prices.

1.3 Commentary on economic outlook (Latin America)

From the previous information, it is evident that all the countries in the region experienced growth in both GDP and GDP per capita during this period, which applies to both the more developed and less developed economies in the region. The former saw absolute growth above 400%, with the sole exception of Venezuela, which grew by 172% compared to its 1980 value. This reflects growth rates above the 5% annual average in some cases, such as Colombia (5.3%) and Chile (5.6%). Growth rates for Argentina, Venezuela, and Mexico were 4.1%, 1.4%, and 2.7%, respectively. Among the less developed economies, Costa Rica reported the greatest absolute growth (1300%), followed by Guatemala (983%) and Paraguay (804%). High growth rates were observed during the period, with Costa Rica showing an average annual growth rate (AAGR) of 6.6%, Guatemala at 5.8%, and Paraguay at 5.3%. In summary, trade liberalization led to a restructuring of the region's economies and an increase in GDP per capita across both the more and less developed countries during the study period.

Regarding foreign direct investment (FDI) in Latin America, the growing importance of foreign capital as a percentage of the GDP in each country is notable compared to its 1980 level. Among the more developed economies, Chile stands out with 4.5%, followed by Mexico with 2.9%, Colombia with 2.8%, and Brazil with 2.6%. These are coincidentally the economies with the most significant activity in the global market. Among the less developed economies, Nicaragua (5.9%) and Costa Rica (3.4%) had the highest levels of foreign investment by the end of the period. The rest reported modest percentages. Again, these are economies that experienced the most growth in their GDP during the period. In conclusion, FDI is a key factor in the growth of the region's economies, especially in the more dynamic ones that have most benefited from trade liberalization during this period.

Finally, the area of investment in science and technology (R&D) presents a discouraging and contrasting picture. With the sole exception of Brazil (1.1%), the other countries report very low levels of spending on science and technology as a percentage of GDP. This results in investments that are mediocre compared to the amounts allocated to this area by developed countries, as we will discuss later. Investment in this area is crucial for greater technological and economic independence and to enhance competitiveness in the global market. The outlook for the less developed economies is even more concerning, showing little importance placed by these

[1] To simplify this exposition, in the following tables of indicators herein reported on the different indicators analyzed (economic, social, and environmental), only data from the beginning and end of the period (1980–2020) are reported, omitting the years of the intermediate decades (1990, 2000 and 2010) for both groups of economies analyzed.

Indicator X	GDP(mms US dls)			GDP pc (US dls)			FDI (% GDP)			R&D (%GDP)		
Country	1980	2020	AAGR (1980–2020)	1980	2020	AAGR (1980–2020)	1980	2020	AAGR (1980–2020)	1980	2020	AAGR (1960–2020)
Mexico	250.1	1090.5	3.7%	2958.0	8655.0	2.7%	1.1	2.9	2.5%	0.2	0.6	2.5%
Brazil	235.0	1476.1	4.7%	1940.0	6924.0	3.2%	0.8	2.6	3.0%	0.8	1.2	1.1%
Argentina	77.0	385.5	4.1%	2738.0	8496.0	2.9%	0.9	1.3	0.9%	0.4	0.5	0.4%
Venezuela	59.1	102.0	1.4%	3853.0	3421.0	−0.3%	0.1	0.1	0.2%	0.7	0.3	−2.5%
Colombia	33.4	270.2	5.4%	1204.0	5304.0	3.8%	0.5	2.8	4.4%	0.3	0.3	0.1%
Chile	29.0	254.1	5.6%	2577.0	17,911.0	5.0%	0.8	4.5	4.4%	0.5	0.3	−1.1%
Guatemala	7.9	77.7	5.9%	1082.0	5061.0	3.9%	1.4	1.3	−0.2%	0.0	0.0	0.0%
Costa Rica	4.8	62.4	6.6%	2022.0	3455.0	1.3%	1.1	3.4	2.9%	0.0	0.0	0.0%
Nicaragua	2.2	12.7	4.5%	674.0	1716.0	2.4%	0.6	5.9	5.9%	0.2	0.2	0.0%
Salvador	3.6	24.9	5.0%	780.0	2345.0	2.8%	0.2	1.6	5.3%	0.1	0.2	2.7%
Paraguay	4.4	35.4	5.4%	1398.0	2067.0	1.0%	0.8	0.3	−2.4%	0.0	0.2	4.9%

Sources: https://datos.bancomundial.org/indicador/EN.ATM.CO2EKT; https://datos.bancomundial.org/indicador/EG.USE.PCAP.KG.OE?name_desc=false; https://datos.bancomundial.org/indicador/NV.AGR.TOTL.ZS?name_desc=false.

Table 1.
Economic indicators (Latam selected countries) (1980–2020).

governments on achieving technological independence from industrialized nations and enhancing competitiveness in the global market. In short, despite the economic growth and prosperity resulting from trade liberalization, Latin American economies, except for Brazil, have placed marginal emphasis on science and technology. The region's investment levels are well below the minimum (1% of GDP) recommended by the United Nations as essential for economic takeoff and reducing technological dependence, as pointed out by I. Torres [11] and Webforum [12].

1.4 Social outlook (selected Latin American economies)

In this regard, the proposed parameters include population (millions of inhabitants), social spending (net social expenditure %GDP), the Gini coefficient as a measure of income distribution among the population, and finally, the Human Development Index (HDI), (IDH in the table above) as a measure the level of general social wellbeing of the population. For this purpose, we refer to **Table 2**, which shows the pertinent information for the study period.

1.5 Commentary on social outlook (Latin America)

The social outlook strongly contrasts with the economic boom experienced by the region during the period, both for the most developed and the most lagging economies in the region, presenting a chiaroscuro panorama.

Firstly, it stands out that the populations of the most developed countries have been growing at a rate of less than 2%, with the rest generally not exceeding this figure. On the lagging side, only three slightly exceed the 2% annual average growth rate (AAGR): Guatemala (2.39%), Paraguay (2.06%), and Costa Rica (2.04%). The rest grew below the 2% annual average. This suggests that only a few countries in the region, except for some lagging economies, saw their population growth exceed 100% during the study period.

This situation reflects a conscious effort by these countries to reduce their population growth rates below the 2% annual average. However, in terms of social spending, the outlook is concerning, both among the most developed and lagging economies in the region. By the end of the period, only three of the more developed economies surpassed 20% of their GDP in social spending: Brazil (38.5%), Argentina (25.9%), and Mexico (22.3%). The rest did not reach this percentage. On the lagging side, two economies stand out for exceeding 20% in social spending: El Salvador (33.9%) and Costa Rica (32.3%). The rest report social spending levels below 20% of their GDP.

Despite the emphasis on social expenditure in these economies, with levels above 20% of their GDP, both highly developed and less developed, this effort pales in comparison to their levels of income distribution (Gini coefficient), as we discuss below.

Concerning income distribution, measured by the Gini coefficient, the situation is troubling. By the end of the period, among the developed economies, inequality ranks as follows: Colombia (53.5), Brazil (48.9), Mexico (45.4), Chile (44.9), Argentina (42.3), and finally Venezuela (41.2). On the lagging side, for this same year, inequality ranks as follows: Guatemala (52.0), Costa Rica (49.2), Nicaragua (45.0), Paraguay (43.5), and Salvador (41.0). Except for Argentina, Nicaragua, and Paraguay, which have seen their levels of inequality worsen compared to 1980, the others, both developed and lagging, have gradually decreased their levels of inequality during the period, albeit very slowly.

Finally, concerning the human development index (HDI), the outlook by the end of the period (2020) shows that HDI levels for both the most developed and the least developed economies do not exceed an HDI index of 0.800. This applies to both

Indicator X	POP (mill. Inhab.)			Net Social Exp (% GDP)			GINI Coef.			Human Develop. Index (HDI)		
Country	1980	2020	AAGR (1980–2020)	1980	2020	AAGR (1980–2020)	1980	2020	AAGR (1980–2020)	1980	2020	AAGR (1980–2020)
México	69.4	126	1.5%	1.7	22.3	6.6%	51.9	45.4	−0.3%	0.680	0.756	0.3%
Brazil	116.7	213.2	1.6%	17.6	38.5	2.0%	57.9	48.9	−0.4%	0.520	0.551	0.1%
Argentina	28.1	45.4	1.3%	19.1	25.9	0.8%	42.8	42.3	0.0%	0.680	0.602	−0.3%
Venezuela	15.3	28.5	1.7%	7.8	14.1	4.6%	55.6	49.5	−0.3%	0.630	0.691	−0.3%
Colombia	27.7	50 9	1.8%	6.0	36.6	4.6%	59.1	53.5	−0.2%	0.560	0.604	0.2%
Chile	11.3	19.3	1.4%	11.9	26.2	2.0%	55.3	44.9	−0.5%	0.640	0.652	0.0%
Guatemala	7.3	16.9	2.4%	7.0	14.9	1.9%	58.3	48.3	−0.5%	0.430	0.461	0.2%
Costa Rica	2.4	5.1	1.8%	16.0	32.3	1.8%	47.5	49.2	0.1%	0.620	0.629	0.0%
Nicaragua	3.2	6.8	2.0%	7.0	16.5	2.2%	46.7	47.7	0.1%	0.460	0.508	0.2%
El Salvador	4.6	6.3	0.9%	6.0	33.9	4.4%	54.0	38.0	−0.9%	0.470	0.546	0.4%
Paraguay	3.2	6.6	2.1%	3.0	18.5	4.6%	40.8	43.5	0 2%	0.540	0.528	−0.1%

Sources: http://data.worldbank.org/indicator/SI.POV.GINI; http://data.bank.worldbank.org/reports.aspx?source=2&series-SI.POV.GINI&country-; http://data.worldbank.org/indicator/SP.POP.TOTL; http://hdr.undp.org/data.center/human-development-index#/indicies/HDI; http://hdr.undp.org/content/human-development-report-2020.

Table 2.
Social indicators (LATAM: Selected countries > (1980–2020).

groups, meaning that most countries in the region are ranked as "medium developed economies." according to the aforementioned international organization.

1.6 Environmental outlook (selected Latin American economies)

In this respect, the parameters considered are carbon dioxide emissions (CO2, in thousands of tons, per capita), electrical energy consumption per capita (KWt, p.c.), and the specific weight of the agricultural sector on the economy (% GDP). Under this premise, we present **Table 3**, which reports this information for the analysis period 1980–2020.

1.7 Commentary on environmental outlook (Latin America)

Here we discuss important climate implications associated with the rapid growth achieved by most of the economies in the region.

Regarding the generation of greenhouse gas (CO2) emissions, among the most developed economies, Brazil, Mexico, and Chile stand out as the major polluters since they almost doubled their emissions during the period. Among the lagging economies, only Guatemala nearly doubled its volume of CO2. These economies have grown at a more accelerated pace of industrialization than others, using fossil fuels in their industrial processes. However, the growth rate of emissions is positive for most advanced economies in the region. In some cases, like Mexico, this rate turns out to be negative, demonstrating the country's efforts to reduce its emissions during the period. Such policies were also adopted by other lagging economies such as Costa Rica, Nicaragua, El Salvador, and Paraguay.

Regarding per capita consumption of electricity in thousands of (KWh p.c.) during the period, the economies with the highest demand for this type of energy by 2020 are Chile (3.9), Argentina (2.7), Venezuela (2.4), Brazil (2.6), and Mexico (2.3). Among the lagging economies, only Costa Rica (1.8) and Paraguay (1.9) stand out for their high per capita electricity consumption by the end of the period.

Finally, regarding the impact of agriculture on the GDP of these economies, a clear trend towards less dependence on agriculture is evident over the period. The exception is Mexico among the more developed economies and Guatemala among the least developed, which increased their dependence on agriculture during the period. By the end of the period, among the most advanced economies, those with greater dependence on the agricultural sector, in order of importance, are Colombia (7.5%), Argentina (6.1%), Brazil (5.7%), Venezuela (5.5%). %), and Mexico (3.8%). Among the lagging economies, Nicaragua (15.7%), Paraguay (11.1%), Costa Rica (4.3%), and Guatemala (4.0%) stand out. Guatemala, Costa Rica, and Mexico have even increased their percentage of dependence on agriculture in their GDP compared to 1980.

1.8 Summary overview of the Latin American region

Firstly, the dynamic growth achieved by Latin American economies associated with trade openness highlights significant resizing in GDP and per capita figures. This growth is partly explained by increased exports and greater foreign investment in the region during this time. Unfortunately, this boom overlooks a greater emphasis by governments on science and technology. By 2020, except for Brazil (1.1% GDP), most nations report low levels of investment in science and technology, below this figure.

In the social realm, despite economic growth, there are high social costs and greater social inequality within the region. Social spending in most countries is low, with

Indicator X	CO2 EMISSIONS (Kton. PC)			ENERGY COMPSUTION (KWH, PC)			Agric. Contrib. (%GDP)		
Country	**1980**	**2020**	**AAGR (1980–2020)**	**1980**	**2020**	**AAGR (1980–2020)**	**1980**	**2020**	**AAGR (1980–2020)**
Mexico	2.4	3.8	1.2%	839.40	2289.1	2.5%	0.9	3.8	3.7%
Brazil	2.1	4.1	1.7%	1005.2	2653.8	2.5%	11.0	5.7	−1.6%
Argentina	1.0	1.5	1.0%	1318.3	2700.4	1.8%	6.4	6.1	−0.1%
Venezuela	0.9	0.7	−0.6%	1794.9	2473.8	0.8%	4.8	3.0	−1.2%
Colombia	0.4	0.8	1.7%	592.0	1430.2	220.0%	19.9	7.5	−2.4%
Chile	2.8	8.5	2.8%	931.6	3995.9	370.0%	7.3	4.0	−1.5%
Guatemala	0.5	0.2	−2.3%	243.0	658.2	250.0%	5.0	9.9	1.7%
Costa Rica	2.5	7.0	2.6%	816.3	1867.1	210.0%	3.3	4.3	0.7%
Nicaragua	2.0	4.6	2.1%	272.8	683.8	230.0%	21.0	15.7	−0.7%
Salvador	2.1	6.4	2.8%	295.6	971.1	300.0%	17.0	5.4	−2.8%
Paraguay	1.6	7.6	4.0%	228.8	1968.4	550.0%	17.0	11.1	−1.1%

Sources: https://datos.bancomundial.org/indicador/EN.ATM.CO2E.KT; https://datos.bancomundial.org/indicador/EG.USE.PCAP.KG.OE?name_desc=false; https://datos.bancomundial.org/indicador/NV.AGRTOTL.ZS?name_desc=false; https://datosmacroexpansion.com/energia-y-medio-ambienete/electricidad-consumo; https://datosmacro.expansion.com/energia-y-medio-ambienete/emisiones-co2?anio=1980; https://www.cia.gov/the-world-factbook/about/archives/2021/countries/germany/; https://www.cia.gov/the-world-factbook/.

Table 3.
Environmental indicators (Latam selected countries) (1980–2020).

only Brazil, Argentina, and Mexico exceeding 20% of their GDP in social expenditure among the developed economies, and Costa Rica (32.3%) and El Salvador (33.9%) among the lagging economies. The rest have levels below this percentage. This imbalance is compounded by troubling levels of inequality, with Gini coefficients levels close to 50 points. For example, the most unequal economies in 2020 were Colombia (53.5), Brazil (48.9), Mexico (45.4), Chile (44.9), and Argentina (42.3).

Regarding HDI, most countries report levels below .800 and are classified as "low to medium development" economies by the United Nations. This indicates uneven and socially exclusive development throughout the region.

In the environmental area, economies that generate pollution, especially greenhouse gases (CO2), and have high energy consumption (KWt, p.c.) based on fossil fuels, stand out. This is a result of accelerated growth but also a gradual move away from mono-exporting countries towards greater diversification of exports in the global market.

Briefly, although economic liberalization has led to economic growth for most countries in the region, both in size and per capita, this growth is unbalanced and unequal. It overlooks population wellbeing, emphasizes insufficient investment in science and technology to boost economic growth and competitiveness, and neglects social spending and income distribution. Additionally, it has produced myriad social costs, externalities, and unsustainable exploitation of natural resources, as a consequence of implementing neoliberal economic strategies.

1.9 Part II. Panoramic outlook (Europe and Scandinavia vs. Latin America: selected countries)

In this second and final part of the analysis, we examine the situation of three European economies—Germany, France, and Spain—and two prominent Scandinavian countries—Sweden and Norway—as examples of balanced, sustainable, and socially inclusive economies. These countries are among the top 16 globally for social wellbeing [12]. On the Latin American side, we consider Brazil, Mexico, Argentina, and Colombia. We use the same set of international indicators applied in the first part of this research.

1.10 Economic indicators (Europe and Scandinavia vs. Latin America: selected countries)

For this aim, we utilize the same set of international indicators used in the first part of this research, as shown in **Table 4**.

1.11 Commentary on economic indicators (Europe and Scandinavia vs. Latin America: selected countries)

To start with, there is a stark asymmetry between some European economies, such as Germany and France, compared to most Latin American countries. This contrast becomes even more dramatic when comparing their per capita and living standards, as we shall see below:

In Europe, Norway and Spain report the highest GDP growth rates during the period, with 4.45% and 4.42%, respectively. However, these figures are overshadowed by the impressive GDP growth rates observed in Latin American during the same period: Colombia (5.37%), Brazil (4.70%), and Argentina (4.11%).

Similarly, the economic resizing of countries in both groups is evident over the analysis years. Within the European group, Scandinavian countries reported higher

Indicator X	GDP(mms US dls)			GDP pc (US dis)			FDI(% GDP)			R&D(%GDP)		
County	1980	2020	AAGR (1980–2020)	1980	2020	AAGR (1980–2020)	1980	2020	AAGR (1980–2020)	1980	2020	AAGR (1980–2020)
Mexico	250.1	1091.0	3.8%	2958.0	8655.0	2.7%	1.1	2.9	2.5%	0.3	0.3	0.2%
Alemania	797.4	3889.7	4.0%	10,170.0	46,773.0	3.9%	0.1	4.1	9.7%	2.1	3.1	1.0%
Francia	690.3	2639.0	3.4%	12,500.0	39,055.0	2.9%	0.5	0.7	0.8%	2.2	2.3	0.1%
España	226.0	1277.0	4.4%	13,415.0	26,960.0	1.8%	0.7	2.9	3.6%	0.8	1.4	1.4%
Suecia	140.1	547.1	3.5%	17,000.0	52,838.0	2.9%	0.2	3.5	7.4%	3.3	3.5	0.1%
Noruega	64.4	367.6	4.5%	11,230.0	68,340.0	4.6%	0.1	1.1	6.2%	2.5	2.3	−0.2%
Brazil	235.0	1476.1	4.7%	1940.0	6924.0	3.2%	0.8	2.6	3.0%	0.8	1.2	1.0%
Argentina	77.0	385.5	4.1%	2738.0	8496.0	2.9%	0.9	1.3	0.9%	0.4	0.5	0.6%
Colombia	33.4	270.2	5.4%	1204.0	6304.0	4.2%	0.5	2.8	4.4%	0.3	0.3	0.0%

Sources: https://datos.bancomundial.org/indicador/EN.ATM.CO2E.KT; https://datos.bancomundial.org//indicador/EG.USE.PCAP.KG.OE?name_desc=false; https://datos.bancomundial.org/indicador/NV.AGR.TOTL.ZS?name_desc=false.

Table 4.
Economic indicators (Europe & Scand vs Latam (selected countries) (1980–2020).

GDP per capita levels by the end of the analysis period: Sweden ($52,838) and (Norway) ($68,340), compared to Germany ($46,773), France ($39,055), and Spain ($26,960). Both the European and Scandinavian groups benefited from economic liberalization, without neglecting investment in science and technology, a fundamental component of their growth and global competitiveness during the analysis period.

Regarding FDI, Europe shows a marked difference compared to Latin America. European countries, except Spain (2.9%) and Germany (4.1%), reported relatively low figures in this area, such as France (0.7%) and Norway (1.1%). Of the Scandinavian countries, only Sweden (3.5%) reported a significant level. This is understandable because most of these countries are capital exporters to the rest of the world. The growth achieved by the European and Scandinavian groups due to economic globalization reflects a significant resizing of their economies and per capita, with less reliance on FDI as a growth component.

Finally, investment in science and technology is notably emphasized in Europe during the period: Sweden (3.3%), Norway (2.9%), Germany (3.1%), France (2.3%), and Spain (1.4%). In contrast, Latin American countries, except Brazil (1.1%), show much lower levels of investment in this area. This indicates that while both groups benefited from trade liberalization, Europeans did not overlook investment in science and technology, which is crucial for maintaining competitiveness in the global market and achieving technological independence. Latin American countries have not taken advantage of their economic boom to emphasize this area, leading to missed opportunities.

1.12 Social indicators (Europe and Scandinavia vs. Latin America: selected countries)

To analyze this aspect, we use **Table 5**, which reports relevant social indicators for both economic groups for the period 1980–2020.

1.13 Commentary on social indicators (Europe and Scandinavia vs. Latin America: selected countries)

Regarding social indicators, there are significant contrasts between the two groups.

Firstly, concerning population growth, most countries in the European group, except France, have grown at a rate of less than 2% annually, with France growing at 2.3% and doubling its population during the period. In contrast, Latin American countries do not exhibit more than double growth in size, with growth rates below 2% annually. Thus, despite the considerable differences between European and Latin American countries, both groups experienced growth rates below the 2% annual average.

Secondly, social expenditure has been notably emphasized in Europe, both in size and growth rate. For example, Germany increased its social expenditure by more than 32.6% from its 1980 figure (1.7%), while France increased it nearly ten times (52.0%), and Spain almost doubled it (41.1%) with average annual growth rates of 7.6%, 5.5%, and 1.9%, respectively. The Nordic countries, despite minimal growth in this period, maintained high levels of social spending compared to 1980 (25–35% of GDP). On the Latin American side, social expenditure presents a mixed outlook. Argentina and Brazil reported levels above 30% of GDP for 2020, with growth rates around 4% over the period. Mexico has quadrupled its expenditure since 2020 to 25% of GDP, though still below Argentina and Brazil. The rest of the countries report

Indicator X	POP (mill. Inhab)		AAGR (1980–2020)	Net Social Exp(%GDP)		AAGR (1980–2020)	GINI Coef.		AAGR (1980–2020)	Human Devel. Index (HDI)		AAGR (1980–2020)
Country	1980	2020		1980	2020		1980	2020		1980	2020	
Mexico	69.4	126.0	1.5%	1.7	22.3	6.6%	51.9	45.4	−0.3%	0.595	0.779	0.7%
Germany	71.0	79.4	0.3%	1.7	32.6	7.7%	51.9	28.2	−1.5%	0.713	0.950	0.7%
France	28.4	58.0	1.8%	6.0	52.0	5.5%	59.1	30.7	−1.6%	0.722	0.903	0.6%
Spain	28.5	38.9	0.8%	19.1	41.1	1.9%	42.8	34.9	−0.5%	0.680	0.905	0.7%
Sweden	8.3	8.6	0.1%	30.0	34.8	0.4%	26.0	28.9	0.3%	0.776	0.947	0.5%
Norway	4.1	4.2	0.1%	31.0	45.4	1.0%	25.0	25.0	0.0%	0.793	0.961	0.5%
Brazil	121.2	150.7	0.5%	17.6	38.5	2.0%	57.9	48.9	−0.4%	0.697	0.754	0.2%
Argentina	28.1	32.6	0.4%	19.1	25.9	0.8%	42.8	42.3	0.0%	0.665	0.845	0.6%
Colombia	27.7	32.6	0.4%	6.0	36.6	4.6%	59.1	53.5	−0.2%	0.557	0.752	0.8%

Sources: https://datos.bancomundial.org/indicador/EN.ATM.CO2E.KT; https://datos.bancomundial.org//indicador/EG.USE.PCAP.KG.OE?name_desc=false; https://datos.bancomundial.org/indicador/NV.AGRTOTLZS?namedesc=false.

Table 5.
Social indicators (Europe & Scand vs Latam (selected countries) (1980–2020).

lower levels. Thus, while Europe emphasizes or at least maintains high levels of social spending, Latin America shows an uneven and less consistent approach to social expenditure during this period.

Thirdly, income distribution, measured by the Gini coefficient, reveals concerning trends. In Europe, Scandinavian countries manage inequality figures slightly above the minimum levels, with Sweden (28.9) and Norway (25.0) being the fairest in wealth distribution worldwide. The rest of the European group reported levels between 30.0 and 35.0. In Latin America, the situation is more worrying. By the end of 2020, the most unequal economies were Colombia (53.5), Brazil (48.9%), Mexico (45.4%), and Argentina (42.3). However, there was a gradual reduction in inequality levels throughout the study period. The stark contrast in income inequality between the two regions highlights the disparity in development outcomes resulting from trade liberalization. In other words, in the face of the economic boom derived from trade liberalization, Europe manages relatively low levels of economic inequality, with Gini coefficients ranging around 31.0–37.0, reflecting a fairer distribution of wealth.

Lastly, regarding HDI, European countries generally reported levels above or very close to 0.900, corresponding to "high or very high development economies." Sweden and Norway, in particular, maintain "world leadership" as "Welfare States". In contrast, most Latin American countries reported HDI levels below 0.800 for 2020, classifying them as "medium-developed" economies according to the United Nations. Thus, while European countries have HDI figures close to or above 0.900, classifying them as "highly or very highly developed economies" like Norway and Sweden, Latin America countries were mostly below the 0.800 level for 2020.

1.14 Environment indicators (Europe and Scandinavia vs. Latin America: selected countries)

To analyze the environmental aspects, we constructed **Table 6** with data from the period 1980–2020.

1.15 Commentary on environmental indicators (Europe and Scandinavia vs. Latin America: selected countries)

Firstly, concerning CO2 emissions for both groups, Europe generally shows a reduction in emissions over the period with a clear downward trend. In contrast, in Latin America, most countries increased their CO2 emissions during this timeframe. For instance, Brazil (1.71%), Colombia (1.65%), México (1.15%), and Argentina (1.03%) all experienced annual growth rates above 1.5%. On the European side, most countries reported a significant reduction in the volume of CO2 emissions in the period, with Spain (5.7%) being a notable exception, showing a 5.7% increase. This reduction in Europe results from deliberate efforts to minimize the environmental impact of these pollutants.

Secondly, regarding electrical energy consumption, a key indicator for understanding the degree of industrialization of an economy and the consumption of consumer goods and electricity (appliances), Europe showed growth in electricity consumption during the period. Spain (1.80%), France (0.89%), and Germany (0.09%) had rates below 2% annually. In Latin America, however, consumption growth exceeded this figure on average, with Mexico (2.54%), Brazil (2.46%), Colombia (2.23%), and Argentina (1.81%) all surpassing a 2.5% annual average. On the European side Sweden stands out on the European side with a remarkable annual

Indicator X	(CO2)Emissions (CO2) (Kton. pc.)		AAGR (1980–2020)	Energy Consume.(KV\A pc)		AAGR (1980–2020)	Agric. Contrib (%GDP)		AAGR (1980–2020)
Country	**1980**	**2020**		**1980**	**2020**		**1980**	**2020**	
Mexico	0.2	0.4	1.2%	839.4	2289.1	2.5%	0.9	3.8	3.7%
Germany	1.1	0.6	−1.6%	5798.0	6003.8	0.1%	1.0	0.8	−0.7%
France	0.5	0.3	−1.6%	4400.5	6282.5	0.9%	3.6	1.6	−2.1%
Spain	0.2	2.0	5.8%	2583.5	5275.5	1.8%	5.0	2.9	−1.4%
Sweden	0.1	0.0	−2.1%	839.4	12,100.3	6.9%	3.6	1.3	−2.5%
Norway	0.0	0.0	−0.3%	18,354.3	23,097.4	0.6%	3.0	1.8	−1.2%
Brasil	0.2	0.4	1.7%	1005.2	2653.8	2.5%	11.0	5.7	−1.6%
Argentina	0.1	0.2	1.0%	1318.3	2700.4	1.8%	6.4	6.1	−0.1%
Colombia	0.0	0.1	1.7%	592.8	1430.2	2.2%	19.9	7.5	−2.4%

Sources: https://datos.bancomundial.org/indicador/EN.ATM.CO2E.KT; https://datos.bancomundial.org/indicador/EG.USE.PCAP.KG.OE?name_desc=false; https://datos.bancomundial.org/indicador/NV.AGR.TOTL.ZS?name_desc=false; https://datosmacroexpansion.com/energia-y-medio-ambienete/electricidad-consumo; https://datosmacro.expansion/energia-y-medio-ambiente/emisiones-co2?anio=1980.

Table 6.
Environmental indicators (Europe & Scand vs Latam (selected countries) (1980–2020).

average increase of 6.9%. This higher consumption in Europe compared to Latin America is attributed to higher industrialization and per capita consumption, leading to higher pollution levels associated with fossil energy. In contrast, Latin America's rising electricity consumption is linked to growing industrialization and increased consumer goods consumption, reflecting the economic benefits from market liberalization.

Finally, regarding the contribution of agriculture to GDP, both European and Latin American countries showed a clear trend towards reduced dependence on agriculture. In Europe, most countries experienced a gradual decline in agriculture's GDP share, with Spain being an exception, maintaining an average annual rate of 2.9% in the period. In Latin America, although countries like Mexico and Brazil showed a reduction, others like Colombia (7.5%), Argentina (6.1%), Brazil (5.7%), and Mexico (3.8%) still displayed notable agricultural contributions by 2020.

In short, European countries exhibited less dependence on agriculture, except Spain, due to greater industrialization and development. Latin American countries, despite a gradual shift away from agriculture, still exhibit considerable dependence, reflecting their ongoing industrialization and diversification from traditional mono-exporting economies. This trend is particularly evident in major economies like Brazil, Mexico, Colombia, and Chile.

1.16 Summary: (II-part): preliminary conclusions

From the comparative analysis of the European (including Scandinavian) and Latin American groups, several preliminary conclusions emerge.

In the economic area, in addition to the observed asymmetries between both groups, it is clear that both are experiencing significant economic growth due to the market liberalization process, both in size and per capita. Notably, while the European group emphasizes growth without neglecting investment in science and technology and the social wellbeing of its population (as measured by social expenditure and income distribution), the Latin American group shows growth with less emphasis on science and technology and unfortunately, on social wellbeing as well (social spending and income distribution). This situation reflects, on one hand, a balanced, socially inclusive, and sustainable development associated with economic globalization in Europe, and on the other hand, an unbalanced, socially exclusive, and unsustainable development in Latin America under the same global environment.

In the social arena, strong asymmetries and contrasts arise between the two groups. Firstly, concerning population growth, except for France (2.3%), most countries in the European group experienced growth of less than double at an average annual rate of less than 2%. Likewise, no country in Latin America grew more than double (2%) annually on average during this period.

In summary, despite the great asymmetry between the populations of European countries and those of Latin America, both groups are growing in population at a rate below the 2% annual average. Meanwhile, in Europe, except for France, the rest of the countries did not manage to double their population during the period. Social expenditure highlights Europe's focus on this indicator, both in size and growth rate, compared to Latin America. For 2020, Argentina, Brazil, and Mexico, among the most developed economies in the region, reported social expenditures above 20% of their GDP, although with growth rates below 1% during the period. Other economies remain well below these levels. In other words, while Europe seeks to increase or at least maintain its levels of social spending amidst economic openness, Latin America remains lagging in this regard, still far from the levels maintained by industrialized economies like those in Europe.

Regarding income distribution, the outlook is troubling. Europe, particularly the Scandinavian countries such as Sweden (27.0) and Norway (25.0), achieves some of the fairest income distribution figures worldwide. The rest of Europe manages levels between 31.0 and 37.0. In Latin America, however, the situation is worrisome, with Colombia (53.5), Brazil (46.9), Mexico (45.4), and Argentina (42.3) reporting high levels of inequality for 2020. Although there is a gradual tendency to reduce inequality levels throughout the study period, some cases like Argentina have worsened.

In summary, income inequality reflects the disparity between development and underdevelopment in a global context and the result of trade openness. While Europe, especially the Nordics, maintains relatively low levels of economic inequality (Gini coefficients in the range of 31.0–37.0), Latin America continues to experience high levels of inequality, with some improvement but also cases of worsening inequality, such as in Argentina.

Regarding HDI in the region, there is an evident economic asymmetry between both groups, both in size and per capita. This disparity is also evident within Latin America, particularly between the most developed economies and the lagging ones. Growth is explainable by greater access to global markets and the corresponding increase in exports during these years. Likewise, we perceive a growing flow of foreign investment towards this region. However, the economic boom achieved by these economies contrasts with the lack of emphasis on science and technology in most Latin American countries, except Brazil, which barely exceeds the 1.1% of GDP recommended by the United Nations for stimulating economic growth. Other countries in the region remain significantly below this figure.

In the social arena, the economic boom resulting from trade openness has led to strong social imbalances, especially between developed and lease developed countries (e.g., Nicaragua, El Salvador, Costa Rica, Guatemala, and Paraguay). Aside from the gap and the existing standards compared to the European economies analyzed. Across the region, low levels of social expenditure and significant income inequality reflect persistent structural problems such as unemployment, poverty, corruption, insecurity, and violence, contributing to social and political tension in many countries.

In environmental matters, the European group stands out as the largest producer of greenhouse gases and electricity consumption, thus being the largest polluter compared to the Latin American group, which shows a marginal dependence on agriculture due to its lower level of industrialization, with obsolete production schemes and generators of greenhouse gases based on fossil energy. However, Latin American economies are expanding and industrializing, moving away from mono-exporting raw materials towards more diversified economies, as seen in Brazil, Mexico, Colombia, and Chile.

In brief, while both groups benefit from trade openness, Europeans, especially the Nordics, show sustainable and socially inclusive economies, whereas Latin Americans show growth that is unbalanced, unsustainable, and socially exclusive, marked by wealth polarization. Achieving a balanced, sustainable, and socially inclusive economic development in Latin America for the 21st century remains a challenge, given the socioeconomic and environmental imbalances observed so far through the current indicators. Despite being supported by hard figures from international organizations, these indicators reveal strong contrasts and imbalances during the market liberalization process, highlighting an evident imbalance between growth and social wellbeing over the last 40 years.

A crucial and unavoidable factor in this analysis is the role of the State in both European and Latin American economies. The State serves as the legal guarantor of balance and the economic watchdog of social interests versus private interests, ensuring the rational and sustainable use of natural wealth. This presupposes a mature democratic regime and a highly participatory and politicized society. In contrast, Latin America often faces incapable, unwilling or just uninterested states in promoting socially inclusive balanced economic strategies, especially under trade liberalization and prone to adopt neoliberal guidelines or "recipes" such as those stemming from the Washington Consensus [3], put into practice by the IMF and other international organizations. Which altogether have led to difficulties in achieving balance under current circumstances. For the fact that these countries have succeeded implement a balanced, socially inclusive and sustainable strategy in their economies, vid Alcalde [13] and managed for instance, to implement in practice a thorough and authentic recycling strategy within their economies vid Alcalde [14].

Furthermore, political regimes in Latin America feature immature or weak democracies, permeated by corruption [15] and facing longstanding structural problems such as unemployment [16], particularly youth unemployment [17], poverty [18, 19], and persistent dependency on foreign debt. These economies are prey to hegemonic interests from global organizations like the IMF [5] and the pervasive influence of transnational companies [20], which steer domestic strategies towards a neoliberal model. This pressure has forced countries like Mexico to abandon nationalist strategies that contributed to its "economic miracle" [21] during Latin American protectionism, a period of four decades of high growth (above 6% annually). The shift to an open economy allowed transnational capital into the country, leading to negative consequences [20, 22], including a 70% deterioration in Mexican's standard of living over 30 years, which has been linked to higher violence rates [23]. The abandonment of the Mexican Miracle strategy has proven disastrous for the country, necessitating its reorientation or "reconversion" due to its "exhausted and surpassed model" [24] and the imperative need for a new approach in line with the country's updated socioeconomic reality, as this author pinpointed more than a decade ago [25].

Fortunately, since the start of the new administration in Mexico (2019–2024), the country's situation has been rapidly changing. President López Obrador's Development Plan (Plan Nacional de Desarrollo) [26, 27] aims to move away from the neoliberal strategy, focusing on wellbeing alongside growth within a socially inclusive and sustainable framework. This approach marks an unprecedented strategy in Latin America. Mexico has diverged from IMF recommendations by refusing external debt to combat the pandemic and revive its economy, instead using internal savings to foster growth through significant public investments in infrastructure projects, particularly in southern Mexico, while minimizing corruption. Strategy which openly contrasts with those implemented most Latin American countries that accepted IMF assistance, leading to regrettable consequences [28].

Mexico's refusal to adopt and follow IMF policies has resulted in unprecedented post-pandemic economic growth, including a nearly 76% increase in its exports [29]. This has led to what some financial sources are calling a new "Second Mexican Miracle" [30] that has given Mexico better regional prospects compared to Brazil and Argentina, as we can observe from international statistics of major economies in Latin America, vid Countryeconomy.com [31]. Overall it results apparent an effective policy management by Mexico with socially inclusive approach that has facilitated productive infrastructure mega-investments, such as the Mayan Train, the Mexican Transismic Corridor, and the Dos Bocas Refinery, particularly in the southwestern region [32]. This strategy's success has been acknowledged by the IMF and many foreign banks and investors [33].

In contrast, the neoliberal "recipes" promoted by the IMF, such as the "Ten Measures of the Washington Consensus" vid, Selfbank [34], which Mexico also adhered to during the neoliberal period before the current administration [35], and implied a host setbacks in critical sectors such as education in the country, vid, OCDE. org [36] causing as well multiple imbalances and recurrent social crises across Latin America. Affecting major economies in this region like Brazil, Mexico [23], Argentina [37], Colombia [38], Venezuela [39], and Chile [40, 41], where high growth came along with unbalanced, unsustainable, and non-inclusive development

Therefore, the IMF's presence and intervention in the region have not only involved managing loans to address debt and economic damage but have also aimed to induce hegemonic policies into domestic strategies, as noted by Ángel Bermúdez [5]. This approach has had catastrophic socioeconomic implications for the region, exemplified by the ongoing crisis in Argentina. Despite this country's debt issues dating back to the beginning of the last decade [37], along with other structural problems and extreme dependence on foreign debt [42], the IMF's recommended measures to address COVID worsened the country's collapse and explain Argentina its current crisis [33]. Furthermore, these measures have exacerbated longstanding social issues such as income inequality, both before and after COVID, as highlighted by Santiago Poy [43].

2. Conclusions

In summary, the achievement of a balanced, sustainable, and socially inclusive economic development strategy for the region by the end of the 21st century, such as that prevailing among European economies and especially the Scandinavian group, must begin with a gradual shift away from the prevailing neoliberal strategies in Latin American states. This shift involves establishing long-term growth paths supported by broad social consensus and tailored to domestic socioeconomic realities. Mexico is currently trying to implement such a strategy successfully, with the Mexican Morena party winning the 2024 presidential elections. Claudia Sheinbaum, the left-wing president-elect, won with an overwhelming majority both for the presidency and in Mexican congress, as reported by the BBC [44].

However, it is essential to recognize and not overlooking the pervasive issue of corruption within the Latin American context, a phenomenon deeply intertwined with the political apparatus in many countries. The political-corruption nexus, as noted by Rogriguez [45], explains the persistence of corrupt governments led by elites typical of immature democracies. These governments often are backed by foreign interests and hegemonic policies, such as the Washington Consensus and transnational capital, which have historically impeded the rise of democratic and honest leaders representing the majority's real interests and wellbeing.

Additionally, another significant factor hindering socioeconomic reorientation and the rise of authentic democratic governance is the low level of social awareness and political consciousness among the populace. This issue is closely related to the historically low educational levels in Latin America, as discussed by Jeffrey Puryear [46].

DOI: http://dx.doi.org/10.5772/intechopen.1005907

Author details

Octavio Luis Pineda
School of Economics (E.S.E), National Polytechnic Institute of Mexico, Mexico

*Address all correspondence to: oluisp@gmail.com

References

[1] Luis-Pineda O. Diversos Enfoques sobre la Globalización: Un Análisis Comparativo. México, D.F: Editorial IPN; 1998

[2] Williamson J. ElConsensode Washington, 2009. 1989. Available from: https://economipedia.com/definiciones/consenso-de-washington.html

[3] Casilda R. América Latina y el Consenso de Washington. In: Boletín Económico. No. 2803. 2004. pp. 19-37. Available from: http://biblioteca.hegoa.ehu.es/system/ebooks/14120/original/America_Latina_y_el_consenso_de_Washington.pdf

[4] Rangel R et al. El Consenso de Washington: la instauración de las políticas neoliberales en América Latina. In: Revista Política y Cultura no. 37 México ene. 2012. 2012. Available from: http://www.scielo.org.mx/scielo.php?pid=S0188-77422012000100003&script=sci_arttext&tlng=en

[5] Bermúdez Á. El FMI en América Latina: el controvertido rol del organismo en grandes crisis económicas en la región y el resto del mundo. BBC News; 2019. Available from: https://www.bbc.com/mundo/noticias-america-latina-50031338

[6] Oro AM. Las economías escandinavas como modelo: ¿Pero qué modelo?. 2011. Available from: http://procesos-de-aprendizaje.libremercado.com/2011/10/13/las-economias-escandinavas-como-modelo-pero-que-modelo/

[7] ABC.com. Definiciónbienestarsocial. 2016. Available from: http://www.definicionabc.com/?s=Bienestar%20Social

[8] Erikson R. The Scandinavian Model: Welfare States and Welfare Research. N.Y: M.E. Sharpe, Inc; 1987

[9] Marín P. España tardará 20 años en volver al nivel de bienestar previo a la crisis. 2012. Available from: http://sociedad.elpais.com/sociedad/2012/12/12/actualidad/1355337410_243341.html

[10] Luis-Pineda O. México y España Frente a la Globalización: Análisis Costo Beneficio e Implicaciones. España: Editorial EAE; 2012

[11] Torres I. América Latina debe evitar dependencia tecnológica para combatir cambio climático. 2008. Available from: http://www.cronica.com.mx/notas/2008/372418.html

[12] Webforum. The 16 Most Socially Advanced Countries in the World. 2019. Available from: https://www.weforum.org/agenda/2019/04/the-16-most-socially-advanced-countries-in-the-world/

[13] Alcalde M. Estado de bienestar: el liderazgo de Escandinavia. 2014. Available from: http://enpositivo.com/2014/07/estado-de-bienestar-el-liderazgo-de-escandinavia/

[14] Alcalde M. Escandinavia una verdadera cultura de reciclaje. 2014. Available from: http://enpositivo.com/2014/06/escandinavia-una-verdadera-cultura-del-reciclaje/

[15] Transparency.org. Corruptionperceptions Index. 2019. Available from: https://www.transparency.org/cpi2019

[16] Garcia V. Desempleo en América Latina. 2010. Available from:

http://coyunturaeconomica.com/noticias/desempleo-en-america-latina

[17] Pesantes G. El desempleo juvenil es alarmante en América Latina. Banco Mundial; 2020. Available from: https://news.un.org/es/story/2020/01/1468672

[18] Botero A. Medimos bien la pobreza en América Latina? Banco de Desarrollo de América Latina; 2016. Available from: https://www.caf.com/es/conocimiento/visiones/2016/02/medimos-bien-la-pobreza-en-america-latina/

[19] ONU. La economía mundial crece poco mientras aumentan los niveles de pobreza en América Latina. Noticias ONU; 2020. Available from: https://news.un.org/es/story/2020/01/1468032

[20] Durand C. et al. Las multinacionales en América Latina: expansión, impacto y resistencias. 2009. Available from: http://www.rebelion.org/noticia.php?id=81765

[21] Carmona F et al. El Milagro Mexicano. Editorial Nuestro Tiempo; 1974. p. 122. Available from: http://www.ccc3m.com/producto/el-milagro-mexicano/

[22] Cifuentes V.R. Transnacionales, saqueo de recursos y conflicto ambiental en Latinoamérica. 2006. Available from: http://rcci.net/globalizacion/2006/fg633.htm

[23] Mayoral I. Mayor desigualdad=Inestabilidad Social. Revista Expansión; 2010. Available from: https://expansion.mx/economia/2010/07/28/violencia-narcotrafico-economia-riesgos

[24] Castellanos A. El modelo económico, agotado; hemos sido lentos para ajustarlo. Apud. Periódico La Jornada, [Sección Economía] México; 30 Jan 2004

[25] Luis-Pineda O. Hacia la reconversión del Modelo Económico Mexicano en el Siglo XXI: Un Imperativo Frente al Nuevo Milenio. México, D.F: Editorial I.P.N; 2008

[26] AMLO.PND. Plan Nacional de Desarrollo prioriza el bienestar y pone punto final al periodo neoliberal: presidente AMLO. In: Plan Nacional de Desarrollo 2019-2024. 2019. Available from: https://lopezobrador.org.mx/temas/plan-nacional-de-desarrollo-2019-2024/

[27] AMLO. Gobierno Mexicano. Plan Nacional de Desarrollo 2019-2024. 2019. Available from: https://lopezobrador.org.mx/wp-content/uploads/2019/05/PLAN-NACIONAL-DE-DESARROLLO-2019-2024.pdf

[28] Luis-Pineda O. Some Externalities Stemming from Mexico's Urban-Regional Imbalance: The Covid's Struggle in the Country's Center Region. 2022. Available from: https://globaljournals.org/GJHSS_Volume22/E-Journal_GJHSS_(E)_Vol_22_Issue_5.pdf

[29] CM.org. Las Exportaciones Mexicanas se disparan un 75.6% luego de haber sufrido una gran caída. Apud. Conociendo México.Org; 2020. Available from: https://youtu.be/PgPLQxHOJQE

[30] Expansion. La logística y-nearshoring el segundo milagro mexicano. 2023. Available from: https://expansion.mx/empresas/2023/06/19/logistica-y-nearshoring-traeran-el-segundo-milagro-mexicano

[31] Countryeconomy.com. Mexico Versus Brasil. 2023. Available from: https://countryeconomy.com/countries/compare/brazil/mexico

[32] CM.org. AMLO: Se Crearán 10 parques industriales en el Corredor

Interoceánico del Istmo de Tehuantepec. Apud. Conociendo México.Org; 2020. Available from: https://youtu.be/o0XcnRjBqgM

[33] Oxford Analytica. Mexican banks face good year despite slower GDP growth. Expert Briefings. 2023. Available from: https://dailybrief.oxan.com/Analysis/ES277952/Mexican-banks-face-good-year-despite-slower-GDP-growth

[34] Selfbank. Washington Consensus. What Is It? and How It Works. 2023. Available from: https://blog.selfbank.es/consenso-de-washington-que-es-y-como-funciona/

[35] Guillén A. México, ejemplo de las politicas antidesarrollo del Consenso de Washington. Estudios Avanzados [online]. 2012;**26**(75):57-76. ISSN 0103-4014. DOI: 10.1590/S0103-40142012000200005

[36] OCDE. OCDE:México, 15 años en el último lugar de educación. Apud.El Universal; 2016. Available from: https://www.eluniversal.com.mx/articulo/nacion/sociedad/2016/12/6/ocde-mexico-15-anos-en-el-ultimo-lugar-de-educacion/

[37] Rebossio A. Argentina se enfrenta a una crisis de deuda que durará al menos hasta 2015. ElPaís: Economía; 2014. Available from: http://economia.elpais.com/economia/2014/08/17/actualidad/1408286907_278770.html

[38] Albisalbis. Crisis Social en Colombia. 2009. Available from: http://co.globedia.com/crisis-social-colombia

[39] Lozano D. Maduro llama a una resistencia histórica. 2016. Available from: http://www.lanacion.com.ar/1904856-4maduro-llama-a-una-resistencia-historica

[40] Gutiérrez E. Bachelet: Nueva Mayoría ha evitado crisis sociales en Chile. 2016. Available from: https://editonline.jornada.com.mx/ultimas/2016/05/26/bachelet-afirma-que-nueva-mayoria-ha-evitado-crisis-sociales-en-chile

[41] Barrios A. Crisis social y política en Chile: la demanda por acceso y cobertura universal de salud. Revista Panamericana de Salud Pública. 2019. Available from: https://journal.paho.org/es/articulos/crisis-social-politica-chile-demanda-por-acceso-cobertura-universal-salud

[42] Féliz M. Beyond the "Pink Tide": Dependent Capitalism in Crisis in Argentina and Lessons to Be Learned for Radical Social Change. Springer; 2019. Available from: https://link.springer.com/chapter/10.1007/978-3-030-13639-0_11

[43] Poy S et al. Rise in Income Inequality before and after the Covid-19 Crisis in Argentina. Rise in income inequality before and after the Covid-19 crisis in Argentina. 2023. Available from: scielo.org.mx

[44] BBC.Apud.BBC. Claudia Sheinbaum arrasa en la elección y será la primera presidenta de México. 2024. Available from: https://www.bbc.com/mundo/articles/c6ppw9572plo

[45] Rogriguez LM. El binomio política-corrupción en América Latina. 2020. Available from: https://dialnet.unirioja.es/servlet/articulo?codigo=7353490

[46] Puryear J. La Educación en América Latina: Problemas y Desafíos. 1997. Available from: https://biblioteca.utec.edu.sv/siab/virtual/interactiva/980000016.pdf

Chapter 7

Green Social Work (GSW): Ensuring Environmental Sustainability

Emad F. Saleh, Wafa Al-Maamari and Magdy M. Mostafa

Abstract

This chapter sheds light on the negative effects of environmental and climate phenomena on the planet and its inhabitants. It also shows the importance of enhancing environmental awareness among residents by highlighting the role of green social work (GSW) in raising this awareness among residents of all social segments to reduce catastrophic climate phenomena and ensure environmental sustainability. The chapter discusses some pivotal concepts, including GSW, environmental awareness, global warming, sustainable development, and environmental sustainability, in addition to local and international efforts to preserve the environment. It also seeks to highlight the role of GSW in developing environmental awareness and maintaining environmental sustainability. This chapter follows the qualitative approach, which is applied through the content analysis method of some field studies, reports, and theoretical literature interested in studying GSW, as well as the relationship between humans and the environment, environmental and climate phenomena, and problems, and their catastrophic effects on the Earth and its inhabitants. It also reviews the role of professional practices in GSW in developing environmental awareness and achieving sustainable environmental goals. It concludes by presenting general outlines of the content of a teaching curriculum for GSW to achieve sustainable development and environmental sustainability.

Keywords: global warming, greenhouse effect, GSW, environment awareness, environmental sustainability, Oman sultanate

1. Introduction

People's relationship with their environment is a necessary and obligatory relationship characterized by strength and mutual interaction that results in mutual influences. The outcome of these influences is determined based on the human ability to understand the environment, their view of it, and how they deal with and appreciate it. The aspects of this mutual relationship are unfolding, and its directions are becoming clearer with the increase in scientific discoveries in all different disciplines and sciences that are concerned with the relationship of human behavior to the environment in which they live. This relationship is characterized by complexity and many facets and patterns. It is always exposed to continuous disturbances, as confirmed by the frequently occurring environmental phenomena that the world suffers from today. Multiple environmental phenomena that afflict the planet and those on it are constantly increasing, and among these phenomena comes change. Climate change, global warming, and associated environmental problems such as hurricanes, floods, and torrents in some areas, drought, and desertification in other areas have harmful effects on the environment and humans.

The occurrence of disturbances in a person's relationship with the environment in which he lives can be explained by his tampering with the necessities of life, as irrational human behaviors have caused pollution of water, air, and soil. While many theories exist to explain the complexity of environmental behavior, the role of individuals' relationships with nature has not yet been fully clarified [1]. Hence, the question that arises is: how do we address the disturbance in the relationship between man and the environment in which he lives, interacts with, influences, and is affected by? Negative human interventions and unsustainable exploitation of natural resources have caused environmental degradation and climate change. Water and air pollution and the use of chemical pesticides and agricultural pesticides have led to soil pollution and decreased food quality. In addition, expanded industrialization and the use of weapons of mass destruction have undoubtedly contributed to environmental destruction and the loss of biodiversity and its sustainability [2].

2. Main topic

The twenty-first century has begun with an economic, environmental, and climate crisis that has led to urgent transformations in various social spheres to ensure the quality of life and care for the environment [3]. The climate change crisis and its effects are considered one of the most important changes that have occurred. Climate warming increases evaporation on land, which can worsen drought and create conditions more susceptible to wildfire and a longer wildfire season. It is also associated with the heaviest precipitation events (rain and snowstorms) through increases in the air's capacity to hold moisture [4]. Many studies indicate that humans may cause these phenomena to a large extent, for example, regarding the recent forest fires in Turkey. According to studies on forest fire risk resources, tourists and local people living in or adjacent to forests cause forest fires due to socioeconomic reasons such as a low level of education, stubble burning, agricultural activities, tourism and recreational activities, illegal utilization, intentional negligence, a lack of knowledge and awareness, etc [5].

El Niño events favor drought in many tropical and subtropical land areas, while La Niña events promote wetter conditions in many places. These short-term and regional variations are expected to become more extreme in a warming climate [4]. The harmful effects resulting from the unbalanced relationship between man and his environment require us to seek to adopt a sustainable approach to development that achieves a balance between human needs on the one hand and respect for the environment and the preservation of its natural resources on the other. Also, working to enhance environmental awareness and encourage sustainable environmental practices in all aspects of life is a necessity that binds all relevant scientific professions and specializations by paying attention to educational curricula, environmental legislation, drawing up environmental policies, and linking them to development and sustainability. In addition, it is important to encourage clean technology and develop

DOI: http://dx.doi.org/10.5772/intechopen.1006739

renewable energy sources as an alternative to relying on fossil fuels that are harmful to the environment.

The importance of GSW appears in introducing new issues into environmental discussions and increasing its centrality in the practice of the profession. It includes mainstreaming environmental considerations and expanding the theoretical and applied base to ensure that social and environmental justice are considered integral parts of any social workers' environmental engagement [6]. In light of the previous presentation, the current chapter focuses on some environmental and climatic phenomena that have negative impacts. In addition to addressing the pivotal role of GSW in enhancing environmental awareness among residents, helping them change their behaviors in dealing with the environment, and training them to adopt sustainable practices that achieve a balance between human needs, protecting the environment, and making good use of its resources to ensure their sustainability. Thus, communication and cooperation must be strengthened between researchers, scientists, local communities, and governmental and non-governmental sectors to develop joint strategies and take effective actions to reduce the effects of catastrophic climate change and preserve the environment. Therefore, solving this complex problem requires coordination of efforts and joint scientific cooperation between all bodies and institutions concerned with the environment at all local, regional, and international levels.

3. Main objectives

The current chapter explains the role of GSW in raising the level of environmental awareness of the population for the purpose of rationalizing human behavior to limit the growth of climate phenomena with catastrophic impacts on the environment and its sustainability. To achieve this goal, it is necessary to discuss the following topics:

1. The main concepts, including GSW, environmental awareness, global warming, climate change, and environmental sustainability

2. International Efforts in Environmental Conservation and Sustainability

3. International Efforts in Environmental Preservation and Sustainability

4. Local efforts in environmental conservation and sustainability (the Sultanate of Oman as a model)

5. The role of professional practices in GSW in Achieving Environmental Sustainability

6. Proposing general guidelines for the content of a teaching curriculum for GSW to activate its role in developing environmental awareness and maintaining environmental sustainability

4. Methodology

The authors have followed a comprehensive scientific methodology to ensure accuracy and coherence in analysis and presentation. The chapter focuses on reviewing and analyzing the contents of some scientific studies and theoretical literature related to GSW. It also focuses on the central points of this concept and defines it precisely. The chapter discusses the role of sustainable development and environmental sustainability in the context of GSW by reviewing research and theoretical literature that indicates how GSW affects sustainable development and how it can contribute to enhancing environmental sustainability across various social, economic, and environmental aspects. The information and data related to this chapter were collected by relying on electronic encyclopedias, open-access publications, scientific journal websites, and statistical centers published on the internet, followed by citation and analysis. For climate phenomena, scientific evidence on climate change and its effects on the environment and society is reviewed. How GSW can adapt to these challenges through multidimensional strategies is also presented. Therefore, from the authors' point of view, this chapter is expected to be based on a solid foundation for a deeper understanding of GSW and its role in ensuring sustainable development and environmental sustainability, based on documented theoretical literature and updated scientific studies that highlight the challenges and opportunities in this growing field.

5. Main concepts

The current chapter seeks to discuss several main concepts, including the following:

5.1 Green social work

GSW is a nascent framework within the social work field that provides insights regarding social workers' engagement in disaster settings [7]. GSW is considered the newest branch of the social work profession, although it was previously practiced under the name environmental social work. It is an approach and perspective in social work that focuses on environmental sustainability and environmental justice in the context of sustainable development (SD) [8]. The relationship between social work and the environment is a strong one that dates back decades. To think of the environment as a universal right is to realize that, early on, it is connected to social work. In the first publication of "Social Diagnostics," social work pioneer Mary Richmond integrates the environment as one of 10 factors to be investigated in a "Family Questionnaire" [9].

The number of publications that highlight the importance of integrating the natural environment into social work practice has grown exponentially [10]. A study of ecological and social work, conducted with 49 postgraduate students at the University of Gävle, Sweden, noted that the last decade has seen the emergence of the idea of GSW, a concept that is now widely recognized in the theoretical literature related to the social work profession [11]. It is increasingly recognized that social work has a vital role to play in shaping effective responses to environmental degradation, climate change, and disasters. Innovative teaching and learning strategies can be used to facilitate understanding of complex, interrelated, and systemic social, economic, and environmental challenges [12].

In light of the previous discussion, the authors present a specific procedural concept of GSW, as it is the latest field of professional practice for social work that is concerned with correcting the course of the relationship between man and his environment by adhering to professional principles and values in applying various methods of practice in order to reach the development of man's environmental awareness and help him reduce the practice of GSW. Therefore, environmental justice can be achieved by preserving the environment and its resources and ensuring their sustainability for future generations.

5.2 Environmental awareness

The number of publications highlighting the importance of integrating the natural environment into social work practice has increased significantly. Historically, many non-Western traditions have integrated the natural environment into their worldview. A systematic review of 70 journal articles focusing on human-environment interactions in the context of natural disasters indicated that environmental status has a significant impact on human health and well-being [13]. In short, environmental awareness can be defined as understanding the environment, the impact of human behavior on it, and the importance of protecting it [14]. From our point of view, the pivotal goals of environmental awareness are to preserve the planet on which we live with other living organisms such as animals and plants, and then to preserve biodiversity by fulfilling the special requirements to reach this awareness. Environmental awareness includes reaching "human awareness that the Earth needs protection for its survival" [15]. Furthermore, Sánchez and Lafuente (2010) define this awareness as an individual's understanding of the environment's fragility and the need to preserve it by rationalizing the use of natural resources and reducing their degradation or pollution [16].

Accordingly, developing awareness among community residents is essential. It aims at creating a desire for joint action to confront the problems of their community and increasing their knowledge and understanding. It also aims to increase their ability to confront the facts related to the problems they live in. This goal is related to forming new orientations or modifying some prevailing ones. Urging individuals and groups in society to confront negativity and indifference is a tool that enables us to achieve the desired change [17]. Based on the belief that only things that can be measured can be managed and directed, measuring environmental awareness based on scientific and objective criteria is becoming increasingly attractive for scholars working in many different fields [18]. Therefore, measuring and developing environmental awareness is the responsibility of all societal professions and specializations to provide people with a sufficient degree of knowledge of environmental and local issues and problems and teach them how to act toward them. Hence, GSW bears the greatest responsibility for developing this awareness.

In conclusion, the researchers derive a procedural concept for environmental awareness, which consists of providing individuals with the following:

- Knowledge related to preserving the environment and environmental phenomena that disturb the ecological balance

- Knowledge that explains the contribution of human behavior to disturbing the environmental balance

- Knowledge about the danger of disturbing the environmental balance on Earth and on living organisms (people, animals, birds, plants, etc.)

- Be convinced of this knowledge and acquire skills that enable them to take actual individual and collective steps to deal positively with the environment and work to sustain it [19].

5.3 Greenhouse effect and global warming

Climate change is part of a natural disaster accompanied by the process of environmental degradation and harms human life holistically. The impact of climate change has become a serious problem experienced by various societal levels, without any exception [20]. Global warming is a term often used interchangeably with climate change, as it is one of the most important measures of global change [21]. Global warming is a natural phenomenon that is necessary for the existence of life, as planet Earth absorbs approximately half of the energy sent by the sun. In order for our planet to benefit from this energy, it must be kept in the rocky, atmospheric, and water atmosphere, and therefore it must be confined as if we were in a greenhouse. However, the envelope of this house is not made of glass or plastic, but from water vapor, clouds, gases, suspended solids, salts, acids, etc. ([22], p. 181).

Global warming occurs as a result of an increase in the temperature of the Earth's surface and the troposphere (the lower layer of the atmosphere) due to the presence of water vapor, carbon dioxide, methane, and some other gases in the air. Among those gases, known as greenhouse gases, water vapor has the greatest effect [23]. With scientific progress, several other gases have been discovered that did not exist in the pre-industrial era. These are greenhouse gases that are considered filters for thermal rays in one direction. Although their percentage is small, these gases are more effective than the effect of carbon dioxide due to their ability to trap heat in the Earth's atmosphere ([24], p. 163). As for the causes leading to global warming, multiple studies find that between 90 and 100% of climate scientists agree humans are causing global warming, with multiple studies converging on a 97% consensus. This position is also endorsed by the Academies of Science from 80 countries and many scientific organizations that study climate science [25].

5.4 Sustainable development and environmental sustainability

Sustainability is the key to ensuring a harmonious coexistence between human society, the environment, and the economy [26]. Especially in light of what modern statistics confirm, the growing population increases human needs, such as those for food, clothing, and housing, but there are limits to how much the world's resources can be expanded to provide. Therefore, population management and control are essential for sustainable development [27]. Therefore, sustainability is very important for a safer and more just future for all current and future generations. Sustainability ensures that our needs are met without compromising the ability of future generations to meet theirs. It also preserves the environment, promotes social justice, and provides opportunities for economic growth in order to reach a better and more equitable world.

Sustainable development was a policy concept formulated to emphasize the importance of rational use of natural resources in meeting basic human needs and fostering environmentally responsible economic development [28]. The concept of sustainable social development plays an important role in the processes of development of society, and in thinking critically about the theory of sustainability [3]. In fact, the process of developing a specific and agreed-upon definition of sustainability and sustainable development is a complex issue, as until now there has been no universally accepted definition. However, sustainable development can be defined as development that meets the needs of the present without compromising the ability of future generations to meet their own needs [29]. Sustainable development is also an approach to social, economic, and environmental planning that attempts to balance the social and economic needs of present and future human generations with the imperative of preserving, or preventing undue damage to, the natural environment [29].

Sustainable development is also known as a process of change in which the exploitation of resources, the direction of investments, the orientation of technological development, and institutional changes are made consistent with future as well as present needs ([27], p. 4). It is also development that meets the needs of the present without compromising the ability of future generations to meet their own needs. Sustainable development can only be achieved if certain principles are followed. However, the economy, environment, and society are usually prioritized when discussing the fundamentals of sustainable development [30]. Mensah's paper finds and argues that the entire issue of sustainable development centers around inter- and intragenerational equity, which is essentially anchored on three distinct but interconnected pillars, namely the environment, economy, and society [31]. Although capitalism, socialism, and communism have been utilized throughout history, their strengths and drawbacks have failed to address sustainable development comprehensively [27]. By highlighting how academics engage in the work of integrating sustainability, this study emphasizes that managers of higher education institutions need to account for the time and additional resources needed to ensure that academics effectively cope with sustainability. Intrinsic motivation may not last if organizational structures and leadership are not supportive on a practical level and in the long run [30].

In conclusion, it can be said that there is a strong closeness and connection between the concepts of sustainable development and environmental sustainability, and one can be used instead of the other. Environmental sustainability is an integral part of the concept of sustainable development. Environmental sustainability means preserving natural resources and managing them in a rational manner that ensures that the needs of current generations are met without compromising their rights or those of future generations. It aims to achieve a balance between three components: economic growth, social development, and environmental protection. In order for sustainable development to be achieved, the links between the environment and development must be studied. Also, it is important to take into account the ultimate goal of development, which is human well-being, as development ideas are based on the concept of human well-being as a central goal for policy discussion, and human well-being is the outcome of development. Human well-being and the state of the environment are also strongly interconnected ([32], p. 44).

Recently, we have found an emerging, almost global consensus on the meaning of the word "sustainability" and a more commonly agreed-upon definition of the "sustainability principle," which is how to achieve sustainability and realize the promise of the sustainability principle through concrete goals and measures [33]. In conclusion, we emphasize that environmental sustainability stems from several basic principles: preserving natural resources, reducing pollution, preserving biological diversity, and protecting species and ecosystems to ensure the continuity of life on Earth. We also rely on renewable energy, innovation, and technology to develop environmentally friendly technologies. It supports sustainability, reduces environmental impact, and promotes environmental justice to ensure a fair distribution of environmental benefits and burdens among all members of society while preserving the rights of future generations. All in order to create a system that can support human and natural life in a sustainable and harmonious manner with the environment.

6. International efforts in environmental preservation and sustainability

In this section, we will highlight international efforts aimed at preserving the environment from human tampering with its resources. Environmental risks have become a concern that worries governments and preoccupies people with the dangers they represent to the environment and to the lives of humankind across the globe. This is confirmed by the results of many specialized scientific studies that have proven that climate change will lead to an increase in volcanic activity and earthquakes, especially if the ice sheets in the Arctic and Antarctica melt [34]. Thus, natural disasters worsen, and we can cite the effects of the severe storm "Sandy" that struck the United States in November 2012. In addition, regardless of the origin of the disaster caused by that storm, its impact will be clear on the lives of living organisms and the natural environment. The degraded environment left by disasters will exacerbate the environmental discrimination felt by the poor [35].

Numerous scientific studies explain that climate change triggers various environmental changes that threaten the survival and welfare of society. Climate change adaptation and mitigation have become the preventive and repressive strategies for controlling climate change [20]. Increasing global risks and the destruction of the environment have driven international organizations to take initiatives for research programs that are both cross-disciplinary and cross-national [36]. Consequently, throughout the past 20 years, there has been a noticeable increase in interest in environmental management and an improved general understanding of environmental issues at all societal levels. It has even grown to be one of the most significant pioneering social objectives, achieving a never-before-seen degree of social and political unity.

The United Nations (UN) is the most important and largest international organization that plays a decisive role in the environmental field. It is committed to addressing global environmental challenges and promoting sustainable development throughout the world. The United Nations Conference on Environment and Population, held in Stockholm in June 1972, issued a warning early to alert the world to the human dangers to the environment and that man is the one who creates and shapes his environment, which gives him sustenance and gives him the opportunity to achieve intellectual, moral, social, and spiritual growth. It was stated in the "Stockholm" Declaration on the Environment and Population: We have reached a stage in history at which we must formulate our actions around the world with more caution, given the environmental impacts they may have. Through ignorance or indifference, we can cause serious and irreversible damage to the Earth's environment, on which our lives and well-being depend. On the contrary, we can do so through more knowledge and action. It is wiser to achieve a better life for ourselves and our future generations in an environment more compatible with human needs [37].

Although progress had been made in developed countries in improving air and water quality, tightening controls on chemicals, and preserving nature's elements, most developing countries were witnessing environmental destruction at a speed and scale never seen before. Many newly industrialized countries have suffered massive environmental degradation. In many regions, environmental destruction is beginning to undermine the prospects for future development and perhaps even the potential for the continued survival of the world. The results of the United Nations Environment Programme report indicate that the accelerating human impact on the Earth is rapidly outpacing the ability of largely existing strategies to respond to and address environmental impacts and the modest and often meager budgets allocated to address them ([38], p. 8).

Global interest in the problem of global warming also began in the late 1980s. At the Earth Summit in Rio de Janeiro in 1992, 160 countries signed a treaty on climate change. It was followed by several conferences that resulted in "The Kyoto Protocol" in 1997, where the international community committed to reducing emissions of heat-trapping gases, especially carbon dioxide ([22], p. 186). The Kyoto Protocol provides several ways for countries to achieve their goals. The first way is to benefit from natural processes known as "sinks," which remove greenhouse gases from the atmosphere, such as planting trees that consume carbon dioxide. The second method is the Clean Development Mechanism (CDM), which encourages developed countries to invest in technology and infrastructure in developing countries to reduce emissions. The investing country has the right to demand an effective reduction in emissions as a credit to achieve its obligations, such as investing in power plants that run on clean gas instead of coal. The third method is emissions trading, which allows countries to buy and sell emissions rights, placing an economic value on emissions. Some European countries have started an emissions trading market to fulfill their obligations under the Kyoto Protocol. Countries that do not meet their targets will pay the emissions differential plus a 30% penalty in the next compliance period in 2012 and will be prohibited from participating in emissions trading until they comply with the protocol. In the future, emissions targets for the post-2012 period will be set in future protocols [39].

International cooperation continues between governments and United Nations environmental organizations in order to reduce the catastrophic effects of climate change. Work was done to reduce the effects of natural disasters and prevent further environmental damage. Disaster risk management strategies have been developed as an attempt to deal with these impacts, and many governments have worked to implement them and build sustainable infrastructure in Asian countries. For example, some countries, such as Laos, Vietnam, and Cambodia, have worked on a flood and drought risk management and mitigation project. In order to reduce the resulting economic losses, the water management structure is combined with improvements in weather forecasting. In 2013, the Philippines' Integrated Disaster Risk Management Fund became the world's first city disaster insurance fund, covering typhoons and earthquakes. Other southeastern countries in Asia have also tried to replicate extreme weather insurance funds. In addition to previous efforts, the road sector databases in "Laos and Cambodia" are integrated into the existing database of the United Nations Development Programme with the aim of tracking disaster losses in vital sectors, and the same project is implemented in the private sector in the "Philippines, Thailand, and Vietnam" [40]. The "Convention on Biological Diversity" initiative is also one of the most important initiatives in this regard and aims to protect threatened species and establish natural reserves [41]. International cooperation also takes place through the United Nations Environment Programme to promote sustainable environmental policies and reduce pollution [42].

In conclusion, recent years have witnessed increasing international efforts to preserve and sustain the environment. Countries and global organizations cooperate to address the increasing environmental challenges. One of the most prominent of these efforts was the "Paris Climate Agreement" in 2015, where countries pledged to reduce carbon emissions to limit rising temperatures [43]. The Sustainable Development Goals launched by the United Nations in 2015 also support environmental initiatives, especially Goal 13 on climate action and Goal 15 on life on Earth [44]. Public-private partnerships are also an essential part of these efforts, as companies work with governments to develop green technologies and reduce the environmental impact of industries. These initiatives demonstrate an increasing global commitment toward a sustainable future, despite the great challenges facing the world in this field [45].

There are also many international organizations that pay great attention to the environment in monitoring, control, awareness, etc., including the Food and Agriculture Organization (FAO), the World Health Organization (WHO), UNESCO, the Scientific Committee on Environmental Problems (SCOPE), the International Academy for Environmental Security (IAES), the International Society for Environmental Toxicology (SECOT), the Organization for Economic Co-operation and Development (OECD), the United Nations Environment Program (UNEP), the United Nations Industrial Development Program (UNIDO), the International Union for Conservation of Nature (IUCN), the United Nations Scientific Committee to Study the Effects of Atomic Radiation (UNSCEAR), and many others, which reflects the extent of contemporary interest in environmental issues [46].

7. Local efforts in environmental conservation and sustainability (the Sultanate of Oman as a model)

Sultan Qaboos bin Said, Sultan of Oman (1970–2020), contributed to establishing and developing the environmental institutional and legislative structure at the national and global levels. Since the beginning of the Renaissance era and taking over the reins of government in Oman in 1970, and at an early stage in the state's life, he laid the foundation stone for organizing and managing environmental work. The Royal Decrees graduated in the environmental sector, confirming the development of the legislative aspect over the past five decades of his rule. The historical starting point was the establishment of the Office of the Environmental Protection Adviser in 1974. In the same year, the Public Authority for Environmental Protection and Pollution Control was established pursuant to Supreme Royal Decree No. 14/79, followed by the establishment of the Environmental Protection and Pollution Control Council in 1979 pursuant to Supreme Royal Decree No. 68/79. In 1984, the Ministry of Environment was established pursuant to Supreme Royal Decree No. 45/84, after which royal decrees continued until the Ministry of Environment and Climate Affairs was announced as an independent institution concerned with the environment by Supreme Royal Decree No. 90/2007 on September 9, 2007 [47].

The Sultanate also moved to draw up an integrated national strategy for environmental diversity, a strategy that focused on the successive five-year development plans that the Sultanate has been implementing since 1976. That strategy focused on renewable and non-renewable resources and emphasized the importance of sectoral integration and the national cooperative approach in the use and management of resources. In addition to monitoring the Sultanate's achievements in the fields of nature conservation and environmental protection, highlighting the obstacles and negatives and emphasizing the role and place of the Omani person are means and goals of development [48]. The Sultan Qaboos Award for Environmental Protection was also launched as the first international Arab award in the environmental field in 1989 to be awarded every 2 years to any individual, group, institution, or organization that contributes to preserving the environment and managing its resources anywhere in the world. The award is considered an expression of appreciation and recognition for the efforts and fruitful contributions made by the candidates and an encouragement to contribute to preserving the environment for others. The award is granted through the International Coordination Office of the Man and the Biosphere Program at UNESCO [49].

In continuation of the Sultanate's approach to caring for the environment and working to preserve it, Royal Decree No. 114/2001 issued the 'Environmental Protection and Pollution Control Law'. In its second chapter, the basic rules and principles for ensuring the safety of the Omani environment are included. Its third chapter contains the penalties related to assaulting or tampering with the environment, with all its environmental contents—natural, animals, birds, plants, and trees [50]. The Sultanate continuously seeks to avoid many of the potential impacts of climate change on natural, economic, and social systems by preparing to manage the risks of climate change by taking measures and preparing national strategies to mitigate greenhouse gas emissions, and others are concerned with adapting to those impacts to confront potential risks in this regard. The Sultanate's interest in environmental issues and climate affairs and its continuous support for these efforts have culminated in achieving many practical measures. Among these measures are the Sultanate's signing of international agreements related to climate change and the issuance of Ministerial Resolution No. 41/2008 regarding the formation of the steering committee to coordinate and supervise the implementation of the capacity building project in the field of climate change, which is implemented by the Ministry in cooperation with the United Nations Development Program [49].

The Sultanate of Oman is also keen to strengthen its international presence in all regional and international forums by participating in global conferences, as His Majesty Sultan Haitham bin Tariq, may God protect him, affirmed in his speech on the occasion of the 32nd anniversary of the establishment of the Sultan Qaboos Award for Environmental Protection: "Protecting the environment and preserving its various natural resources are among the priorities of the development plans in the Sultanate of Oman, based on our firm conviction that caring for the environment and its resources is a global responsibility that is not limited by the political borders of countries. Hence, this is what the international community should be keen to do by providing all available capabilities to maintain the system, preserving its natural resources, developing action plans to achieve sustainable development goals at the national, regional, and international levels, spreading awareness of the importance of preserving the balance of the environment, and investing in its resources to the extent that achieves development and ensures its sustainability for future generations" [51].

Moving toward a sustainable environmental future, the Sultanate's Vision 2040 includes a main axis related to the priorities associated with the sustainable environment. This axis includes many goals that work to achieve sustainability, such as:

- Reaching an environment that achieves a balance between environmental and economic

- Working with the rules of sustainable development, renewable energy, diverse sources, and rationalization of consumption to achieve energy security and social requirements

- High-quality, pollution-free environmental environments, a green and circular economy that responds to national needs and is in line with global orientation

- Food and water security based on renewable resources and advanced technologies

- Optimal exploitation of the strategic location and biodiversity of the Sultanate of Oman

- Environmental awareness in the effective application of sustainable consumption and production rules

- Sustainable use and investment of natural resources and wealth to ensure the achievement of high-added value [52].

8. The role of green social work in achieving environmental sustainability

Climate change is an issue of social justice and a basic human right. Therefore, it requires the attention of the social work profession based on the international organizations' emphasis that human rights and the environment are interconnected. A person cannot enjoy his rights without a safe, clean, and healthy environment, and safe environmental sustainability cannot exist without establishing and respecting human rights [53]. GSW calls for working to rationalize the human relationship with the environment through cooperation between various professions and specializations concerned with the environment. The profession of social work has a role to play in greening and sustaining the environment [54]. It claims to be an innovative approach to environmental issues as practiced in the social work profession [55]. This innovative view expands to include all living organisms on Earth, such as humans, animals, and plants, and the impact of these organisms on the ecosystem. In order to emphasize the relational support between them for the purpose of preserving the status and privacy of the planet and everyone who lives on it, environmental problems receive a degrée of attention at all levels of professional practice in social work (micro, mezzo, and macro levels). However, we look forward to more of this attention, as the planet is constantly witnessing more dangerous environmental phenomena that are accompanied by harmful effects.

As environmental concerns have increased, national and international social work organizations have called on social workers to incorporate issues of the environment into their professional practice. Although there is a small body of literature related to social work and the environment, the profession has not fully embraced the need to incorporate these issues into social work education or practice [56]. Hence, concerns are increasing at all levels of professional practice and academic institutions, which requires taking proactive steps to rationalize human behavior in dealing with turbulent environmental situations. To address today's complex problems, social work must inform itself of advances in the knowledge base of the natural and social sciences, particularly those that explain the changing construction of different systems and the way the world works. These explanatory theories offer a relevant understanding of the social, economic, and ecological contexts within which social workers practice [57]. Accordingly, it is important to assess social workers' knowledge and attitudes regarding the environment ([56], p. 5). This is for the purpose of identifying what they lack in environmental knowledge and information that can help them understand and explain the behaviors of individuals that cause environmental phenomena. That knowledge also improves their professional interventions with individuals and communities affected by environmental disturbances and problems.

Professional social work organizations must realize that the world today lives in the current "social-ecological crisis" that demands a major social transition to a sustainable society that touches all dimensions of our lives. So, social work cannot escape this process [58]. This requires viewing the environment as one of the elements of the professional practice process, and therefore the focus of practice is on strategies for development, empowerment, and the formation of social and environmental capital as strategies to assist in the processes of transformation and the desired social and political change. Therefore, GSW must rely on environmental planning and achieve local sustainable development through increased reliance on the active participation of affected communities. Hence, residents' participation in development plans, consensus building in environmental conflict situations, and residents' participation in the design of local facilities are also required [59]. Dominelli's study (2014) argues for the inclusion of environmental justice in contemporary social work practice as one way of promoting inclusionary social work that meets some of the challenges of the twenty-first century [60].

The social work profession's keen interest in human rights, social justice, and community building also creates a sound base from which to engage in the collaborative, creative, and interactive processes required for environmental practice [61]. Based on the above, social workers must be called upon to take the necessary measures to address these environmental issues as a strong and unified profession that works side by side with the deprived and marginalized, who do not receive their share of global resources, despite their limited environmental footprint. Proposing a new model of practice rooted in environmental justice that promotes the well-being of people, plants, animals, and the ecosystem is imperative [35].

The historical involvement of social workers in women issues such as poverty alleviation, domestic violence, displacements, social inequality, and destitution, among other challenges faced by rural women, is the ambient platform for social workers to embrace their crisis intervention skills to infuse climate dynamics change into their interventions [62].

In this regard, a study confirms that environmental social workers are indispensable in empowering society, especially in encouraging environmental improvement to remain sustainable for the prosperity of society [20]. Another study presents a model based on experiences in southern Africa to provide some guidance for social workers on how best to mainstream gender dimensions in climate change interventions. This would foster social and environmental justice, social resilience, equal participation in the climate change discourse, capacity building, and adaptive capacity for rural women [62]. Ogunbode and Arnold's study suggests that given the current socioeconomic situation in Nigeria and limited knowledge of environmental issues, the most effective strategy for promoting environmentally responsible behavior and positive attitudes regarding less obvious issues such as biodiversity loss might simply be to increase exposure to direct messages emphasizing their relevance [63]. This would enhance empowerment and build capacity to deal with environmental sustainability issues. Within the framework of professional interventions, social workers face many contemporary challenges: the difficulties faced in supporting human rights, social justice, and active citizenship; the difficulties related to affirming environmental justice and preserving the planet; and the practice of social work and addressing climate change and other disasters [35]. With the successful implementation of a holistic approach to sustainability, students will have better insights and understanding of both themselves and the surrounding society, laying the groundwork for an inclusive future society [30].

In conclusion, the authors believe that GSW can play a pivotal role in developing environmental awareness and achieving sustainable environmental goals for the population in general by providing direct professional services in multiple professional practice institutions, such as universities, schools, sports and youth clubs, and others. Green social workers can also practice their professional roles by performing the following tasks:

1. Organizing targeted awareness campaigns and programs that target all segments of society (children, adolescents, young people, and the elderly) introduces them to environmental issues, their importance, and the role of the population in preserving the environment and its natural resources. These campaigns can be implemented through organized training programs, workshops, seminars, lectures, and media releases to raise the level of environmental awareness among the population.

2. Enhancing community participation in environmental activities such as volunteer camps in schools and youth centers, collective cleaning campaigns for beaches and streets, recycling projects, and afforestation not only reflects the improvement and protection of the environment but also develops the spirit of cooperation and teamwork among residents in local communities.

3. Developing and implementing environmental programs and projects will spread environmental awareness and achieve sustainability. Green social workers can provide support and advice to individuals and communities to adopt environmentally friendly practices, tend to rely on clean sources of energy, use renewable energy, and turn to organic agriculture, all of which can contribute to achieving environmental sustainability.

4. Green social workers can exercise their professional roles as intermediaries between local communities and government agencies, helping to achieve effective environmental policy goals. In this regard, they can participate in meaningful community dialogs and positively influence officials to help them make appropriate environmental decisions in order to ensure access to sustainable environmental policies.

5. Green social workers adopt work to strengthen environmental education in schools and educational institutions, with the aim of instilling the values of environmental sustainability in current generations. This education adopts integrated curricula and practical activities that develop awareness of environmental issues and work to sustain them.

The outcome of the above is that professional practices of GSW contribute significantly to developing environmental awareness and achieving sustainable environmental goals through many paths, such as environmental awareness, community participation, participation in drawing up environmental policies and strengthening environmental education curricula at all educational levels.

9. General guidelines for the content of a teaching curriculum for green social work

This curriculum aims to provide students with the knowledge and skills necessary to activate their role in developing environmental awareness and maintaining environmental sustainability through the practice of GSW. The proposed curriculum begins by shedding light on the concept of GSW and its importance in the environmental context and is expected to include five basic units, which are:

9.1 Unit one: basics of environment and sustainability

This unit addresses basic environmental concepts such as biodiversity, climate change, global warming, and natural and industrial environmental disasters, in addition to explaining the principles and goals of sustainable development.

9.2 Unit two: professional roles of the green social worker

It is concerned with clarifying the role of the social worker in promoting environmental awareness and applying sustainability practices, training students in effective listening skills, communication, environmental education, and running workshops, seminars, and all other tools and means of professional work.

9.3 Unit three: community work and environmental initiatives

This unit includes helping students develop and implement community projects aimed at improving the local environment, such as afforestation and recycling campaigns, and enhancing cooperation with civil society organizations and government institutions to support environmental initiatives.

9.4 Unit four: environmental education and awareness

This unit aims to help students design awareness programs to increase environmental awareness in all professional practice institutions, such as schools, universities, clubs, youth centers, etc., and how to benefit from social media networks in environmental education and awareness.

9.5 Unit five: environmental policies and legislation

This unit focuses on helping students know national and international environmental laws and legislation and how social workers participate in formulating environmental policies at all local, regional, and international levels.

The proposed approach concludes by addressing methods for evaluating the effectiveness of environmental programs and developing them to ensure sustainability and positive impact. By studying this curriculum, students are expected to be able to play an effective role in promoting environmental awareness and contributing to preserving the environment through GSW practices.

10. Conclusion

This chapter discusses in detail the negative influences of environmental and climate phenomena on our planet and its inhabitants. The chapter highlights the urgent need to raise environmental awareness among all segments of society, emphasizing the pivotal role played by GSW in promoting this awareness. By addressing key concepts such as environmental awareness, global warming, and sustainable development, along with environmental conservation efforts at the local and international levels, the chapter demonstrates the significant contributions of GSW to achieving environmental sustainability. By adopting a qualitative approach, which includes content analysis of field studies, reports, and theoretical literature, it highlights the deep links between human activities, environmental challenges, and their disastrous consequences. Furthermore, the chapter discusses the importance of professional practices within GSW in promoting environmental awareness and achieving sustainability goals. The chapter concludes by presenting an outline of the content of a GSW educational curriculum aimed at promoting sustainable development and ensuring a sustainable environmental future.

Author details

Emad F. Saleh*, Wafa Al-Maamari and Magdy M. Mostafa
Sultan Qaboos University, Oman

*Address all correspondence to: emadf@squ.edu.om

References

[1] Braito MT, Böck K, Flint C, Muhar A, Muhar S, Penker M. Human–nature relationships and linkages to environmental behaviour. Environmental Values. 2017;**26**(3):365-389

[2] Saleh EF, Al-Maamari WS. Green social work and its role in developing environmental awareness and ensuring environmental sustainability. Scientific Journal of Social Work. 2023;**4**(2):68-95

[3] Josemanuel L-N, Tobón S, Juárez-Hernández LG. Sustainability-based on socioformation and complex thought or sustainable social development. Environment and Sustainability. 2020;**2**:100007

[4] The Royal Society. Climate Change: Evidence and Causes [Online]. London, United Kingdom: The Royal Society; 2020. Available from: https://royalsociety.org/news-resources/projects/climate-change-evidence-causes/question-13/

[5] Gültekin YS, Gültekin P. Forest Fire Risk Management at the Country Scale: The Case of Turkey, in Fire Hazards: Socio-Economic and Regional Issues. Cham: Springer International Publishing; 2024. pp. 43-52

[6] Dominelli L, Ku HB. Green social work and its implications for social development in China. China Journal of Social Work. 2017;**10**:3-22

[7] Breen K, Greig M, Wu H. Learning Green social work in global disaster contexts: A case study approach. The Social Science. 2023;**12**(5):288

[8] Rambaree K. Environmental social work: Environmental social work, international journal of sustainability in higher education. International Journal of Sustainability in Higher Education. 2020;**21**:557-574

[9] Rocha HB. Social work practices and the ecological sustainability of socially vulnerable communities. Sustainability. 2018;**10**(5):1312

[10] Ramsay S, Boddy J. Environmental social work: A concept analysis. The British Journal of Social Work. 2017;**47**(1):68-86

[11] Rambaree K. Environmental social work: Implications for accelerating the implementation of sustainable development in social work curricula. International Journal of Sustainability in Higher Education. 2020;**21**(3):557-574

[12] Drolet J, Wu H, Taylor M, Dennehy A. Social Work and Sustainable Social Development: Teaching and Learning Strategies for 'Green Social Work' Curriculum, in Environmental Justice. 1st ed. London: Routledge; 2016

[13] Wu H, Greig M, Bryan C. Promoting environmental justice and sustainability in social work practice in rural community: A systematic review. Social Sciences. 2022;**11**(8):336

[14] Pachamama Alliance. Environmental Awareness [Online]. San Francisco, United States: Pachamama Alliance; 2024. Available from: https://pachamama.org/environmental-awareness?utm_term [Accessed: July 9, 2024]

[15] The Heart Water Foundation. Environmental Awareness [Online]. Austin, Texas, United States: The Heart Water Foundation; 2024. Available from: https://drinkheartwater.com/blog/environmental-awareness [Accessed: July 9, 2024]

[16] Sánchez M, Lafuente R. Defining and measuring environmental consciousness. Revista Internacional de Sociología (RIS). 2010;**68**(3):731-755

[17] Khalifa MM. Social Work Practice: Studies in Planned Change. الثاني ed. Alexandria: University Knowledge House; 1992

[18] Ham M, Mrčela D, Horvat M. Insights for measuring enviornmental aweareness. Ekonomski Vjesnik. 2016;**29**(1):159-176

[19] Saleh EF. The mechanisms of the social work profession in developing university students' awareness of the phenomenon of the greenhouse effect. Studies in Social Work and Human Sciences Periodical. 2010;**11**(31):5157-5214

[20] Ramdani J, Gunawan B, Darwis RS. The contribution of green social workers in addressing the impacts of climate change in Indonesia. Jurnal Pendidikan dan Pemberdayaan Masyarakat. 2022;**9**:50-60

[21] Ramkumar S. What is Climate Change? [Online]. 2021. Available from: https://studentenergy.org/influencer/climate-change/?

[22] Al-Mukhtar B. Population and Environmental Resources: A Geographical Reading. 1st ed. Muscat: Scientific Publishing Council, Sultan Qaboos University; 2008

[23] Britannica. Greenhouse effect [Online]. 2024. Available from: https://www.britannica.com/science/greenhouse-effect

[24] Al-Ajmi DN. Climate change and its impact on the environment. Alam Al-Fikr. 2008;**37**(2):157-183

[25] Skeptical Science. The 97% Consensus on Global Warming [Online]. 2017. Available from: https://skepticalscience.com/global-warming-scientific-consensus-intermediate.htm

[26] Reyes M. 3 Principles of Sustainability: Building a Greener Future [Online]. 2024. Available from: sustainabilityeducationacademy.com

[27] Hariram NP, Mekha KB, Suganthan V, Sudhakar K. Sustainalism: An integrated socio-economic-environmental model to address sustainable development and sustainability. Sustainability. 2023;**15**(3):10682. DOI: 10.3390/su151310682

[28] Tennberg M. Sustainable Development. Massachusetts, United States: The Blackwell Encyclopedia of Sociology; 2018

[29] Jarvie ME. Encyclopedia Britannica. Chicago, Illinois, United States: Brundtland Report; 2016

[30] Argento D, Einarson D, Mårtensson L, Persson C, Wendin K, Westergren A. Integrating sustainability in higher education: A Swedish case. International Journal of Sustainability in Higher Education. 2020;**21**:1131-1150

[31] Mensah J. Sustainable development: Meaning, history, principles, pillars, and implications for human action: Literature review. Cogent Social Sciences. 2019;**5**(1):1-27. Available from: https://www.tandfonline.com/doi/epdf/10.1080/23311886.2019.1653531?needAccess=true

[32] United Nations Environment Programme. Global Environment Outlook for Environment for Development. Nairobi, Kenya: United Nations Environment Programme; 2007

[33] Crawford C. Sustainability principle. In: Cremades J, Hermida C,

editors. Encyclopedia of Contemporary Constitutionalism. Cham: Springer; 2022. pp. 1-14

[34] Pearce F. Could a Climate Change Set Off Volcanoes and Quakes? [Online]. 2012. Available from: http://e360.yale.edu/feature/could_a_changing_climate_set_off_volcanoes_and_quakes/2525/

[35] Dominelli L. Environmental justice at the heart of social work practice: Greening the profession. International Journal of Social Welfare. 2013;**22**:431-439

[36] Takala M. Environmental awareness and human activity. International Journal of Psychology. 1991;**26**(5):585-597

[37] Al-Tal S. Greenhouse effect. Alam Alfikr. 2008;**37**(2):47-98

[38] United Nations Environment Programme. Good Environmental Management. New York: United Nations Environment Programme; 2001

[39] Britannica T. Greenhouse effect [Online]. 2023. Available from: https://www.britannica.com/science/greenhouse-effect

[40] Özgür U. Integrating Disaster Risk Reduction (DRR) and Ensuring Sustainable Infrastructure in South and East Asia. London, United Kingdom: Sustainable Development Commission (SDC); 2023

[41] Convention on Biological Diversity. The Convention on Biological Diversity; 2020. [Online]. Available from: https://www.cbd.int/ [Accessed: September 9, 2024]

[42] UNEP. United Nations Environment Programme. Nairobi, Kenya: UNEP; 2020

[43] United Nations. Paris Agreement. New York, United States: United Nations; 2015

[44] United Nations. Sustainable Development Goals. New York, United States: United Nations; 2020

[45] World Economic Forum. How Public-Private Partnerships are Essential for Tackling Climate Change. Geneva, Switzerland: World Economic Forum; 2020

[46] Al-Makkawi AM. Human, Environment and Health, a Study in Sociology. Cairo: Dar Al-Nasr for Publishing and Distribution; 2002

[47] Oman Environment Authority. Founder of Environmental Action [Online]. Muscat, Oman: Oman Environment Authority; 2023. Available from: https://www.ea.gov.om/ar/the-authority/about-authority/the-founder-of-environmental-work/?csrt=16891787683487641375

[48] Al-Samsamiya N. The Lofty Thought of Sultan Qaboos gave the Environment its Due Rights Through Five Decades of Attention and Care [Online]. 2020. Available from: https://www.omandaily.om/

[49] Eumanna. Sultan Qaboos Award for Environmental Protection [Online]. 2023. Available from: https://omanportal.gov.om/wps/portal/index/gov/environment/SultanQaboosEnvironmentAward

[50] Ministry of Justice and Legal Affairs. Environmental Protection and Pollution Control Law, the Official Gazette, no. 707. Muscat, Oman: Ministry of Justice and Legal Affairs; 2001

[51] Tariq S, HB. Booklet for the 32nd Anniversary of the Establishment of the UNESCO-Sultan Qaboos Prize for Environmental Conservation [Online]. Paris, France: UNESCO;

2023. Available from: https://onc.om/unesco-sultan-qaboos-prize/

[52] Oman Vision 2040. Implementation Follow-Up Unit. Muscat, Oman: Sustainable Environment Focus [Online]; 2023. Available from: https://www.oman2040.om/mahawer-4.html

[53] Anderson R. Why is climate change a pertinent issue for social work and how can social workers contribute to efforts to address it? Social Work & Policy Studies: Social Justice, Practice and Theory. Sydney Open Journals. 2021;**4**(1). Available from: https://journals.indianapolis.iu.edu/index.php/advancesinsocialwork/article/view/1340/1811; https://openjournals.library.sydney.edu.au/SWPS/article/view/15004

[54] Lucas-Darby ET. The new color is green: Social work practice and service-learning. Advances in Social Work. 2011;**12**(1):113-125. Available from: https://journals.indianapolis.iu.edu/index.php/advancesinsocialwork/article/view/1340

[55] Dominelli L. Green social work in theory and practice, a new environmental paradigm for the profession. In: Dominelli L, Nikku BR, Ku HB, editors. The Routledge Handbook of Green Social Work. Vol. 1. London: Routledge; 2018. p. 628

[56] Shaw TV. Is social work a green profession? An examination of environmental beliefs. Journal of Social Work. 2013;**13**(1):3-29

[57] Green D, McDermott F. Social work from inside and between complex systems: Perspectives on person-in-environment for today's social work. The British Journal of Social Work. 2010;**40**(8):2414-2430

[58] Peeters J. The place of social work in sustainable development: Towards ecosocial practice. International Journal of Social Welfare. Jul 2012;**21**(3):287-298. Available from: https://onlinelibrary.wiley.com/doi/epdf/10.1111/j.1468-2397.2011.00856.x

[59] Selman P. Social capital, sustainability and environmental planning. Planning Theory & Practice. 2010;**2**(1):13-30

[60] Dominelli L. Promoting environmental justice through green social work practice: A key challenge for practitioners and educators. International Social Work. 2014;**57**:338-345

[61] Schmitz CL, Matyók T, Sloan LM, James C. The relationship between social work and environmental sustainability: Implications for interdisciplinary practice. International Journal of Social Welfare. Jul 2012;**21**(3):278-286. Available from: https://onlinelibrary.wiley.com/doi/epdf/10.1111/j.1468-2397.2011.00855.x

[62] Nyahunda L. Social work empowerment model for mainstreaming the participation of rural women in the climate change discourse. Journal of Human Rights and Social Work. 2021;**6**:120-129

[63] Ogunbode CA, Arnold K. A study of environmental awareness and attitudes in Ibadan, Nigeria. Human and Ecological Risk Assessment: An International Journal. 2012;**18**(3):669-684

Chapter 8

Socio-Cultural Environment: Spotting Implication of Globalization on Sustainability

Noha Essam Khamis

Abstract

The phenomena of globalization are intricate and multifaceted. It is the process of international integration that results from global exchanges of ideas, goods, worldviews, and other cultural aspects, leading to the global interchange of national and cultural capitals. The current wave of globalization is the result of technological innovation and progress, as well as the interaction of numerous stakeholders at different economic levels. The rapid production, trade, and consumption of material products in previously unheard-of amounts has been encouraged by globalization, which has both positive and negative features. On the positive side, globalization is often associated with an increase in international trade. This has been influenced by the global ecological footprint of human activities. The effects of globalization on the environment are still hard to measure. One of the primary environmental issues brought on by globalization is climate change. Above all, globalization has impacted the environment as a whole at various levels and in a range of dimensions over the years, and it continues to do so. This chapter covers the socio-cultural effects and environmental aspects of globalization, which is a complex phenomenon in and of itself.

Keywords: globalization, ecological footprints, climate change, socio-cultural aspects, technological innovation

1. Introduction

The globe has grown increasingly more interconnected and dependent on one another in recent decades [1]. The trade of goods and services between nations and continents marked the beginning of this growth millennia ago. Exporters had to look for more effective means to deliver their goods overseas as the supply of these foreign goods rose along with the demand in their home regions. Globalization is largely due to innovations, which have led to breakthroughs in transportation and technology and have made it feasible for people to exchange much more than simply goods and services ([2], p. 14). Globalism is defined as "a state of the world involving networks of interdependence at multicontinental distances." According to Keohane and Nye [3], p. 4, "The links arise through flows and influences of capital and goods, information and ideas, people and forces, as well as biologically and environmentally relevant

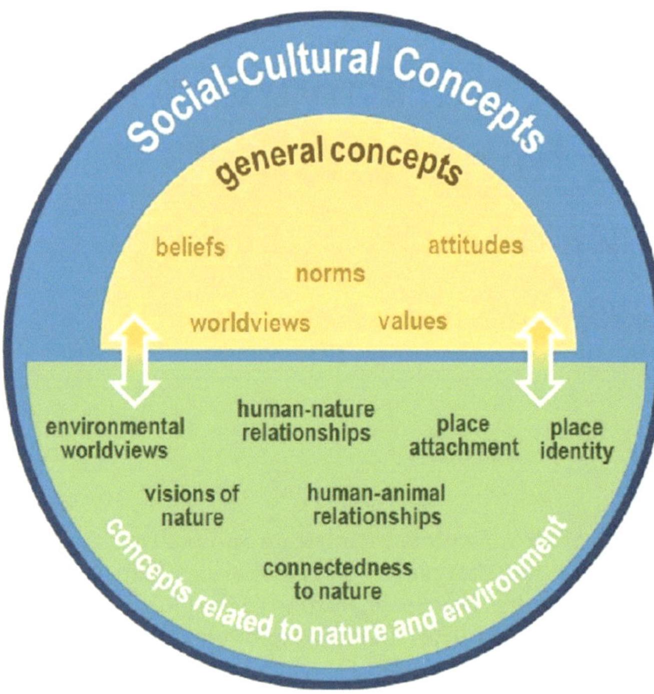

Figure 1.
Overview of representative social-cultural concepts in the context of human-nature relationships.

substances." Accordingly, globalization is the expansion of globalism and the fortifying of ties between all nations by the development of networks and the cross-border transfer of each country's tangible and intangible resources ([3], p. 6). Globalization, however, is a multifaceted notion with a wide range of frequently unforeseen impacts and ramifications on the planet and its ecosystem, economy, and human population; in other words, Sustainable Development. Globalization has been beneficial to the world in numerous capacities, but it has also brought forth new and serious issues that have caused controversy. One of the most significant problems with globalization today is sustainable development (SD), which is defined as "development that meets the needs of the present without compromising the ability of future generations to meet their own needs". Sustainability is also referred to as the "Triple Bottom Line" as there are three components to it: economy, environment, and equity (social). Studies have demonstrated that globalization has varying effects on SD, with some being detrimental ([4], p. 19; [5], p. 3) and some being beneficial ([4], p. 15; [6] p. 104; [7], p. 31). However, because nations differ, what works for one may not be beneficial for another. The UN is the largest supporter of SD, and its main objective is to create an environment where global peace and equality can be fostered. Over the past years, authorities and the general public have also come to understand and accept SD more globally (**Figure 1**).

2. Implication of globalization on sustainability

In recent times, the negative consequences of globalization have been recognized not only for developing countries but also for industrialized nations. A cohesive and mutually beneficial solution will require the cooperation of governments, individuals, corporations, IGOs, and NGOs at various levels of the global scale. While many scholars, policymakers, and supranational authorities have differing ideas on how to "solve" the globalization problem, it must be understood that each party has widely varying motives ([8], p. 5; [9], p. 977). Making concessions and gaining information and awareness will be necessary for this endeavor to reach a consensus that will benefit all parties involved and the globe at large.

2.1 Problem statement

Sustainable development and globalization have been working against each other, leaving the world in an unequal state of poverty and inequality while straining the earth beyond its limits ([5], p. 4). Some claim that the greatest challenge with making globalization and SD work is inequality, which is preventing developing countries from adopting sustainable methods, while others claim it lies in the policies and regulations of global governance, which should be better regulated [1]. A number of approaches have been put up to address this issue, such as the UN's SD efforts, citizen involvement in global governance, and various frameworks for incorporating SD into corporate operations. In spite of these initiatives, poverty persists, and environmental pollution levels are rising, making SD an increasingly pressing issue [10, 11]. What is the reason for this, what is generating these issues, and how can they be resolved? Assessing the impact of globalization on sustainable development will offer a foundation for comprehending the problems impeding the worldwide attainment of sustainable development objectives and opportunities for improvement going forward.

2.2 Objectives

This chapter's primary goal is to present a comprehensive and up-to-date analysis of how globalization has affected sustainable development. There are differing opinions about how globalization affects SD. Thus, in order to arrive at a fair conclusion, it will be crucial to consider a wide range of viewpoints. Highlighting significant ideas and subjects from the scholarly literature would also be crucial. Following the identification of the effects of globalization on sustainability, it will be feasible to ascertain the state of some SD-related issues, thereby illuminating the current state of advancement or regression that the globe is experiencing.

2.3 Methodology

The study methodology follows a descriptive-analytical approach that includes correlational data on globalization and sustainability. A systematic literature review will be conducted using scholarly journal articles that have been published in the last 10 years. The selected articles will be categorized and ordered to reveal the most popular themes and topics of globalization and Sustainable development currently under discussion in the academic literature. After highlighting the most common themes in the literature, the findings will be reviewed, and the main challenges facing globalization and Sustainable Development today will be summarized. This study concludes with a discussion on how to initiate a change toward Sustainable Development using globalization as a catalyst. The globalization section will include conceptualizing, defining, and explaining globalization and its various dimensions: economic, environmental, social, and cultural ([3], pp. 4–6).

3. Globalization

Globalization is a difficult term to define as it has multiple dimensions that overlap, relate, and affect one another. As a result, definitions of globalization widely vary, as many scholars define globalization in the context of a single dimension. Author Martell states "you can't understand globalization without looking at its economic and political dimensions, or by analyzing cultural and social spheres instead of, or separately from, politics and economics" ([12], p. 2). He explains that if one dimension of globalization is assessed without consideration of the others, then the effects that these dimensions have on the other dimensions are overlooked. Therefore, this section will begin by conceptualizing globalization. Following the concept, the definitions of globalization will be compared, and the various dimensions of globalization will be described. This study will be based on defining globalization and its dimensions with a number of different foci. In the modern world, globalization has grown to be a significant and well-liked topic of discussion. Because of its many facets, viewpoints, and opinions on the subject are widely varied in the literature, leading to intense debate on the subject lack of a consensus or definition ([12], p. 12; [13], p. 15). For example, Martell [12] offers four pluralist interpretations of globalization to counteract too westernized, homogenizing, and monocausal interpretations of the term. He contends that perspectives on globalization that are pluralist, hybrid, and multidimensional see it functioning at several levels, ranging from the political to the cultural.

Multi-causal, multilevel, hybrid, and localized are the characteristics of Martell's [12] pluralist views on globalization ([12], p. 12). Multi-causality suggests that there are multiple primary causes of globalization. Five levels are classified by multilevel classification: political, cultural, military, economic, and environmental. The hybrid perspective incorporates a variety of East/West/North/South inputs. Last but not least, the shape that globalization takes differs according to the context; for this reason, a localized perspective is crucial ([12], p. 12). The goal of this paper is to present the concept of globalization in an objective and equitable manner by upholding a pluralistic perspective on it.

3.1 Globalization of the Zeitgiest

Owing to the intricacy of globalization, in order to completely comprehend the subject matter, it is imperative to elucidate the notions, definitions, and diverse aspects of globalization. Since comprehending the origins of the term's definition is essential, this section will start with the concept of globalization. Over the past 20 years, a number of researchers have conceptualized globalization; despite all the changes the world has experienced, some of these notions overlap, revealing important themes in the literature.

Held et al. [13] draw attention to the difficulty in defining globalization and explain that it exists on a continuum with local, national, and regional elements making up the several levels at which globalization manifests itself. The social and economic networks and relations that are organized locally and/or nationally make up one end of this globalization continuum, while the social and economic networks and relations that crystallize on the larger scale of regional and global interactions are found on the other end ([13], p. 15). Held and McGrew [14] further explain this continuum, stating that globalization does not displace or take precedence over local, national, or regional orders of life but that the local becomes more embedded within more expansive sets of interregional relations and networks of power ([14], p. 3). Martell [12] uses his idea of globalization to highlight these multiple levels and uses "guidelines" to set the phenomenon apart from other comparable concepts. The significance of these "guidelines" lies in the fact that they provide a means of differentiating globalization from internationalization, regionalization, or transnationalization in the absence of geographical referents ([14], p. 4). Since "the global cannot be interpreted literally, as a universal phenomenon, then the concept of globalization lacks specificity," Martell's [12] principles seek to create a cohesive and universal understanding of globalization ([14], p. 4). These four principles will be explained in the sections that follow, together with the major ideas that comprise the notion of globalization. According to Martell [12], one aspect of globalization is the compression of space, meaning that distance no longer plays as much of a role in information, communication, and mobility as it once did ([12], p. 12). In their paper, Keohane & Nye [3] also discuss the importance of space in their understanding of globalization (pp. 4–7). Held et al. [13] stated that by connecting and extending human activity across regions and continents, "globalization can be taken to refer to those spatio-temporal processes of change which underpin a transformation in the organization of human affairs" ([13], p. 15).

1. Therefore, globalization must be global, meaning that it needs to reach all continents and majority of those continents ([12], p. 6).

2. Another 'global' factor is that globalization needs to have inputs from across continents and multiple countries within them ([12], p. 12). According to Martell [12] and Held et al. [13], relations need to be stable and regular to establish a structure or system in order to be global in scale.

3. Furthermore, Martell [12] claims that globalization needs to involve the elites as well as the masses, and a global consciousness must exist so that people have an awareness of the globe as a single place ([12], p. 12).

(1) Consequently, globalization has to be worldwide, which means it must affect all or most of the continents ([12], p. 6). (2) Inputs for globalization must come from several countries on different continents, which is another "global" component ([12], p. 12). According to Held et al. [13] and Martell [12], relations must be regular and stable in order to create a structure or system that is global in scope. (3) In addition, Martell [12] argues that in order for people to perceive the world as a single location, globalization must involve both the masses and the elites ([12], p. 12).

By going over the several definitions and dimensions of globalization, we can show why it is important to evaluate each one and also highlight the areas where they overlap. However, the idea of globalization clarified how the phenomena function, defining it and highlighting the significance of its material components.

3.2 Dimensions of globalization and socio-cultural environment

This section aims to introduce several aspects of globalization, encompassing economic, environmental, social, and cultural globalization. Prior to discussing the dimensions, it is important to note that advanced and developing economies are affected by globalization in two different ways. The different parties and activities engaged in the globalization process are introduced through economic globalization. The section on economic globalization will be the largest and include a variety of topics, such as trade, foreign direct investment, and short-term capital flows. It will also discuss the role that multinational corporations (MNCs) have had in the process of globalization (**Figure 2**).

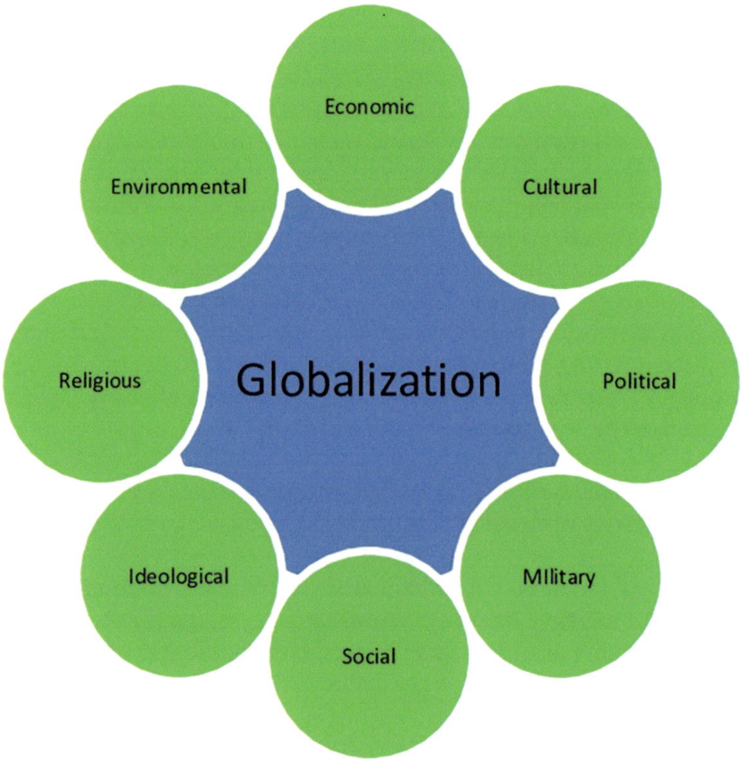

Figure 2.
Dimensions of globalization and socio-cultural environment.

3.2.1 Economic globalization

The structuring of the processes that connect the long-distance flows of capital, products, and services is referred to as economic globalism ([3], p. 4). A more specific definition of economic globalization is from Bhagwati, who defines it as the "integration of national economies into the international economy through trade, FDI, short-term capital flows, international flows of workers and humanity generally, and flows of technology" ([15], p. 439).

3.2.1.1 Consumerism

Free trade agreements necessitate negotiation and compromise among the parties around mutually agreed-upon norms. Concerns pertaining to trade and consumer protection center on the food, product, and service standards that a nation has put in place to safeguard its citizens. Not every nation has the same criteria.

3.2.1.2 Environmental and social responsibility

Labor and environmental protection regulations differ from nation to nation, particularly between industrialized and developing nations, much like consumer protection issues. Choosing which nation's labor and environmental protection regulations to apply when producing items for export is difficult due to the differences between developed and developing nations.

3.2.2 Environmental globalization

The increase in the global flow of people, capital, products, and services was demonstrated by the economic aspect of globalization. The production and movement of commodities between continents are growing along with these activities. As science and technology have advanced and given rise to inventions like the Internet, telephones, ships, and airplanes that have accelerated global communication and travel, they have also brought to light the detrimental effects that industrial and capitalistic activities have been having on the environment. Since poverty is a key source and impact of global environmental concerns, environmental challenges are linked to globalization and development. "Ecology and economy are becoming ever more interwoven locally, regionally, nationally, and globally into a seamless net of causes and effects," according to a UN report on Our Common Futures (1987). The idea of environmental limits originated in 1972 at the Stockholm Conference when the Club of Rome unveiled the concept of "Limits to Growth," is one of the core ideas of environmental globalism ([16], p. 107).

The majority of the world's population expansion is predicted to occur in impoverished countries' urban regions over the next 30 years, making rapid, unplanned, and unsustainable urban development patterns the main source of new environmental and health risks [10, 11]. The World Health Organization [10, 11] states that socio-environmental factors are a major contributing factor to the burden of death, sickness, and disability, especially in poor nations. Poor water quality, availability, and sanitation; vector-borne infections; poor indoor and outdoor air quality; hazardous compounds; and global environmental change are the main sources of risk [10, 11]. Unsafe water, indoor smoke from solid fuel use, malaria from poorly designed irrigation and water systems, inadequate housing, poor waste disposal, and water storage, deforestation and loss of biodiversity, traffic, lead exposure, climate change, and unintentional poisonings from improper use of toxic chemicals and pesticides are the most significant environmental-related causes or conditions that lead to global deaths [10, 11]. Therefore, the poor and vulnerable populations in developing countries are most adversely affected by environmental concerns in terms of health [10, 11]. However, it is also possible that the poorest people are more reliant on natural resources for their livelihood and well-being, making them more vulnerable to the effects of unsustainable resource extraction and depletion [10, 11, 17].

3.2.2.1 Air pollution

The World Health Organization (WHO) lists the following as the main human activity-related sources of air pollution: • Fuel combustion from motor vehicles, such as cars and heavy-duty vehicles • The production of heat and power (such as boilers and oil and coal power plants). • Industrial facilities, such as mines, oil refineries, and factories that manufacture goods. • Residential cooking, heating, and lighting with polluting fuels; • Municipal and agricultural waste sites and waste incineration/burning [10, 11].

3.2.2.2 Differing standards

Developing countries still struggle to meet their most basic requirements, including those for food, water, energy, hygienic conditions, healthcare, and education, and lack the resources, capital, and organizational capacity to put environmental, labor, and health norms into effect.

3.2.2.3 Global warming

There is another essential component of environmental pollution, aside from the detrimental impacts of air pollution on human health. This is the case with CO_2 emissions and other greenhouse gas emissions (GHG), which are significantly influencing the Earth's atmosphere's CO_2 concentration.

3.2.3 Social and cultural globalization

Globalization of the environment has a significant impact on humanity, yet a nation's culture and traditional values also constitute a form of social capital. Keohane and Nye define social and cultural globalism as the long-distance flow of people, ideas, information, and images ([3], p. 5).

One crucial component of social globalism is isomorphism, which is the replication of a society's structures and practices by another ([3], p. 5). Srivastava and Khan define cultural globalism as the spread of ideas, meanings, and values around the world by extending and deepening social ties [18], which is similar to Keohane and Nye's concept of social and cultural globalism. According to Srivastava and Khan [18], on page 18, the common consumption of cultures disseminated through the Internet, popular culture media, and foreign travel is indicative of the progress of cultural globalization.

On the one hand, without the global movement of capital, goods, and services that facilitate cross-cultural interaction, social and cultural globalism would not be feasible.

3.2.3.1 Culture clash

According to a number of scholars ([2], p. 28; [18], p. 20), globalization has a detrimental impact on culture. Globalization frequently suggests that two traditional cultures will conflict as a result of new cultures interacting. Since a nation's culture defines its identity, this has a negative impact on that nation's social capital. The isomorphism associated with social and cultural globalization may be the cause of this. More local cultures will be exposed to foreign cultures if social and cultural globalism disseminate concepts, data, pictures, and people over great distances. Local cultures get increasingly diluted as a result of isomorphism—the adoption of other cultures by local people.

The powerful October 18 Brooke Baldwin 48/148 colonies overran the weak, imposing their culture, political beliefs, and institutional framework on the citizens of the weak colonized countries. This is comparable to what happened during the colonial era: "Psychologically, the superior attitude of the colonist who viewed the local culture and practices as inferior, local religions as barbaric, and local institutions as inefficient not only created a new Westernized ruling class among the native in the country but also had a psychological impacting the local self-esteem and sense of identity" ([2], p. 28).

3.2.3.2 Knowledge of labor

The foundation of a company's competencies is knowledge, which is increasingly produced and maintained by the company and travels with it wherever it does business. In terms of global economic efficiency, labor flows represent the movement or flow of people worldwide. Labor moves around the world as certain regions become more productive than others. This has the potential to be both detrimental and advantageous at the same time. Although it may result in an influx of labor into areas that are understaffed, it may also cause a migration of young, skilled people from nations that depend on them for economic growth.

3.2.3.3 Controversial events

The Syrian Civil War in 2011 caused a massive exodus of immigrants into Europe in 2015, which in turn sparked the global refugee crisis. This massive influx of migrants and asylum seekers put a significant burden on public authorities in multiple Member States and even had consequences in the United States.

4. Sustainability and globalization

In order to create a world of peace, prosperity, and justice for all people, sustainable development (SD) aims to create a society in which social, economic, and environmental issues are shared by all people and ultimately resolved. According to Blewitt [19], p. 29, SD "is about promoting social equity and a degree of economic equality within and between nations as well as protecting and conserving the planet's natural environment."

With global ecological changes, an ever more integrated global economy, and other global trends, political activity increasingly takes place at the global level (**Table 1**).

4.1 Globalization and sustainability in the shades of the NEO-normality

Some authors contend that the UN's most recent set of sustainable progress goals (SDGs) has facilitated progress by encouraging collaborative efforts toward shared global objectives, while others believe that they have limited popular acceptance.

Critics of globalization contend, on the one hand, that it has resulted in a widening disparity between the rich and the poor and increased social systemic imbalances ([20], p. 1; [21], p. 15; [22], p. 5). However, proponents of globalization contend that without it, there would not have been a rise in living standards or a decline in poverty brought about by increased trade and capital flows, idea sharing, and the expansion of democratic institutions across international borders ([23], p. v100).

Ecological scale	Socio-cultural scale
Biosphere	World
Biome-Type	Supranational regions
Biome	State
Landscape	Region
Ecosystem	Locality 1: City, Town,
Community	Locality 2: Village, community, neighborhood
Population	Household
Organism	

Table 1.
Sustainability spatial scale (ecological scale, socio-cultural scale).

4.1.1 Income inequality

In developing nations, where competition and technology make it difficult for them to compete in the global market and where living standards are low for the environment, labor, and society, income inequality is especially difficult to overcome ([7], p. 30; [24], p. 67; [25], p. 553).

4.1.2 Technology trends

Technology today is evolving at a rapid pace, enabling faster change and progress, causing an acceleration of the rate of change. However, it is not only technology trends and emerging technologies that are evolving; a lot more has changed. Technology has made it easier to conduct cross-border trade and interactions. Similarly, technologies that have made air flight more efficient have helped increase the flow of people around.

4.2 Sustainable development guiding principles

In fact, during the past few decades, sustainability strategies have garnered a great deal of attention and favor in the business world. Businesses have been implementing sustainability initiatives more frequently as a result of the ongoing awareness of SD issues and the detrimental effects that corporate operations have on the environment ([8], p. 4).

The following describes these eight SD driving principles:

Nature is not subject to systematically increasing in a sustainable society.

1. Concentrations of materials taken out of the crust of the Earth (such as metals and fossil carbons).

2. Concentrations of materials created by civilization (e.g., protective measures, restricted production, and thoughtful molecular designs).

3. Physical degradation (such as excessive fishing and forest harvesting).

4. Health (e.g., hazardous working circumstances or inadequate post-work relaxation).

5. Influence (e.g., by stifling free expression or disregarding viewpoints).

6. Competence (e.g., through educational barriers or a lack of opportunities for personal growth).

7. Impartiality (e.g., through unfair hiring practices or discrimination).

8. Meaning-making (e.g., by stifling cultural expression or putting up barriers to the joint production of conditions that serve a goal).

4.3 Sustainability and responsible innovation

According to Voegltin and Scherer [26], "the idea of responsible innovation is that new products should not harm the general public's or consumers' health, new processes should be safe for all parties involved, and neither of these should pollute or harm the environment in any way." Three dimensions make up responsible innovation. These are as follows: (1) innovations prevent harm to people or the environment; (2) innovations "do good" by providing new goods, services, or technological advancements that support sustainable development; and (3) global governance frameworks are in place to support innovations that both "do" good and prevent harm.

5. Results and discussion

A broad range of issues and solutions pertaining to globalization and SD were discussed in the papers evaluated across the academic literature. The literature has made it abundantly evident that globalization has both positive and negative effects on SD. This suggests that the results of research conducted by Das and Das [6], Fura et al. [9], Ertrugul et al. [25], and Grossman and Helpman [23] also support the idea that globalization can help advance the sustainable development agenda, especially with certain modifications. These changes, however, are not straightforward; rather, they call for coordinated and cooperative decision-making and execution within as well as across national borders ([9], p. 978; [27], pp. 27–29).

The damage caused by environmental and economic globalism, as well as global governance and policy coordination, were the primary causes of income disparity. Good global governance and policy coordination, according to many authors, will help limit the negative effects of economic and environmental globalism ([5], p. 2; [6], p. 110; [8], p. 5; [16], p. 101; [27], pp. 27–29; [28], p. 379). There is a strong correlation between these two topics.

Global governance, policy coordination, and the damage caused by economic and environmental globalism were the primary causes of income disparity. Strongly correlated with these two topics are the opinions of numerous authors who believe that enhanced global governance and policy coordination will help prevent the negative effects of globalization on the environment and economy ([5], p. 2; [6], p. 110; [8], p. 5; [16], p. 101; [27], pp. 27–29; [28], p. 379).

Increased capital and control in developed countries, coupled with persistently low social standards (health, work, living conditions) in developing nations, are the common outcomes ([20], p. 3; [25], p. 553; [29], p. 26). Furthermore, foreign direct investment (FDI) undermines local enterprises and competitiveness in developing nations, which makes it harder for those nations to grow their own economies ([22], p. 4; [30], p. 29).

To compete in the global market, developing nations must continue to entice international investment despite the negative impacts of foreign direct investment (FDI) ([25], p. 543; [31], p. 1142). Once more, the outcome is increased money gain and control for industrialized countries and increased suppression of developing countries. Consequently, the role of global governance regimes is to supervise international operations and support national governments that are incapable of operating independently.

As a result, national governments in both rich and developing nations perceive a loss of sovereignty [32]. Thus, we have new types of global governance, such as corporate governance across borders and citizen engagement. The choice between the higher standards of the developed countries and the lower standards of the developing countries becomes difficult when it comes to the coordination of policies between these countries. Furthermore, although to a lesser extent than previously, transnational corporate governance seeks to raise the social standards of nations that participate in international commerce ([24], p. 66; [29], p. 26; [33], p. 31).

6. Conclusion

Global networks of interdependence made up of nations, businesses, and individuals characterize modern globalization. These networks facilitate the metaphorical shrinkage of space and the exchange of goods, services, capital, and knowledge at a rate and frequency that is both fast and frequent ([3], p. 5; [12], p. 12; [13], p. 15). The corporate climate has altered as a result of globalization, enabling corporations to look for and locate in regions of the world that provide them the most financial opportunities ([34], p. 132). In the twenty-first century, wealth and mass production have increased like never before because to the unrestricted movement of capital and corporations throughout the globe, technological advancements, and the Internet ([4], p. 13). However, income disparity persists despite rising income levels, indicating that wealth is not being dispersed equitably throughout countries. Furthermore, the stresses placed on the environment are constantly rising ([10, 11]; [16], p. 103).

Technology, foreign direct investment (FDI), and competition have been identified as the main causes of the negative effects ([7], p. 30). Knowledge transfer and spillovers, higher income levels, increased support in supply chain networks, increased collaboration and coordination, increased awareness of SD, increased ESD, and increased support of SD are some of the positive effects of globalization on SD that have been identified ([6], p. 101; [9], p. 977; [23], p. 100; [25], p. 553; [35], p. 1). However, an intriguing discovery was how some of the positive effects of globalization on SD, especially for poor nations, can be used to counteract the negative consequences of competition, FDI, and technical advancement. For example, Das and Das [6] discussed how technical advancement, international commerce, and investment may globalize an economy and raise living standards in nations such as India ([6], p. 101).

The results of the systematic literature study further emphasized the importance of key stakeholders' cooperation, coordination, and support for the SD goal. This was underlined in a number of settings across the scholarly literature. First, in terms of the SDGs, Fura et al.'s [9] paper clarified how the 28 EU members have been cooperating to achieve the objectives of the Europe 2020 Strategy.

In summary, globalization has both negative and positive consequences on SD. Nonetheless, the results of the systematic literature assessment offer hope for overcoming a number of the current problems that globalization and SD are confronting. Global progress toward sustainable development is shown in the growing acceptance and endorsement of the SD agenda by national governments, academic institutions, developing nations, and global governance frameworks. However, because of the study's narrow focus, many consequences of globalization on SD have not been taken into account despite the fact that there are thousands of scholarly books and papers on the subject.

Additionally, this study's broad data search may have produced more general conclusions than if it had focused more on certain features of SD or types of globalization. Nonetheless, the study's findings might offer a broad summary of the principal subjects and ideas covered in the most current scholarly literature, serving as a solid base of information for later investigations, especially in light of the three primary obstacles that globalization and sustainable development must overcome: (1) efficient global governance and policy coordination, (2) winning over society to support SD, and (3) winning over business to support sustainable development.

Author details

Noha Essam Khamis
College of Architecture and Design, Effat University, Jeddah, KSA

*Address all correspondence to: nkhamis@effatuniversity.edu.sa;
noha.essam8511@gmail.com

References

[1] UN. Office for ECOSOC Support and Coordination. UN; 2018. Available from: http://www.un.org/esa/coordination/globalization.htm [Accessed: July 5, 2018]

[2] Gopinath C. Chapter 2: What is new about globalization? In: Gopinath C, editor. Globalization: A Multidimensional System. CA, USA: Sage Publications; 2008. pp. 7-31

[3] Keohane RO, Nye JS. Globalization: What's New? What's not? And so what. In: Krug JA, Daniels JD, editors. International Business and Globalization: The Growth, Consequences, and Future of Globalization. Vol. 1. 2007. p. 3

[4] Brondoni SM. Global capitalism and sustainable growth. From global products to network globalization. Emerging Issues in Management. 2014;**1**:10-31

[5] Holden E, Linnerud K, Banister D. The imperatives of SD. SD. 2016:1-14

[6] Das S, Das SK. Knowledge, attitude and practice of stroke in india versus other developed and developing countries. Annals of Indian Academy of Neurology. 2013;**16**:488-493. DOI: 10.4103/0972-2327.120431

[7] Jaumotte F, Lall S, Papgeorgiou C. Rising income inequality: Technology, or trade and financial globalization? IMF Economic Review. 2013;**61**(2)

[8] Almeida C, Agostinho F, Gianetti BF, Huisigh D. Integrating cleaner production into sustainability strategies: An introduction to this special volume. Journal of Cleaner Production. 2015;**96**:1-9

[9] Fura B, Wojnar J, Beata K. Ranking and classification of EU countries regarding their levels of implementation of the Europe 2020 strategy. Journal of Cleaner Production. 2017;**165**. DOI: 10.1016/j.jclepro.2017.07.088

[10] WHO. Environment and health in developing countries. UNEP. The Health and Environment Linkages Initiative; 2018. Available from: http://www.who.int/heli/en/ [Accessed: July 5, 2018]

[11] WHO. Air Pollution. World Health Organization; 2018. Available from: http://www.who.int/airpollution/en/ [Accessed: July 5, 2018]

[12] Martell L. Introduction: Concepts of globalization. In: Martell L, editor. The Sociology of Globalization. Cambridge, UK: MPG Books Group; 2010. pp. 1-14

[13] Held D et al. Global Transformations. Politics, Economics, Culture. Stanford, CA: Stanford University Press; 1999

[14] Held D, McGrew A. The Global Transformations Reader. Vol. 2nd. Malden, MA, USA: Polity Press; 2003. pp. 1-603

[15] Bhagwati J. Anti-globalization: Why? Journal of Policy Modeling. 2004;**26**:439-463

[16] Waisman H-D, Cassen C, Hamdi-Cherif M, Hourcade JC. Sustainability, globalization, and the energy sector Europe in a global perspective. Journal of Environment & Development. 2014;**23**(1):101-132

[17] UNDP. Human Development Report - Consumption for Human Development. New York and Oxford; 1998

[18] Srivastava S, Khan A. Globalization and development in contemporary India: Cultural perspective. International

Journal of Social Science Studies. 2016;**4**:18. DOI: 10.11114/ijsss.v4i3.1375

[19] Blewitt J. Understanding SD. Vol. 2. Devon, UK; 2015

[20] Beck M. The risk implications of globalisation: An exploratory analysis of 105 major industrial incidents (1971-2010). International Journal of Environmental Research and Public Health. 2016;**13**(309):1-21

[21] Herrschel T. Competitiveness and sustainability – Can 'smart city regionalism' square the circle? Urban Studies. 2013;**50**(7):1-17

[22] Firat AF. Orbits of contemporary globalization. Markets, Globalization & Development Review. 2017;**2**(3):1-8

[23] Grossman GM, Helpman E. Globalization and growth. American Economic Review. 2015;**105**(5):100-104

[24] Bennett EA. Voluntary Sustainability standards: A squandered opportunity to improve workers' wages. SD. 2018;**26**:65-82

[25] Ertugrul HM, Cetin M, Seker F, Dogan E. The impact of trade openness on global carbon dioxide emissions: Evidence from the top ten emitters among developing countries. Ecological Indicators. 2016;**67**:543-555. DOI: 10.1016/j.ecolind.2016.03.027

[26] Voegtlin C, Scherer AG. Responsible innovation and the innovation of responsibility: Governing sustainable development in a globalized world. Journal of Business Ethics, Springer. 2017;**143**(2):227-243

[27] Biermann F, Kanie N, Kim RE. Global governance by goal-setting: the novel approach of the UN SD goals. Current Opinion in Environmental Sustainability. 2017;**16**:26-31

[28] Frugoli PA, Almeida CM, Agostinho F, Giannetti BF, Huisingh D. Can measures of well-being and progress help societies to achieve SD? Journal of Cleaner Production. 2015;**90**:370-380

[29] Brandi CA. Sustainability standards and SD – Synergies and trade-offs of transnational governance. SD. 2017;**25**:25-34

[30] Ampuja M. Globalization and neoliberalism: A new theory for new times? Education and Policy Research. 2015:17-31

[31] Ahmed K, Rehman MU, Ozturk I. What drives carbon dioxide emissions in the long-run? Evidence from selected South Asian Countries. Renewable and Sustainable Energy Reviews. 2017;**70**:1142-1153

[32] Tucker T. How to fix the most controversial elements of trade deals. The Agenda. 2016. Available from: https://www.politico.com/agenda/story/2016/09/fix-isds-trade-deals-000204 [Accessed: November 20, 2017]

[33] Barkemeyer R, Holt D, Preuss L, Tsang S. What happened to the 'development' in SD? Business guidelines two decades after Brundtland. SD. 2014;**22**(1):15-32

[34] Gopinath C. Globalization: A Multidimensional System. Michigan, US: Sage Publications; 2008. pp. 1-255

[35] Lozano R, Ceulemans K, Alonso-Almeida M, Huisingh D, Lozano FJ, Waas T, et al. A review of commitment and implementation of SD in higher education: Results from a worldwide survey. Journal of Cleaner Production. 2015;**108**:1-18

Chapter 9

Stakeholders Perspectives on Ecotourism in Biosphere Reserves in Morocco and Lebanon: The Case of Shouf Biosphere Reserve and Intercontinental Biosphere Reserve of the Mediterranean in Morocco (IBRM)

Faiza El Mejjad, Yamina El Kirat El Allame and Laoucine Amzil

Abstract

This chapter investigates the perspectives of local stakeholders involved in implementing strategies to enhance ecotourism activities within two distinct biosphere reserves: The Intercontinental Biosphere Reserve of the Mediterranean (IBRM) in Morocco and the Shouf Biosphere Reserve (SBR) in Lebanon. The study adopts a qualitative approach using in-depth interviews to uncover both commonalities and differences in the stakeholders' viewpoints regarding ecotourism within these two protected areas. Despite the commonly shared goals, the results reveal nuanced distinctions in the strategies, initiatives, and promotional methods used to promote ecotourism in the two reserves. The research highlights the roles of local stakeholders and their unique contributions to the advancement of ecotourism in these regions.

Keywords: ecotourism, biosphere reserves, stakeholders, Lebanon, Intercontinental Biosphere Reserve of the Mediterranean (IBRM), Shouf Biosphere Reserve (SBR), Morocco

1. Introduction

Environmental challenges, such as ozone depletion, the greenhouse effect, declining animal populations, global warming, desertification, deforestation, loss of biodiversity, and improper waste disposal, represent some of humanity's most critical issues. These challenges predominantly arise from human activities, leading to habitat destruction and ecosystem harm.

The attainment of the United Nations Sustainable Development Goals (SDGs) requires a collaborative endeavor to tackle environmental challenges and establish a sustainable global society, as emphasized by [1]. At the heart of the Sustainable Development Goals lies the crucial task of protecting the Earth's ecosystems and biodiversity to ensure the well-being of both current and future generations.

Moreover, the phenomenon of globalization has intensified the interconnectivity of environmental issues on a global scale. As economies become increasingly intertwined, so too do the environmental challenges they confront. However, globalization also presents an opportunity for international collaboration and cooperation to effectively address and tackle these challenges, as noted by [2].

The importance of international cooperation is especially evident in the realm of ecotourism. Ecotourism represents a sustainable model within the tourism industry that emphasizes environmental preservation, cultural conservation, and community involvement. Consequently, it acts as a driver for progress across multiple Sustainable Development Goals, including those addressing environmental sustainability, poverty reduction, and community empowerment.

The sustainability of our planet and the welfare of its inhabitants face significant threats stemming from the global human-ecological crisis. This crisis, largely precipitated by the exponential growth of the world's population, is evidenced by escalating ecological footprints, intensified consumption patterns, and heightened pollution levels. Human expansion into natural ecosystems further compounds these challenges, contributing to increased carbon dioxide emissions and loss of biodiversity. Effective mitigation of this crisis requires immediate and concerted international action, entailing collaborative efforts across borders and sectors to formulate sustainable solutions.

This global crisis at the intersection of human and environmental concerns has unprecedentedly elevated the urgency of sustainability and biodiversity preservation [3]. Biosphere reserves are crucial in enhancing our planet's ecological resilience, serving as defenders of diverse ecosystems and their myriad species [4].

Sustainable development involves adopting various strategies to protect resources for the well-being of future generations. As articulated by Keeble [5], sustainable development entails pursuing progress that meets current needs without jeopardizing the ability of future generations to fulfill their own requirements [6]. The objectives delineated in the 1987 report "Our Common Future" by The World Commission on Environment and Development underscore the imperative of achieving sustainable development and ensuring the enduring availability of resources.[1]

The growing global demand for ecotourism, particularly after the pandemic, has incited nations like Morocco and Lebanon to adopt strategies that aim at promoting sustainable practices in their biosphere reserves. The Intercontinental Biosphere Reserve of the Mediterranean (IBRM) in Morocco and the Shouf Biosphere Reserve in Lebanon are both dynamic areas that are vital for preserving biodiversity and promoting socioeconomic development. Morocco and Lebanon have strategically placed

[1] Brundtland, G.H. (1987) Our Common Future: Report of the World Commission on Environment and Development. Geneva, UN-Document A/42/427. http://www.un-documents.net/ocf-ov.htm.

their biosphere reserves as major contributors to environmental conservation and sustainable development due to the increased interest in ecotourism (**Figure 1**).

Cognizant of the constraints inherent in direct ecotourism initiatives, stakeholders in Morocco have embraced a comprehensive strategy within the Intercontinental Biosphere Reserve of the Mediterranean (IBRM). This approach places a strong emphasis on ecological preservation and the conservation of biodiversity. The strategies adopted are highlighted through collaboration with international partners and active involvement with the local community. Projects that promote sustainable practices in accommodation, agriculture, and cultural preservation are further financed by government programs led by the Department of Agriculture and the Ministry of Tourism [7].

Similarly, Shouf Biosphere Reserve (SBR) in Lebanon stands out as an example of a successful ecotourism site despite the challenges related to security incidents and a decrease in visitor numbers. Stakeholders in the Lebanese context prioritize environmental preservation, improved quality of life for local communities, and enhancements of tourism products and services. The link between tourism and the biosphere reserve is becoming increasingly vital, with an evolving mutual dependency. The Shouf Biosphere Reserve reveals the effectiveness of programs, partnerships, and a well-defined tourism strategy that not only promotes the reserve but also contributes to the economic development of local communities [8].

The present study adopts a qualitative approach through in-depth interviews, to elucidate both the convergences and divergences in stakeholders' viewpoints on ecotourism in the IBRM and SBR. The focus is on the diverse roles of local stakeholders and their contributions to the promotion of ecotourism. The study also explores how these initiatives can be used to conserve biodiversity.

Figure 1.
Caricatural diagram of the components of the SBR (EDUBIOMED project). Source: Fanny Didou, 2021, dans le cadre du projet EDUBIOMED.

2. Literature review

Biosphere reserves are crucial in implementing several strategies and initiatives to maintain ecotourism development. Hence, ecotourism itself can be a reason for promoting sustainability in such areas. Countless definitions are given to the concept of "ecotourism". They do, however, cover the same viewpoints such as protecting the environment and helping people to improve their income from their surroundings [9]. The concept of ecotourism and the concept of biosphere reserves go together in many ways.[2]

Many scholars have defined ecotourism as a special form of tourism that improves the well-being of local communities, protecting biodiversity, and providing tourists with a satisfying eco-experience [10].

Ceballos-Lascurain provided one of the earliest definitions of ecotourism, stating that it is a type of tourism focusing on visiting natural environments that are relatively undisturbed or uncontaminated. Ceballos-Lascurain [11] defined ecotourism as the pursuit of observing, appreciating, and enjoying the landscapes, wildlife, and plant life in these areas. He emphasized the importance of including all cultural manifestations, both historical and contemporary, as a crucial component of the ecotourism experience in these regions [12].

In 2002, Quebec officially acknowledged ecotourism as a manifestation of sustainable tourism principles, taking into consideration its impacts on the economy, society, and the environment.[3] Furthermore, specific principles distinguishing ecotourism from the broader concept of sustainable tourism were identified. During the same year, the World Tourism Organization (WTO) and the United Nations Environment Program (UNEP) issued guidelines, as documented by [13, 14].[4]

Scholars assert that the success of ecotourism hinges upon substantial engagement, effective communication, and shared knowledge among both leaders and participants [15]. Unlike many other forms of tourism, ecotourism experiences are deemed to have a comparatively lower long-term impact on the environment and biosphere. Consequently, ecotourism emerges as a potent means of promoting the tangible values inherent in a biosphere reserve.

The exploration of ecotourism within biosphere reserves is recognized as a pivotal component of sustainable tourism, seamlessly integrating conservation efforts with responsible economic development. The synergy between the concepts of ecotourism and biosphere reserves is evident, as they uniquely combine the preservation of ecosystems with sustainable human activities. This amalgamation allows ecotourism initiatives in these areas to effectively strike a balance between environmental preservation and socioeconomic advancement [16].

---

[2] Ecotourism and Sustainable Development in Biosphere Reserves: Experiences and Prospects; Workshop Summary Report, Quebec City, Canada, May 24–25, 2002." UNESCO Bibliothèque Numérique, n.d. Include the URL: https://unesdoc.unesco.org/ark:/48223/pf0000127757. Accessed 31 Dec. 2023.

[3] United Nations Environment Program. "Québec Declaration on Ecotourism Defines Basis for Its International Development." UNEP, 23 May 2002, https://press.un.org/en/2002/unep113.doc.htm. Accessed 27 December 2023.

[4] Jonathan Tardif, "Écotourisme et développement durable," VertigO - la revue électronique en sciences de l'environnement, Volume 4, Numéro 1 (May 2003), online since 01 May 2003, accessed on 13 January 2024, URL: http://journals.openedition.org/vertigo/4575; DOI: https://doi.org/10.4000/vertigo.4575.

The United Nations World Tourism Organization (UNWTO) has underscored the positive impact of ecotourism in protected areas, lauding its contributions to fostering environmental awareness and instilling a sense of responsibility for biodiversity conservation. Ecotourism emerges as a proactive means of safeguarding local flora and fauna, significantly enhancing the sustainability of these delicate ecosystems. Furthermore, engagement in ecotourism activities not only benefits the environment but also contributes to economic growth through job creation and the development of local communities.

Ecotourism, as defined by the International Ecotourism Society, embodies responsible travel practices geared toward preserving the environment and enhancing the well-being of local communities [16]. Also referred to as "nature tourism", ecotourism endeavors to minimize the ecological impact of tourism activities.

The management of biosphere reserves garners significant international attention, with robust recommendations and regulations established globally [17, 18]. However, within the jurisdiction of UNESCO Member States, particularly in countries of the Global South like Morocco and Lebanon, biosphere reserves often receive limited recognition due to a primary focus on international cooperation. Yet, these reserves are pivotal for maintaining eco-socio-systems and balancing natural resource needs amidst environmental decline [19]. The sustainable use of biosphere reserves is the subject of extensive recommendations and restrictions at the international level [17, 18]. However, the importance of the international jurisdiction of the biosphere coincides with a modest or even marginal presence at the level of the jurisdiction of UNESCO Member States, particularly the countries of the South. The case of the jurisdiction of Morocco and Lebanon barely recognize the existence of biosphere reserves given their strategic interest in international cooperation. However, Biosphere Reserves are much more than areas for protecting biodiversity, they are vital areas for the eco-sociosystem and the continuity of the balance between the needs for natural resources and the potential offered by a natural substrate in strong regression. Various UNESCO summits and agreements emphasize this equilibrium [20–22], highlighting sustainable tourism as crucial for promoting awareness and maintaining balance [23]. Studies underscore the benefits of ecotourism within biosphere reserves, reshaping stakeholder perceptions and enhancing sustainability [19, 24, 25].

The literature on ecotourism in Morocco and Lebanon remains sparse, with few studies dedicated to the topic. One significant contribution is Selkani's [26] investigation titled "Management of Protected Areas and Ecotourism: Exploring the Relationship", focusing on Toubkal National Park. Selkani examines ecotourism as a nature-based tourism form, stressing its role in environmental conservation and local community welfare. The study underscores protected area management objectives, including biodiversity conservation, sustainable development promotion, and awareness-raising. Strategies for managing Morocco's protected areas are discussed, highlighting their benefits in conserving natural and cultural resources and fostering sustainable local development. Selkani highlights ecotourism's potential as a sustainable development tool, emphasizing sustainable practices and integration into broader development frameworks. The study cites Toubkal National Park as a successful example, showcasing its biodiversity, cultural significance, and tourism appeal. However, while insightful, Selkani's [26] study lacks detailed methodology disclosure. This study will try to check these findings against the data from IBRM and the Shouf reserve.

Azzam's [27] study of Shouf reserve in Lebanon investigates the management of the Shouf Biosphere Reserve (SBR) in Lebanon, where urbanization often takes precedence over environmental conservation. Azzam develops an evaluation framework in his master's thesis to assess the governance structure of the SBR, blending preservation and planning components specific to biosphere reserves. The study adopts a qualitative approach making use of interviews and participant observation to evaluate governance, management, and planning strategies. However, the study lacks detailed insight into its methodology. The findings highlight the need for policy interventions to support decision-makers in managing biosphere reserves, both locally in Shouf, Lebanon, and globally. Azzam's research underscores the necessity for an integrated management model incorporating spatial planning principles to address challenges faced by protected areas like the SBR. By integrating spatial planning perspectives, the study contributes to the understanding of biosphere reserve management and evaluation, contributing to knowledge of regional governance and planning modalities essential for effective management of biosphere reserves like the Shouf Biosphere Reserves (BR) in Lebanon. The present study will check some of the key findings reported by Shouf reserve against its data.

Two recent studies contribute significantly to the understanding of the topic under investigation. Sethy and Senapati [28] conducted a comprehensive study on stakeholder perceptions regarding ecotourism development in the Chilika coastal wetland ecosystem in Odisha. Their research aimed to uncover the challenges and benefits associated with ecotourism in the region, as well as stakeholders' willingness to invest in ecotourism development and biodiversity conservation. Utilizing semi-structured interviews, the study engaged 250 informants from various adjacent areas, including Balugaon, Barakul, Pathara, and Khalikot. The contingent valuation method and the Double Hurdle Model were employed to analyze stakeholders' willingness to contribute financially. The findings underscore the critical importance of environmental awareness and education among local communities and tourists for biodiversity conservation in Chilika Lake. Furthermore, the study highlights the significance of considering socioeconomic and demographic characteristics in formulating sustainable ecotourism practices. It emphasizes the necessity for environmental education and awareness initiatives, as well as improved tourism infrastructure, to effectively conserve biodiversity in the region. The current investigation will examine how the conditions in the IBRM and Shouf reserves compare with those observed in the Chilika case.

Esparza et al. [29] conducted an investigation titled "Perceptions of Stakeholders and Challenges Faced by Ecotourism Enterprises in the Bosque de Pómac Historic Sanctuary, Lambayeque, Peru", which delved into the perceptions and challenges encountered by key stakeholders representing both public and private entities engaged in the protection and management of the sanctuary. Employing an exploratory qualitative methodology centered on semi-structured, in-depth interviews, the research aimed to elucidate the primary drivers and obstacles pertaining to the economic, social, and environmental dimensions of sustainability within this context. The findings of the study furnished valuable insights into the sustainability concerns confronting local ecotourism enterprises, thereby providing pragmatic recommendations for bolstering economic viability, fostering social cohesion, and advancing environmental preservation endeavors within analogous ecotourism locales. This study will investigate whether the circumstances observed in the cases under examination align with those documented in the Bosque Pómac Historic Sanctuary case.

Mondino and Beery's [10] study, titled "Ecotourism as a Learning Tool for Sustainable Development: The Case of Monviso Transboundary Biosphere Reserve, Italy", investigates the role of ecotourism in advancing sustainable development, specifically within the Monviso Transboundary Biosphere Reserve. Utilizing qualitative methods, such as interviews with local stakeholders, the research explores their perspectives on ecotourism's potential as an educational mechanism. The primary focus lies in elucidating ecotourism's capacity to facilitate the integration of conservation efforts with sustainable development initiatives. The study reveals a complex picture: while ecotourism demonstrates promise in fostering collaboration among stakeholders, it encounters significant challenges, notably persistent negative perceptions likely stemming from insufficient environmental education [10]. Consequently, the study underscores the necessity of enhancing environmental literacy programs and strengthening governmental support to optimize ecotourism's effectiveness as an educational tool. Ecotourism in the Monviso Transboundary Biosphere Reserve serves as a catalyst for stakeholder cooperation, yet confronts perceptual barriers due to inadequate environmental education. This study will examine these findings in relation to the pertinent data collected for the contexts under examination.

3. Methodology

The present study adopted a qualitative approach, using semi-structured interviews as the primary data collection tool. Nineteen informants participated in the interviews, selected through a snowball sampling technique to capture diverse perspectives from stakeholders associated with the Intercontinental Biosphere Reserve of the Mediterranean (IBRM) in Morocco and the Shouf Biosphere Reserve in Lebanon. The snowball sampling method commenced with interviews conducted with key stakeholders identified through established networks. Subsequently, each participant was requested to nominate additional individuals deemed relevant, thus enabling the identification of further participants within the target population.

The semi-structured interviews were designed to elicit comprehensive responses from stakeholders regarding their perceptions, experiences, and perspectives on ecotourism within the biosphere reserves. The questions aimed to probe stakeholders' comprehension and conceptualization of ecotourism, the extent of its promotion and implementation within the reserves, specific initiatives and strategies employed to bolster ecotourism, and the engagement and cooperation of local communities in ecotourism endeavors. Furthermore, stakeholders were prompted to provide tangible examples of successful ecotourism initiatives, delineate the economic benefits accrued by local communities from ecotourism development, and identify challenges confronting ecotourism within the reserves, along with potential solutions and future prospects.

The data analysis employed Iramuteq[5] to conduct statistical analyses on text corpora and tables containing individuals/words, with a specific emphasis on the discourse emanating from local actors within biosphere reserves. Employing discourse analysis techniques facilitated the identification of prominent themes, patterns, and perspectives pertaining to ecotourism and environmental concerns in the South Mediterranean region, as gleaned from stakeholders' testimonies. Guided by the study's objectives, which sought to elucidate stakeholders' worldviews and perceptions regarding ecotourism, as well as their awareness of broader environmental issues, the analysis centered on identifying recurrent themes within the discourse. Particularly, the section focusing on the necessity and perception of collaboration within the local community entailed an analysis of the frequency of co-occurrence

[5] Iramuteq [RAMUTEQ is a GNU GPL (v2) licensed software used for the statistical analysis of text corpus and tables composed by individuals/words] is used to analyze the discourse of the local actors in the biosphere reserves.

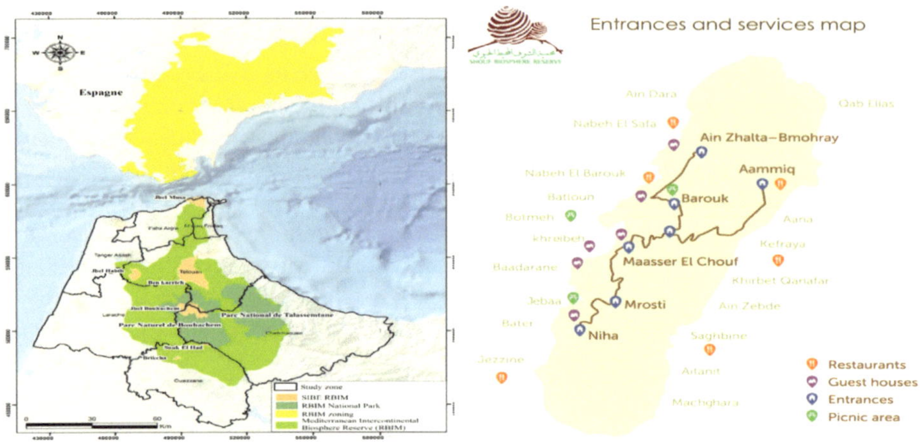

Figure 2.
Geographical location of IBRM and SBR. a. Source: Elaborated by EL MEJJAD F, 2020. b. Source: Shouf Biosphere Reserve website.

of specific terms pertinent to ecotourism and biosphere reserves. This analytical approach aimed to elucidate the interrelationships between these terms and discern salient themes in the discourse of local actors.

A comparative analysis approach was used to juxtapose viewpoints from Lebanon and Morocco, aiming to discern potential cultural variations. The objective of this methodological approach is to identify and scrutinize the cultural determinants that shape perceptions of ecotourism and environmental sustainability within each context. Moreover, qualitative methodologies such as in-depth interviews were employed to delve into individual perspectives, thereby offering insights into the multifaceted interactions among culture, ecotourism, and environmental attitudes.

The study focuses on Shouf Biosphere Reserve (SBR) and Intercontinental Biosphere Reserve of the Mediterranean in Morocco (IBRM). The two biospheres are located in the South Mediterranean region, namely the Northern part of Morocco and Southern Lebanon[6] (see **Figure 2**). More details about the two reserves are presented in the data analysis and discussion section. The choice of the two sites is motivated by the intricate richness of their landscapes intertwined with a myriad of cultural systems, rendering them highly attractive destinations within the framework of sustainable tourism practices. Indeed, the two reserves serve as a focal point for cultural tourism, offering authentic experiences in immediate proximity to human settlements and natural environments. IBRM and SBR also align with the prevailing international trends in the countries' sustainable tourism market.

4. Informants' profiles

This study involved people who are actively involved in the Shouf Biosphere Reserve (SBR) and the Intercontinental Biosphere Reserve of the Mediterranean in Morocco (IBRM). The main objective of the study is to capture diverse perspectives

[6] EduBioMed—Capacity Building For Education And Applied Research In Mediterranean UNESCO's Biosphere Reserves.

Informant	BR	Gender	Level of education	Age	Position
1. TA	IBRM	Male	BA	63	ATED Association Pres
2. ALB	IBRM	Male	BA	40	Tour guide
3. ASB	IBRM	Male	Master	27	Tour Guide
4. FH	IBRM	Female	Master	48	Hostel Owner
5. HJ	IBRM	Female	Master	49	Director, Sayida El Hourra
6. AI	IBRM	Male	Doctorate	67	Association president and professor
7. MD	IBRM	Male	Doctorate/Engineer	45	Head of Department (ANEF)
8. SA	IBRM	Female	Doctorate	32	Manager in Tangier-Tétouan Tourism Delegation
9. MA	IBRM	Male	BA	38	Educational Facilitator
10. AK	IBRM	Male	BA	59	Charge the office
11. JM	IBRM	Male	Master	44	Head of Partnership and Animation Department Management (ANEF)
12. OA	SBR	Male	BA in Advertising and Marketing	39	Ecotourism Manager
13. GN	SBR	Male	BA	35	Owner of the project Beil Leila
14. GF	SBR	Male	BA in interior design	34	Owner blate Atiq
15. LS	SBR	Female	BA in political sciences	64	Development Coordinator
16. NH	SBR	Male	PhD in Ethnobotany	46	General Manager SBR
17. JK	SBR	Female	BA in music education	48	Tour Guide
18. CA	SBR	Male	MA in animal science	39	Farm Owner, Maasir Donkey Farm
19. SN	SBR	Female	BA in business Marketing	35	Communication Officer at the Reserve

Source: field survey, interviews with IBRM and SBR stakeholders, El Mejjad F, 2021–2023.

Table 1.
Informants profiles.

from both Lebanese and Moroccan actors. The informants include ecotourism managers, tour guides, and hostel owners. Due to geographical constraints, online interviews were conducted to ensure the participation of both Lebanese and Moroccan informants. The educational backgrounds of the informants ranged from a Bachelor's degree to Ph.D. qualifications, which indicates a range of expertise relevant to the focus of the study, involved in managing natural reserves and practicing ecotourism. The study favored the informants with professional experiences within the biosphere reserve and integrated management. **Table 1** lists the informants from the two biosphere reserves with some specifications of their profiles.

5. Data analysis and discussion

To gain deep insight into the topic under investigation and achieve a comprehensive understanding of the matter, the data underwent an initial coding process. This involved a thorough review of the transcripts to identify key patterns and themes before proceeding to analyze and interpret the stakeholders' testimonies.

The analysis was executed with the research questions outlined in the study, placing particular emphasis on the viewpoints of the stakeholders. A meticulous examination of the coded segments was done to identify the recurring patterns and ensure a precise alignment with the research objectives to capture pertinent insights.

The analysis also involved a comparison of the perspectives reported for IBRM and SBR. The systematic study of the two biospheres stakeholders' responses highlights the similarities and differences in their testimonies. The analysis focused on underlining the best practices observed in both contexts, emphasizing the successful strategies and identifying the challenges or the practices that require improvement. This comparative analysis helped to gain valuable insights into the distinct dynamics at play in each biosphere, contributing to a comprehensive understanding of the focal areas under study. The subsequent subsections present the data analysis.

Drawing upon the researchers' extensive field experience, intimate familiarity with the study settings, and visits to both contexts, supplemented by data sourced from relevant websites and on-site observations, a comprehensive exposition is offered regarding the distinctive attributes of the two biosphere reserves, complemented by photographic documentation. This information is systematically categorized under the overarching theme titled "IBRM and SBR: Exploring Diverse Ecological Potentials with Ecotourism Value".

5.1 IBRM and SBR: Exploring diverse ecological potentials with ecotourism value

The geographic expanse encompassed by the two studied reserves boasts a rich array of biological, ecological, and landscape diversities of considerable heritage significance. These features, including landscape characteristics, flora and fauna compositions, and the presence of historical and cultural assets, serve as fundamental pillars for fostering ecotourism initiatives within the two biosphere reserves.

The Shouf Biosphere Reserve, accorded UNESCO designation in 2005, spans an area of 29,540 hectares within Lebanon's Chouf region. Renowned for its remarkable biodiversity and distinctive landscapes, this reserve constitutes a critical habitat supporting a plethora of plant and animal species. In contrast, the Intercontinental Biosphere Reserve of the Mediterranean, established in 2006, covers a vast expanse of 1 million hectares. Encompassing territories within the Province of Chefchaouen and various regions within the Wilaya of Tétouan, alongside the Provinces of Fnidek, Fahs-Anjra, and Larache, this expansive reserve epitomizes the diverse ecological and cultural heritage intrinsic to the Mediterranean region. As such, it assumes a pivotal role in both conservation endeavors and the promotion of ecotourism development. In what follows, photos illustrating each biosphere landscape characteristics, flora, and fauna compositions are provided.

The geological landscape of the Intercontinental Biosphere Reserve of the Mediterranean (IBRM) represents a continuation of the intricate Alpine Chain, characterized by picturesque and panoramic geological formations.[7] The Rif Mountains, within the IBRM, present a distinct landscape featuring an ancient

[7] The High Commission for Water and Forests and the Fight against Desertification, currently the National Agency for Water and Forests. Data from 2018 to 2021.

dolomitic-limestone ridge dating back to the Triassic-Lias period. The physical attributes and climatic conditions of the Rif, such as the substantial thickness of carbonate substrates in the ridge, intense tectonic activity, high humidity levels, and dense forest cover, have sculpted a karstic landscape of exceptional beauty, accentuated by awe-inspiring gorges. Notably, locations like Aqchour in the Talassmtane Park feature towering walls flanking deep valleys, offering opportunities for hiking and guided excursions catering to nature and geology enthusiasts. This geological terrain harbors notable caverns and caves conducive to both recreational and scientific spelunking activities.

The vegetation pattern within the IBRM comprises a mosaic primarily dominated by cork oak trees, influenced by the region's humid climate, and adorned with one of the last remaining fir forests in Africa. The IBRM encapsulates Mediterranean biogeographic zones ranging from humid to semi-arid, fostering a unique flora characterized by rich diversity and significant endemism. Noteworthy among its botanical treasures is the presence of one of Africa's two remaining fir forests, covering an approximate area of 3000 hectares. This natural legacy has attracted considerable interest from a diverse array of visitors, ranging from enthusiasts to ecotourism specialists. The floral composition within the IBRM exhibits richness and diversity, encompassing approximately 4800 rare species, subspecies, or endemic taxa.

The IBRM territory encompasses ecosystems spanning both continental and littoral/marine realms, leading to the establishment of Natural Parks and Sites of Biological and Ecological Interest, notably the Talassmtane and Bouhachem Parks. The region's high biodiversity value, coupled with bioclimatic contrasts, has facilitated the development of distinct and diverse fauna populations within the IBRM, as the following photos show (**Figures 3** and **4**).

The Shouf Biosphere Reserve encompasses approximately 5% of Lebanon's total land area, extending along the ridge of the western Mount Lebanon chain from Dahr-el Baidar in the north to Jezzine in the south.[8] Situated between the Bekaa Valley to the east and the Shouf Valley to the west, the reserve comprises various ecologically significant zones, including the Shouf Cedar Nature Reserve, established in 1996. In July 2005, UNESCO designated the Shouf Cedar Nature Reserve as a biosphere reserve, encompassing approximately 50,000 hectares. This designation also includes the Ammiq Wetland, located east of the Shouf in the Beqaa Valley, recognized as a

Figure 3.
Significance of biodiversity: excerpts from the Talassmtane eco-museum. Photos taken by Amzil 2020.

[8] Shouf Biosphere Reserve, "Lebanon." UNESCO, https://en.unesco.org/biosphere/arab-states/shouf. Last updated: October 2018.

Figure 4.
Botanical diversity of IBRM reserve. Source: https://www.clorofila.ma/ and other websites.

Ramsar site and one of the few remaining wetlands in the Middle East. Additionally, 24 surrounding villages contribute to the biosphere's cultural and ecological diversity, located on the eastern and western flanks of the Barouk and Niha mountains.

Renowned for its towering mountains, the Shouf Biosphere Reserve is a vital ecosystem that provides crucial environmental services. It boasts the largest stands of Lebanese Cedars globally, celebrated as iconic trees of significant cultural and ecological importance. The reserve's diverse flora includes numerous medicinal plants with notable economic potential, while its varied wildlife presents opportunities for conservation efforts, particularly through the reintroduction of species such as Ibex, Gazelle, and Vultures. Positioned as a pivotal crossroads for bird migration between Europe, Africa, and West Asia, the reserve plays a pivotal role in avian biodiversity preservation. Encompassing 50,000 hectares, it stands as one of the largest protected areas in the Middle East and East Mediterranean, with its cedar forest alone covering 620 hectares, safeguarding a quarter of Lebanon's remaining cedar forests. The reserve hosts an impressive botanical diversity, boasting 520 plant species, including 25 internationally and nationally threatened species, along with 48 plants endemic to Lebanon. Furthermore, the reserve supports over 250 bird species, with many reliant on the Ammiq Wetland for habitat within the reserve (Shouf Biosphere Reserve, Lebanon, n.d.). The following photos illustrate the reserve's flora and wild life (**Figure 5**).

The Shouf Biosphere Reserve (SBR) serves as a crucial refuge for biodiversity and sustains approximately 70,000 residents inhabiting its surrounding areas. The reserve's territory encompasses 24 distinct municipalities across three districts, with a primary focus on rural development. The local community, predominantly led by

Figure 5.
Vegetal species found in Shouf reserve.

women, actively participates in the production of 70 different products using traditional methods. These products, along with their accompanying recipes, are made available for purchase in visitor centers managed by the Reserve, thereby contributing to the economic prosperity of the region. Regarding ecotourism management, the reserve prioritizes interpreter/guiding services, visitor safety measures, organization of ecotourism activities, and provision of guest houses for accommodation. Environmental education initiatives are also emphasized and aimed at fostering awareness about conservation significance and promoting sustainable lifestyles among both visitors and the local community.

The Cedars of Chouf Reserve boasts an impressive array of fauna and flora, including approximately 500 plant species, 200 bird species, and 32 species of wild animals such as wolves, lynxes, gazelles, and hyenas (see **Figure 6**).

5.2 Stakeholders' perspectives on ecotourism at IBRM and Shouf reserve

One of the key questions addressed during the interview concerned the definition of the term "ecotourism". The analysis of the data collected from the stakeholders from Morocco and Lebanon showed both similarities and differences. In Morocco, "ecotourism" is viewed as a concept that requires the preservation of nature and

Rock Hyrax (Procavia capenis syriaca)

Griffon Vulture *Gyps fulvus*

Striped hyaena (Hyaena hyaena syriaca)

Mouse-eared bat (Myotis myotis)

Figure 6.
Natural habitats of Shouf Biosphere Reserve. Source: Shouf Biosphere Reserve website.

heritage. For example, AL. B, a 40-year-old male tour guide in IBRM, stated, "*When I hear the term ecotourism, I think of a form of tourism focused on nature preservation and minimizing environmental impact*".

Similarly, FH, a 48-year-old female; hotel owner in Tallasmatan believes in protecting nature and local traditions, while also improving travel experiences. She declared that "*Ecotourism brings to mind the preservation of nature and heritage while providing responsible travel experiences*".

The President of Asayida Elhourra association in Chefchaouen, a 49-year-old female with an MA, discusses the ways to help both the environment and people through projects such as cooperatives. She did not provide any definition of the term, but she directly referred to the initiatives of the association as she understood the meaning of ecotourism. She affirmed

> *"We work with women on environmental issues and natural resources to generate income that promotes their empowerment. We have created cooperatives focusing on environmental issues. Our main objective is the socio-economic development of women".*

Stakeholders in the IBRM and SBR highlighted the importance of preserving natural landscapes and cultural heritage in ecotourism. Local guides emphasized educating visitors on environmental conservation and cultural authenticity. In contrast, stakeholders in a Peruvian natural area prioritized economic benefits over environmental concerns, viewing ecotourism mainly as an income source. While both regions face challenges with infrastructure and visitor management, the IBRM and SBR stakeholders are concerned about balancing tourism growth with environmental protection; those in Peru struggle with resource scarcity for remote ecotourism sites.

Professor AI, a 67-year-old mathematics professor and president of the "Patrimoine Développement Citoyenneté association in Tetouan, argued that ecotourism does not always have to play a role in the responsibility of the tourism industry. He stressed the fact that there are projects that do not respect nature and involve external investors, which is not considered responsible […] *not all ecotourism is responsible* […]".

The above testimonies reveal that ecotourism can have beneficial effects on individuals, nature as well as local communities.

Consistent findings have emerged across studies conducted in diverse ecotourism destinations. Sethy and Senapati's [28] investigation of the Chilika Lake conservation area in Odisha, India, revealed a spectrum of stakeholder perceptions toward ecotourism, highlighting the necessity to harmonize conservation objectives with socioeconomic development. However, notable distinctions between the two contexts persist, which are largely attributed to the distinct environmental challenges and community dynamics inherent to each locale.

The present study serves as a complementary addition to the existing body of literature, aligning with insights from the IBRM and SBR. It elucidates stakeholders' perspectives on contemporary ecotourism issues and benefits, emphasizing the pivotal roles of environmental awareness, educational initiatives, and enhanced tourism infrastructure in biodiversity conservation efforts. Furthermore, stakeholders' willingness to invest in ecotourism development and biodiversity conservation within the Chilika coastal wetland ecosystem of India resonates with the broader objective of promoting sustainable ecotourism practices observed in both the IBRM and SBR.

The findings derived from the Chilika coastal wetland ecosystem in Odisha mirror the collective sentiments of stakeholders in the IBRM and SBR, emphasizing the critical importance of responsible tourism practices aimed at safeguarding environmental integrity and enhancing community welfare.

The statements from stakeholders in the Shouf Biosphere Reserve (SBR) revealed both shared perspectives and distinctive elements in their understanding of ecotourism within this distinct regional context. In Lebanon, ecotourism is generally perceived as a conscientious form of travel that prioritizes the preservation of both natural landscapes and the cultural heritage of the region.

JK, a 48-year-old female tour guide from Lebanon, aligns with this perspective, asserting that her involvement in this form of tourism is driven by the goal of educating people about the significance of nature preservation and environmental conservation.

The significance of generational experiences in the local environment and the commitment to environmental sustainability is underscored by GN, a 35-year-old proprietor of residence in Maasir Shouf. GN perceives ecotourism as a duty that encapsulates both cultural authenticity and ecological preservation, aligning seamlessly with the overarching objectives of the Shouf Biosphere Reserve. In his words, "*Ecotourism, to me, is a responsibility – a commitment to environmental sustainability and preserving our cultural authenticity* […]".

The Shouf Biosphere Reserve's General Manager, NH, 46-year-old male with a Ph.D, in Ethnobotany, presents a complete definition of ecotourism. He identifies three essential components: "*supporting protected areas through fees, helping local communities to enhance their income, and raising awareness among visitors about nature, biodiversity, and protected areas*".

The interpretations provided by Lebanese stakeholders underscore a commitment to environmental sustainability and cultural authenticity, emphasizing a distinctive focus on the symbiosis between tourism and nature. The interconnection among stakeholders from both biosphere reserves reveals a shared definition of ecotourism as a responsible practice dedicated to the preservation of both natural ecosystems and cultural heritage. Lebanese stakeholders, in particular, stress the socioeconomic dimensions of ecotourism, indicative of a more comprehensive approach to community development. A Word cloud was generated using a free online word cloud generator to sort out the most recurrent terms used in the definition of the term "ecotourism". The figure is presented below.

The viewpoints expressed by stakeholders in both the SBR and the IBRM align with the conceptualization of ecotourism delineated by Selkani [26]. According to Selkani, ecotourism is characterized as a paradigm emphasizing the dual objectives of preserving natural environments and cultural heritage, and concurrently advocating for conscientious and sustainable travel practices.

The depicted figure encompasses all lexical items employed by the informants in articulating the concept of "ecotourism". The font size and spatial arrangement of each term within the figure serve to visually convey its recurrence in the narratives. A lexicometric tool, this diagram systematically represents the frequencies of lexical forms by correlating them with their size, specifically for the terms "ecotourism", "IBRM", "SBR", and "environment" (refer to **Figure 7**). Notably, the word cloud

Figure 7.
Word cloud showing the stakeholders' views about ecotourism in BRs. Figure Designed by F. Elmejjad 2023.

illuminates additional semantically associated terms such as "initiative", "local", "biosphere", and "development", offering a comprehensive linguistic snapshot of the informants' perspectives.

The following section presents and discusses the stakeholders "roles and initiatives for the promotion of ecotourism in IBRM and SBR".

5.3 Local stakeholders' roles and initiatives in the promotion of ecotourism at IBRM and Shouf reserve

One of the key questions directed toward stakeholders during the interviews pertained to their respective roles and initiatives in fostering ecotourism within the biospheres under investigation. The Intercontinental Biosphere Reserve of the Mediterranean (IBRM) emerges as a propitious terrain for tourism, accentuating ecological affluence and catalyzing a socioeconomic dynamism in the region, as elucidated by Abdul Malak et al. [30]. Within this context, tourists are presented with an array of tourist circuits, offering a nuanced exploration of the local culture. The IBRM holds the inherent potential to champion environmental sustainability, ecotourism, and programs that could yield dividends for the inhabitants of the IBRM territory, concurrently contributing to the socioeconomic advancement of rural communities, as posited by Boushaba [31].

An analysis of the initiatives spearheaded by the ATED association and ANEF (l'Agence Nationale des Eaux et Forêts) illuminates their pivotal role in enhancing the local economy and fortifying the tourism infrastructure within the region. The president of the ATED association, AT, 63 years old with a BA underscores the organization's proactive initiatives, encompassing the establishment of accommodations, the creation of designated circuits, and educational endeavors. He elucidates:

> *"We've also been working on setting up tourist circuits, with an emphasis on signage. In addition, we are focusing on training in the management of rural houses and cottages covering areas such as communication, reception, gastronomy, and administrative and financial management".*

Similarly, JM, a 44-year-old male engineer affiliated with ANEF, underscores collaborative efforts with GIZ to advance ecotourism-related initiatives. These efforts include the establishment of tourism cooperatives and the restoration of lodges, aligning with the broader goal of fostering the economic growth of the region. Both entities express a shared commitment to ecotourism, emphasizing environmental stewardship and fostering economic development within local communities.

The participants' testimonies collectively underscore a dedicated commitment to the progression of ecotourism, with active involvement from stakeholders, including government representatives. However, AI, a 67-year-old professor and president of an association, introduces a nuanced perspective, highlighting that while existing initiatives are noteworthy, their scope remains limited compared to the vastness of the territory.

This testimonial indicates that despite the limited direct ecotourism initiatives, stakeholders in the IBRM diligently focus on ecological preservation and biodiversity, indirectly aligning with the fundamental tenets of ecotourism. Additionally, biosphere reserve stakeholders initiate comprehensive endeavors linked to conservation, diversity, the promotion of ecotourism, and economic and social development. Their collaborative efforts with international partners, coordination strategies, and encouragement of local community involvement stand as a testament to their commitment. The IBRM also implements educational programs, guided eco-friendly tours, and partnerships with local communities to ensure the enduring sustainability of the territory.

Stakeholders within the Intercontinental Biosphere Reserve of the Mediterranean (IBRM) purport to embrace diverse strategies for the development of ecotourism, encompassing conservation, education, collaboration with international investors, and community engagement. Despite these aspirations, the actualization of projects encounters substantial challenges, marked by coordination issues, a lack of awareness, and inadequate communication among the local populace. Notably, the collaborative ethos among local stakeholders within the IBRM remains relatively sporadic, primarily in response to specific projects that bring stakeholders together. The absence of a consistently proactive approach, where stakeholders actively initiate contact and propose joint ventures, impedes the establishment of a robust collaborative network, as underscored by MD, a 45-year-old male engineer affiliated with ANEF and the coordinator of IBRM.

Within the context of Morocco's biosphere reserves (BRs), some participants expressed apprehension regarding the limited communication surrounding these sites. The dearth of communication channels and the inadequate dissemination of pertinent information contribute to a diminished promotion of the significance of ecotourism. The lack of a systematic communication strategy not only restricts public awareness of the ecotourism concept but also diminishes the influx of visitors to these areas. AB, a 27-year-old male tour guide, affirmed that "*Sometimes, while certain associations undertake environmental awareness efforts to protect resources, their endeavors alone prove insufficient*".

Throughout the interviews, stakeholders within the Intercontinental Biosphere Reserve of the Mediterranean (IBRM) expounded on a myriad of actions and activities they have undertaken. These include the completion of comprehensive studies delving into the cultural, natural, and heritage potential of the region. Notably, they have identified ecotourism circuits, laying the groundwork for the future Bouhachem Natural Park, Tallasmtane National Park, and the prospective Jbel Moussa Park. The clarity and tangible evidence of these efforts are underscored by the meticulous identification of these circuits.

While numerous initiatives have been proposed under the umbrella of ecotourism in the IBRM, some have already been set in motion. Certain associations have proactively devised and implemented thematic routes designed to captivate tourists. The concerted support from international cooperation agencies has bolstered endeavors aimed at enhancing the overall quality of the tourist experience. Specific instances, such as the ecotourism cooperatives and accommodations in Moulay Abdessellam within the Bouhachem region, serve as exemplars of successful initiatives that have contributed to the enhancement of the natural, cultural, historical, and spiritual heritage of Bouhachem. These ecotourism activities seamlessly integrate accommodation, catering, hiking, and animated outings, all facilitated by local guides. This example illuminates the multifaceted dimensions of both the natural and cultural heritage intrinsic to the region.

An illustrative instance of an ongoing initiative is the collaborative program with GIZ, the German International Development Cooperation Agency, presently progressing into its second phase. This program is specifically designed to fortify ecotourism activities and support cooperatives, facilitating their transition into more professionally run entities. This underscores the local agency's unwavering dedication to fostering the ecological development of the region and leveraging its natural and cultural potential to the fullest. Despite these commendable efforts, the tourism sector has yet to reach a level where it can generate substantial economic benefits. While certain activities such as accommodation, catering, guided tours, hiking, kayaking, and mountain biking are already underway, there is a need for further development and organization within the industry to maximize its potential. Additionally, the ongoing biosphere reserve project is positioned as a viable alternative to combat cannabis monoculture in the Moroccan area, as noted by scholars [32].

Stakeholders associated with the Shouf reserve accentuated their commitment to promoting sustainable tourism practices through initiatives such as training sessions for visitors, particularly focused on the preservation of the cedar forest. General Manager, NH, a 46-year-old male, General Manager SBR, introduced innovative programs like the adoption of cedar trees, allowing individuals to adopt a tree with a personalized sign indicating the adopter's name, along with certification and a 10-year membership card for free visits to the reserve. Moreover, extensive training sessions for cedar tree preservation have been instituted in the Shouf Biosphere Reserve, contributing significantly to the advocacy of sustainable tourism. Ecotourism Manager, OA, a 39-year-old male Ecotourism Manager, emphasized the cultural practices within the biosphere reserve, including beekeeping and sustainable agriculture, aligning with the dual objectives of providing unique experiences and educational opportunities. The reserve actively welcomes visitors from various regions to explore its ancient trails, fostering mutual benefits for both visitors and the Shouf Biosphere Reserve community.

Furthermore, NH, A 46-year-old male, General Manager SBR elucidates diverse strategies employed for marketing ecotourism, encompassing both online and traditional channels. The array of strategies includes online advertising through social media platforms, websites, and electronic media, alongside traditional methods such as television, radio, conferences, meetings, and newspapers. To enhance the biosphere reserve's visibility and attract visitors, NH underscores the importance of organizing diverse programs, events, and festivals.

JK, a 48-year-old female tour guide, provides valuable insights into the ecotourism initiatives within the Shouf Biosphere Reserve. Notably, several training sessions have been conducted, specifically focusing on forest soil preservation, contributing to the promotion of sustainable tourism. The reserve has a well-defined environmental conservation plan, led by a dedicated team committed to preserving sustainable tourism. In Beirut, despite the various environmental challenges, residents actively strive to maintain a healthy lifestyle.

While both SBR and IBRM share a commitment to environmental conservation, SBR's initiatives place a distinct emphasis on cultural experiences and community engagement in specific activities within the reserve. In contrast, IBRM adopts a broader approach, addressing legal matters, fostering collaborations with external entities, and prioritizing economic assistance for local communities.

5.4 Necessity and perception of collaboration within the local community

The importance of collaboration between actors has been highlighted by all stakeholders in both IBRM and the Shouf Biosphere Reserve (SBR). Indeed, local communities play a crucial role in the development and continuity of ecotourism within biosphere reserves. A complex interplay of challenges and successes is revealed through the varied perspectives of professionals, educators, and local communities. The maintenance of sites of ecotourism interest, the sustainability of traditional practices and know-how as well as the promotion and reception on site, are the basis of the intervention of grassroots communities. Thus, the co-management and protection of landscape and heritage resources must be shared between reserve managers and the local community. There seems to be a positive development in community involvement, with an increase in interest in ecotourism initiatives and the achievement of economic advantages.

However, the perception of the actors interviewed regarding the role of the local community is different for the two biosphere reserves. While the local community in the IBRM-Morocco fears the control over the resources of the reserve, its role in the development of ecotourism is limited, "the sustainability and success of the projects depends on the awareness and involvement of the local communities and international cooperation rather than exclusive action by the State" (interview AI, a 67-year-old male professor expert and associative actor). The involvement of the local community is the only subject of discourse by managers of the IBRM without real participation in decision-making around development actions and the promotion of ecotourism. One of the informants, MD, a 45-year-old male from the National Agency for Water and Forests, declared:

> *"I would say that there is sometimes personal involvement with initiatives of NGO and cooperatives...some people organize hikes, visits, excursions, also the provision of tourist accommodation...there is a certain collaboration with administrations in particular National Agency for Water and Forests...however, I think we can do better. There are more opportunities to explore and promote in the field of ecotourism".*

Moreover, in the IBRM, there are some exceptions to the commitment of the local community. The President of ATED association, TA, a 63-year-old male, discusses a substantial shift in the local community engagement in ecotourism since the 2000s. Residents are actively participating through heightened awareness and demonstration efforts, which is evident in the increased requests for lodging projects. Furthermore, local community involvement exhibits positive evolution, with a growing interest in promotional activities and collaborative programs with both local and international partners, as reported by JM, a 44-year-old male, Head of Partnerships and Activities at ANEF.

On the other hand, the Shouf Biosphere Reserve is a case in point, where local actors have collaborated and supported each other to achieve economic growth and maintain cultural and environmental authenticity. Stakeholders' testimonies from the Shouf Biosphere Reserve (SBR) unveil a synergistic collaboration between ecological preservation and community engagement, shaping the framework for sustainable ecotourism. The owner of houses in Maasir Shouf, GF, a 34-year-old male, explains how local communities play an active role in showcasing agriculture. Local guesthouses in the Shouf Biosphere collaborate with farmers, establishing a network that mutually benefits from the existence of the biosphere reserve. This collaborative model has developed over the years.

The engagement of local actors in both the IIBRM and Shouf Biosphere Reserve signifies a gradual shift toward community-centered ecotourism. Testimonies from various stakeholders emphasize the pivotal role of local communities in shaping these initiatives, reflecting an increasing awareness of the interconnectedness between ecological preservation and community well-being. The long-term success and sustainability of ecotourism in these biospheres hinge upon the support of this collaborative approach.

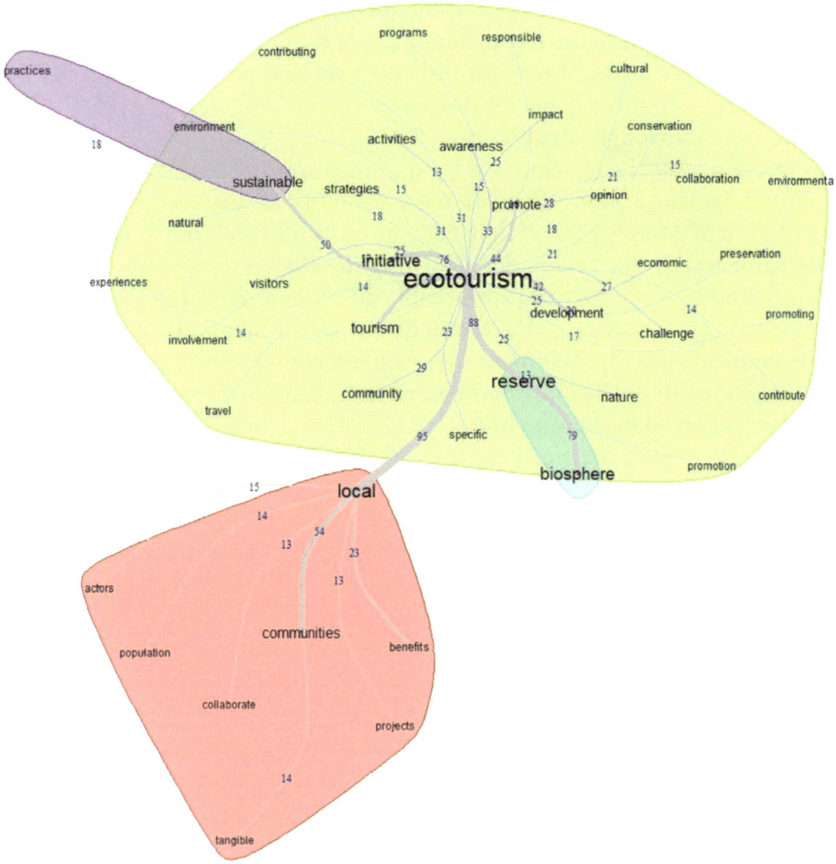

Figure 8.
Similarity analysis of local communities in BRs. Source: Field survey, interviews with IBRM and SBR stakeholders, El Mejjad F, 2021–2023.

Figure 8 illustrates the frequency of co-occurrence of specific terms, with thicker lines indicating a higher frequency of co-occurrence. Notably, the terms "Local" and "communities" exhibit a strong connection with the term ecotourism, occurring 95 times. Also, the relationship between "Ecotourism" and "biosphere" is comparatively strong, with 88 occurrences. This visual representation serves to elucidate the relationships between these terms, emphasizing "Local community" as a central term and delineating varying strengths in connections with other terms in the analysis of actors' discourses.

5.5 Ecotourism in IBRM and Shouf reserve: Strengths and weaknesses of the strategies adopted

Based on the analysis and interpretation of stakeholders' testimonies in both biospheres, a comparative approach was conducted to evaluate the situations in these areas. The analysis brought to light the strengths and weaknesses of the strategies employed. According to the perspectives of Moroccan stakeholders, the Intercontinental Biosphere Reserve of the Mediterranean (IBRM) boasts notable strengths, including its proximity to the European continent [33] and the active involvement of international non-governmental organizations. This involvement has significantly contributed to the implementation of various development projects in recent years, such as the establishment and enhancement of lodges, improvement of infrastructure, initiation of businesses and cooperatives, and the exploration of potential resources through ecotourism.

However, it is essential to acknowledge that the promotion of the IBRM is considerably limited, lacking actions to enhance the visibility of the territory through the implementation of a territorial mark in signage or initiatives. Such measures could prove instrumental in conserving and elevating the shared natural and cultural resources, fostering economic and sustainable development for the local inhabitants.

In contrast to the situation in the Moroccan Intercontinental Biosphere Reserve of the Mediterranean (IBRM), Lebanese stakeholders are actively engaged in the promotion and development of the biosphere reserve in Lebanon. Participants emphasized the crucial roles played by a multitude of local, national, and international stakeholders, acting as supporters, partners, contractors, beneficiaries, and providers of technical and financial assistance in the development of the Shouf Biosphere Reserve.

Mondino and Beery's [10] examination of ecotourism within the Monviso Transboundary Biosphere Reserve in Italy reveals a significant divergence in stakeholders' collaboration levels when compared with the IIBRM Biosphere Reserve in Morocco. Their findings indicate a constructive trend of stakeholder collaboration in the Monviso Transboundary Biosphere Reserve, akin to observations in Lebanon. In contrast, within the IBRM Biosphere Reserve, the study underscores a deficiency in collaborative endeavors among stakeholders.

While Lebanon demonstrates a proactive approach to environmental education, the situations observed in Morocco and Italy underscore the pressing need for enhanced educational initiatives. These efforts are crucial for bridging knowledge gaps and fostering meaningful stakeholder engagement in ecotourism endeavors geared toward sustainable development. Azzam [27] goes further and highlights the need for policy interventions to support decision-makers in managing biosphere reserves in Shouf, Lebanon. The study underscores the necessity for an integrated management model incorporating spatial planning principles to address challenges faced by protected areas like the SBR. Azzam [27] argues that the integration of spatial planning perspectives could improve biosphere reserve management and governance as well as the planning modalities essential for effective management of biosphere reserves like the Shouf BR in Lebanon.

The collaborative efforts extend to the entire community, encompassing women, men, youth, students, local authorities, local partners, agricultural cooperatives, and women groups, as articulated by NH, 46-year-old male General Manager: "Our program involves everyone". Despite the challenges posed by the political events causing destruction in Lebanon and a significant reduction in the number of visitors to the SBR, the reserve has achieved a positive impact on socioeconomic and conservation fronts.

The substantial role of women within the Shouf Biosphere Reserve was highlighted by participants, emphasizing their involvement in cultural practices and the agricultural sector. The Ecotourism Manager, OA, 39-year-old, acknowledged their diverse roles, including farming, beekeeping, guesthouse ownership, handicraft production, traditional food manufacturing, leadership and membership in agricultural cooperatives and associations, as well as engagement in various professions. The president of the association, Assayida El Hourra, HJ, a 49-year-old male, underscored the crucial role of women in promoting cultural development, preserving traditions, and contributing to the natural product landscape in both the Shouf Reserve and IBRM. Stakeholders in

Shouf Reserve expressed a commitment to friendly management practices integrated into the existing community on the reserve, further emphasizing practices and policies that promote the full participation of local communities and empower women.

The analysis reveals that local actors in Lebanon embody a harmonious blend of elements contributing to sustainable development within the biosphere reserve territory. This combination notably involves economic development achieved through the promotion of local and specialized products, goods, and services, coupled with the sustainable management of natural resources and biodiversity. The Shouf Biosphere Reserve, in particular, benefits from the active involvement of numerous associations and organizations, consistently engaging in activities that stimulate social and productive actions to elevate the level of socioeconomic development.

Significantly, the Shouf reserve management team comprises exclusively staff members who are residents of the Shouf region. This intentional composition aims at fostering a deeper connection with the local community and, consequently, contributing to the overall development of this biosphere.

5.6 Future prospects and opportunities

The exploration of future prospects and opportunities for the two biosphere reserves, SBR and IBRM, reveals their distinctive potential, encompassing rich historical, cultural, and spiritual heritage coupled with exceptional geographical locations. Both initiatives share the overarching goal of environmental preservation and the promotion of biodiversity value.

In the context of IBRM, stakeholders identify ongoing programs, notably the second phase of the collaboration with GIZ, aimed at enhancing ecotourism activities and supporting cooperatives committed to the ecological development of the region. These endeavors leverage the natural and cultural potential of the area, emphasizing sustainability in the ecotourism industry.

IBRM's future plans involve a strategic focus on local collaborations, engagement with the Spanish network, development of tourist circuits, promotion of local products, and the pursuit of labels to increase visibility. This multifaceted approach aims not only to attract tourists but also to ensure sustainable and locally beneficial practices within the ecotourism sector.

The IBRM region, with its significant potential for ecotourism, manifests a commitment to ongoing programs and strategic initiatives. However, existing challenges necessitate proactive efforts to strengthen collaboration among stakeholders. The outlined perspectives and strategies propose an approach emphasizing local cooperation and international networking for the successful development of ecotourism in IBRM.

Shouf Biosphere Reserve envisions a future for ecotourism centered on encouraging visitors to actively participate in the preservation of trees and animals. JK, a 48-year-old female tour guide, emphasized the desire for people to engage in this form of tourism for its perceived health benefits and ecological importance. The aim is for Shouf Biosphere Reserve to become a pioneering biosphere reserve in the Mediterranean, implementing a balance between nature conservation, social well-being, and economic activities in the surrounding area.

Stakeholders at Shouf reserve note a gradual increase in foreign visitors, particularly during the spring period. Efforts are being made to maintain the reserve's condition and benefit to local residents. SN, a 35-year-old female Communication Coordinator at Shouf biosphere reserve, emphasizes the successful future of ecotourism, attributing it to a growing interest in experiences beyond simple visits.

Public sector stakeholders and local NGOs in both Shouf reserve and IBRM are actively contributing to the socioeconomic development of local populations. They promote economic sectors, employ new communication strategies, and support local cooperatives, emphasizing activities like beekeeping and the valorization of aromatic and medicinal plants. Sensitization efforts in Shouf Reserve play a crucial role in encouraging public involvement in the planning and decision-making process. Similar contributions are observed in IBRM, highlighting a shared commitment to socioeconomic development through strategic initiatives.

6. Conclusion

The objective of this study was to investigate stakeholders' perspectives on ecotourism in biosphere reserves in Morocco and Lebanon, with a specific focus on Shouf Biosphere Reserve in Lebanon and Intercontinental Biosphere Reserve of the Mediterranean in Morocco (IBRM). The study aimed to unveil stakeholders' understanding of environmental issues and the development of ecotourism activities, along with the implementation of ecotourism to foster economic development in the respective territories. Additionally, it sought to identify the roles played by stakeholders in the two biospheres, examining both strengths and weaknesses.

Data analysis revealed that IBRM holds significant ecotourism potential owing to its rich landscape and traditional use of natural resources. Challenges such as limited accommodations, weak coordination among stakeholders, and a need for increased public awareness were identified. Despite these challenges, opportunities exist for improvement through collaboration with international non-governmental organizations and the implementation of specific projects to enhance tourist attractions.

IBRM faces difficulties due to its vast territory and limited accommodations, with threats posed by weak coordination of ecotouristic initiatives and monoculture. However, proximity to the European continent and the involvement of international organizations present opportunities that should be capitalized upon.

In contrast, Shouf Biosphere Reserve in Lebanon exhibits a clearer institutional structure and better coordination of ecotouristic initiatives. The utilization of media for promoting public awareness and strong stakeholder engagement contributes positively to sustainable agriculture and the creation of new circuits. Additionally, Shouf reserve has the potential to develop new services for visitors.

In sum, both biosphere reserves face challenges, yet they possess inherent strengths and opportunities for improvement. While both reserves encounter management challenges, issues related to public awareness, and stakeholder engagement, the SBR exhibits more effective management practices and higher levels of public awareness compared to the IBRM. Collaborative endeavors, educational outreach programs, and diversification of sustainable agricultural practices emerge as potential strategies to address these challenges and enhance the sustainability of both reserves.

Recognizing these aspects and learning from stakeholders' perspectives can guide future policies, strategies, and initiatives to further enhance the development and promotion of ecotourism in these regions. Despite limitations related to the number of informants, this study serves as a foundational exploration of ecotourism development in Morocco and Lebanon, urging ongoing efforts to address identified weaknesses and maximize existing strengths. Further research should delve into the roles of investors, NGOs, and associations in shaping and impacting ecotourism within biosphere reserves.

Other researchers [26, 28, 29] provide invaluable insights and strategies that could enrich the development of ecotourism within biosphere reserves, particularly in the contexts of IBRM and SBR. A comprehensive approach to ecotourism development in these reserves is essential, encompassing awareness campaigns, educational initiatives, habitat conservation efforts, and the enhancement of local products and services. Effective promotion of these reserves at both national and international levels necessitates the implementation of efficient and robust communication strategies and the establishment of partnerships with universities and communication experts. By adopting these measures, biosphere reserves can not only attract environmentally conscious tourists but also serve as pivotal hubs for environmental education and research, thereby significantly contributing to the sustainable development of these invaluable ecosystems.

Furthermore, it is imperative to underscore the critical role of environmental education and stakeholder coordination in the ongoing management and promotion of biosphere reserves. Biosphere reserves can play a crucial role in advancing sustainability and stewardship by integrating environmental education into school curricula and fostering collaboration among government agencies, local communities, and non-profit organizations. This collaborative endeavor not only enhances the visitor experience but also ensures the long-term preservation of biodiversity and ecosystem services within these protected areas.

The present study represents a modest addition to the scholarly discourse concerning biosphere reserves and ecotourism in the South Mediterranean region, with particular emphasis on Morocco and Lebanon. Further research endeavors are warranted to explore additional facets pertinent to this subject matter. Of particular significance is the examination of local communities' perspectives, given their direct involvement in ecotourism initiatives. Additionally, a comparative analysis of stakeholders' viewpoints in the North Mediterranean region, encompassing both the Northern segment of the IBRM and its Southern counterpart, holds considerable relevance.

Investigating topics such as biosphere reserve management, collaboration among various stakeholders, communication strategies, and public perceptions of biosphere reserves and ecotourism in these two countries would provide valuable insights. These areas merit comprehensive examination to better understand the complexities surrounding biosphere reserve governance and to identify opportunities for enhancing collaboration and communication within the ecotourism sector.

Acknowledgements

The authors would like to thank the administrative team of Shouf Biosphere Reserve, participants, and all the institutions involved in the management of the IBRM for providing the needed information and documentation that was of great help for the elaboration of the study. Also, the authors would like to express their appreciation for the EduBioMed project and their support in strengthening, improving, and enhancing academic activity related to biosphere reserves.

Additional information

This chapter includes some results of the research undertaken within the framework of the ERASMUS-EDUBIOMED project.

Author details

Faiza El Mejjad, Yamina El Kirat El Allame* and Laoucine Amzil
Mohammed V University in Rabat, Morocco

*Address all correspondence to: yelkirat@gmail.com

References

[1] Opoku A. Biodiversity and the built environment: Implications for the sustainable development goals (SDGs). Resources, Conservation and Recycling. 2019;**141**:1-7. DOI: 10.1016/j.resconrec.2018.10.011

[2] Zhang L, Xu M, Chen H, Li Y, Chen S. Globalization, Green economy and environmental challenges: State of the art review for practical implications. Frontiers in environmental Science. 2022;**10**:4. Available from: https://www.frontiersin.org/articles/10.3389/fenvs.2022.870271

[3] Holden E, Linnerud K, Banister D. The imperatives of sustainable development. Sustainable Development. 2017;**25**(3):213-226. DOI: 10.1002/sd.1647

[4] Ferreira AF, Zimmermann H, Santos R, von Wehrden H. Biosphere reserves' management effectiveness—A systematic literature review and a research agenda. Sustainability. 2020;**12**(14):Article 14. DOI: 10.3390/su12145497

[5] Common future. Medicine and War. 1988;**4**:17-25. DOI: 10.1080/07488008808408783

[6] Omoloso O, Mortimer K, Wise WR, Jraisat L. Sustainability research in the leather industry: A critical review of progress and opportunities for future research. Journal of Cleaner Production. 2021;**285**:125441. DOI: 10.1016/j.jclepro.2020.125441

[7] Lelloltery H, Puttileihalat MMS, Sitanala M, Seipalla BB, Kunda RM. The role of stakeholders in ecotourism development in Marsegu Island nature Tourism park (MINTP) in Western Seram regency, Eastern Indonesia. IOP Conference Series: Earth and Environmental Science. 2021;**800**(1):012056. DOI: 10.1088/1755-1315/800/1/012056

[8] Abou A, Hady S. Ecotourism and sustainability: Practices of the Lebanese nature-based operators. Menadzment u Hotelijerstvu i Turizmu. 2019;**7**(1):11-23. DOI: 10.5937/menhottur1901011A

[9] Wiesli TX, Hammer T, Knaus F. Improving quality of life for residents of biosphere reserves and nature parks: Management recommendations from Switzerland. Sustainability: Science, Practice and Policy. 2022;**18**(1):601-615. DOI: 10.1080/15487733.2022.2100128

[10] Mondino E, Beery T. Ecotourism as a learning tool for sustainable development. The case of Monviso transboundary biosphere reserve, Italy. Journal of Ecotourism. 2019;**18**(2):107-121. DOI: 10.1080/14724049.2018.1462371

[11] Ceballos-Lascurain H. The future of 'ecotourism'. Mexico Journal. 1987:13-14

[12] Higham J. Ecotourism: Pondering the paradoxes. Critical Issues in Ecotourism. 2007. Available from: https://www.academia.edu/16936757/Ecotourism_pondering_the_paradoxes

[13] Tardif J. Écotourisme et développement durable. VertigO - la revue électronique en sciences de l'environnement. 2003;**4**(1). DOI: 10.4000/vertigo.4575

[14] United Nations Environment Programme and World Trade Organization (WTO). Making Tourism more Sustainable: A Guide for Policy Makers. Knowledge Repository - UNEP. UNEP; 2005. Available from:

https://wedocs.unep.org/20.500.11822/8741 [Accesed: May 14, 2024]

[15] Fennell DA. A content analysis of ecotourism definitions. Current Issues in Tourism. 2001;**4**(5):403-421. DOI: 10.1080/13683500108667896

[16] Stojanović V, Demonja D, Mijatov M, Dunjić J, Tišma S. Concept of ecotourism development in unesco biosphere reserves: case studies from croatia and serbia koncepcija razvoja ekoturizma u unesco rezervatima biosfere: studije slučaja iz hrvatske i srbije. Šumarski list. 2022;**5-6**:243-253. DOI: 10.31298/sl.146.5-6.6

[17] Amos B, Bridgewater P, Green MJB, Phillips A. Biosphere Reserves and the IUCN System of Protected Area Management Categories. Canberra: Australian Nature Conservation Agency; IUCN; UNESCO, Man and the Biosphere Programme (MAB). 1996. Available from: https://portals.iucn.org/library/sites/library/files/documents/1996-081.pdf

[18] United Nations Educational, Scientific and Cultural Organization (UNESCO). Designation and Review Process. 2023a. Available from: https://www.en.unesco.org/biosphere/designation

[19] Botha E. Stakeholder perceptions of sustainability and possible behaviour in a biosphere reserve. Sustainable Development. 2023;**31**(5):3843-3856. DOI: 10.1002/sd.2629

[20] United Nations Educational, Scientific and Cultural Organization (UNESCO). Biosphere Reserves, About. 2023b. Available from: https://www.en.unesco.org/biosphere/about

[21] Lima action plan for UNESCO's man and the biosphere (MAB) programme and its world network of biosphere reserves (2016-2025). In: Conference: International Coordinating Council of the Programme on Man and the Biosphere, 28th, Lima. 2016 [27]. Document code: SC-16/CONF.228/11. Collation: 13 pages. Language: English

[22] UNESCO. Biosphere Reserves: The Seville Strategy and the Statutory Framework of the World Network. Paris; 1996

[23] United Nations Educational, Scientific and Cultural Organization (UNESCO). Management Manual for UNESCO Biosphere Reserves in Africa: Field Version. 2015. Available from: https://www.unesco.de/sites/default/files/2018-01/Manual_br_field_.pdf

[24] Barukchieva EI. The relationship between tourism and the biosphere reserve status: The Danube Delta—If the Danube is "the sustainable highway" of Europe, then the Danube Delta should be the sustainable gate to the Black Sea. International Journal of UNESCO Biosphere Reserves. 2017;**1**(2):21-55. DOI: 10.25316/IR-7

[25] Gu XP, Lewis BJ, Niu LJ, Yu DP, Zhou L, Zhou WM, et al. Segmentation by domestic visitor motivation: Changbai mountain biosphere reserve, China. Journal of Mountain Science. 2018;**15**(8):1711-1727. DOI: 10.1007/s11629-017-4708-1

[26] Selkani I. Management of protected areas and ecotourism, what relationship? Austin Journal of Business Administration and Management. 2020;**4**(1):1045

[27] Azzam Norma Marwan. Evaluating the Management of Biosphere Reserves in Lebanon: The Shouf Biosphere Reserve (SBR) as Case Study. Master's Thesis, American University of Beirut,

Maroun Semaan Faculty of Engineering and Architecture, Department of Architecture and Design, April 2022

[28] Sethy MK, Senapati AK. Perceptions towards ecotourism practice and the willingness to pay: Evidence from Chilika coastal wetland ecosystem, Odisha. International Journal of Geoheritage and Parks. 2023;**11**(3):497-513. DOI: 10.1016/j.ijgeop.2023.08.001

[29] Esparza R, Caicedo Y, Flores C, Romo Román P, Mougenot B. Perceptions of stakeholders and challenges faced by ecotourism management in a natural protected area in Peru. Environment, Development and Sustainability. 2023. DOI: 10.1007/s10668-023-03501-9

[30] Abdul Malak D et al. The intercontinental biosphere reserve of the Mediterranean. In: Adapting to Climate Change, SpringerBriefs in Environmental Science. Cham: Springer; 2017. DOI: 10.1007/978-3-319-51680-6_2

[31] Boushaba K. Relations sociétés-nature et stratégies intégrées de conservation et du développement: Cas de la Réserve de Biosphère Intercontinentale de la Méditerranée [Thèse de doctorat]. Faculté de l'Aménagement, Université de Montréal; 2017

[32] Moore HM et al. Environmental challenges in the Rif mountains, northern Morocco. Environmental Conservation. 1998;**25**(4):354-365

[33] Barbosa A et al. Cost-effective restoration and conservation planning in green and blue infrastructure designs. A case study on the intercontinental biosphere reserve of the Mediterranean: Andalusia (Spain) – Morocco. Science of The Total Environment. 2019;**652**:1463-1473. DOI: 10.1016/j.scitotenv.2018.10.416

Chapter 10

Perspective Chapter: Rethinking Tourism for the Benefit of All in a Postnormal between Space

Michael McAllum

Abstract

There is considerable evidence that many of the systemic features and issues of the present global condition—including how we consider both globalization and sustainability—are problematic in a postnormal world that is complex, chaotic and contradictory. The numerous uncertainties generated by the postnormal suggests that humanity and our social institutions are now decoupled from a dominant, Westernized narrative that finds it difficult to navigate an indeterminate between space. Nowhere is this failure of shared narrative and institutional disfunction more evident than in the global tourism complex. It seduces us into accepting the unsustainable premises of low cost and high volume that underpin its offerings, whilst ensuring that the vast majority of benefit for the same goes to technological platforms and those that package and transport. The consequence is a world of 'overtourism', which continues to destroy habitats, marginalize communities and minimize benefit to local workforces, all the while shifting infrastructure costs to local governments. This chapter will explore how tourism-dependent societies can create sustainable benefit for both habitat and the environment, whilst at the same time deliver better visitor experiences. It will assert the case for both systemic enquiry and navigation-centric practice in a future of pluriversal world making.

Keywords: overtourism, unsustainable models, postnormal, hyperobjectvity, radical uncertainty, pluriversalism, post globalisation, centrality of narratives, way finding, warm data

1. Introduction

To many, what we think of as 'globalisation' and even 'sustainability' seems abstract. As ideas they are difficult to define, complex in nature and seemingly disconnected from our everyday lives. Further, more precise definitions of these particular human social arrangements are complicated by the situation in which they are being discussed, and more often than not, influenced by the world views of those implicated and impacted by the conversation at hand. This level of abstraction and multidimensionality of understanding makes it difficult to coherently ringfence the ecological, economic, societal and social issues that they raise, unless they are more specifically linked to particular manifestations of the effects and issues in question.

In this sense, both the abstract (the models and narratives being expounded) and the particular (the confrontation of particular instances) need addressing together. It is the intent of this chapter to try and do both, whilst recognising that, because we, as humans, have created the essentially westernised conditions, as well as the issues around both, we cannot (in a sense) stand apart from them, because we are completely immersed in the systems and system effects they have created. In other words, we need to try and resolve what are potentially existential issues for humanity through dealing with what at best can only be described as a partial understanding of particular instances.

Based on these premises, and without pretending that what follows is in any way a complete or even partial solution, this chapter proposes, as an instance of the particular, to explore issues of globalisation and sustainability (or the lack thereof) as they relate to the global tourism industry complex. It is defined and complex because it has many facets and it operates at many levels with inconsistencies and differences that create layers of contradiction within the complexity. Through several case studies, this chapter suggests potential options that might go some way to addressing issues of globalisation and sustainability as they relate to this modernised technological packaging of tourism; or what used to be called 'visitation'. It begins by exploring a number of critical conceptual foundations whilst unpicking a number of implicit assumptions embedded in common, though rarely examined, definitions. It then briefly canvases the sources for a number of critical arguments. These arguments include assertions that there is a misunderstanding and misreading of the context in which we must address tourism; that many of the problems that exist in what is now called 'overtourism' [1] are because economic, technological and social arrangements have decoupled from each other [2]; that issues in the tourism complex cannot be resolved if we simply maintain expectations, orientations and systems that created the problems in the first place; and finally, that in conditions of ever-increasing uncertainty, new models, offerings and narratives require both development and rapid deployment, in the process creating new meanings about what we call 'globalisation', all the while slowly changing the trajectory of unsustainable activity, upon which we have become dependent.

Based on these arguments, it is suggested that tourism—as is true for many other parts of the human condition—needs a new narrative of transformational systemic change. This should inform a rapid migration from current 'universal and western-centric models' that concentrate benefit in the hands of the few, to 'pluriversal' offerings (the making of many worlds) [3] that deliver enduring sustainable benefit for both people and environments. This will necessitate an abandonment of the culture of end-focused and growth-obsessed goal setting, which privileges universalism and a scarcely undifferentiated sameness that destroys flora and fauna, cultures and community dignity. This chapter asserts that, in its place, a new 'learning by navigation' narrative is needed; one that blends both local offerings and the entire global tourism sector into a sustainable network within which each part, through the quality of its relationship with the others, looks to add value and sustainable benefit, rather than competing—or even outcompeting—with each other.

2. Foundational intelligence

In arguing a case for change, it is important to set out the reasons as to what it is that needs resolving at what might be termed a 'second order change level'. 'First order change' is that which can be accommodated into an existing system. 'Second order' is that which requires change to the system itself. This rationale might be called 'foundational intelligence'. After exploring in more detail the nature of what has been described as the tourism complex, and why change is required in that sphere (among others), there are three other important concepts that require canvassing. Firstly, it is asserted that the context and conditions that created global tourism—and on which its current business model depend—no longer exist. In their place is what has come to be known as the 'postnormal' [4]. Secondly, it is asserted that what we understand by 'contemporary globalisation' is a narrative that is unravelling in the face of meta shocks, and as a set of human arrangements and institutions, is proving incapable of solving the multiple problems it is creating. This includes significant issues of sustainability that cannot be resolved through local action alone. Thirdly, all of this is set in a situational chaos that we are deeply immersed within. In reality, we can never stand aside from it. This inability to be truly objective (one of the foundations of both science and policy development) is what Timothy Morton has described as 'Hyperobjectivity' [5]. Hyperobjectivity completely changes the manner in which we think about how change might, and could, occur, and is perhaps the most important piece of intelligence we can possess. It requires us to think about things in a systemic—rather than simply a systematic—way and in so doing, challenges problem solving as we have understood it. The value of Hyperobjectivity (I argue) is that its foundational enquiry makes visible the implicit assumptions and meanings that words or concepts discreetly carry; meanings that can potentially get in the way of both understanding the scale at which enquiry might occur, and/or the potential resolution of critical issues.

2.1 The global tourism complex

In many senses, the global tourism system and its co-dependent twin, the global aviation sector, is an evident poster child of contemporary globalisation practice and the broader sustainability challenges that threaten humanity. What makes confronting the many issues it contributes to is that both suppliers and customers are deeply implicated and invested in its continuation. Now, on a post-Covid recovery path, the World Travel and Tourism Council estimates its value as north of US$10 trillion; and that it accounts for around 10% of all jobs globally [6]. More importantly, its dominant narrative is that it will continue to grow—and indeed, it needs to. There is foundational to a narrative that produces an unhealth dependence and over reliance for many economies, including small islands, resource poor communities and emerging middle income nations. This dependence creates a fragility that relies on the current model. Egypt, for instance, has plans to triple the size of the tourism sector, which would inflate tourism's share of the nation's GDP to (by some estimates) more than 60% [7]. In the Maldives, tourism is already 82% of GDP, with much of the labour being supplied by migrants, who make up around one third of the Maldives population in the process insulating them and isolating them from the wider communities who have little contact with the tourist islands. Because these migrants are seen as outsiders or others they become vulnerable to unacceptable levels of exploitation in communities indifferent to their plight [8]. Further complicating the issues around tourism are the emergence of Russian and Chinese diaspora communities that thrive in conclaves, delivering little benefit to the local workforce, economies and communities in question [9]. However, the sheer scale and impact of the sector cannot be ignored, and any change to the system and models will be problematic.

2.2 Fragility and meta shocks

What the recent COVID crisis made visible was the high level of dependence that the tourism complex has on the present model of globalisation (i.e., the free movement of goods, service and people without hinderance). There are some commentators, though, who maintain that, just as with other phases of globalisation, this phase is also coming to an end. If this is the case then the arteries that facilitate tourism flows may not be available to facilitate the growth that is expected.

Changes in the global flow of services through changes in technology could also have an impact especially for those known as global nomads. A recent Cato Institute argues that globalisation is morphing away from goods to services [10], but that begs the question of how and when AI will reframe the services model. The Future of Life Institute, for instance, frames the problem as follows:

The trajectory of AI development thus points to the emergence of asymptotically comprehensive, superintelligent-level AI services that— crucially—can include the service of developing new services, both narrow and broad, guided by concrete human goals and informed by strong models of human (dis)approval. The concept of comprehensive AI services (CAIS) provides a model of flexible, general intelligence in which agents are a class of service-providing products, rather than a natural or necessary engine of progress in themselves [11].

If this is the case and AI becomes a counter to the low-cost labour model, then the whole system of goods and services as we now understand it will be reframed. When this fast-moving AI disruption is added to the unexpected consequences of severe disruptive events on a planet approaching, or probably past, 1.5 degrees of climate shift, then it is reasonable to assume that both globalisation as we understand it, and sustainability as we practice it, are fragile propositions upon which to build future success for anything and that includes tourism.

2.3 Unsustainable models

Notwithstanding the potential meta shocks outlined above, most of the present activity is based on a low-cost, high-volume model facilitated by technological platforms and other intermediary devices (bank fees and usurious exchange rates, IP charges from global hotel chains) which have three important consequences. The first is that the reach of these platforms is completely pervasive in a globalised world reliant on connected devices. This 'platforming of access to facilities and experiences' enables the vast majority of the profits to be channelled to source countries, leaving the bare minimum in destinations to supply the services offered. By some estimates, more than 30% of every dollar (and sometimes more) spent by onshore tourists in many destinations is repatriated offshore. In a recent World Trade Organisation Symposium it was reported that, through what is known as leakage, the "average leakage for most developing countries today is between 40 and 50 percent of gross tourism earnings for small economies and between 10 and 20 percent for most advanced and diversified developing countries" [12]. The second is that this same process lends itself to the packaging of experiences in ways that are disconnected from each other. For instance, there is no way for tourists who visit Cairo's Museum of Antiquities to know that one of the world's oldest potteries is a mere 400 metres away, or that some of the world's oldest bars are right across the road. In other words, the technology privileges the singular packaging of experience rather than the development of an 'ecology of experiences' that are both less structured and likely to bring increased benefit to a more diverse community. The third consequence is that the cost/volume model is driving what is known as 'overtourism', which is undermining both the quality of experiences and, more importantly, many of the habitats upon which particular offerings depend.

2.4 The emergence of the postnormal

As COVID graphically illustrated, the global tourism complex is reliant upon a relative stability in source countries and destinations, and in its infrastructure. Any kind of disruption, and the uncertainty it creates, shifts the market's attention to seemingly more stable offerings. The critical theorist Zia Sardar argues that, in fact, the conditions of certainty are becoming rare in the face of global conditions that are increasingly complex, chaotic and contradictory [13]. He asserts that these might be described as an ever-present postnormal to which humanity must now become accustomed. He is not alone in this view. Recently, the eminent economists King and Kay wrote that we now face a time of radical uncertainty, where we cannot rely on the models and templates that were constructed in more certain times, and that the only way forward is to develop and build new shared narratives [14]. Similar views also emerge in the works of Notwotny [15] and Gowing [16]. As canvassed earlier, it is possible that even the present era of globalisation is at the beginning of its end and that other disruptions similar to COVID might occur. If that was to come to pass, then the levels of uncertainty would increase astronomically.

2.5 What is globalisation?

Given that the concept of globalisation is one of the central themes of this book, it is perhaps useful to spend a little time considering what it is that we actually mean by the term. For some it has economic connotations and is generally used to describe flows of goods and services based on a neoclassical approach to economy, with competition based on low-cost labour at its heart. The consequence has been an insidious form of economic neocolonialism, where the majority of benefit flows to wealthier, so-called developed, nations. Facilitating this vast, globalised economic expansion has been a dramatic reduction in communication costs through, firstly, a digital revolution in the 1990s, and more latterly (circa 2007 onwards), through both a mobile technology and network revolution. It is worth noting that both of these might well be upended by the emergent AI revolution, which will essentially destroy the advantage of low-cost labour in many communities that have benefitted from the current phase of globalisation. All of this has culminated in the rise of mega cities—tourist destinations in their own right— that are essentially cosmopolitan complexes with enormous ecological and energy footprints that function not only as centres for the flow of goods, but also for both information and people. As their power and influence increases, they act to preserve their own interests, and in so doing, undermine the nation state of the prime unit of organisation. There are in addition many other kinds of globalisation—of pandemics and diseases; of knowledge of cultures; and of niche groups, perversions, criminal activity and opinion—each being amplified in its own technological echo chamber. If there is a common thread in these many manifestations of globalisation, it is that conceptually and in actuality, narratives and institutions that have supported previously accepted definitions of success and globalisation are unravelling quickly in front of us.

2.6 Beyond a sustainability of convenience

The concept of the Anthropocene argues that humans have modified the planet at an ecological scale in the same way the planet was previously modified by extreme weather, asteroids or other factors [17]. It is what might be called a globalised sustainability crisis, in which the inhabitants of this unguided spaceship are destroying the very things that keep it functioning for the benefit of its human inhabitants [18]. Many eminent climate change and ecological theorists would also assert that humanity is in postnormal conditions as a consequence, because the life systems upon which humans in our present socio-economic condition depend have either crossed, or are about to cross, thresholds that will make them something else entirely [19]. Of concern in the science and among scientists are the unknown (but mostly negative effects) of interacting feedback loops and the variability in ecological time spans that govern these systems. All of this confuses and confounds what we mean by 'sustainability'. Firstly, one of the initial definitions of sustainability, articulated by the Bruntland Commission (*meeting the needs of the present without compromizing the ability of future generations to meet their own needs* [20]) is unavailable, because even if the world was to meet its 2050 CO2 emission reduction targets (which currently seems unlikely), life systems have already changed to an extent that it is not possible to make available the same types and levels of resources that the current generation have enjoyed. Secondly, particularly in the case of both aviation and tourism more generally, it is not possible to promote both a cost/volume growth model *and* to be sustainable at the same time. In other words, the journey towards sustainability will require the tourism complex to be completely reinvented complete with levels of change that few who are invested in its continuation care to contemplate.

2.7 Hyperobjectivity

The idea of hyperobjectivity is, in some ways, linked to the concept of complexity at the core of the postnormal proposition. It draws from, and is part of, a wider philosophical argument that suggests that the basis for considering the relationship between things, and people and things, should be to see all as objects in relation to the other [21]. As many recent philosophers have suggested [22], this Object Oriented Ontology, or OOO as it is sometimes known, marks a radical departure from the Cartesian and Kantian self/other premises upon which the western knowledge system is based. Hyperobjectivity suggests that many systems that affect both human and life-giving arrangements are so large and so complex that we can never truly stand outside of them, and thus we cannot see them objectively. The nature of the cosmos, the interaction of life systems on earth, the global socio-political relationships of nations, and the global economic system are examples of these complex hyperobjects. One might argue that many megacities also could be considered that way, as could the global tourism complex. Just as we experienced with COVID, in hyperobject situations we can only deal with what appear to be local manifestations of issues that arise for humans; we really have no idea what the result of our actions or solutions might be.

A recent scientific discovery provides a useful illustration as to how problem solving changes if and when we can stand outside of any system. In the late 1970s two satellites—Voyager I and II—were launched by NASA. However, it wasn't until 2012 that Voyager, just before it crossed into what we now call interstellar space, sent signals back of unusual invisible disturbances. These data helped to prove the theory of the heliosphere, or solar waves protecting the solar system. Until that point, no human controlled device had ever been outside of the solar system [23].

The sustainability journey in the tourism complex is still very much at the edges of sustainability consciousness. Indeed, it seems to be almost unthinkable and unspeakable except where egregious localised instance of poor practice can no longer be ignored. Dave Keating, in a recent opinion piece [24], noted that there were no tourism specific speakers at the recent COP 28 conference in Dubai, nor were the any at the previous COP 27 event. This is despite global tourism being an estimated 10% of the global economy and the knowledge that a staggering 80% of tourists visit only 10% of destinations (overtourism). He argues that this is in part because tourism issues and impacts are cross sectoral in nature and therefore not obvious and easy to tackle. In that sense he could be arguing about sustainability more generally.

If we consider the tourism complex with all its interrelationships to be a hyperobject the challenge both for policy makers and institutional strategists become obvious. Firstly, few have models or approaches that can cope with the fluidity and ambiguity that hyperobjectivity necessitate, nor do they have conversational frameworks that even consider the conversation that hyperobjective situations require. These for the most part sit outside most institutional experience. They require what Helga Novotny described as the embrace of the cunning of uncertainty; "a wholesome counterforce to the false uncertainties induced by hubris and over-reliance on the assumptions that people think they know" [25]. The recognition of this situation is a necessary precondition to the emergence of a sustainable tourism model.

3. Understanding the kinds of knowledge we need and the nature of the change that is required

Where the core tenets of foundational intelligence collide together in the tourism complex space we can find a number of propositions. These include the following.

1. *Moving beyond disciplinary thinking*. The resolution of the issues that are immediately evident cannot be resolved through the application of any one knowledge base or disciplinary approach. This requires a bias towards transdisciplinary practice and the use of what both Gregory and Nora Bateson describe as warm data. This is

   > *The interrelational processes between and among systems. It can involve recognizing how patterns repeat and reflect each other among multiple contexts and across multiple systems – recognizing that many of these systems' maintenance and renewal is critical in the coming decades. The underlying premise is to address and experiment with how we perceive [26].*

2. *Creating new narratives in a time of radical uncertainty.* It has been argued that the Tourism complex, as with many other systems, cannot sustain itself in its current format unless it wishes, in the next few decades, to be completely overwhelmed by the unsustainable outcomes of its prosecution. This is because, for the most part, the type of "successful" change being advocated for and pursued in the current mechanistic model only concerns itself with actions that require measurable adjustment and adaptation to systems that already exist. The current plans for transition to renewable fuels in aviation are a useful example of this [27]. This incrementalism is insufficient for the level of renewal that is required.

3. *Developing a navigation rather than a destination mindset in the Between Space.* There is now an emerging understanding that many of the systems upon which humanity has relied are insufficient for the challenges at hand. Yet, while recognising a cynical and broad trend that appeals to base instincts and nostalgia [28] and despite the many advocacies of potential transition such as those of James Martin [29] or Lester Brown [30], to name but two, most societies have yet to agree on 'where to next'. This inability to see the 'other shore' is disconcerting to those that believe in the abilities of humans to solve everything through technology, and tourism is not immune from that effect. However, in this between space which may exist for a short, or for a very long time, all societies and sectors must learn to navigate what seems to be emerging and perhaps learn something about what the future might be like in the process. Thus, the very definition of 'change' and 'change theory' is reframed. It moves from considering change as a process of adapting to broadly agreed external forces, to one where the focus is on strengthening the capacity and ability of any entity or network to navigate. It requires constantly adjusting and coexisting (perhaps by navigating away from) alongside the manifestations of uncertainty, asymmetry and volatility that now define our existence.

3.1 Pluriversal rather than universal thinking

Perhaps the most difficult proposition to accept is that the hyperextension of the western model has created the situation that humanity now faces. By hyperextension, I and others [31] mean that universally we have over-privileged individualism at the expense of all other kinds of relational forms; put society in the service of economy rather than the other way around; created unnecessary dualisms that establish false dichotomies; and, as Morton argues, wrongly sustained a belief that rationalism and technology can solve everything. The prosecution of this model has made mostly invisible other ways of thinking, and on occasion, colonised and adopted it (think yoga as practiced in the western model) in ways that make it almost unrecognisable. This is not to argue that everything western is somehow wrong, but rather that it should play *a* role—not *the* role—in the contest of ideas as we contemplate the place of unknowing; what the Greeks described as abyss, in front of us. This will allow for what Arturo Escobar [3] calls pluriversal thinking: the making of many worlds rather than just clones of one world. In a slight recasting of Raworth's *Doughnut Economics* [32], and as indicated in **Figure 1**, this many-world making must foster the rebuilding of ecologies on the planet and ensure that people and communities receive an equitable share.

3.2 An ecology of minds

Given this vast interconnectedness and the levels of dependence many people and societies have upon it, the trajectory of tourism requires a new ecology of minds. In Batesonian terms this means:

Changes of thought that will impact upon our government, economic structure, educational philosophy and military stance because the old premises are deeply built into all these sides of our society. [While] nobody can predict what new patterns will emerge from these drastic changes, we hope that the period of change may be characterised by wisdom rather than by either violence or the fear of violence [33].

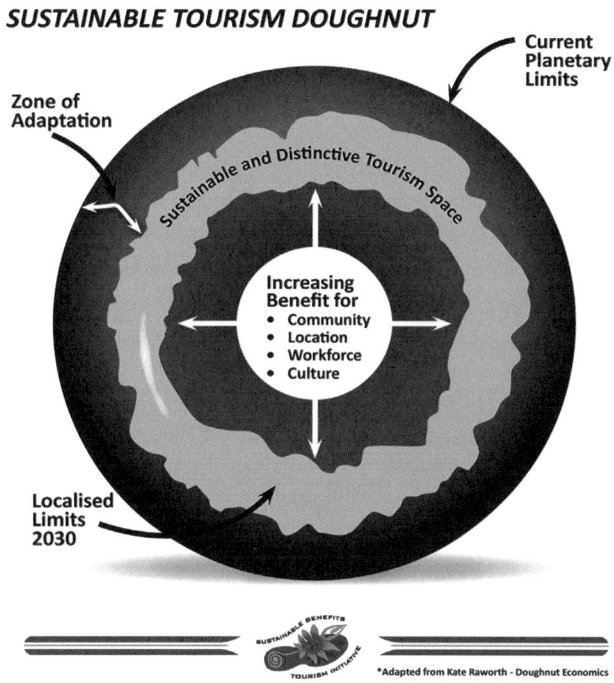

Figure 1.
The sustainable tourism doughnut (source: K Raworth, doughnut economics).

4. Navigational possibilities and potential trajectories for the tourism complex

In this 'between space' there are no certainties about the way forward. Navigating will require a clarity of intent, a preparedness to learn and a high level of knowledge humility. Core to the process will be a recognition that overtourism is in no-one's interest, and therefore any viable future requires different ways or models of tourism that ensure that the benefits of tourism flow both equitably and sustainably to all those that participate, and all those things that are involved. However, none of this can occur unless there is a new kind of resilience built into the tourism complex. This is necessary to facilitate the management of increasing uncertainties (as was the case with COVID) and other kinds of volatility (as is the case with both the Ukraine invasion and the current Palestinian conflicts). This will require the development of new pluriversal narratives that move away from a world of cloned offerings; that inform the design of ecologies of experiences and build new kinds of systems that marginalise the dark underbelly of the global tourist system.

Before exploring what kind of system transformation interventions might be necessary, it is perhaps useful to summarise the many differences between the present system and what a more sustainable system might look like. As **Table 1** shows, the difference is stark and one that can be distinguished from the kinds of sustainability practices that can be contemplated inside the current system with in any meaningful way changing is unsustainable effects. This kind of quasi sustainability practice includes lower emissions travel and tourism practices, reusable and recycling systems and perhaps more support for localism and more sustainable food practises. While

	Overtourism model	Sustainable benefit tourism
Business model	Low cost/high volume/extractive	Fair cost/value constrained/pluriversal distributed/beneficial world making
Dominant world view	Neo classical economic growth dependency	Doughnut Economics (Raworth)
Information technology facilitation	Global platforms dominant choice and access	Global—local AI networks showcase experiences with different benefit flows
Travel technology facilitation	Cost & volume in air travel comes from lower cents /passenger KM. In cruising cost is created through scale with very high ecological footprints	More richly curated experiences that come together to create more isolated visits in concentrated locations.
Policy orientation	Increased dependency on tourism and foreign exchange at expense of more diverse economies	Tourism is additive to, integrated and facilitative of other local economic community and cultural experiences
Tourism activity effects	Measurable and detrimental reduction of urban, rural and costal spaces. Tourism induced (and more recently long term stay induced) real estate speculation	Restorative practice funded through tourism of urban, rural and coastal spaces. Constrained tourism footprints.
Long term impacts	Reduction of local quality of life, community cohesion (which increases tourismphobia) and loss of cultural distinctiveness	Primacy of local distinctiveness shapes tourism choices Pro community orientation (recoupling) Cultural re-invigoration.

Table 1.
Characteristics of current globalised tourism complex and a potential post normal sustainable system.

this are worthy and perhaps useful they are not sufficient to change the trajectory of the system.

While concerns about excessive visitation are not new there seems to be a reluctance to address root causes and rethink in a systemic way. As economies increase rather than decrease their reliance on tourism, with all the benefits and issues that this brings the belief has been that "all problems can be solved by exerting greater effort and demanding greater efficiency within the status quo of continued growth and consumption" [34]. Scant regard is paid its monocultural nature, the destruction of other forms of more integrated economic activity, its dislocation and isolation from local communities and the fact that the reality of any growth focused system will always be the bottom line. The consequence is that, despite a few stand out cases where sustainable tourism and tourism policy has been shown to a better and more enriching kind of tourism, management and policy making with respect to the complex is "unable to reduce tourism related tensions without a profound shift in the underlying structure of the global tourism economy" [35].

The necessity to explore systemic change is not just a good thing to do; it is as vital to tourism as the shift from fossil fuels is to the global energy system.

4.1 Recoupling the tourism complex with ecology and humanity

It was asserted earlier that one of the reasons why systems fail to work is that they are decoupled. The alternative is to develop tourism offerings and models that recouple ecology, society, culture and economy. There are two important design considerations here. The first is to treat the non-human as having as much right to benefit (or perhaps even more) as humans. In this instance, for example, what would a tourism experience look like from the point of view of a world heritage forest? This is not as fanciful as it sounds. Some argue that, indeed, forests can think, and this opens up the way for an anthropology that goes beyond the human [36]. Interestingly, in New Zealand there is by law already a river and a mountainous area that have standing in the courts as persons in their own right [37]. The second is to make visible where current benefits are flowing to, as the basis for redesigning those flows. Again, there are small case studies everywhere that point to this possibility; visitor fees for restored historical sites, levies on sports people such as fly fishermen and an increasing intolerance of accommodation practices and cruise ships that promote tourism at the expense of community (e.g. Air BNB in Barcelona) and their habitats (as in Venice).

4.2 Beyond decoupling

Decoupling is at the core of the tourism system that increasingly offers cloned offerings that are increasingly isolated from the communities in which they situate themselves. It privileges the tourism experience over everything else. In places like Bali, Phuket in Thailand and the Maldives this manifests itself in an astonishing number of almost identical hotel resorts where many 'vacationers' relax without ever leaving the premises. In what was a small fishing town in Cambodia, Shianoukville is now completely overwhelmed by casino-based offerings owned by and run for Chinese gamblers [38]. The US$7.7 billion global cruise ship industry (with its estimated growth of over 11% per annum) represents another extension of the model; one that perhaps takes decoupling to the extreme. It creates a kind of 'Disneyland effect' where, if there *is* interaction with communities—'the strange other'—then it is at a distance and on a schedule dictated by the ship. The fact that the industry expects some 58% of first-time vacationers to use this offering in the next few years should be a cause for alarm [39].

4.3 Redesigning benefit flows

The challenge is to counter this trend with the design of experiences which are more attractive whilst also perhaps developing stronger policy approaches that change the benefit flows. In Phuket, for instance, there is a refuge for mistreated elephants where the interest of the elephants is at the centre of the offering, and the interests of the tourists are designed around it. This allows for experiences such as safely walking with giants, where those involved see and learn how elephants in their habitat actually forage, and on occasion, the invasion of the bar by one or two of the inhabitants who enjoy the company of humans [40]. The second challenge is to design situations and experiences where the unplanned, the accidental and the unexpected enhance the experience. This design ethos is presently being built into what is known as gastronomy, or food, tourism where participants not only experience great food but also mingle with locals in the process. Popular motorbike tours of street food in Vietnam and the development of locally-oriented cooking classes, are great examples. Current research shows that authentic experiences rank highly with Millennials, who are now emerging as the largest cohort of tourists [41]. Given that they are still very young, authenticity should now be fundamental to design.

As **Figure 2** demonstrates, without such information metrics, it is difficult for the system to make decisions about itself and its future.

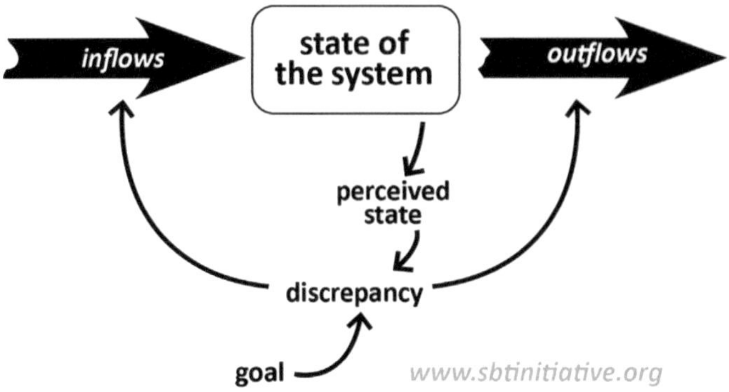

Figure 2.
Information metrics (source: www.sbtinitiative.org).

4.4 New place-spaces, new stories

While there is a need for meta narratives in the global tourist complex, there is also a need for more localised narratives. Nowhere is this more evident than in the competitive market of large city tourism; a set of places, spaces and activities that are increasingly vital to the economies of many. The challenge for them all is that previous key differentiators were the nature (and price) of the retail experience, the food experience and perhaps the night life. Now, in a world of online shopping, globalised sourcing and ubiquitous branding, with perhaps the exception of nightlife, none of these are key differentiators. While some are building identity around their role as a key gateway with all the overtourism challenges that presents, few are clear about the image they want to project, and thus, the story they want told about them. These narrative images are likely to be a set of specialised offerings, each being a fragment of the future image.

4.5 Curation of difference

Again, using food as an example of curated specialisation, in the present system there are a few indicators of possibility that point to a new kind of city tourist design. While the island state of Singapore is curating its image as an epicurean melting pot, it is doing so through creating and highlighting interesting urban spaces that both reflect vitality and incorporate the culture of the people living there [42]. In Singapore's case, it is the love of food that is central to Singaporean culture and community. Encouraging tourists to experience its diversity in place and space creates interactive spaces wherein both the tourist and the community are present and share common interests. The conception of a marketplace of experiences is a powerful antidote to the packaged approach. It brings together spatial perceptions that are psychologically, sociologically and aesthetically functional, providing both a foundation for identity and a new narrative [42].

However, positive perceptions of this kind are difficult to sustain in those tourist nations where control by the state, oligarchs and/or criminals influences interaction and activity at a micro level.

4.6 Resilience requires the system to understand how it takes decisions about itself

In order to design, project and promote this new kind of tourism system, it will also be necessary in the transition to develop intelligence, strategies and new kinds of networks in order to cope with increasing levels of uncertainty and volatility, particularly at local levels. Typically, sector information is historical and is mostly about arrivals departures, spend and economic contribution. As **Figure 3** shows, significantly more nuanced information flows are required.

4.7 Paradigm shifts

The above discussion on potential trajectories are summarised in **Table 2**, They underpin the argument that sustainability cannot be achieved in the current business model and the system that it supports. From this, three important principles emerge: firstly, that the focus and conversation must be systemic not systematic (the latter is where the focus on measurables often takes us); secondly, that potential trajectories need to coexist and synergise at international, national and local levels; and thirdly,

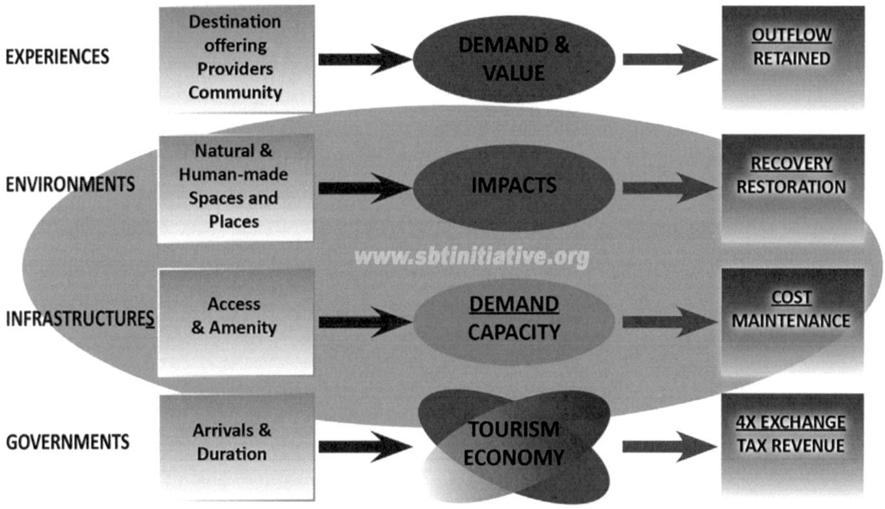

Figure 3.
Tourism stocks and flows (source: www.sbtintiative.org).

Frames for framing and shaping Potential System transformation
Create shared meta, national and localised narratives about the role of tourism and place -spaces
Redesigning the tourism complex to rebalance and equally benefit the ecology and humanity
Decouple tourism from monocultural (or cloned practice), over reliance on its economy and its global intermediaries
Recouple tourism to better integrate it into the lived experiences of its peoples and communities and design it for restorative purposes
Change the benefit flows throughout the system to better reflect how and where value is delivered
Create conversations around shared space-maps of what a system looks like and where and how one might intervene.

Table 2.
Summary of system acupuncture points that can change the nature of the tourism complex system.

that what we might change, and what we need to change, is something we need to collectively learn about. The intent should be to create an everchanging portfolio of learning options and an increasing number of viable, actionable and scale-able options that will facilitate the change we think needs to be made.

5. Social system visuospatial space maps

To enable a shared view of what is occurring and what is required, the conversations among all stakeholders is greatly enhanced if they have a shared map of the nature of the system they are looking at. Nowhere is this more important than in those circumstances where the imperative for transformation and renewal is significant. Remembering that the map is never the territory, these visuospatial maps, even if they are not completely accurate, act as a shared reference point for exploring what might exist in any given space, and where possibility might lie. These representations provide a common, or shared, understanding of 'what is going on here.' Similarly, as **Figure 4** shows, they can also be used in any tourism complex dialogues alluded to above. In this diagrammatic example, taken from the work of the Chôra Foundation [43], in a number of post-COVID tourism system interventions—both tourist-lived experiences and the available resources, both human made and natural—were used to create the horizontal axis. A vertical axis was then designed around how any human system takes decisions about itself and how it looks to intervene in that system. Contextual influences or system shapers were then mapped and wrapped around the system itself. In sum, the graph provides understanding of who is (or should be) involved, what is involved, how change might occur and what outside influences affect any system.

Having established a representation of the system space, enquiry then needs to turn to what is *inside* that space. These might be called 'areas of interest', 'learning spaces' or even warm data locations. They are the places in a system where the necessary level of transformation can or should occur. Some of the potential warm data

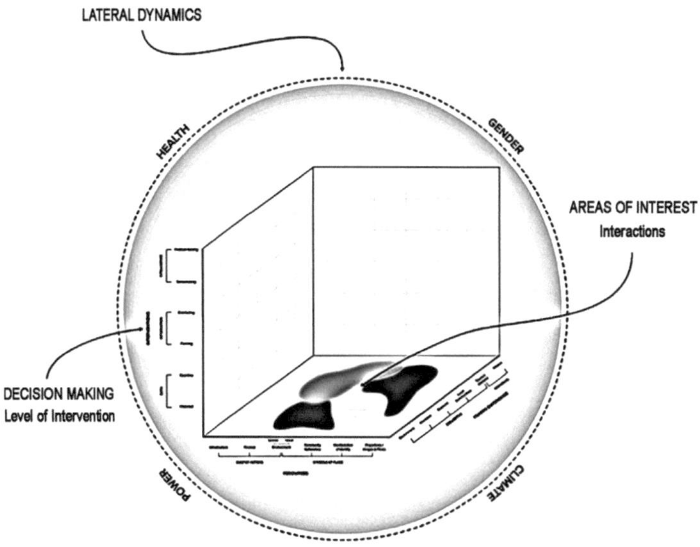

Figure 4.
Illustration of a tourism stencil or space map (source: Chôra foundation. www.chôra.org). Note: the figure presents an imagination of the tourism space that can be contextualised with location-specific data.

locations have been canvassed earlier in this chapter. Many of the potential learning spaces have also been described earlier in this chapter. Detailed visualisations can then be developed for every component of the space in ways that can be location-specific, and that will go on to inform a portfolio of learning options.

6. Final thoughts

Seeing a way to navigate towards a truly sustainable tourism future without compromising the ability of future generations to enjoy the level of experiences that have been available to the present generation seems both bewildering and problematic, given the multiplicity of interconnected ecological, socio-economic and cultural issues that now confront us all.

This chapter has asserted on a number of occasions the systemic nature and the difficulties of transition for policy makers, suppliers and indeed the tourists themselves. The 2023 Bookings.Com Sustainability Report [44] is illustrative. It highlights both the debate about transformation and the initiatives being taken to achieve that transformation. This is in line with that organisations stated goals to get to net zero emissions by 2040. Whether that includes the activities of customers using the platform is not clear. The report found that over 75% of its customers were conscious about the need to take positive action to save the planet now and that, as a consequence, they wanted to travel more sustainably. They also noted that over 50% believed that sustainability choices were difficult to find and that they were also unacceptably more expensive when available. These findings highlight the global dilemma and perhaps gives new meaning to the 'between space.' On the one hand, most are now acutely aware that time has run out for remedial action and that adaptive strategies are now the only option. But, and it is a big but, few are prepared to accept the costs and make short term changes to their lifestyles to facilitate that adaptation and are not prepared to support at any level of government the policies and framings that are needed. The consequence is a world in stalemate, where the costs of inaction continue to rise inexorably in both short-term and long-term ways while continued, and arguably cynical, investment in the existing model makes it increasingly likely that the worst-case sustainability scenarios come to pass. In that sense, the global tourism complex is both what we have become and what we are becoming. It mirrors what we are.

Yet in the unexpected discontinuity of COVID, the global tourism sector has stared into the abyss of a post-globalised world that undermines its business model. It is still recovering and fervently hoping that nothing will derail a business-as-usual recovery. The mantra of its various peak bodies simply amplifies that view. At the same time, it is also acutely aware of the mounting evidence that its practises rarely align with sustainable maintenance or restoration. Still, it chooses to pursue future paths that deny both, instead of usefully instigating effort and time to rethink, resign and reposition itself. While these may seem like harsh words, there is little evidence to the contrary. If the tides of change are truly upon us and humanity is now in an in-between space, then this is a dangerous assumption upon which to build a future, and this might be regarded as an absence of stewardship and leadership for the many people and things that depend on its continuance.

If indeed the tourism complex is by definition a hyperobject the challenge is considerable because the knowledge and approaches required are for the most part foreign to those invested in the model. Those invested in potential futures might perhaps adopt a curiosity and knowledge humility that comes with that when travelling to places where what is known provides almost no compass. In such circumstances it is hard to set aside the accumulation of supposed wisdom in favour of relentlessly pursuing the difficult conversations that must be had.

It will be difficult, given the entrenched and vested interests of many stakeholders, to agree on any kind of future goal, or even a shared narrative about what it might become. Perhaps the alternative may be found in navigating a way through and past the many issues that confront the sector. This wayfinding begins with bringing alternatives into focus through considered questions and conversations that make visible all kinds of knowledge in a transdisciplinary fashion. It will be greatly aided if there are visuospatial maps of what it is we think we are talking about. It continues with the redesign of benefit flows, places, spaces and experiences to create an authenticity that is often absent in current cloned models, which tend to create separateness rather than togetherness. This authenticity should also embrace the principle that the natural resources that are used as attractors are not there simply to be used and abused. Rather, they need nurturing, and where necessary, be provided restorative care. This 'reanimation' is a relational reorientation that is fundamental to the sustainability dialogue.

Over time this wayfinding provides the fragments of narratives of difference; possibilities both tourists and providers can draw upon as the sector reaches limits that cannot be sustained anymore in any shape or form. These narratives should show those whose entire lives are dependent on sustainable tourism a sense that things can continue, albeit in a slightly different way.

The journey to rethink, redesign and remake the tourist experience is simply one component of the wider transition that humanity must undertake because of the fundamental shifts to life systems that have been caused by neglect, disinterest and cynical exploitation. It mirrors what we have become and what we are becoming. It is helpful to keep it in focus, because it is so visible; if we care to look.

Author details

Michael McAllum[1,2,3]

1 University of the Sunshine Coast Buderim, QLD, Australia

2 Chôra Foundation, Haarlem, Netherlands

3 Global Foresight Network, VIC, Australia

*Address all correspondence to: michael.mcallum@globalforesight.net

References

[1] Mahalic T. Conceptualizing overtourism: A sustainability approach. Annals of Tourism Research. Elsevier. 2020;**84**(Sept):103025

[2] Polanyi K. The Great Transformation: The Political and Economic Origins of our Time. 2nd Beacon Paperback ed. Vol. xli. Boston, MA: Beacon Press; 2001. 317 p

[3] Escobar A. Designs for the Pluriverse: Radical Interdependence, Autonomy, and the Making of Worlds. Durham: Duke University Press; 2018

[4] Sardar Z. On the nature of time in Postnormal times. Journal of Future Studies. 2021;**25**(4):17-30

[5] Morton T. Hyperobjects: Philosophy and Ecology after the End of the World. Minneapolis: University of Minnesota Press; 2013

[6] World Travel and Tourism Council Economic Research. Available from: https://wttc.org/research/economic-impact

[7] Egypt's National Strategy for Sustainable Tourism 2030. Available from: https://beta.sis.gov.eg/en/media-center/strategies/egypts-national-strategy-for-sustainable-tourism-2030/#:~:text=The%20strategy%20aims%20to%20increase,vision%20for%20sustainable%20development%202030.

[8] Use of Migrant Labour in the Maldives. Available from: https://www.ilo.org/wcmsp5/groups/public/---dgreports/---inst/documents/publication/wcms_776391.pdf

[9] Russian Only Business in Phuket. Available from: https://www.aljazeera.com/economy/2023/9/6/russian-only-businesses-in-thailands-phuket-spark-backlash

[10] Cato Institute. Available from: https://www.cato.org/commentary/its-end-globalization-we-know-it-thats-probably-fine-part-1

[11] Future of Life Institute. Available from: https://www.fhi.ox.ac.uk/reframing/

[12] UNCTAD Paper to a World Trade Organization Symposium. Available from: https://www.google.com/url?sa=t&rct=j&q=&esrc=s&source=web&cd=&ved=2ahUKEwibq87EyN6DAxXj3TgGHVTuAxwQFnoECBEQAQ&url=https%3A%2F%2Fwww.wto.org%2Fenglish%2Ftratop_e%2Fserv_e%2Fdiaz.doc&usg=AOvVaw08mPQmpyb_jcWjXIszi7Tu&opi=89978449

[13] Sardar Z. Welcome to postnormal times. Futures. 2010;**42**(5):435-444

[14] King MA, Kay JA. Radical Uncertainty: Decision-Making for an Unknowable Future. London: Little, Brown Book Group; 2020

[15] Nowotny H. The Cunning of Uncertainty. Cambridge UK: Polity Press; 2015

[16] Gowing NLC. Thinking the Unthinkable. A New Imperative for Leadership in the Digital Age. London: Chartered Institute of Management Accountants; 2015

[17] Steffen WC, McNeill PJ. The Anthropocene: Are humans now overwhelming the great forces of nature? Ambio. 2007;**36**(8):614-621

[18] Fuller RB. Operating Manual for Spaceship Earth. Carbondale: Southern Illinois University Press; 1969

[19] Steffen W, Rockstrom J, Richardson K, Lenton TM, Folke C, Liverman D, et al. Trajectories of the earth system in the Anthropocene. Proceedings of the National Academy of Sciences of the United States of America. 2018;**115**(33):8252-8259

[20] Brundtland Commission Definition of Sustainability. Available from: https://www.un.org/en/academic-impact/sustainability

[21] Harman G. Object-Oriented Ontology: A New Theory of Everything. London: Pelican Books; 2018. p. 295

[22] Ricœur P. The Course of Recognition. Cambridge, Mass: Harvard University Press; 2005

[23] Discovery of the Heliosphere. Available from: https://manyworlds.space/2022/06/07/our-heliosphere-home/#:~:text=Despite%20the%20fact%20that%20it,he%20termed%20the%20solar%20wind

[24] Keating D. COP 28: Is the UN Climate Process Ignoring the Tourism Sector? Energy Monitor. Available from: https://www.energymonitor.ai/sectors/transport/cop28-is-the-un-climate-process-ignoring-the-tourism-sector/?cf-view

[25] Nowotny H. The Cunning of Uncertainty. UK: Polity press; 2015. Loc 210

[26] Source: The Bateson Institute. Available from: https://batesoninstitute.org/warm-data/

[27] Status of Renewable Fuels in Aviation. Available from: https://www.iea.org/energy-system/transport/aviation

[28] Latour B. Down to Earth: Politics in the New Climatic Regime. Cambridge, UK: Polity Press; 2018

[29] Martin J. The Meaning of the 21st Century: A Vital Blueprint for Ensuring our Future. New York: Riverhead Books; 2006

[30] Brown LR. Plan B 4.0: Mobilizing to Save Civilization. New York: W.W. Norton; 2009

[31] Gebser J. The Ever-Present Origin. Athens, Ohio: Ohio University Press; 1984

[32] Raworth K. Doughnut Economics: Seven Ways to Think like a 21st-Century Economist. London: Random House Business; 2017

[33] Bateson G. Steps to an Ecology of Mind. University of Chicago Press ed. Vol. xxxii. Chicago: University of Chicago Press; 2000. 533 p

[34] Hall C. Constructing sustainable tourism development: The 2030 agenda and the managerial ecology of sustainable tourism. Journal of Sustainable Tourism. 2019;**27**(7):1044-1060

[35] Bianchi R. In: Milano C, Novelli M, Cheer J, editors. Travel and Tourism in an Age of Overtourism. Oxon: Routledge; 2021

[36] Kohn E. How Forests Think: Toward an Anthropology beyond the Human. Vol. xiii. Berkeley: University of California Press; 2013. 267 p

[37] River Granted the Same Rights as a Human Being. Available from: https://www.theguardian.com/world/2017/mar/16/new-zealand-river-granted-same-legal-rights-as-human-being

[38] The Guardian. News, Sport and Opinion from the Guardian's Global Edition | The Guardian [Internet]. The Guardian; 2015. Available from: https://www.theguardian.com/

[39] Grandview Research Cruise Market Report. Available from: https://www.grandviewresearch.com/industry-analysis/cruise-market-report#:~:text=The%20global%20cruise%20market%20was,USD%2018.30%20billion%20by%202030

[40] Laws, Scott, Font & Kowdolski, editor. Interview with John Roberts as Chapter 1 Managing the Anatara Golden Triangle Elephant Camp in The Elephant tourism Business. CABI; 2021

[41] Kim S, Lee CK, Jung T. Millennials' vacations: Types and preferences. In: Millennials: An In-depth Look into the Travel Behavior of Millennials. ResearchGate [Internet]. 2017 [cited 2024 Sep 25]. Available from: https://www.researchgate.net/figure/Millennials-Vacations-Types-Source-Millennials-An-In-depth-Look-into-the-Travel_fig1_328275976/download

[42] Ronan O'Connell. World Nomads. Available from: https://www.worldnomads.com/explore/southeast-asia/singapore/asias-top-food-destination

[43] Rozman Cafuta M. Framing the tourist spatial identity of a city as a tourist product. International Journal of Tourism Cities. 2022. Available from: https://www.emerald.com/insight/content/doi/10.1108/IJTC-06-2022-0151/full/html

[44] Booking.com. Sustainable Travel Report. 2023. Available from: https://globalnews.booking.com/download/31767dc7-3d6a-4108-9900-ab5d11e0a808/booking.com-sustainable-travel-report2023.pdf

Chapter 11

Demystifying Technology as the Basis of Education for Sustainability

Damir Purković

Abstract

The chapter discusses the meaning and current understanding of societal sustainability and resilience, particularly in relation to the development of technologies that can or should enable the sustainability of any community and society. It presents the problems associated with the sustainability of small communities and societies, that is, the challenges they face in line with global trends and efforts that fundamentally undermine their resilience and sustainability. In doing so, it highlights education for sustainability and sustainable development as one of the most important prerequisites for the sustainability of any society and critically analyses the trends to which it is exposed. Arguments are then presented for the need to demystify technology for young generations as a crucial segment for the development of their critical thinking, which is largely neglected in such education. Such demystification is necessary for the younger generations to understand the causes of disruption, but also the needs of society and the economy at a local and global level. This is the only way to ensure their critical approach to solving the problems of the present and the future and to motivate them to take transformative action aimed at solving the problem of sustainability and sustainable development.

Keywords: demystification of technology, education for sustainable development, sustainability, resilience, technology education

1. Introduction

Today's world is changing rapidly, and as the population grows, so does the demand for essential resources such as energy, food, and water, while the amount of waste we release into the environment increases every day. All of this has led to a serious violation of planetary boundaries [1, 2], which is best shown by the climate disruption we have faced in recent decades. Despite these and numerous other indicators pointing to the catastrophic impact of human activities on the environment, compromising the sustainability of life on Earth, the human needs for more, better, stronger, more lucrative, and more profitable, continue to override all these indicators and common sense. In order to reduce the pressure on natural resources and the environment in the coming decades, the prosperity and development of any modern society should be based on "intelligent" management of its own potential and resources. This should ultimately ensure the sustainability and sustainable development of society, but also its resilience to the challenges it faces.

Therefore, sustainability as the ability to do something continuously over a long period of time [3] remains a primary goal of any society, and the only question is how to achieve it. As sustainability is not a static state, sustainable development is about maintaining productivity by replacing depleted resources with resources of equal or greater value without compromising or jeopardizing natural biotic systems [4]. One could say that sustainable development is development that meets the needs of the present generation in a way that does not jeopardize the ability of future generations to meet their needs. Although this definition of sustainable development is inadequate because needs can be interpreted in different ways, it expresses the concept in a simple way. Such development also presupposes the resilience of society, which is an important component of sustainability. Today, society's resilience is threatened by numerous challenges from natural disruptions such as earthquakes to disruptions caused by human activity such as climate disruption, economic and financial disruption, to threats from recent pandemics or current wars. Resilience is the ability of a system, community, or society exposed to hazards to withstand, absorb, adapt, transform, and recover from the effects of hazards in a timely and effective manner, including maintaining and restoring its basic structures and functioning through risk management [5]. Basically, it is about socio-ecological resilience, that is. the ability to adapt or transform under conditions in which we are confronted with unexpected changes in socio-ecological systems to ensure continuous support for human well-being [6, 7]. Sustainability, that is. sustainable development, today generally encompasses three basic dimensions: the ecological (environmental), economic, and social dimensions [8], to which the institutional (political), cultural, and technical dimensions are increasingly being added [9, 10]. Although it is clear from this broad understanding of sustainability that sustainability and sustainable development are a cultural, traditional, linguistic, spiritual, political, and national (patriotic) phenomenon, it is often overlooked that it is still primarily an energy-ecological, economic, and labor-social issue [10]. At the same time, technology is an indispensable segment that should be given more attention. Indeed, if society does not ensure stability in these crucial economic, technological, and social issues, any other stability, including cultural sustainability and sustainable development, is highly questionable and without fundamental prerequisites for development. For this reason, the resilience of a society is measured by its capabilities, which include the ability to predict, absorb, adapt, and transform [11]. Predictive capacity refers to the ability of the system to predict and mitigate the effects of shocks through preparedness and planning, and absorptive capacity refers to the ability to absorb and manage the effects of shocks and stress [11]. Adaptation is the ability of human systems to change and learn in response to multiple, long-term, and future risks and to adapt after the shock has occurred, while transformability is the ability to take deliberate steps to change systems that cause risk, vulnerability, and/or inequity [11]. From the above characteristics and the definition of sustainability and resilience, it is clear that human capabilities are an extremely important pillar of society that should be given special attention. For although the global community, as well as local communities and states, are trying to counter the aforementioned disruptions with various regulations, it is already clear today that the problems and challenges of sustainability will have to be solved by future generations. These are problems that we are not yet able to articulate clearly and with technologies that we do not yet know. Thus, in addition to natural resources and infrastructure, the human component of sustainability and sustainable development also plays a major role as the concept of fairness encompasses several manifestations concerning the sustainable development of human society [12]. This also means that significantly more attention should be paid to learning and educating future generations for sustainability and sustainable development than is the case today, and that the focus should be on actions that lead to a real improvement in socio-ecological problems and are carried out consciously and critically [10]. At the same time, fostering critical thinking, creativity, and ingenuity among younger generations and raising awareness among the wider population about the challenges of sustainability, resilience, and sustainable development [13] remains one of today's priorities for tomorrow's world. The success of future generations in solving problems will certainly depend on what kind of generations we educate and raise today. Therefore, it is very important whether we produce generations of new unscrupulous and greedy people who will accelerate the further decline of the Earth and lead humanity to ruin, or whether we produce generations of conscious and empathetic critical thinkers and creatives who will find solutions for the benefit of humanity. For this reason, the main problem that this paper seeks to solve is the question of what the Education for Sustainable Development (ESD) of the future should look like and what role technology and technology education should play in it.

1.1 Challenges and problems of sustainability and resilience of small communities and societies

As stated in the introductory chapter, there is no alternative to sustainability and resilience if we are to balance the natural environment and allow future generations to develop unhindered. To some extent, this is confirmed by the recent IPCC report on assessing the impacts of climate change on ecosystems, biodiversity, and human communities at global and regional scales [14]. This report identifies the vulnerabilities and limits of nature's and society's ability to adapt to climate change and notes that current trends do not point to progress, but also warns that the prospects for resilience to climate change are very limited unless work to reduce greenhouse gas emissions begins immediately [14]. However, the possibilities for reducing greenhouse gases require radical and very rapid changes, in technological, economic and social terms. This is particularly difficult to achieve in populous and poor countries and can be very devastating for small countries and their economies. The reason for this is often insufficient capacities, mechanisms and technologies for rapid and dynamic change. Although the Paris agreement provides for developed countries to provide financial, technical, and capacity-building support to countries that need it [15], the question is whether this will be enough to reduce greenhouse gas emissions on a global scale and stop global warming. At the same time, it is questionable to what extent National Determined Contributions (NDCs) can be considered relevant indicators of sustainability and resilience. The large and economically powerful countries and their economies, which are often also the biggest polluters, are already finding it very difficult to reach a consensus on the most important issues of sustainable development as this undermines their economic sustainability in many aspects. Nevertheless, there is a fundamental agreement between large and powerful states, communities, and economies, which has so far led to numerous agreements, as well as directives and regulations to stop further global warming and normalize the natural environment in due course. It is obvious that such measures favor the industrialized countries, which use their influence (superiority) to harmonize key documents according to their own economic resources and technological development. These countries also have sufficient financial and human resources to drive the necessary change. Small countries, whose contribution to the effects of climate change is often negligible, must also adapt to such regulations and measures as otherwise, they will lack the promised financial support for the transformation. Due to modest financial resources, limited human capacity, weak influence at the global level and, above all, technological inferiority (because technology ultimately enables change), small countries are often forced to buy technology, but also goods that meet the standards set. In this way, much of the financial support received from the industrialised countries flows back to these countries, while the small and poor countries still do not achieve the expected results because they try to adhere to something that is not adapted to them. All this hinders their own development and leads to economic decline, technological backwardness, social discontent, and political instability, which ultimately encourages the migration of their populations to countries with better living conditions. In such a context, it is difficult to expect the strengthening of sustainability, resilience, and sustainable development of small countries and peoples, which could undermine their social and cultural survival in the near future. It can be concluded that many societies and communities today are not resilient and their sustainability is severely compromised because they are unable to respond adequately to the many challenges and needs that lead society to sustainability, resilience, and sustainable development. One of these challenges is meeting the needs of the labor market for competent personnel with different profiles and levels of education, including a large number of people from the fields of engineering and technology. The other capacities of small countries and communities required to predict, absorb, adapt and transform into a sustainable and resilient society are often inadequate, to say the least. This can be seen in energy dependence, dependence on imports of all kinds of products (including food), dependence on imports of (often inadequate or low-skilled) human resources, but also financial dependence on rich countries. Hence, the need to intensify learning and education for sustainable development, especially in small and poor countries, but it must be significantly different from what is implemented today. Since the resilience and sustainability of society, that is. the sustainable and environmentally sound production of energy, food, and other goods, is the basis for the resilience and sustainability of the economy and society and thus the prerequisite for the cultural and demographic development of any country, it is undeniable that technology, engineering, and accompanying sciences should be more strongly represented in education for sustainable development. This is particularly important for small communities and countries whose path to sustainability certainly cannot follow the examples of large and powerful countries and cannot afford to be technologically inferior [16] because only technological ingenuity and creativity can lead them to sustainability.

2. Education for sustainable development (ESD) and technology

The accelerated degradation of the natural environment, population growth, and the associated increase in needs, as well as scientific progress, has led to the recommendation to integrate the values of sustainability and sustainable development into all aspects of learning [17–20]. This is understandable as learning is an integral part of almost all conceptualizations of sustainable development, either implicitly or explicitly [2]. Sustainable development is, therefore, not only a product of learning but also a learning process, that is. a way of thinking and acting that is open to learning and change [2, 21–23]. However, despite the efforts of the global community and the fact that ESD is now an integral part of youth education, it has not mitigated the negative effects and has had a weak effect in promoting the value of sustainable development at the global community level. It happens although the Education for Sustainable Development (ESD) approach is attempting to transform from an idealistic to an Education for Sustainable Citizenship, which is a more realistic approach to change concerning sustainability and sustainable development [24]. In other words, insufficient guidelines, false idealism, and censorship of critical ideas and content [10, 24] led to the need to focus education for sustainable development on local communities (countries, regions) as this is the only way to resist global corporate interests. Nevertheless, there are numerous unresolved problems, which stand in the way of successful education and learning for sustainable development today. Some of these issues are highlighted by Boström et al. [2] and relate to the problem of dialogue with those who have the power to prevent change, structural inertia, the value of different knowledge in the context of sustainability, the relevance of perspectives and decision-makers, resistance to learning, the problem of climate change denial, and the impact on social tensions and conflicts [2]. Some authors consider that the approach to ESD requires the renewal of so-called citizen pedagogy [25], which aims to educate people to think critically about political, environmental, economic, and social issues of sustainability and sustainable development. However, ecopedagogy [25–27] does not provide for the development of knowledge through which an individual could understand the background processes important for sustainability and sustainable development. At the same time, Global Education for Sustainable Citizenship (GESC), which is taught within the framework of ecopedagogy includes the scale dimension — insight into global society and mechanisms of action, the ethical dimension — sustainability as a normative concept, the relational dimension — the social construction of sustainability, and the political dimension — social and societal issues of justice [24]. Certain issues of sustainability, including ESD, are seen as a problem of human rights and the community's right to real or simulated participation and decision-making on these issues [24]. All the mentioned dimensions and issues of sustainability and sustainable development are significant, but they cannot be sufficient to understand the real challenges of sustainable development because they exclude or do not explicitly name the crucial dimension, namely the technological one. According to its starting points, ESD should be inclusive and interdisciplinary, which means that this education should provide young generations with observation skills, critical thinking, creativity, adaptability, and skills necessary for a good life where they live, and give them the "tools" to build healthy and resilient communities [28]. This means that without knowledge about existing and sustainable technologies and the technological-ecological and energy problems of today, it is not possible to truly understand the problems, and therefore not possible to develop the critical thinking of the young generation. In fact, the development of critical thinking and criticism through "ecological indoctrination" have nothing to do with each other. The term "ecological indoctrination" implies various influences that radically shape the consciousness of younger generations in such a way that they regard any technology a priori as a danger to people and the environment [10]. Such ecologically indoctrinated people do not really understand the important issues and problems of sustainability and are easy prey for corporate interests and neoliberalism because they readily accept the "greening of the capitalist way out of the crisis" [24]. This favors the shifting of responsibility to corporations rather than changing current patterns of production, consumption, and distribution in the interest of greater environmental and social justice. Learning for sustainability, as well as education for sustainable development, often omits the dimensions necessary for understanding the actual cause-effect processes of human activities, which have caused the current state of the planet, namely the scientific and especially the technological dimensions [13]. Although it is often assumed that the scientific and technological dimensions are included in ESD through the ecological dimension in reality, this is often not the case and is left to the preferences of educators who often do not feel ready for such challenges [10]. However, without the inclusion of these dimensions in learning and ESD, it is not certain that acceptable solutions to the problems of sustainability and sustainable development can be found today, let alone in the future. Against this background, the question arises as to how the technological dimension can be integrated into learning and education for sustainable development and whether the approach to such integration can be unique. Thus, the approach of learning for sustainable development is not only the development of environmental literacy [29] or universal critical thinking [30], but necessarily includes the action skills needed to implement changes [31] that lead the local community, region or country toward sustainability. ESD should therefore be much more than the development of environmental awareness [32, 33] and take into account the environmental, social, political, but also human, economic and technological context of the reasons why something is used and the consequences it has for people, society and nature. At the same time, the focus should be on the changing context of the environment, whereby the active participation of young generations in shaping sustainable solutions or solving community problems makes this education transformative [13] to achieve transformative change toward resilience and sustainability [33–36]. This requires teaching and learning processes that help students construct and appropriate new and revised meanings of acquired experiences [37]. The perspective of transformative learning assumes that learning and change are anchored in practice and institutions that it is a process of reviewing, questioning, and revising previously taken-for-granted assumptions, that it comprises cognitive, social, moral, and affective components, and that the social context, social relationships, conflicts, and power relations play a central role as an integral part of the learning process [2]. Boström et al. [2] present an approach to further develop the concept of transformative learning by integrating three additional dimensions, namely institutional structures, social practices, and conflict perspectives, which should be taken into account in a stronger integration of the technological dimension in ESD. However, although the authors critically note that the issues of education for sustainable development cannot be resolved solely within the framework of the social sciences, their proposals do not fundamentally abandon this context. In these considerations, the dimension of technological knowledge is once again marginalized, although an understanding of technology is a prerequisite for understanding the actual causes (and sources) of problems and challenges of sustainable development and a starting point for solving these problems in the future. So, it is not a question of training people who will only fight for sustainability through social initiatives, actions, and policies, but critical thinkers who understand today's technology to be able to change it in the future, or who understand the arguments of those who will do so. Therefore, this education should focus on actions that lead to a real improvement of socio-environmental problems carried out consciously and critically. At the same time, the problem should always be approached systematically so that the individual can see the problem as a whole, understand the importance of the interrelationships of the individual elements of the whole, and act accordingly. This requires not only an understanding of the whole system but also an understanding of how things are connected [34, 38–40], which requires the development of the individual's systemic and critical thinking. In order for the younger generations to change the world of the future for the better, they must, therefore, become critical thinkers.

3. Critical thinking as an important goal of ESD

Critical thinking is an intellectually disciplined process of actively and skillfully conceptualizing, applying, analyzing, synthesizing, and/or evaluating information gathered or generated through observation, experience, reflection, contemplation, or communication as a guide to beliefs and actions [13, 41]. This means that in the process of developing such an opinion, information and facts should be systematically collected and then analyzed and evaluated in terms of their clarity, accuracy, relevance, depth, breadth, and logic. This includes the individual's ability to process and synthesize information in such a way that they can apply it meaningfully to tasks, make informed decisions, and solve problems effectively [42]. By its very nature, this requires recognizing that all reasoning occurs within viewpoints and frames of reference that it proceeds from specific goals that it has a relevant information base, that all data, when used in reasoning, must be interpreted, that interpretation involves concepts, that concepts involve assumptions, and that all basic reasoning conclusions have implications [43]. Given that the Sustainable Development Goals are relatively clear, learning and education for sustainable development need to consider all viewpoints and frames of reference (environmental, social, economic, and other), but the information base must include everything necessary to present valid interpretations, concepts, and conclusions. Of course, this inevitably includes technological information (knowledge). In the process, the individual develops the skills of analysis, interpretation, explanation, argumentation, evaluation, and self-regulation [44], which are used in parallel in an authentic context [42]. Underpinning this development is questioning that focuses on the fundamentals of thinking and reasoning, whereby the individual should be able to articulate thinking that reflects basic mastery of the intellectual dimensions of reasoning. Considering that the information base of sustainable development encompasses numerous domains (knowledge) from scientific, technological, and social to economic, political, and humanistic domains, diminishing the importance of any knowledge will not lead to adequate development of critical thinking. In this sense, technology and engineering, that is. technical education, are often considered as a discipline that involves critical thinking and the application of knowledge and skills because research shows that knowledge is effectively developed through interdisciplinary connections of the real world with content or practices [45]. Indeed, the experience of technical education cannot occur without integration with other disciplines and the application of prior knowledge and skills to solve problems [46], which is a comparative advantage of this education in the context of sustainability and the development of critical thinking. At the same time, research shows that engineering (technical) design activities can help to foster various forms of so-called "soft" skills such as creativity, critical thinking, collaboration, and communication [47]. Due to the development of students' ability to deal with the problems of sustainability and sustainable development in the future, the development of critical thinking is now seen as an important goal of any education and at any level so it is reasonable to expect such development to include a holistic technological understanding of the world. The previous mastery of basic skills is, therefore, no longer sufficient because work, learning, and citizenship in the twenty-first century require that we all know how to think, argue, analyze, weigh evidence, and solve problems [48], which is particularly important in learning and education for sustainable development.

4. Demystifying technology in the context of ESD

In the background of today's progress and the maintenance of the quality of life of every community and society is technology. It provides everything that many people take for granted today such as energy, food, and water. However, in addition to its contribution, every technology also has negative consequences for the natural environment and by-products that damage this environment. However, it is not possible today to meet people's growing needs without technology. Due to the noticeable negative consequences for the environment, it is, therefore, clear that technology must be further developed and improved in order to be sustainable in the future. At the same time, the question arises as to who will improve it if we educate children today as if it did not exist. How will these same children develop technology tomorrow if they do not know it? Some studies point to a significant relationship between the inclusion of technological knowledge in the college curriculum and sustainable development [49]. However, these and similar studies often emphasize information technology and digital skills. Although this is also an important technology that can contribute to informing individuals, research, and the development of environmental awareness, it will not lead to truly innovative solutions in the future. At the same time, it is extremely important that the wider population is also aware of sustainability and the Sustainable Development Goals [50], precisely so that they can recognize and support initiatives that offer sustainable solutions. Research shows that although the general public supports the goals of sustainable development, they are generally not sufficiently informed about the real problems and ways to solve them [51–55]. Some studies also show that young people are largely aware of the Sustainable Development Goals [56, 57], but that they often do not have acceptable attitudes when it comes to linking these goals to technology [10]. In other words, young people today, especially in industrialized countries, do not know what and how they are supplied daily with electricity, various foods, housing, drinking and sanitary water, and who and how to clean their waste. At the same time, they have no insight into the amount of different goods that someone has to provide daily so that they can live at an acceptable level of civilization. Finally, they are aware of the need to protect the natural environment and the environmental trends propagated by the media or school, but nothing further than that. The reason for this is the current structure of ESD, which ignores the technological dimension, which is why the ecological dimension of this learning and education is often reduced to the aforementioned ecological indoctrination. In ESD learning activities, participants are often not co-creators (they do not engage in research or design activities), but only performers of something that may be interesting to them, but they do not understand the real purpose of the activity or the cause-effect relationships behind such activities [10], i.e. the context that connects such activities to their understandable reality [10]. One of the reasons for this is that such activities have nothing to do with real issues of sustainability and sustainable development of the local community, so students do not develop a deeper understanding of these issues and problems. Another reason is the lack of competent educators and people who are considered capable of delivering education for sustainable development [58], which requires systematic training of ESD educators. For this reason, technological (and engineering) knowledge, which is in the background of cause-effect relationships of sustainability, often remains unclear in the minds of the participants.

Although each of the 17 Sustainable Development Goals [15] can be linked to technology and engineering and education in this field, some goals are more closely related and to achieve them it is necessary to develop the young generation's awareness and critical thinking skills from elementary school [13]. These are goals in the areas of energy, food, water, and waste [13]. Addressing these problems and future challenges is important for the long-term sustainability of the community, as well as for resisting influences that may jeopardize this sustainability. It is, therefore, extremely important to sensitize the young generation to these problems but also to educate them to become critical thinkers who are able to look at problems systematically and holistically and find appropriate solutions [13]. The Sustainable Energy Development Goal should ensure access to affordable, reliable, sustainable, and modern energy for all, new economic opportunities and jobs, empower women, children and youth, enable better education and health, more sustainable, equitable, and inclusive communities, and better protection against and resistance to climate change [15]. The goals in the area of nutrition not only include the eradication of hunger, food security, and improved nutrition but also the promotion of sustainable agriculture [15], which is linked to the empowerment of smallholder farmers, rural development, healthy lifestyles, and the fight against climate change. The goal related to water refers to the fact that all people have the right to access to drinking water in a quantity and quality that meets their basic needs [15], but also to the obligation to protect the water world and the environment, which ensures the sustainability of ecosystems and biodiversity. Waste-related objectives include the application of transparent and science-based risk assessment procedures for the use of chemicals, leading to a reduction of adverse effects on human health and the environment and the environmentally sound management of hazardous waste [15]. This goal is closely linked to sustainable production and consumption, as well as sustainable tourism. Ultimately, technology and engineering, as well as science and capacity building, are the main pillars of the means to implement these and other sustainable development goals [15], recognizing the research, development, implementation, and dissemination of environmentally friendly technologies as an important segment of sustainability, including education for sustainable development. Such technologies can only be created by people who can arrive at solutions through critical thinking. However, technological knowledge remains hidden, unknown, and unfathomable during learning and education for the majority of the population, especially for young people. Perhaps the reason for this is a kind of hidden curriculum that suits countries with developed technology or large corporations that build their profits on the ignorance of the general public, or perhaps just the incompetence of education systems or educators' ignorance of technology, especially sustainable technology. In any case, technology and technological knowledge, which are important from a sustainability and sustainable development perspective, often remain hidden or mystified (unclear) in the minds of younger generations, which is why it is important to demystify technology, especially that which is in the background of today's sustainability issues. This is necessary so that the young generation gets to know the causes of the problems, the consequences, and the possibilities well enough and develops the skills they need to find solutions for the future, which, in turn, are technological in nature. This is also necessary so that they can think critically about sustainable development issues. These processes include the formation of concepts, principles and understanding, which are necessary for the acquisition of new knowledge, but also the processes of problem solving, decision making, testing and constructing, which are necessary for the application of knowledge, as well as the mastery of oral discourse, which is necessary for the further acquisition and application of knowledge [59]. In this way, they develop the skills of focusing (on a problem), analyzing, gathering information, creating (solutions), remembering, organizing, integrating, and evaluating [59, 60], which is important from the point of view of sustainability and is also compatible with the mental processes that take place during the activities of individuals in technology and engineering education [13].

4.1 The concept for demystifying technology (CDT)

From all that has been said in the previous chapters, it can be concluded that today's ESD concept is mainly based on the (mostly negative) consequences of human actions and on "miracle technologies" (which do not exist), which many believe will solve sustainability problems, as well as on social initiatives and/or policies that will influence communities, societies, or states through various pressures (individual, social and economic) to become sustainable. At the same time, technology today is mostly considered in the areas that are attractive or currently in the spotlight for society, academia, and professionals in general such as digital technologies, AI, information and computing technology, etc. Although these technologies are important today and form an integral part of everyone's life, they will neither ensure the survival of nature nor provide people with what they need to live such as food, energy for heating and transportation, housing, necessary products, drinking water and sanitation water, healthy air, and clean environment, etc. Therefore, the concept for demystifying technology (CDT) (**Figure 1**) is presented here to complement today's incomplete pathways of learning and education for sustainable development. CDT is considered here primarily from the starting point of the necessary needs and goals of sustainability, without which neither humans nor nature can live, and it continues with a "deep" insight into the resources and technologies that ensure this and ends with an insight into the consequences and perspectives for the future. The concept is based on the operationalization of the development of critical thinking in the context of the Sustainable Development Goals [13],

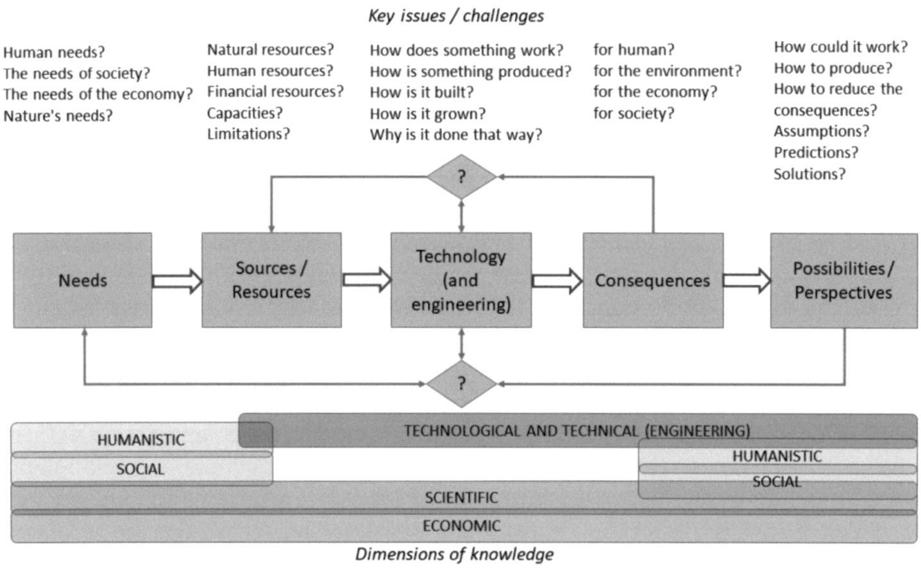

Figure 1.
The Concept for demystifying technology (CDT) in ESD.

which considers the mentioned Sustainable Development Goals as the most important for the general population and as necessary and suitable for lifelong learning.

In this concept, technology forms an insurmountable step (segment) on the path from the needs of humans, nature, society, and the economy to the consequences and prospects for the future. Of course, everything starts with needs that the general public, especially young people, are often unaware of. That is why learning and ESD should always begin with an understanding of how much water, air, food, living space, and energy, how much and what kind of waste needs to be cleaned up on a daily, monthly, or annual basis. This insight is also important at a local, national, and even global level as the younger generations, especially in industrialized countries, often do not yet have a strong awareness of this. At the same time, they take for granted everything they need daily, which someone or something has to provide or make possible, and all this also has its economic, ecological, and scientific background. Therefore, this knowledge should be considered from the perspective of the economic, social, and scientific dimensions of knowledge. In order to satisfy these needs, resources and sources are needed, which are developed and exploited and which are not inexhaustible, which is the second section (step) of this concept. This step is directly related to technology and technological knowledge as the exploitation and exploration of resources are enabled by a specific technology. Nevertheless, most of the basic needs mentioned are covered by a specific technology, which is the central or basic segment of this concept. In order for an individual to understand what needs they are meeting or should be meeting for themselves, society and the economy, a "deep" insight into technology is required. This includes not only knowledge of how something is made or produced today but also insight into the technical, scientific, and economic reasons why a particular technology is used. In the context of the selected core objectives of sustainability, this refers to technologies for the production and distribution of energy and energy sources, technologies for the production, cultivation, and processing of food (agriculture, livestock, and industry), and production in general, technologies for the management of water resources (drinking and industrial water), construction technologies and housing, and technologies for waste management. In the absence of this insight and experience, individuals can only have their own opinion, which cannot serve as a basis for arguing positions or for a reasoned defense of their own solutions, so such an opinion will certainly not be critical [13]. Every technology is determined by needs, but also by the development opportunities and prospects, and at the same time has certain consequences for people, society, the environment, and the economy. Insight into the consequences is closely related to technological knowledge as this knowledge is necessary for the individual to truly understand the emergence of positive and negative consequences. Understanding the consequences of technology in this concept includes not only all contributions but also the negative impacts and learning about existing technologies that attempt to mitigate them. In the context of the above objectives, these are technologies and standards to reduce fossil fuel pollution and impacts on nature in energy production, technologies and means to reduce natural pollution in food production, technologies to ensure the purity of drinking water and the water cycle in nature, and to preserve biodiversity, technologies to reduce the impact of pollution of seas, oceans and coastal areas and technology to reduce the impact of excessive and inappropriate waste disposal (prolonging the life of products, recycling, recovery, reuse, rational use, recovery, etc.).). It is clear from the above that the consequences are linked simultaneously to all the other dimensions of knowledge presented. Although every technology brings significant benefits, it also has some negative consequences for people, society, nature, and the economy. In other words, individuals need to understand what a particular technology brings to them and the community, what it means for society and the economy, what a particular technology disrupts and to what extent, and what the scientific reasons for these consequences are. Such multidimensional knowledge is necessary to understand possibilities and perspectives, as the next segment of this concept. Although possibilities and perspectives include humanistic, social, and economic dimensions, they are primarily determined by scientific and especially technical and technological knowledge. In other words, individuals should gain insight into which technologies can already ensure better sustainability and how they could be improved. With this in mind, individuals should familiarize themselves with new, still insufficiently developed but promising technologies. This concerns energy production (alternative, non-conventional and renewable energy sources and technologies for their use), food production (technologies for sustainable agriculture and breeding, environmentally friendly processing methods, preservation of original seeds and species, etc.), sustainable construction and housing, new technologies for the management of water resources (extraction and treatment of drinking water, technologies for wastewater management, methods and technologies for the protection of coastal areas, seas and oceans, etc.), and modern and new technologies for waste management and disposal. At the same time, the individual should also gain an insight into the scientific assumptions about the development of new ways and technologies that could ensure sustainability, what forecasts are associated with the prospects of such technologies, and think about possible solutions for the future, first in the local and then in the wider context of their application. Although modern digital technologies can be used for research and learning in the previous segments of this concept, the opportunities and perspectives segment is particularly suitable for learning about modern technologies and their application in management, decision-making, research, and development. At the same time, advanced systems such as the application of advanced information and computer systems, AI or IoT can also be learned, but this depends primarily on the structural and organizational form of learning, the level and type of training, the dispositions of the individual, the learning conditions, etc. This segment of learning should give individuals the creative freedom to create solutions that are not sustainable at first glance, but this can influence the growth of radical innovations [61] that can bring us contributions in the future that we do not see today. In the area of opportunities and perspectives, certainly, not every individual will make an equal contribution during learning and education for sustainability, but this area is a real "training ground" for those who are creative and capable, for whom it represents a real challenge and/or stimulus for future transformative action in the local and wider social community.

4.2 Limitations and assumptions for the implementation of the CDT

The presented CDT concept tries to convey in a vivid way that knowledge from technology and engineering is essential for learning and education for sustainable development and that it is very important for the development of critical thinking skills and also for the transformative behaviour of an individual. The concept also shows that technological knowledge cannot be assumed in isolation within a discipline or field, but that the humanistic, social, and economic dimensions of knowledge must also be taken into account alongside the already traditional scientific dimension. This knowledge forms an important context for technology because technology and technological knowledge are not and cannot be an end in themselves. It is essentially the needs, sources, consequences, opportunities, and challenges of the concept presented as a context without which technological knowledge cannot exist, nor can the context be truly recognized without technological knowledge. Therefore, learning and education for sustainability and sustainable development must be kept in the original context as a teaching and learning process that helps the individual to construct and appropriate new and revised meanings of the acquired experiences [37]. In other words, the individual should be confronted with technology and the background and consequences of this technology in the learning process as directly as possible or through contextualized material to develop an incentive for him or her to critically question possibilities and perspectives. It is important to point out that the aforementioned social dimension of knowledge should also include insights into the politics, social tensions, and structures that guide and manage processes, regulation, and the economy, which are relevant from the point of view of sustainability and sustainable development. Indeed, societal change will not occur simply on the basis of advances in knowledge if the institutional, social, and conflict-related dimensions of learning are not taken into account [2]. Moreover, social learning, where communities work together to find solutions to local problems, is a key factor in the transition to sustainability [33]. The scientific dimension of knowledge in this concept encompasses not only the scientific knowledge which relates to understanding the workings of the living world and the human activities that disrupt that world, but also the knowledge that enables and limits the application of a particular technology. In this concept, the humanistic dimension of knowledge refers first of all to the individual (egoistic) dimension from which every human being starts, even when learning. Given the fact that people's needs often determine their interests, learning should begin with human needs, continue with the individual's role in these processes and insight into the consequences for his or her quality of life, and end with the opportunities and perspectives for personal engagement in the direction of transformative behavior. This dimension should also include assumptions and predictions of different scenarios that may occur at the individual level as an undesirable consequence if sustainability is not ensured. All that has been said so far in this chapter are preconditions for the possible successful implementation of the proposed concept in some form of learning or education for sustainability and sustainable development. However, there are also limitations that point to the complexity of implementing such a concept in the learning and education system.

Regardless of whether and how it is implemented, such a process is in itself lengthy and requires exceptional synergy and coordination of actors from different sectors, as well as competent, motivated, and broadly trained educators. Furthermore, the concept is certainly not complete, which is an additional limitation. Many other knowledge and experiences are important to achieve transformative behavior that leads to sustainability and sustainable development. For example, knowledge about cultural and traditional sustainability is particularly important for small nations and countries, which is often not recognized clearly enough in policies, recommendations, and regulations at the global level. As a result, communities that have lived sustainably for centuries may simply disappear under the onslaught of ruthless global, corporate, and political trends. Political stability and nonviolence also have a positive impact on the sustainability of a society [62], so young generations should acquire a comprehensive knowledge of this and develop an awareness of the importance of maintaining the political stability of the country. Ultimately, the point is that the world needs to achieve transformative changes that are not yet in sight and that are envisioned by the ESD learning objectives [63]. Indeed, one of the recent situations review reports presented by the European Commission points out that implementation of the Sustainable Development Goals is lagging behind and even regressing in key areas such as poverty reduction, addressing inequality and solving the climate, biodiversity, and pollution crises [64]. Therefore, it is difficult to expect that the concept presented, even if implemented on a global scale, could bring about transformative change in a short period. However, the concept is not intended for the people of the present, but for future generations who will solve problems that do not yet exist. So, if we educate large parts of the young generation to become critical thinkers who know how people, nature, society, and technology work and become conscientious, thrifty, humble, and empathetic people, then the world has a prospect of achieving sustainability. Otherwise, we will embark on an irreversible path of self-destruction.

5. Conclusions

It is certain that despite the basic agreement on what should be done to stop or reduce the negative consequences of human activities, the efforts of the global community will not bring the desired results any time soon, and it is unlikely that this will happen in the long-term plans. In addition to the numerous interests of large corporations and economies that may pose an obstacle, there are also technological challenges that still need to be overcome in the future. These are challenges and technologies that do not yet exist and will have to be faced by future generations. Therefore, it is certain that the young generation should be sensitized in this sense, which requires the development of their critical thinking toward what exists today. If technology is the cause of many undesirable consequences for sustainability and sustainable development, and if some solutions could contribute to solving these problems and challenges today and soon, the question arises as to how the young generations should develop critical thinking on these issues if they are not familiar with these technologies? Technology and engineering and their societal necessity and justification are part of the whole system, which means that a critical, systemic, and holistic way of thinking and acting cannot be developed without the learning and education in this field.

This paper, therefore, proposes the concept of CDT, which integrates technology and engineering into learning and education for sustainable development, thus demystifying them in the minds of the younger generations. The concept is based on learning segments that represent a unique process and consists of a segment of insight into needs, a "deep" insight into the resources and technologies used to meet these needs, an insight into the consequences of technology use, and the discovery of opportunities and perspectives that will contribute to mitigation and sustainability in the future. The combination of knowledge with humanistic, social, scientific, and economic knowledge presented in this concept shows that technological knowledge is multidimensional and should be learned in the process of ESD in the context of related knowledge dimensions. However, since technological knowledge is necessary to understand the cause-effect relationships of sustainability and to develop critical thinking, it is also fundamental knowledge without which there is no ESD. Although the concept is only intended for learning about key sustainability issues and goals such as energy, food, water, construction, and waste management, it can be applied to all other goals where technology plays an important role.

From all of the above, it can be concluded that the concept for demystifying technology (CDT) must be an integral part of ESD for individuals to understand the cause-effect relationships of sustainability, and that such ESD must be an integral part of learning and education for the wider population to develop critical thinking at a level suitable for transformative action in the future.

Author details

Damir Purković
University of Rijeka, Rijeka, Croatia

*Address all correspondence to: damir@uniri.hr

References

[1] Steffen W, Richardson K, Rockström J, Cornell SE, Fetzer I, Bennett EM, et al. Planetary boundaries: Guiding human development on a changing planet. Science. 2015;**347**:6223. DOI: 10.1126/science.1259855

[2] Boström M, Andersson E, Berg M, Gustafsson K, Gustavsson E, Hysing E, et al. Conditions for transformative learning for sustainable development: A theoretical review and approach. Sustainability. 2018;**10**(12):4479. DOI: 10.3390/su10124479

[3] Onions Charles T. The Shorter Oxford English Dictionary. Oxford: Clarendon Press; 1964. p. 2095

[4] Kahle LR, Gurel-Atay E. Communicating Sustainability for the Green Economy. 1st ed. New York: Routledge; 2014. DOI: 10.4324/9781315705491

[5] HLPF. HLPF thematic review: Transformation towards sustainable and resilient societies - Building resilience. [Internet] 2018. Available from: https://sustainabledevelopment.un.org/hlpf/2018 [Accessed: May 12, 2021]

[6] Schlüter M, Biggs R, Schoon ML, Robards MD, Anderies JM. Reflections on building resilience – interactions among principles and implications for governance. In: Biggs R, Schlüter M, Schoon ML, editors. Principles for Building Resilience: Sustaining Ecosystem Services in Social-Ecological Systems. Cambridge: Cambridge University Press; 2015. pp. 251-282. DOI: 10.1017/cbo9781316014240

[7] Folke C, Biggs R, Norström AV, Reyers B, Rockström J. Social-ecological resilience and biosphere-based sustainability science. Ecology and Society. 2016;**21**(3):41. DOI: 10.5751/ES-08748-210341

[8] UNESCO. Sustainable development. [Internet]. 2015. Available from: https://en.unesco.org/themes/education-sustainable-development/what-is-esd/sd

[9] Purvis B, Mao Y, Robinson D. Three pillars of sustainability: In search of conceptual origins. Sustainability Science. 2019;**14**(3):681-669. DOI: 10.1007/s11625-018-0627-5

[10] Purković D, Kovačević S, Runko Luttenberger L. Attitudes of Croatian pupils on the relationship of environmental issues and sustainable development with technology and engineering. International Journal of Technology and Design Education. 2023;**33**(4):1285-1307. DOI: 10.1007/s10798-022-09779-6

[11] UN-ADB-UNDP. Transformation towards sustainable and resilient societies in Asia and the Pacific: Theme report for the Asia-Pacific Forum on Sustainable Development, 2018. Bangkok, Thailand: United Nations; 2018. DOI: 10.22617/TCS189274-2

[12] Duran DC, Gogan LM, Artene A, Duran V. The components of sustainable development-a possible approach. Procedia Economics and Finance. 2015;**26**:806-811

[13] Purković D. The importance of technical and technological knowledge for the development of students' critical thinking. In: Mrnjaus K, editor. VIII. International European Congress on Social Sciences - Full Text Book. Rijeka: IKSAD; 2022. pp. 326-338

[14] IPCC. In: Pörtner H, Roberts D, Tignor M, et al., editors. Climate Change

2022: Impacts, Adaptation, and Vulnerability. Contribution of Working Group II to the Sixth Assessment Report of the Intergovernmental Panel on Climate Change. Cambridge: Cambridge University Press; 2022. Available from: https://www.ipcc.ch/report/ar6/wg2/

[15] UN. Transforming Our World: The 2030 Agenda for Sustainable Development Resolution. New York: United Nations; 2015

[16] Purković D. Realiteti tehničke kulture. Rijeka: University of Rijeka, Faculty of Humanities and Social Sciences; 2015

[17] UNESCO. The DESD at a Glance. Paris: UNESCO; 2005 [Internet]. Available from: http://unesdoc.unesco.org/images/0014/001416/141629e.pdf

[18] UNESCO. Bonn Declaration of the World Conference on Education for Sustainable Development. Paris: UNESCO; 2009 [Internet]. Available from: http://www.esd-world-conference-2009.org/fileadmin/download/ESD2009_BonnDeclaration080409.pdf

[19] UNESCO. Shaping the Education of Tomorrow – Abridged Version. Paris: UNESCO; 2012 [Internet]. Available from: http://unesdoc.unesco.org/images/0021/002166/216606e.pdf

[20] UNESCO-DESD. Shaping the future we want: UN decade of education for sustainable development (2005-2014) – Final Report. UNESCO-DESD. 2014 [Internet]. Available from: https://unesdoc.unesco.org/ark:/48223/pf0000230171

[21] Hjorth P, Bagheri A. Navigating towards sustainable development: A system dynamics approach. Futures. 2006;**36**(1):74-92. DOI: 10.1016/j.futures.2005.04.005

[22] Henry AD. The challenge of learning for sustainability: A prolegomenon to theory. Human Ecology Review. 2009;**16**(2):131-140 [Internet]. Available from: http://www.jstor.org/stable/24707537

[23] Smulders-Dane S, Smits T, Fielding G, Chang Y, Kuipers K. Learning from regional sustainable development in The Netherlands: Explorations from a learning history. Sustainability. 2016;**8**(6):527. DOI: 10.3390/su8060527

[24] Huckle J, Wals AEJ. The UN decade of education for sustainable development: Business as usual in the end. Environmental Education Research. 2015;**21**(3):491-505. DOI: 10.1080/13504622.2015.1011084

[25] Laker J, Naval C, Mrnjaus K. Introduction — Civic pedagogies in higher education: Teaching for democracy in Europe, Canada and the USA. In: Laker J, Naval C, Mrnjaus K, editors. Civic Pedagogies in Higher Education. Palgrave Studies in Global Citizenship Education and Democracy. London: Palgrave Macmillan; 2014. DOI: 10.1057/9781137355591_1

[26] Kahn R. Critical Pedagogy, Ecoliteracy and the Planetary Crisis: The Ecopedagogy Movement. New York: Peter Lang; 2010

[27] Misiaszek GW. Educating the Global Environmental Citizen: Understanding Ecopedagogy in Local and Global Contexts. New York: Routledge; 2017

[28] Robertson M. Sustainability Principles and Practice. New York: Routledge; 2017

[29] Orr D. Ecological Literacy: Education and the Transition to a Postmodern World. New York: State University of New York; 1992

[30] Huckle J. Chapter 3: Environmental education and sustainability: A view from critical theory. In: Fien J, editor. Environmental education: A pathway to sustainability. Geelong: Deakin University Press; 1993. pp. 43-68

[31] Jensen BB, Schnack K. The action competence approach in environmental education. Environmental Education Research. 1997;**3**(2):163-178. DOI: 10.1080/1350462970030205

[32] Tilbury D, Wortman D. How is community education contributing to sustainability in practice? Applied Environmental Education and Communication. 2008;**7**(3):83-93. DOI: 10.1080/15330150802502171

[33] Aguayo C, Eames C. Promoting community socio-ecological sustainability through technology: A case study from Chile. International Review of Education. 2017;**63**:871-895. DOI: 10.1007/s11159-017-9685-7

[34] Sterling S. Sustainable Education: Re-visioning Learning and Change. Bristol: Green Books; 2001

[35] Sterling S. Learning for resilience, or the resilient learner? Towards a necessary reconciliation in a paradigm of sustainable education. Environmental Education Research. 2010;**16**(5-6):511-528. DOI: 10.1080/13504622.2010.505427

[36] Uzzell D, Räthzel N. Transforming environmental psychology. Journal of Environmental Psychology. 2009;**29**(3):340-350. DOI: 10.1016/j.jenvp.2008.11.005

[37] Pavlova M. Teaching and learning for sustainable development: ESD research in technology education. International Journal of Technology and Design Education. 2013;**23**:733-748. DOI: 10.1007/s10798-012-9213-9

[38] Armstrong J. En'owkin: Decision making as if the sustainability mattered. In: Stone M, Barlow Z, editors. Ecological literacy: Educating Our Children for a Sustainable World. San Francisco: Sierra Club Books; 2005. pp. 11-17

[39] Capra F. Speaking nature's language: Principles for sustainability. In: Stone M, Barlow Z, editors. Ecological Literacy: Educating our Children for a Sustainable World. San Francisco: Sierra Club Books; 2005. pp. 18-29

[40] Orr D. Systems thinking and the future of cities. Solutions. 2014;**5**(1):54-61 [Internet]. Available from: http://www.resilience.org/stories/2014-05-30/systems-thinking-and-the-future-of-cities [Accessed: Feb 17, 2021]

[41] Scriven M, Paul R. Defining critical thinking. In: Eighth Annual International Conference on Critical Thinking and Education Reform. [Internet]. Available from: http://www.criticalthinking.org/pages/defining-critical-thinking/766 [Accessed: May 21, 2020]

[42] Heard J, Scoular C, Duckworth D, Ramalingam D, Teo I. Critical thinking: Definition and structure. Camberwell: Australian Council for Educational Research; 2020:1-7 [Internet]. Available from: https://research.acer.edu.au/ar_misc/38

[43] Paul RW, Elder L, Bartell T. California Teacher Preparation for Instruction in Critical Thinking: Research Findings and Policy Recommendations. Foundation for Critical Thinking. Sacramento (USA): California Commission on Teacher Credentialing; 1997 [Internet]. Available from: https://files.eric.ed.gov/fulltext/ED437379.pdf

[44] Thomas K, Lok B. Teaching critical thinking: An operational framework. In: Davies M, Barnett R, editors. The

Palgrave Handbook of Critical Thinking in Higher Education. New York: Palgrave Macmillan; 2015. pp. 93-105. DOI: 10.1057/9781137378057_6

[45] Schwartz MS, Sadler PM, Sonnert G, Tai RH. Depth vs. breath: How content coverage in high school science courses relates to later success in college science coursework. Science Education. 2009;**93**:798-826

[46] Spicer Y, Informal M. Out-of-school technology education. In: De Vries MJ, editor. Handbook of Technology Education. Cham: Springer International Publishing AG; 2018. pp. 267-279. DOI: 101007/978-3-319-44687-5

[47] Hathcock SJ, Dickerson DL, Eckhoff A, et al. Scaffolding for Creative Product Possibilities in a Design-Based STEM Activity. Research in Science Education. 2015;**45**:727-748. DOI: 10.1007/s11165-014-9437-7

[48] Wagner T. The Global Achievement Gap: Why Even Our Best Schools Don't Teach the New Survival Skills Our Children Need–And What We Can Do About It. New York: Basic Books; 2008

[49] Alshammari AM, Alshammari FF, Thomran M, Altwaiji M. Integrating technological knowledge into higher education curricula: An essential measure for attaining sustainable development in Saudi Arabia. Sustainability. 2023;**15**:15956. DOI: 10.3390/su152215956

[50] Bain PG, Kroonenberg PM, Johansson LO, et al. Public views of the sustainable development goals across countries. Nature Sustainability. 2019;**2**:819-825. DOI: 10.1038/s41893-019-0365-4

[51] Dlamini S, Block EK, Mathibela N. Knowledge of and attitudes towards sustainable development goals among adults in South Africa. South African Geographical Journal. 2023;**105**(3):384-401. DOI: 10.1080/03736245.2022.2148723

[52] Maciejewski G, Lesznik D. Consumers towards the goals of sustainable development: Attitudes and typology. Sustainability. 2022;**14**(17):10558. DOI: 10.3390/su141710558

[53] Guan T, Meng K, Liu W, Xue L. Public attitudes toward sustainable development goals: Evidence from five Chinese cities. Sustainability. 2019;**11**:5793. DOI: 10.3390/su11205793

[54] Schneider SH, Gleser SH, Bruder M. Public opinion of the 2030 agenda: Results of the DEval opinion monitor for development policy 2018. DEval Policy Brief. 2018;**6** [Internet]. Available from: https://www.deval.org

[55] Omisore AG, Babarinde GM, Bakare DP, Asekun-Olarinmoye EO. Awareness and knowledge of the sustainable development goals in a university community in Southwestern Nigeria. Ethiopian Journal of Health Sciences. 2017;**27**(6):669-676. DOI: 10.4314/ejhs.v27i6.12

[56] Minimol MC, Francis S. Young people's awareness, perception, and attitude towards sustainable development goals in India. Visions for Sustainability. 2023;**21**:1-20. DOI: 10.13135/2384-8677/8415

[57] Balakrishnan B, Tochinai F, Kanemitsu H. Perceptions and attitudes towards sustainable development among Malaysian undergraduates. International Journal of Higher Education. 2020;**9**(1):44-51. DOI: 10.5430/ijhe.v9n1p44

[58] Vukobratović J, Rončević N. Education for sustainable development

as the educational imperative of the teaching profession: “Personal example is more important than all the teaching materials”. In: Ilić BĆ, Buchberger I, editors. Suvremeni Trendovi i Izazovi Nastavničke Profesije. Rijeka: University of Rijeka, Faculty of Humanities and Social Sciences; 2020. pp. 95-116. [Internet]. Available from: http://izdavastvo.ffri.hr/wp-content/uploads/2020/05/Pedagogija_Zbornik_Suvremeni-trendovi-i-izazovi-nastavnicke-profesije_FINAL_Online_izdanje.pdf [Accessed: Feb 23, 2021]

[59] Marzano RJ, Brandt RS, Hughes CS, Jones BF, Presseisen BZ, Rankin SC, et al. Dimension of thinking: A framework for curriculum and instruction. Alexandria, VA: Association for Supervision and Curriculum Development; 1998

[60] Johnson SD. A framework for technology education curricula which emphasizes intellectual processes. Journal of Technology Education. 1992;**3**(2):26-36

[61] Verter V, Boyaci T, Galbreth M. Design for reusability and product reuse under radical innovation. Sustainability Analytics and Modeling. 2023;**3**:100021. DOI: 10.1016/j.samod.2023.100021

[62] Agheli L, Taghvaee VM. Political stability effect on environment and weak sustainability in Asian countries. Sustainability Analytics and Modeling. 2022;**2**:100007. DOI: 10.1016/j.samod.2022.100007

[63] UNESCO. Education for sustainable development goals: Learning objectives. Collectivité auteur: UNESCO. 2017 [Internet]. Available from: https://unesdoc.unesco.org/ark:/48223/pf0000247444. ISBN: 978-92-3-100209-0

[64] EC. Report from the commission to the European parliament, the council, the European economic and social committee and the committee of the regions. 2023 [Internet]. Available from: https://ec.europa.eu/info/law/better-regulation/

Chapter 12

Environmental Activism for "Common Home": Through the Documentary "My Octopus Teacher"

Sonia Martins Felix and Sikunder Ali

Abstract

This chapter presents an overview of knowledge processes interconnected with two main perspectives, the so-called Euro-American scientific way and the indigenous ways. Taking these two different perspectives, we will revisit the concept of environmental education. We explore how environmental education can benefit from indigenous ways of knowing. While doing so, we will draw on a documentary film-based example, "My Octopus Teacher," to articulate activism for environmental education. Here, activism could be enhanced and interpreted from the critical thinking dispositions perspective. Activism is then considered a stimulus for environmental education and education for sustainable development issues. This documentary exemplifies by viewing how "to know," as verb-based knowledge integral to indigenous ways of being, as particular ways of living with nature, can provide deeper meanings to activism in education for sustainability.

Keywords: environmental education, critical thinking dispositions, education for sustainable development, indigenous knowledges, activism

1. Introduction

Questions that are not necessarily answered but serve as guidance in this chapter to articulate our position of verb-based knowledges, such as (i) *What is scientific knowledge? or* (ii) *How is scientific knowledge entangled as an integral part of being with nature as a way to achieve sustainable living?*.

In doing so, we intend to elaborate scientific discourses around environmental issues as can be seen in action *in situ* (with the lens of activism) [1, 2]. This chapter provides a brief background on ways of knowing in science while juxtaposing the Euro-American scientific way and indigenous ways. After that, we provide an overview of environmental education in international reports/documents on environmental education and education for sustainable development [3]. The role of environmental activism has been presented from indigenous knowledge perspective.

This perspective is based on actions (verb-based) in contrast with noun-based interpretation within Eurocentric knowledge orientation.

In the discussion, we elaborate on how verb-based, indigenous knowledge or "to know" perspectives can be considered a form of activism [4] with an illustration of the documentary "My Octopus Teacher." That is, the chapter specifically addresses the question: *how can indigenous ways of knowing exemplified through "My Octopus Teacher" serve the purpose of activism by relating to education for sustainability issues?* In addition to this, we consider the possible role of critical thinking as dispositions (in contrast to skills) and how it can contribute to sensitizing and engaging with environmental concerns with principles of activism. Here, the importance of the activism role lies in the importance of transcending the "instrumental view" of environmental and sustainability issues. This can encourage a theoretical discussion focused on the "culture-dependent collective rationale perceiving of reality" [5] that can serve a "our common home" [6] premised in environmental education.

2. Ways of knowing in science

The origins of science can be traced back to ancient philosophies; however, it is claimed that the major transformations in science happened in Europe. The main reason is that certain movements that have influenced the world, such as the Renaissance movement that occurred in the seventeenth century, influenced the knowledge system as a whole. A group of scientists/mathematicians (e.g., Galileo, Kepler, Descartes, Leibniz, Newton, and so on) created the need for natural science to be established as a knowledge system predicated on the authority of empirical evidence and rational deliberations as opposed to church and royal authority [2]. The figure of a scientist/mathematician becomes an embodiment of scientific authority based on a claim to objective/rational knowledge.

The Industrial Revolution changed the *status quo* associated with knowledge/power through technologists, especially since it has institutionalized scientific practice in terms of establishing scientific staff as professionals who are paid to use knowledge for the benefit of their institutions [2] (*idem*). In addition, the ideology of positivism had a big impact on shaping the nature and form of Eurocentric sciences up until the 1960s [7] and is still present in school sciences by portraying science as an act of measurement in terms of creating stable entities (whether material or conceptual) as a thing where the distinction between living and nonliving being is valorized and established.

Here, positivism follows the separation between the knower from the known. In contrast, traditions of indigenous knowledges are based on and developed on the idea of relationality in terms of considering the monism perspective where there is no distinction between animate and inanimate beings. According to these indigenous knowledge practices, the organic and evolving connection cannot be captured with the act of measurement, but is felt and sensitized with feelings and emotions. This way of knowing is considered as knowing with the verb "to know" as action with others (includes both animate or inanimate as one), as opposed to the" knowing" (action on others—a noun-based knowing) as knowledge way based in Eurocentric sciences [2], where the distinction between knower and known is maintained. In professional science settings, the prevailing perspective is that there is a *Eurocentric* science [2] (*idem*) framed in the historical-social context that is dated since the seventeenth century, where physical sciences establish a foundation of knowledge on the act of measurement and turning phenomena as knowable through empirical measurement and logical and rational adjudication. For example, Galileo Galilei emphasizes the mathematical foundation of physics through experimental observations with telescopes.

3. Juxtaposition between Euro-American scientific way and indigenous ways

In juxtaposing with Eurocentric science, the indigenous perspectives provide possibilities to relate human action as an integral connection with nature (as living with nature), which valorizes wisdom in action. Here, indigenous knowledge can be understood as a way of understanding, skills, and philosophies held by local societies with their long stories and integral experiential interaction within their social, cultural, and natural environments [8]. According to Aikenhead and Ogawa [2], this might offer a problematic nuance when indigenous wisdom is seen through a Eurocentric perspective, a noun-based knowledge that cannot be easily translated into verb-based Indigenous languages and integral ways of being with nature, where the purpose of knowledge is harmonic experience with nature. The main point of difference between Eurocentric knowledge structure and indigenous ways of knowing can be seen in verb-based knowledge in opposition to noun-based knowledge, while similarities can be found in the way they give importance to interpreting natural phenomena. When comparing Eurocentric science knowledge with indigenous knowledges, there is often a misunderstanding between distinction and foundation. Here, foundational dichotomies between Eurocentric knowledge and indigenous knowledges are again maintained, despite efforts on the part of Aikenhead and Ogawa [2] by proposing three new categories: (1) Indigenous ways of knowing (plural), (2) neo-indigenous ways of knowing nature, and (3) Eurocentric sciences (plural) [2] (see Van Eijck, 2007 *in* Aikenhead & Ogawa, 2007). The foundation can be perceived as a legal entity or authoritarian body. Taking this perspective, foundation as a legal entity is often used to describe knowledge, and the problem arises when in education, knowledge is often calibrated from the temporal-spatial scaling of life [9]. This foundational perspective complicates a comparison between Eurocentric knowledge and indigenous knowledges. This means that Eurocentric knowledge uses the advantage of having been defined, centuries ago, and indigenous knowledge has been characterized recently; that is, Eurocentric knowledge is a consequence of compression of time and space [2, 10] (Appadurai, 1996; Brandt, 2007 in Aikenhead & Ogawa, 2007). This points out a false dichotomy, since indigenous knowledge versus science, framed through colonial discourse with its subtext of winners and losers [2] (Macedo, 1999 *in* Aikenhead & Ogawa, 2007), which is in alignment with Eurocentric sciences, has a privileged position and other ways of knowing were rarely ever broached or validated on equal footings [2] (Brandt, 2007 in Aikenhead & Ogawa, 2007). This differential treatment of indigenous knowledges can be seen in how it is judged with criteria of validity and legitimacy as invoked through a foundational perspective.

One misunderstanding is that often the underlying meaning is that there are differences between Eurocentric knowledge and Indigenous knowledge that can be characterized by establishing distinctions. However, these characteristics often fall in the "foundational" category, without being properly addressed from the "distinction" perspective and that might be the misinterpretation referred to by critical comments of Van Eijck [2] (2007 in Aikenhead & Ogawa, 2007) on the differences that Aikenhead and Ogawa [2] illustrate through distinctions of Eurocentric and indigenous knowledges. One example of this type of false distinction can be seen in the ways the word "tracker" is applied to a profession where people are following wild animals. The word "tracker" might be interpreted from both Eurocentric and Indigenous perspectives. This example is presented in the documentary "My Octopus Teacher," where the "tracker" profession is presented as someone who is following animal traces, with all the senses (verb-based description) and described as someone who is following subtle signs of animals as they are part of "inside of the natural world." From the Eurocentric perspective (and this is not explicit in the documentary), tracking wild animals would probably be enhanced from taxonomic arguments. The taxonomic argumentation is useful for organizing knowledge, but the point is that it is entangled in binary ways of sorting by characteristics as modeled on the so-called classificatory value grids of Eurocentric knowledge (whose prime example was Linnaean classifications of biological kinds).

So, the foundational (verb-based vs. noun-based) is here misunderstood from the "distinguish" facet. Here, the verb-based or indigenous way of knowing is equally capable of fixing the "solution" despite not using the "taxonomic arguments." These misunderstandings in the act of "distinguishing" both types of knowledge aspects may be the reason why Ogawa [11] proposes that every culture has its science, and this means that science is culture-dependent and framed in a collective rationale for perceiving reality through each culture's situational specific collective frame.

4. Education for sustainable development and critical thinking (as dispositions)

UNESCO's last documents seem to carry a less instrumental view of education concerning actions related to environmental issues and education for sustainable development [3]. After 30 years of formal international efforts to promote environmental education, mainly through the UNESCO-UNEP International Environmental Education Program (1975–1995), environmental education has entered a new phase of institutionalization. Environmental education has a broad scope, embracing social, political, economic, technological, and cultural aspects [12]; as compared to earlier instrumental views, "discourse has been reduced to a set of problems to be solved and a pool of resources to be managed in more efficient ways" ([3], p. 40). First introduced in the early 1970s, the concept of Sustainable Development was to "meet(ing) people's needs today without diminishing opportunities for future generations" ([13], p. 43). At that time, this report of the World Commission on Environment and Development received the title "Our Common Future," but there was an absence of a relationship between people and the environment since it did not integrate the base of human development and action [3]. Probably this happened because of financial/economic motives that have been the shortcut to serving environmental purposes. It is interesting to note that critical thinking or reflective thinking was not mentioned, for example, at the United Nations Conference on Environment and Development until 1993 [3]. In other words, the need to create means to reflect on environmental and developmental realities was absent. Another interesting point is that education tends to be replaced by learning, over the years. Possibly, economic approach was dominant and overshadowed human relations with the world. But later, education for sustainable development is characterized by three fundamental pillars economic, social, and environmental protection as pillars of sustainable development [14]. Moreover, sustainable development has been replaced by education for sustainable development, which highlights educational action related to sustainability. Furthermore, it has been more recently stated that education for sustainable development is a learning opportunity that should provide learners of all different age levels the knowledge, skills, values, and agency to address interconnected global challenges [15]. In other words, educating for sustainability requires citizens to make informed decisions, in individual and collective settings for achieving sustainable goals. Mainstreaming education for sustainable development is seen as one of the core strategies to impact critical thinking competencies among the future generation while helping in achieving SDG-4 – Quality Education [16, 17]. In this way, education for sustainable development has been prescribed as the central dimension of all educational projects where critical thinking plays a crucial role in promoting active sustainability citizens. It can be interpreted that critical thinking is the individual's ability to engage with education for sustainable development [17]. Critical thinking can be defined as "reasonable reflective thinking focused on deciding what to believe or do" [1, 18]. In this context the focus on critical thinking, dispositions are defined as something that implies a willingness [19] or a tendency to do something, given certain conditions [18], which can provide a better understanding of education for sustainable development in practice. However, identifying peoples' dispositions also offers some challenges. Ennis [18] provides a curious metaphor about dispositions: "We cannot see that glass is brittle simply by looking at it" (p. 166). This means that dispositions are not easily quantified since they can be considered nondeterministic. One assumption is that critical thinking dispositions as a tendency to do something given certain conditions can improve education for sustainable development actions. Here, the policy turn to education for sustainable development became an occasion to broaden what had been assessed to be a rather narrow fact-based or normative-based environmental education in favor of a more critical, political, and pluralistic meaning approach [20, 21]. Education for sustainability should then be defined as embedded in the local and cultural contexts, emphasizing quality of life and capacity-building for communities [22].

4.1 The role of activism

Activism can enhance moral concerns by bringing focus on values in terms of science, technology, society, and environmental education [4, 23]. In the social context, the focus on values helps to incorporate the critique of science and social institutions. For example, the collective action stimulated by activism can be seen as a tool through documentaries, which can be inspired by "verb-based knowledge" [2]. Through stories, and especially through drama, students are stimulated to address issues and events from the perspective of *others*, explore and develop understanding, establish new relationships, and consolidate existing ones [4]. In other words, social context can help to stimulate issues and events from the *other's* perspective [24]. Here, self (as *I*) presupposes necessarily the existentiality of *otherness.* The underlying idea is that the critical accounting of the *I* with others has a particular meaning through the sociocultural contexts [25]. Here, we will rely on activism in education as a key component that can activate engagement among youth through the promotion of democratic activism [4]. Although the focus here is not on learning, the case that the social context in which the student is located outside school is likely to be a major factor impacting learning [4]. The notion of the *self* to others provides an important relational aspect through the sociocultural contexts. Considering the school arena, the curriculum needs to focus on two "currents." On one hand, students need to learn how to participate (with *otherness*), and on the other hand, they need to experience participation that can develop their critical thinking dispositions (through *self*). Perkins et al. [26] have offered a triadic theory of critical thinking dispositions as an effort to characterize them. According to these authors, disposition as a generative way of approaching issues has three components: inclination, sensitivity, and ability. Trying to understand how "dispositions" can be pursued as actions that are (i) in relation to the self, (ii) in relation to others, and (iii) in relation to the world [27] might contribute to a further understanding about people's engagement with environmental concerns. So, the point here is that contextualized in the environmental concern or education for sustainability, the way one can experience watching documentaries can entail dispositional attitudes related to activism principles.

5. Sociocultural collective action in environmental concerns

Education for sustainable development can be defined as a reaction or response to the urgent and dramatic challenges that the planet faces [28]. The United Nations General Assembly proclaimed the 2030 Agenda for sustainable development goal for a prosperous and inclusive future for all citizens to accomplish solid sustainability in the pursuit of 17 SDGs and 169 targets [29]. One of the arguments is that Western modern science is often seen as an icon of prestige, power, and progress, and Western culture permeates the culture of those who engage in science [2]. How can activism be developed in the intersection between "to know" and the so-called collective action in environmental concerns [4]? For example, the animistic beliefs of Amazonian people are documented through the work of the anthropologist Eduardo Kohn about the shared characteristics of themselves and the animals that live around them: "They are selves ... that have a point of view. This is what makes them animate and this animation enchants the world" ([30], p. 97). Kohn's work is an example of how Amazonian people recognize personhood in nonhuman animals [31]. This pinpointed the need to create a sense of community and develop an awareness of ties to others that can be developed through a shift of how the animals and other nonhumans are presented. In this connection, "My Octopus Teacher" can enhance the purpose of activism while helping us to exemplify our connection with animals as an essential part of the environment.

5.1 Drawing on the documentary example: "My Octopus Teacher"

Education for sustainable development from the knowledge perspective can be challenged through critical thinking approaches, while taking indigenous ways of knowing as more disposed to challenge scientific rationality. Indigenous ways of knowing are held by local societies with experience of interaction with their cultural and natural environments [8]. Here, we want to draw on one documentary example to consider (i) possible activism-role and (ii) identifying the implicit knowledge.

This is particularly interesting how some social media are challenging the prevalent rationale. After 1 year of interacting with an octopus in a South African kelp forest, the filmmaker Craig Foster presents "My Octopus Teacher" [32]. The film shows the bond developed between Foster and the eight tentacled creature that helps to develop a deeper understanding of the sea creature and her environment and relays the impact of this experience on the humans' lives. Foster himself starts the documentary, by stating how much he has been inspired by the indigenous ways of knowing. Here, he refers to his experience in Central Kalahari where he met people who were trackers of wild animals. He emphasized that they were "inside of the natural world" as they could follow subtle signs of the animals. During the documentary and from critical thinking dispositions, Foster's interaction with the octopus can be interpreted as an action that arises in relation to the self [27]. Critical thinking disposition might indicate the vital role in relation to *otherness* as a tendency to do something in relation to the world [27]. Taking that as the desirable approach to relating environmental issues is to work on relationships between people and the environment, which can be considered as the basis of human development and action [3]. The documentary provides an experience that can be comparable to the experience of "to know," present from the indigenous experiential perspective. Observing the interaction between the octopus and Foster offers a verb-based knowledge experienced directly related to environmental concerns. The interaction between Foster and the octopus relates to each other's ways of interaction, which can be characterized as "to know" experiences. One could also claim that scientific knowledge from "knowing" happens after a disposition, as a tendency to interact with something as a "to know" perspective. That is what critical thinking as disposition has been outlined by Ennis [1, 18, 19].

6. Discussion

Here, we wish to discuss how activism can be experienced in environmental education or education for sustainability in a verb-based knowledge. This can be even experienced as a tipping point in the way one considers education for sustainability from an exclusively Eurocentric way of knowing. Although this paper focuses on the "culture-dependent collective rationale perceiving of reality" [5] that can serve a common good premise in education for sustainability, some limitations are also emerging through a possible conflict from the "nature" of the "knowledge" perspective. The compression of time and space that has come with globalization and new transnational flows of people and information [10] has influenced the "to know" perspective [2]. That is, this luxury of time (with slow and patient dialog/interactions between Octopus and Foster), effort, and commitment that Foster demonstrates in coming closer to the world of Octopus and in turn how Octopus enhances sensitivities and frame of viewing of Foster has created a holistic view that indigenous knowledge promotes through verb-based knowledge. In the revisitation of knowledge of science, when juxtaposing the science knowledge and indigenous knowledges, there is a misunderstanding between distinction and foundation [2] (comments of Van Eijck, 2007 in Aikenhead & Ogawa, 2007). Often, indigenous knowledge is being described from the foundation perspective. However, when operationalizing the education for sustainability action and considering, for example, critical thinking to make decisions, the role of the dispositions, as a tendency to care about, is inevitably part of any of them. That invites for questioning the foundational claims of knowledge-based—whether noun- or verb-based—actions [2] (as is well argued by comments of Van Eijck 2007, in Aikenhead & Ogawa, 2007).

Firstly, it brings attention to whose purpose of education for sustainability serves. This is integraly linked with the social and cultural preferences of the population. Secondly, critical thinking as "reasonable reflective thinking focused on deciding what to believe or do" [1, 18] means that dispositions imply a willingness or a tendency to do something, given certain conditions [19]. So, despite the initial idea that there is an opposition between "scientific" and "indigenous knowledges", in an environmental context, both have a space to coexist. Ultimately, the individual effort in collective actions embedded in every life effort matters in the sustainability world [33]. Therefore, critical thinking dispositions have a crucial role in environmental activism. Here, considering holistic pluralities of indigenous knowledge can provide rich resources to bring total commitment to seeing us humans as an integral part of the dynamic system of nature, where humans and nature are tied together in a mutually sustaining way. Sheila Jasanoff, the 2022 Holberg prize[1] winner, mentioned Pope Francesco's efforts regarding his vision of sustainability, during the SDG 2023 conference in Bergen, Norway. Having a vital role in the religious sphere, as well as in the global world, the understanding that humanity is sharing a "common house" [6] might help to dilute the responsibility toward *otherness.* Pope Francesco points out different "common goods" [13], reminding us that natural resources are not unlimited, and while doing so, he is trying to minimize the *otherness* effect while putting all of *us* in the "common house."

7. Conclusion

To conclude, educators might have a particular role through activism, while inviting students and the population in general, to see all the available positions, which can easily be accessed through the documentary's examples such as "My Octopus Teacher". This knowledge-based experience presupposes a verb-based knowledge, inviting to describe and deconstruct action-based experiences that characterize the knowledge. Thus, it invites a non-instrumentalist perspective and pluralistic view of education for sustainability while providing the integration of the positions of different disciplines. The critical thinking role in education for sustainability is here enhanced from the disposition perspective, highly intertwined with verb-based knowledge. Dispositional action arises in relation to others and by consequence in relation to the world [27], and these are ways of engaging with activism in the sustainability context.

Environmental education can be enhanced as ways of activism in different ways, and documentaries can serve as a goal to develop interactions between different actors, promote diversity of opinions, and critically engage in different arguments [20]. Considering the positions of different actors is a challenge that can be manifested using critical thinking dispositions. In this way, the previous sustainability motto of "common good" [13] can be seemingly replaced by Pope Francesco's [6] proposal of sustainability as the "common home."

8. Possible future implications

As octopus is a popular ingredient used in many cultures, with many tons of this molluscan being captured per year, the fact that the food processing corporation "Nueva Pescanova" aims to build the world's first outdoor octopus farm in Gran Canaria is a real concern [34]. The presented documentary "My Octopus Teacher" serves the purpose of offering a way to relate with the water-dwelling body as a way to relate with this nonhuman animal, which after all, creates an interaction with a human. Given the circumstances that sustainability as a concern is a fact and a major force in today's world, the interaction between humans and nonhumans offers the possibility to develop a dispositional attitude as a tendency to reconsider the human consumerism related to the habit of eating octopus. Here, science education

[1] Available from: https://holbergprize.org/en/holbergprisen/holberg-prize.

concerning teachers can have a critical role in deconstructing "to know" through a "knowing" experience.

Acknowledgements

We would like to thank Helena Bichão, Trude Nordal, and Gerd Johansen for the comments in the early version. We are also grateful to the suggestions provided during the review process.

Author details

Sonia Martins Felix* and Sikunder Ali*
Department of Teacher Education, Norwegian University of Science and Technology (NTNU), Trondheim, Norway

*Address all correspondence to: sonia.f.martins@ntnu.no and sikunder.ali@ntnu.no

References

[1] Ennis R. Critical thinking: A streamline conception. In: Davies M, Barnett R, editors. The Palgrave Handbook of Critical Thinking in Higher Education. Vol. 2015. London, UK: Palgrave Macmillan; 2015. pp. 31-48

[2] Aikenhead GS, Ogawa M. Indigenous knowledge and science revisited. Cultural Studies of Science Education. 2007;**2**(3):539-620

[3] Sauvé L, Berryman T, Brunelle R. Three decades of international guidelines for environment-related education: A critical hermeneutic of the United Nations discourse. Canadian Journal of Environmental Education (CJEE). 2007;**12**(1):33-54

[4] Hodson D. Becoming part of the solution: Learning about activism, learning through activism, learning from activism. In: Activist Science and Technology Education. Netherlands: Springer; 2014. pp. 67-98

[5] Ogawa M. Science education in a multi-science perspective. Science Education. 1995;**79**:583-593

[6] Pope Francis. Laudato Si': On Care for our Common Home [Encyclical]. The Vatican; 2015

[7] Ziman J. An Introduction to Science Studies: The Philosophical and Social Aspects of Science and Technology. Cambridge: Cambridge; 1984

[8] Zidny R, Sjöström J, Eilks I. A multi-perspective reflection on how indigenous knowledge and related ideas can improve science education for sustainability. Science & Education. 2020;**29**:145-185. DOI: 10.1007/s11191-019-00100-x

[9] Nespor J. Educational scale-making. Pedagogy, Culture and Society. 2004;**12**(3):309-326

[10] Appdurai A. Modernity at Large: Cultural Dimensions of Globalization. Vol. 1. Minnesota: University of Minnesota Press; 1996

[11] Ogawa M, Ogunniyi MB, Jegede OJ, Yandila CD, Oladele FK. Nature of worldview presuppositions among science teachers in Botswana, Indonesia, Japan, Nigeria, and the Philippines. Journal of Research in Science Teaching. 1995;**32**(8):817-831

[12] UNESCO-UNEP. The Belgrade Charter: A Global Framework for Environmental Education. Belgrade: International Environmental Education Workshop; 1976

[13] World Commission on Environment and Development. Our Common Future (So-Called Brundtland Report). Oxford: Oxford University Press; 1986

[14] UN. Report on the World Summit on Sustainable Development. Johannesburg: World Summit on Sustainable Development; 2002. Available from: https://documents-dds-ny.un.org/doc/UNDOC/GEN/N02/636/93/PDF/N0263693.pdf?OpenElement

[15] UNESCO. Reimagining our Futures Together: A New Social Contract for Education. UN; 2022. Available from: https://unesdoc.unesco.org/ark:/48223/pf0000379707

[16] UNESCO. SDG4-Education 2030, Incheon Declaration (ID) and Framework for Action. For the Implementation of Sustainable Development Goal 4, Ensure Inclusive and Equitable Quality Education and Promote Lifelong Learning Opportunities for all, ED-2016/WS/28. UNESCO; 2015

[17] Taimur S, Sattar H. Education for sustainable development and critical thinking competency. In: Leal Filho W, Azul Brandli L, Özuyar P, Wall T, editors. Quality Education. Encyclopedia of the UN Sustainable Development Goals. Cham: Springer; 2019. Available from: https://link.springer.com/referenceworkentry/10.1007/978-3-319-69902-8_64-1

[18] Ennis R. Critical thinking dispositions: Their nature and assessability. Informal Logic. 1996;**18**(2):165-182

[19] Ennis R. Critical thinking: What is it? In: Proceedings of the Forty-Eighth Annual Meeting of the Philosophy of Education Society Denver, Colorado, March 27-30. Denver, Colorado: The Philosophy of Education Society; 1992

[20] Rudsberg K, Öhman J. Pluralism in practice – Experiences from Swedish evaluation, school development and research. Environmental Education Research. 2010;**16**(1):95-111. DOI: 10.1080/13504620903504073

[21] Sandell K, Öhman J, Östman L. Education for Sustainable Development. Nature, School and Democracy. Lund: Studentlitteratur; 2005

[22] Pavlova M. Teaching and learning for sustainable development: ESD research in technology education. International Journal of Technology and Design Education. 2012;**23**:733-748

[23] Pedretti E, Nazir J. Currents in STSE education: Mapping a complex field, 40 years on. Science Education. 2011;**95**(4):601-626

[24] Butler J. Conclusion: From Parody to Politics. Gender Trouble-Feminism and the Subversion of Identity (181-190). Oxford: Routledge; 2001

[25] Butler J. Giving an Account of oneself. NY: Fordham University Press; 2005

[26] Perkins DN, Jay E, Tishman S. Beyond abilities: A dispositional theory of thinking. Merrill-Palmer Quarterly. 1993;**39**(1):1-21

[27] Davies M, Barnett R, editors. The Palgrave Handbook of Critical Thinking in Higher Education. NY, USA: Palgrave Macmillan; 2015

[28] UNESCO. Roadmap for Implementing the Global Action Programme on Education for Sustainable Development. UNESCO; 2014. Available from: http://sdg.nuk.edu.tw/data/230514e.pdf

[29] UN. Transforming our world: The 2030 agenda for sustainable development. In: Resolution Adopted by the General Assembly on 25 September 2015. Vol. 42809. UN; 2015. pp. 1-13. DOI: 10.1007/s13398-014-0173-7.2

[30] Kohn E. How Forests Think. California: University of California Press; 2013

[31] Holway W. Towards an eco-psycho-social analysis of climate change. In: Frosh S, Vyrgioti M, Walsh J, editors. The Palgrave Handbook of Psychosocial Studies. Cham: Palgrave Macmillan; 2022. DOI: 10.1007/978-3-030-61510-9_28-1 [Accessed: February 20, 2023]

[32] Ehrlich P, Reed J. My Octopus Teacher. South Africa: Netflix; 2020. Available from: https://www.netflix.com/watch/81045007?source=35 [Accessed: February 27, 2023]

[33] O'Brien K. You Matter More than you Think: Quantum Social Change for a Thriving World. Oslo: Change Press; 2021

[34] Hamilton L. Why the Ethics of Octopus Farming Is So Troubling. The Conversation. York, UK; 2023 (theconversation.com) [Accessed: March 27, 2023]

Chapter 13

Perspective Chapter: Empowering Sustainable Development through Education – A Perspective on the Educational Landscape and Its Impact on Sustainability

Khalid Aada

Abstract

Over the past decade, there has been a compelling discourse on transitioning from a rational paradigm, steeped in functionalist and utilitarian models, to a paradigm more aligned with the preservation of natural resources and life. This shift is crucial for addressing the urgent need to integrate sustainable principles into the educational sphere, with the aim of instigating a genuine cultural transformation. Recognizing the global impact of social, economic, and environmental challenges, numerous international entities advocate for an ambitious, complex, and transformative approach to sustainability in education. This scholarly work explores the intricate relationship between education, cultural change, and sustainability. It provides a comprehensive overview of the educational landscape, critically examining objectives, foundations, challenges in environmental education, and avenues for improvement. The narrative highlights aspects that positively influence conceptual development and educational practices. Moreover, it delves into innovative approaches for addressing environmental issues and steering society toward a more sustainable future. The work concludes by proposing pragmatic solutions to propel the envisioned changes forward.

Keywords: collective responsibility, conceptual development, education practices, educational landscape, environmental stewardship, humanity and the planet, sustainability

1. Introduction

In a world at the edge of environmental, social, and economic challenges, the pressing need for sustainable development has become undeniable. This urgency stems from the recognition that our planet is facing a delicate balance, with potential consequences that demand immediate attention. At the heart of the solution to these challenges is the powerful influence of education, acting as a guiding light on the path to fostering a harmonious coexistence between humanity and the planet [1].

The metaphor of the world teetering on the precipice paints a vivid picture of the precarious situation we find ourselves in, emphasizing the gravity of the issues at hand. The call for sustainable development is not just a suggestion but an imperative, signifying that action is crucial to avert impending crises.

Education takes center stage in this narrative. It is presented not merely as a tool for knowledge dissemination but as a transformative force capable of shaping perspectives and driving meaningful change. Kathleen McLean, a teacher at Great Falls Public School in Great Falls, Montana reveals that she takes her students to places where they can see evidence of environmental problems "I am inspired by their creativity and persistence in finding solutions." [2].

The implication is that education has the potential to empower individuals and societies to navigate the complexities of sustainability, influencing behaviors and fostering a collective responsibility for the well-being of both humanity and the planet.

This text endeavors to explore and elucidate the profound nexus between education and sustainable development, dissecting the various facets of the educational landscape and its consequential impact on the broader tapestry of sustainability.

At its core, education instills a sense of environmental stewardship, fostering a generation that understands the delicate balance between human activities and the health of the planet. By integrating sustainability into curricula, educational institutions become incubators for eco-literate individuals equipped with the knowledge and skills to address pressing environmental challenges.

The first facet of this profound nexus lies in the integration of sustainability into educational curricula. From primary schools to universities, a paradigm shift is needed to infuse sustainability principles across disciplines. A holistic approach to education, incorporating environmental sciences, social studies, and ethical considerations, cultivates a comprehension understanding of the interconnectedness of economic, social, and environmental systems. As an example, in primary schools, incorporating modules on environmental conservation into science classes allows students to learn about ecosystems, climate change, and biodiversity. By integrating sustainability into the curriculum, students gain not only theoretical knowledge but also develop a holistic understanding of their role in preserving the environment [3].

As we embark on this exploration of the profound nexus between education and sustainable development, we recognize the urgency of our times. The world stands at the crossroads of unprecedented challenges, demanding immediate action and a collective commitment to sustainable practices. The metaphorical edge upon which our planet teeters, underscores the gravity of the situation, emphasizing that the call for sustainable development is not a choice but a necessity [4].

In the lines that follow, we will delve into the transformative power of education, guided by the belief that knowledge is not just a tool for understanding but a catalyst for change. Through the lens of dedicated educators and the integration of sustainability into educational curricula, we witness the potential for education to shape a generation that comprehends not only the intricacies of environmental stewardship but also actively engages in preserving the delicate balance between humanity and the planet.

This journey into the heart of education's influence on sustainable development is a call to action, an invitation to explore the interconnected threads that weave together knowledge, responsibility, and meaningful change. As we turn the pages of this chapter book, let us embark on a voyage of discovery, uncovering the ways in which education can be the guiding force toward a future where harmony between humanity and the planet is not just an aspiration but a tangible reality.

1.1 Fostering a green mindset: education's role in sustainable development

To fully elucidate the nexus between education and sustainable development, we must embrace experiential learning and practical application. Field studies, community projects, and internships provide students with hands-on experiences that bridge theoretical knowledge with real-world challenges. This enhances not only the educational experience but also empowers individuals to become agents of positive change within their communities. High school students engaged in a community garden project learn not only about sustainable agriculture but also witness firsthand the impact of responsible land use. Through planting, cultivating, and harvesting, they gain practical skills and a deeper appreciation for the connection between food production, ecological health, and sustainable living.

It's also worth to highlight that:

> *[...] supporting research asserts that environmental education and outdoor programs utilizing explorative learning can drastically improve the student's learning attitudes. Additionally, when these outdoor learning experiences are implemented by trained teachers in familiar settings, students are more engaged and experience greater science achievement [5].*

Education molds not only professionals but also global citizens with a heightened sense of social responsibility. By fostering inclusivity, diversity, and a deep understanding of global interdependencies, education becomes a driving force behind sustainable development. Ethical considerations and a commitment to social justice are integral components of an education that aims to create a sustainable and equitable world [6]. University students participating in international exchange programs could experience different cultures and societal structures. This exposure fosters a global perspective, encouraging them to consider the social and environmental implications of their actions on a global scale. This firsthand experience contributes to the development of socially responsible global citizens.

A university research team focuses on developing sustainability energy solutions. Through their work, they create innovations such as efficient solar panels or bioenergy technologies. The knowledge and technological advancements generated by such research contribute not only to academic excellence but also offer practical solutions for a more sustainable energy future.

The nexus between education and sustainable development is further exemplified through the role of innovation and technological advancement. Education acts as a catalyst for research and development, inspiring breakthroughs that address environmental challenges. The integrations set an example for the broader society, influencing industry practices, and shaping a sustainable future.

We see education as a catalyst for societal progress because it plays a pivotal role in shaping the perspectives, values, and actions of individuals. It serves as the crucible where the seeds of awareness are sown, cultivating minds that are attuned to the complexities of our interconnected world. By fostering a holistic understanding of the environment, society, and economy, education lays the groundwork for sustainable development by instilling a sense of responsibility and stewardship.

Imagine education as a gardener cultivating a vibrant garden of knowledge. It begins by planting seeds of awareness in the fertile soil of young minds. This is the first step in the journey toward societal progress.

Consider the classroom as the greenhouse where these seeds germinate and grow. Through lessons on history, literature, and science, students develop a holistic understanding of the environment, society, and economy. For instance, they might study environmental issues, historical events, and economic systems, gaining insights into the interconnectedness of these elements [7].

Now, let us talk about the catalyst effect. Education acts as the sunlight that fuels the growth of these young minds. It illuminates perspectives, shaping values, and influencing actions. Just like how exposure to diverse literature or discussions about global issues broadens a student's worldview, the sunlight of education fosters a holistic understanding.

Engaging students in hands-on, real-world projects rather than traditional workbook exercises has been shown to significantly enhance their science self-efficacy. Researchers conducted pre- and post-surveys to gauge students' beliefs and experiences throughout this study, revealing compelling results. By immersing students in focused environmental science investigations that are relevant to their local context, a notable improvement in their confidence levels was observed. This heightened confidence empowered not only students to explore and address local environmental challenges but also substantially bolstered their overall scientific inquiry skills. The study highlights the transformative impact of practical, community-centered learning experiences on students' abilities to apply scientific knowledge to real-world scenarios, and fostering a sense of efficacy and competence in scientific endeavors [8].

The pre- and post-surveys administered to the students revealed a marked increase in science self-efficacy. The hands-on projects, which focused on local environmental issues, captured not only the students' interest but also provided them with a tangible connection to the scientific concepts being taught.

For instance, one project involved students investigating the impact of urbanization on local ecosystems. They conducted fieldwork, collected data, and analyzed the results, allowing them to see the direct relevance of scientific principles to their community. The post-survey responses indicated a significant rise in the students' confidence in their ability to understand and contribute to environmental science discussions [9].

As these educated individuals graduate into the wider world, they become stewards of knowledge. It's like sending out well-prepared gardeners into the world, equipped with the tools of responsibility and stewardship. They understand the delicate balance of the ecosystem—akin to the balance of a well-tended garden—and are committed to sustainable development.

So, education is not just a process; it's a journey of planting seeds, nurturing growth, and ultimately harvesting a crop of responsible, informed individuals who contribute to societal progress [10].

2. Education's canvas and pedagogical performance

The foundation of sustainable development is deeply embedded in the ethos of education. From the early formative years to the halls of higher learning, educational institutions bear the weighty responsibility of nurturing a generation that comprehends not only the intricacies of sustainability but also embodies its principles. The curriculum becomes a canvas upon which the brushstrokes of ecological consciousness, social equity, and economic resilience are painted. Subjects such as environmental science, social studies, and sustainable business practices form the tapestry of knowledge that equips students with the tools to address the multifaceted challenges facing our world.

To dive into the layers of this educational foundation for sustainable development, let us picture education as the bedrock upon which the pillars of sustainable development are erected. Starting from the earliest stages of learning and extending through the corridors of higher education, institutions play a pivotal role in sculpting a generation attuned to the nuances of sustainability. Think of it as crafting architects of a sustainable future.

Now, envision the curriculum as an expansive canvas, stretched wide to capture the essence of sustainable living. Each subject is a brushstroke contributing to the masterpiece of ecological consciousness, social equity, and economic resilience. Consider subjects like environmental science as the vibrant green strokes symbolizing a profound understanding of our planet's delicate balance. Social studies, in turn, adds hues of empathy and awareness, weaving threads of understanding into the fabric of societal interconnectedness. Sustainable business practices are like bold strokes of economic paint, creating a resilient foundation that withstands the tests of time [11].

As students advance through this educational journey, they become artisans of sustainable knowledge. They grasp not only the intricacies of sustainability but also embody its principles. It's akin to watching artists skillfully apply paint, turning a blank canvas into a vivid representation of a harmonious and sustainable world.

In essence, education transforms into a workshop where the tools of awareness, empathy, and practical skills are forged. These tools are crucial for addressing the multifaceted challenges that our world faces—challenges that require a nuanced and sustainable approach. So, from the foundational strokes of early education to the intricate details of higher learning, the canvas of education becomes a tapestry that equips individuals with the brushes and colors to paint a sustainable future [12].

Moreover, the pedagogical approach itself must undergo a metamorphosis, transcending traditional paradigms to embrace experiential, interdisciplinary, and collaborative learning. The integration of real-world scenarios and practical applications into the educational milieu enhances the relevance of sustainability concepts, fostering a dynamic engagement with the material. Through project-based initiatives, field studies, and community involvement, students grasp not only theoretical frameworks but also develop the skills and aptitudes necessary for active participation in the sustainable development agenda [13].

Imagine the pedagogical approach as a script undergoing a rewrite, breaking free from the constraints of traditional scenes to embrace a dynamic narrative of experiential, interdisciplinary, and collaborative learning. It's like transforming a classic play into an interactive and engaging performance that captivates the audience.

Now, think of the educational landscape as a stage, with the integration of real-world scenarios and practical applications as the special effects that bring the script to life. The classroom becomes a theater where sustainability concepts take center stage, creating a dynamic and immersive experience. It's not just teaching; it's a performance where students actively participate and engage with the material.

Enter project-based initiatives, field studies, and community involvement as the plot twists that add suspense and depth to the educational storyline. Projects are the unexpected turns that challenge students to apply theoretical knowledge in creative ways. Field studies are the on-location scenes that transport students from the theoretical setting to the real-world stage of sustainability. Community involvement is the subplot that connects students to the broader narrative of societal development.

As students become the actors in this educational play, they do not just memorize lines—they develop the skills and aptitudes necessary for an Oscar-worthy performance in the sustainable development agenda. It's like transforming students from passive spectators into active participants, ready to take on roles that contribute meaningfully to the unfolding drama of a sustainable future.

2.1 Extending the impact of education on sustainable development

In our current century, we cannot deny recognizing the integration of technology that has become increasingly indispensable in our lives. When educators leverage technology to impact environmental education concepts, they are equipping students with a valuable life skill. A notable study, Project Engaging At-Risk Populations Outdoors, Digitally (EARPOD) [14], delved into the impact of utilizing digital technology in outdoor education experiences. This comprehensive program engaged not only underserved students in environmental education but also sought to assess the effectiveness of incorporating technology into their learning journeys. The finding revealed significant outcomes as students reported a heightened inclination toward using technology outdoors. Furthermore, they demonstrated an increased awareness of available technologies and expanded their proficiency with diverse technological tools. This surge in technology-specific knowledge translated into a notable boost in students' confidence when utilizing technology, emphasizing the transformative potential of integrating digital tools into environmental education [15].

The impact of education on sustainable development extends beyond the classroom, permeating the fabric of societal norms and behaviors. Educational institutions, as bastions of enlightenment, wield considerable influence in shaping cultural narratives and societal values. By promoting inclusivity, diversity, and ethical considerations, education becomes a transformative force that transcends individual minds, permeating the collective consciousness and fostering a culture conducive to sustainable practices.

What would happen if we were in a bustling city where educational institutions stand as towering pillars of enlightenment? In this context, let us focus on a university that actively promotes inclusivity and diversity, and picture students from various backgrounds coming together, sharing experiences, and engaging in dialog that transcends cultural boundaries. This university becomes a microcosm of a society that values differences, fostering an environment where individuals learn not only from textbooks but also from the rich tapestry of perspectives around them.

Another relevant aspect for the mental health benefits standpoints, is revealed by a study investigating the association between greenness around schools and symptoms of ADHD [ADHD stands for attention deficit hyperactivity disorder. It is a neurodevelopmental disorder that affects individuals' ability to focus, control impulses, and regulate their energy levels. ADHD is typically diagnosed in childhood, although symptoms can persist into adulthood], data showed that greater greenness levels were significantly associated with lower incidences of ADHD symptoms [16]. Building upon this insight, additional research underscores the intrinsic value of nature in mitigating psycho-physiological stress. Exposure to natural environments is identified as a potent mediator, actively countering the adverse effects of stress by reducing negative mood states and, most importantly, amplifying positive emotions. The therapeutic impact of spending time in nature extends to eliciting calming responses, facilitating the recovery of both physiological stress and mental fatigue, and even contributing to the restoration of cognitive performance. These findings collectively emphasize the pivotal role of green environments not only in addressing ADHD symptoms but also in promoting holistic well-being and cognitive resilience through a symbolic relationship with nature [17].

As these students graduate and step into the wider community, they carry with them the seeds of inclusivity and diversity planted during their educational journey. In the workplace, they advocate for fair hiring practices, ensuring that opportunities are extended to individuals from diverse backgrounds. This, in turn, influences the cultural narrative within the corporate world, challenging traditional norms and creating an atmosphere that thrives on the strength of varied perspectives.

2.2 Education's role in fostering informed choices and cultural transformation

Now, let us shift our focus to the ethical considerations integrated into the curriculum of a high school. Here, students engage in discussions about responsible consumerism and sustainable business practices. As they delve into real-world case studies, they develop a keen understanding of the ethical implications of their choices. These students, armed with ethical awareness, become informed consumers who prioritize sustainable products and support businesses that align with their values.

As this ethical consciousness permeates the societal fabric, it prompts businesses to reconsider their practices. Picture a scenario where a local company, influenced by the demand for ethical and sustainable products, adopts eco-friendly manufacturing processes. The ripples of this change extend beyond the business realm, influencing the cultural narrative by elevating ethical considerations in the community's collective mindset [18].

In essence, education acts as a catalyst for change, influencing not only individual minds but also shaping the broader cultural narrative. By promoting inclusivity, diversity, and ethical considerations, educational institutions become beacons of transformation. They sow the seeds of awareness and ethical consciousness that, when carried into society, foster a culture conducive to sustainable that extends far beyond the classroom, weaving a narrative of positive change throughout the fabric of societal norms and behaviors.

In addition to the cognitive dimensions of education, the emotional and ethical dimensions must be nurtured. Empathy, compassion, and a sense of global citizenship are integral components of a sustainable mindset. Educational initiatives that cultivate these qualities contribute to the development of responsible and conscientious individuals who recognize the interdependence of all life on Earth. Through fostering a sense of interconnectedness, education becomes a powerful instrument in dismantling the barriers that perpetuate environmental degradation, social injustice, and economic inequality.

Our perception changes when we look at education as a three-dimensional puzzle, with cognitive, emotional, and ethical dimensions interlocking to form a comprehensive understanding of the world. In this puzzle, cognitive aspects represent the intellectual knowledge and skills gained through learning, while emotional aspects encompass the ability to understand and manage one's emotions and empathize with others. Ethical dimensions, on the other hand, involve the development of a moral compass, guiding individuals to make decisions that align with principles of fairness, justice, and global responsibility [19].

The cultivation of empathy, compassion, and global citizenship is crucial for the development of responsible and conscientious individuals. It's not just about what we know, but also about how we feel and the ethical principles that guide our actions. Students, through educational initiatives focusing on empathy, learn to understand and share the feelings of someone from a different cultural background. This emotional intelligence, coupled with ethical considerations, forms the foundation of a mindset that values diversity and promotes inclusivity.

Now, let us explore the idea of a sustainable mindset. Enriched with empathy and a sense of global citizenship, this mindset acknowledges the interdependence of all life on Earth. It's like realizing that each puzzle piece has a role in forming the bigger picture of a sustainable and interconnected world. Education, in this context, is not just about acquiring knowledge but about fostering a deep understanding of the intricate web of relationships that bind humanity, nature, and the global community. By nurturing a sense of interconnectedness, education becomes a potent instrument for change. It's as if education is dismantling the barriers that perpetuate environmental degradation, social injustice, and economic inequality. Consider a ripple effect where individuals, equipped with a sustainable mindset, actively work toward breaking down these barriers. They engage in environmentally conscious practices, advocate for social justice, and contribute to economic systems that prioritize equality.

The role of educational institutions is not confined to the dissemination of knowledge alone; they must also serve as exemplars of sustainability. Campuses should evolve into living laboratories where eco-friendly practices, renewable energy sources, and waste reduction strategies are implemented. By embodying the principles they espouse, educational institutions become living testimonies to the feasibility and desirability of sustainable living [20].

A university instead of just teaching environmental ethics, becomes a testament to these principles. Solar panels adorn rooftops, providing a substantial portion of the energy needed to power classrooms and facilities. This reduces not only the institution's carbon footprint but also serves as a tangible demonstration of the viability of renewable energy sources.

Furthermore, the university incorporates waste reduction strategies by implementing comprehensive recycling programs and composting initiatives. Waste bins are strategically placed, accompanied by educational signage encouraging students and staff to actively participate in the recycling process. As a result, the campus functions as a "living laboratory" showcasing how waste reduction can be seamlessly integrated into daily life.

In terms of infrastructure, the university embraces eco-friendly designs and construction materials. Green spaces and gardens are interspersed throughout the campus, promoting biodiversity, and serving as outdoor classrooms. Rainwater harvesting systems are in place, providing irrigation for these green areas and reducing reliance on external water sources [21].

This holistic approach extends beyond the physical environment to educational practices. Sustainable living is woven into the curriculum, fostering a mindset of environmental stewardship among students. A biology class does not only teach the theory of ecosystems but also involves students in hands-on projects to restore local habitats or monitor wildlife populations on campus. In essence, that setting becomes a dynamic, real-world example of sustainable living. Students learn not only about these principles but actively engage with them in their daily lives. The university, in embodying the sustainability it preaches, becomes a living testimony to the feasibility and desirability of integrating eco-friendly practices into our broader society.

3. Education for climate action: UNESCO's role in catalyzing global transformation

UNESCO [22] organization has been mobilizing education to address climate change. It underscores the pivotal role of education, particularly within the framework of education for sustainable development (ESD), in addressing the complexities of climate change highlighting it as a central theme with ESD, acting as a catalyst to enhance people's understanding and response to the impacts of the climate crisis. The aim is to empower individuals with a comprehensive set of tools, including knowledge, skills, values, and attitudes, transforming them into proactive agents of change.

The global recognition of the significance of education and training in the context of climate change is emphasized by its inclusion in key international agreements. The UN Framework Convention on Climate Change, the Paris Agreement, and the associated Action for Climate Empowerment agenda all underscore the imperative for governments to educate, empower, and engage diverse stakeholders and major groups in policies and actions related to climate change. This collective call reflects a shared understanding that addressing climate change requires a broad and informed participation from all sectors of society.

UNESCO, through its ESD program, actively contributes to making education a more prominent and integral component of the global response to climate change. The organization plays a multifaceted role in this regard. It generates and disseminates knowledge on climate change and sustainable practices, ensuring that the latest insights are accessible to educational institutions and policymakers. Additionally, UNESCO provides crucial policy guidance and technical support to countries, aiding them in the development and implementation of effective climate change education strategies.

This engagement makes us to see education from different lenses. First, education is a fundamental human right and cornerstone for building peaceful, inclusive, and sustainable societies. Second, challenges like climate change are inherently interconnected, spanning environmental, social, and economic dimensions. By prioritizing ESD, we acknowledge that addressing these complex challenges requires a holistic and integrated approach that transcends disciplinary boundaries.

The emphasis on climate change education aligns with the belief that informed and educated individuals are better equipped to contribute meaningfully to sustainable development. By providing people with knowledge, skills, and attitudes related to climate change, we aim to make informed decisions, adopt sustainable practices, and actively participate in the collective efforts to mitigate and adapt to climate change.

Third, the inclusion of education in key international agreements reflects a broader consensus on its importance to address global challenges, which led to the urgency of translating the global mandates into actionable initiatives, ensuring that education remains a central component of international strategies to combat climate change.

Fourth, education is viewed as a means of building individual and societal capacity to adapt to changing circumstances. We may need to enhance resilience at both the individual and community levels, preparing people to navigate the evolving challenges posed by climate change. But this multifaceted approach also needs the participation of spanning knowledge generations, policy guidance, technical support, and practical projects, which would reflect an understanding that systemic change requires a comprehensive strategy to catalyze transformative shifts in educational systems, societal attitudes, and sustainable practices [23].

In essence, the decade of ESD represents a concerted effort to foster a global consciousness with education, aiming to cultivate a nuanced understanding of the world's current state. The overarching goal is to instill responsible attitudes and commitments in learners, equipping them with the capacity to make informed decisions crucial for achieving cultural pluralism, social equity, and ecological sustainability. This educational paradigm transcends traditional anthropocentric perspectives, emphasizing the need for more holistic and intelligent models of interaction with ecosystems.

It is crucial to recognize that the ESD movement is far from monolithic; it exhibits a diverse range of meanings, conceptual expressions, and practices, often giving rise to tension among proponents [24]. However, certain fundamental elements have persisted since its inception, actively championed by its foremost advocates. Among these elements, critical practice stands out as a pivotal pedagogical approach that integrates both reflection and action [25].

The notion of critical practice within ESD prompts two overarching questions that place culture at the heart of educational endeavors. Firstly, within the content of a critical practice and amidst the backdrop of globalization, which frequently engenders unsustainable values and practices within communities, what role should education assume in the pursuit of sustainable development and the resilience of human-natural systems? Secondly, the question arises regarding the content and methodology of teaching—what should be taught, and how should it be taught to effectively fulfill the overarching purpose of sustainable development? These inquiries underscore the imperative for a thoughtful and adaptive educational approach that imparts not only knowledge but also cultivates the critical thinking skills necessary for navigating the complexities of our interconnected global society [26].

4. Action plan for sustainable development in educational institutions

Crafting an action plan for sustainable development within educational institutions involves a multifaceted approach encompassing curriculum, operations, and community engagement. A comprehensive guide to such a plan may include:

- Integrating sustainability into the curriculum. In other words, developing and incorporating modules on sustainable development across disciplines; establishing interdisciplinary courses that encourage holistic thinking about ecological, social, and economic sustainability; and fostering critical thinking by including real-world case studies and global examples of sustainability practices.

- Green campus initiatives by conducting energy audits to identify areas for efficiency improvements; by implementing renewable energy sources such as solar panels and promoting energy conservation practices; and by ensuring waste reduction through recycling programs and waste segregation.

- Eco-friendly transportation through encouraging the use of public transportation, cycling, or walking; and implementing carpooling initiatives and providing incentives for sustainable commuting practices.

- Sustainable infrastructure that will include a design and construction of buildings with eco-friendly materials and energy-efficient technologies; and a creation of green spaces on campus, promoting biodiversity, and serving as educational tools.

- Community engagement by collaborating with local communities to address shared sustainability challenges; and, by organizing workshops, seminars, and awareness campaigns on sustainable living for students and community members.

- Partnership and networking by fostering partnerships with organizations and businesses committed to sustainable practices; and engaging in knowledge exchange and collaborative projects with other educational institutions pursuing sustainable development.

- Ethical procurement prioritizes purchasing products with eco-friendly certifications and from socially responsible suppliers; and establishing guidelines for ethical procurement practices across departments.

- Student involvement and leadership by creating student-led sustainability committees or clubs; and encouraging student participation in decision-making processes related to sustainable initiatives.

- Professional development providing training and workshops for faculty and staff on integrating sustainability into teaching and daily operations; and encouraging research and publications on sustainable practices within the educational context.

- Regular assessment and reporting through establishing key performance indicators (KPIs) to measure and track progress; and regularly assessing the impact of sustainable initiatives and communicating findings to stakeholders.

- Cultivate a culture of sustainability fostering a sense of responsibility for sustainable practices among students, faculty, and staff; celebrating and recognizing achievements in sustainable development to reinforce positive behaviors.

- Continuous improvement by regularly reviewing and updating the action plan based on feedback and changing sustainability goals; and embracing a culture of continuous improvement to adapt to emerging challenges and opportunities.

5. Conclusion

The symbiotic relationship between education and sustainable development is the linchpin upon which the future of our planet hinges. As we stand at the crossroads of unprecedented challenges, education emerges as the catalyst for transformative change, empowering individuals to become stewards of a sustainable and equitable world. Through a recalibration of curricula, pedagogical methodologies, and institutional practices, education can transcend its conventional role and emerge as a formidable force for positive change. The journey toward sustainable development is, at its core, an educational odyssey—one that demands our unwavering commitment to fostering enlightened minds capable of navigating the intricate tapestry of a sustainable future.

In concluding our exploration of "Empowering Sustainable Development through Education: A Perspective on the Educational Landscape and its Impact on Sustainability," we find ourselves at the intersection of knowledge and action. Throughout this chapter book, we have traversed the vast terrain of education's transformative power, weaving a narrative that unfolds like a tapestry of interconnected concepts and initiatives.

From envisioning education as a garden, where seeds of awareness are planted and nurtured, to portray it as a dynamic stage, where students actively engage with the material through experiential and collaborative learning, we have embraced diverse metaphors to capture the essence of education's role in sustainable development.

Our journey has taken us through the multifaceted dimensions of education, exploring its cognitive, emotional, and ethical aspects. We've witnessed how education acts as a catalyst for societal progress, shaping the perspectives, values, and actions of individuals. It is not merely a process but a continuous journey of planting seeds, nurturing growth, and ultimately harvesting a crop of responsible, informed individuals ready to contribute to societal progress.

The chapter book unfolded the canvas of education, where brushstrokes of ecological consciousness, social equity, and economic resilience were painted. We delved into the technological integration that empowers students with essential life skills and the ethical considerations that shape responsible consumer choices. We examined how education extends its influence beyond the classroom, permeating the fabric of societal norms and behaviors.

UNESCO's pivotal role in mobilizing education to address climate change highlighted the global recognition of education's significance in combatting environmental challenges. The multifaceted approach to sustainable development within educational institutions, as outlined in our action plan, underscores the comprehensive strategy required to catalyze transformative shifts in educational systems, societal attitudes, and sustainable practices.

As we stand at this juncture, it's essential to emphasize the collective responsibility we bear in translating the insights gained into actionable initiatives. Our understanding of education's potential to drive sustainable development necessitates a commitment to continuous improvement and adaptation to emerging challenges. The tapestry we have woven together is a dynamic, living entity that requires nurturing, revisiting, and expanding.

In the chapters ahead, we invite you to delve deeper into specific aspects, projects, and initiatives, as we collectively embark on a journey toward a more sustainable and equitable future. "Empowering Sustainable Development through Education" is not just a chapter book; it's a call to action, an exploration of possibilities, and a testament to the transformative impact education can have on shaping a more sustainable world. The journey continues, and with each step, we move closer to realizing the vision of a harmonious coexistence between humanity and the planet.

6. Source selection criteria

In curating the literature for this chapter, a methodical approach was employed to ensure relevance, credibility, and comprehensiveness. A thorough review of peer-reviewed journals, academic publications, and reputable sources was conducted to identify literature that contributed significantly to the understanding of sustainability in education. Emphasis was placed on selecting sources that provided current insights and perspectives on the role of education in fostering sustainability and cultural transformation. Additionally, sources were evaluated based on their alignment with the overarching themes and objectives of the chapter, ensuring that they provided valuable insights into the multifaceted dimensions of sustainable development within educational contexts.

The criteria for source selection encompassed various factors aimed at ensuring the quality and relevance of the literature included in this chapter. Primarily, sources were evaluated based on their authority and credibility within the field of sustainability education. Peer-reviewed journals, academic publications, and reputable organizations were prioritized to ensure that the information presented was grounded in rigorous research and scholarship. Additionally, the recency of the sources was considered to incorporate the latest insights and developments in the field. This criterion aimed to provide readers with up-to-date perspectives on the evolving discourse surrounding sustainability education.

Another key criterion was the thematic alignment of the sources with the overarching themes and objectives of the chapter. Literature that addressed the role of education in fostering sustainable development, promoting informed choices, and catalyzing cultural transformation was prioritized. By selecting sources that directly contributed to these thematic areas, the chapter aims to provide a comprehensive exploration of the intersection between education and sustainability. Additionally, sources that offered diverse perspectives and insights, including theoretical frameworks, empirical studies, and case examples, were included to provide readers with a holistic understanding of the topic.

Furthermore, the accessibility and availability of the sources were considered to ensure that readers could access and engage with the literature cited in the chapter. Whenever possible, open-access publications and widely available sources were prioritized to promote inclusivity and facilitate further exploration of the topics discussed. Overall, the criteria for source selection were designed to uphold the highest standards of quality, relevance, and accessibility, thereby enriching the reader's understanding of sustainability education and its transformative potential.

Acknowledgements

In the pursuit of knowledge and the creation of this chapter book, I am grateful for the unwavering support that has enriched every step of the journey. To my beloved wife, Faviola, your enduring encouragement, and unwavering support that fuels my passion and understanding have been my rock.

To my children, Samia and Ismael, your commitment, understanding, and boundless joy have been the heartbeat of our home. During the moments when Dad is immersed in his work, your laughter and love have made every endeavor more meaningful.

A special thank you goes to Mrs. Sarah Munoz, our dedicated department assistant. Your guidance through administrative processes has been invaluable, and your unwavering commitment to efficiency has made this journey smoother.

I extend my gratitude to the committee of enhancement funds for their willingness to support the publication of creative works. Your belief in the value of knowledge and creative endeavors contributes significantly to the broader intellectual landscape.

A heartfelt acknowledgment to Mrs. Helena Recic, the Publishing Process Manager of IntechOpen publisher, for her professional follow-up. Her expertise and dedication in managing the publishing process are crucial to the successful realization of this project.

To all those who have been part of this chapter book's journey, your contributions, support, and encouragement have been instrumental. This work is a culmination of collaborative efforts, and I am deeply appreciative of the collective spirit that has fueled its creation.

Author details

Khalid Aada
Literature and Cultural Studies Department, College of Liberal Arts, University of Texas Rio Grande Valley, Brownsville, Texas, USA

*Address all correspondence to: khalid.aada@utrgv.edu

References

[1] Anyolo E, Kärkkäinen S, Keinonen T. Implementing education for sustainable development in Namibia: School teachers' perceptions and teaching practices. Journal of Teacher Education for Sustainability. 2018;**20**(1):64-81. DOI: 10.2478/jtes-2018-0004

[2] McLean Kathleen In NEEF, What We Do; l-12 Education; Benefits of Environmental Education [Internet]. 2023. Available from: https://www.neefusa.org/what-we-do/k-12-education/benefits-environmental-education

[3] Carban E, Fisher D. Sustainability reporting at schools: Challenges and benefits. Journal of Teacher Education for Sustainability. 2017;**19**(1):69-81. DOI: 10.1515/jtes-2017-0005

[4] Cantrell DC. Alternative paradigms in environmental education research: The interpretive perspective. Alternative Paradigms in Environmental Education Research. 1993;**8**:81-104

[5] Corney G. Education for sustainable development: An empirical study of the tensions and challenges faced by geography student teachers. International Research in Geographical and Environmental Education. 2006;**15**(3):224-240. DOI: 10.2167/irgee194.0

[6] Dimenäs J, Alexandersson M. Crossing disciplinary borders: Perspectives on learning about sustainable development. Journal of Teacher Education for Sustainability. 2012;**14**(1):5-19. DOI: 10.2478/v10099-012-0001-0

[7] Gayford, C. Learning for Sustainability: From the Pupils Perspective. 2003. Retrieved December 3, 2017. Available from: http://www.se-ed.co.uk/sites/default/files/resources/WWF%20Learning_for_sustainability.pdf

[8] Yang B-Y, Zeng Z-W, Markevych I, Bloom MS, Heinrich J, Knibbs LD, et al. Association between greenness surrounding schools and kindergartens and attention-deficit/hyperactivity disorder in children in China. JAMA Network Open. 2019;**2**(12). DOI: 10.1001/jamanetworkopen.2019.17862

[9] Hougham RJ, Nutter M, Graham C. Bridging natural and digital domains: Attitudes, confidence, and interest in using technology to learn outdoors. The Journal of Experimental Education. 2018;**41**(2):154-169. DOI: 10.1177%2F1053825917751203

[10] van Gejeka N. Research into learning process in a vocational secondary school in the context of sustainable development. Journal of Teacher Education for Sustainability. 2013;**15**(2):67-77. DOI: 10.2478/jtes-2013-0012

[11] Hungerford HR. Environmental education (EE) for 21st century: Where have we been? Where are we now? Where are we headed? Journal of Environmental Education. 2010;**41**(1):1-6. DOI: 10.1080/00958960903206773

[12] Majumdar S. Integrating sustainable development in TVET curriculum. In: 11th UNESCO–APEID International Conference on "Reinventing Higher Education: Toward Participatory and Sustainable Development"; Bangkok, Thailand. Berlin, Germany: Walter de Gruyter GmbH; 2012

[13] Mweti I, van Wyk H. NSSC Development Studies. Module 2.

Cambridge: Cambridge University Press; 2005

[14] Rios JM, Brewer J. Outdoor education and science achievement. Applied Environmental Education & Communication. 2014;**13**(4):234-240. DOI: 10.1080/1533015X.2015.975084

[15] Barnett M, Vaughn M, Strauss E, Cotter L. Urban environmental education: Leveraging technology and ecology to engage students in studying the environment. International Research in Geographical and Environmental Education. 2011;**20**(3):199-214. DOI: 10.1080/10382046.2011.588501

[16] NEEF. What We Do; l-12 Education; Benefits of Environmental Education [Internet]. 2023. Available from: https://www.neefusa.org/what-we-do/k-12-education/benefits-environmental-education

[17] Berto R. The role of nature in coping with psycho-physiological stress: A literature review on restorativeness. Behavioral Science. 2014;**4**(4):394-409. DOI: 10.3390/bs4040394

[18] Ololube NP, Egbezor DE, Kpolovie PJ. Education policies and teacher education programs: Meeting the millennium development goals. Journal of Teacher Education for Sustainability. 2008;**9**:21-34. DOI: 10.2478/v10099-009-0016-3

[19] Olsson D, Gericke N, Chang Rundgren S-N. The effect of implementation of education for sustainable development in Swedish compulsory schools—Assessing pupils' sustainability consciousness. Environmental Education Research. 2016;**22**(2):176-202. DOI: 10.1080/13504622.2015.1005057

[20] Sund L. Facing global sustainability issues: Teachers' experiences of their own practices in environmental and sustainability education. Environmental Education Research. 2016;**22**(6):788-805. DOI: 10.1080/13504622.2015.1110744

[21] Velazquez L, Munguia N, Sanchez M. Deterring sustainability in higher education institutions: An appraisal of the factors which influence sustainability in higher education institution. International Journal of Sustainability in Higher Education. 2005;**6**(4):383-391. DOI: 10.1108/14676370510623865

[22] UNESCO. UNESCO at COP28: Making Education the Long-Term Solution to the Climate Crisis [Internet]. 2023. Available from: https://www.unesco.org/en/articles/unesco-cop28-making-education-long-term-solution-climate-crisis#:~:text=UNESCO%20is%20accelerating%20climate%20change,for%20sustainable%20development%20(ESD)

[23] Nikolopoulou A, Abraham T, Mirbagheri F. Education for Sustainable Development: Challenges, Strategies and Practices in a Globalizing World. 1st ed. Thousand Oaks, CA, USA: SAGE Publications Pvt. Ltd.; 2018

[24] Huckle J, Sterling S. Education for Sustainability. London: Earthscan Publications; 1996. p. 278. DOI: 10.4324/9781315070650

[25] Tilbury D. Emerging issues in education for sustainable development. In: Bhandari BB, Ab O, editors. Education for Sustainable Development in Nepal: Views and Visions, Japón. International Institute for Global Environmental Strategies (IGES). Kamiyamaguchi, Hayama, Kanagawa, Japan [Internet]; 2003. Available from: https://www.iges.or.jp/en/system/files/publication_documents/pub/policyreport/749/999-report9_full.pdf

[26] Stevenson RB, Brody M, Dillon J, Wals AEJ. International Handbook of Research on Environmental Education. New York, London: Routledge Taylor & Francis Group; 2013. ISBN 9780415892391

Secondary sources consulted

Summers M, Childs A, Corney G. Education for sustainable development in initial teacher training: Issues for interdisciplinary collaboration. Environmental Education Research. 2005;**11**(5):623-647. DOI: 10.1080/13504620500169841

Uitto A, Saloranta S. Subject teachers as educators for sustainability: A survey study. Education Sciences. 2017;7(8). DOI: 10.3390/educsci7010008.10.3390/educsci7010008. Available from: https://www.mdpi.com/2227-7102/7/1/8

Wals AEJ. Learning our way to sustainability. Journal of Education for Sustainable Development. 2011;**5**(2):177-186. DOI: 10.1177/097340821100500208

Willis GB. Analysis of the Cognitive Interview in Questionnaire Design. Understanding Qualitative Research. Oxford: Oxford University Press; 2007

Yavetz B, Goldman D, Pe'er S. How do preservice teachers perceive 'environment' and its relevance to their area of teaching? Environmental Education Research. 2014;**20**(3):354-371. DOI: 10.1080/13504622.2013.803038

Yin RK. Case Study Research: Design and Methods. 4th ed. Thousand Oaks: Sage; 2009. Retrieved December 3, 2017, Available from: https://manyebooks.org/download/Yin-case-study-research.pdf

Yin RK. Applications of Case Study Research. 3rd ed. Thousand Oaks: Sage; 2012

Zhang Y, Wildemuth BM. Qualitative analysis of content. Applications of social research methods to questions in information and library science. Libraries Unlimited, 2009. pp. 1-12. Scientific Research, Academic publisher, Open access. 2009. pp. 308-319. Available from: https://www.scirp.org/reference/ReferencesPapers?ReferenceID=675664

Chapter 14

Perspective Chapter: SDG 4, Educational Strategy and Awareness for Social Innovation in Sustainable Development

Rodolfo Martinez Gutierrez, Gaudencio Lucas Bravo, Angel Ernesto Jimenez Bernardino and José Daniel Padilla De la Rosa

Abstract

The education sector at the local level around the world has great challenges regarding global awareness actions in the student population, to creatively convey the importance of environmental protection as an action of social responsibility toward future generations. That is why, SGD 4 of the 2030 goals represents a key link to encourage students at different educational levels to develop initiatives to take care of natural resources and empathy for well-being, health, and sport as key factors for better living conditions in the population. Beyond conceptual understanding, academic practices and research are required to manage knowledge of sustainable development.

Keywords: 2030 goals, social innovation, sustainable development, SDG 4, sectoral competitiveness

1. Introduction

The great global challenges to raise awareness among the population about the importance of actions for sustainable development reflect local, regional, and national efforts, with systemic initiatives promoted by different sectors of society, especially the 2030 Agenda for Sustainable Development Goals (SDGs) established by the United Nations (UN), have become an international benchmark for setting commitments, targets, and indicators in each of the 17 SDGs [1]. SDG 4: Quality Education is the objective of the analysis in this document as a contribution to the State of the Art of Development from the Higher-Level Education Sector in the City of Tijuana, Baja California, Mexico, with national and international linkage actions. The methodological emphasis of the research is the awareness of Social Innovation for Sustainable Development [2].

The research begins with the analysis of the regulations of the Mexican Government for the training and updating of teachers, which includes the focus on empowerment and management of academic activities, linkage, social service, and development of projects through the SDGs in the classes of different educational programs. Methodologically, social innovation is influenced to improve sustainable development, and thus, sectoral competitiveness. Analysis was carried out on the social impact of progress in the implementation of the SDGs. Follow-up was carried out on Educational Social Service Activities to support the development of Node Programs to develop the Social and Solidarity Economy [3].

Initiatives are taken to develop TecNM Scientific Research Projects to promote sectoral linkages through scientific dissemination activities. Likewise, the activities of national and international linkage through the productivity of the Academic Body (ITTIJ-CA-12) and the National Network for Research in Social and Solidarity Economy (RIESS) of the National Technological Institute of Mexico [4]. The research addresses the following topics, from the conceptual approach, and the evidence focused on SDG4, quality education, is the main objective of research:

1. Guidelines of the Program for the Professional Development of Teachers (PRODEP)
2. Fifth Systemic Helix Methodology (QHS) and Dictionaries of Sectoral Competencies (DCS) for Sectoral Development and Social Innovation
3. Municipal Development Agenda (MDA)
4. Center for Research and Documentation of the Public, Social, and Cooperative Economy—Mexico (CIRIEC)
5. TecNM Scientific Research Projects
6. TecNM's NODESS Tijuana Project
7. Commission for the Promotion of the Social and Solidarity Economy
8. Training Program in Social and Solidarity Economy (SSE) (Diploma, Master's, and Doctorate)
9. TecNM Academic Body ITTIJ-CA-12
10. Voluntary Report on Progress on the SDGs in Cities
11. TecNM Social and Solidarity Economy (SSE) Research Network

2. Methods and materials

According to the State of the Art of Bibliometrics of Scopus Article Databases, Web of Science on the topic of SDG 4, and Quality Education, there are great challenges and critical roles that depend on the teaching practices of teachers [5], to influence motivation and awareness of the elements and actions that make up the quality of life from the student phase to the professional phase [6], being the teacher who plays a key role as a mentor of good practices, tasks, projects, social service, professional practices, essays, theses, publications, and the development of critical thinking through the development of scientific research and development initiatives [7–9].

The great expectation of a professor and researcher in the development of an article, or research report, is to disseminate the findings, product of teaching, and research activities, as well as the link with the different sectors of society as part of the efforts of full-time professors and researchers with national accreditation. This chapter presents the efforts of good practices to develop a roadmap that can be replicated in a local, national, and international systemic context, as a framework of reference for activities from academia and applied research.

In Mexico, full-time professors must accredit by regulation a certification named PRODEP (Desirable Profile), which includes the need to empower the 2030 Agenda and develop actions within their classes on the understanding and awareness of each SDG [10].

With the focus of the Fifth Helix Methodology (QHS in Spanish), the importance and benefits of linking with the representatives of the society are illustrated by government, companies, academia, associations—business chambers, and consultants, to influence actions to raise awareness among the population from the local level in a city, with small efforts and constant actions and seek the relevance and belonging of the society with each SDG, as well as the affinity in the sense of social responsibility. In the case of INAFED (National Institute for Federalism and Municipal Development), the 2030 Agenda is promoted in public policy actions in the municipalities in the Municipal Development Plans, in coordination with the academic sector to train society in public spaces and civic squares, as well as through mass media such as social networks and giant screens for public dissemination of information on the SDGs. In the specific case of the city of Tijuana, Baja California, efforts have been made to provide the necessary resources. For example, in Avenida Revolución, the most famous street in the city, there is a mega screen with the capacity to view information from more than 85,000 people daily. From the Tijuana Institute of Technology of TecNM, different efforts have been developed, as is the case of the research project developed between the TecNM Campus Tijuana and the Municipal Institute of Citizen Participation of

Objectives		Feedback/goals
1	Diagnosis and municipal training in the necessary competencies of higher education students on the Sustainable Development Goals (SDGs) of the 2030 Agenda.	TecNM, at the national level, has more than 600,000 students throughout Mexico, with social service activities, dissemination through digital media, social networks (TikTok, YouTube, and Facebook), in squares and public spaces, the goal of 1,000,000 people empowered and aware of the importance of the SDGs can be achieved.
2	Diagnosis and development of workshops for methodological competencies, to encourage the design of research projects of higher education students on the SDGs.	There is low participation of students in TecNM Research Project Calls, so it is necessary to train and motivate them to develop the scientific research profile, emphasis on the 2030 Agenda, priority issues for the country, and regional and local development.
3	Evaluation and design of a methodological proposal to encourage the development of subnational voluntary reports in cities on the progress of the SDGs.	Currently, Mexico has only presented national reports with statistical averages of the progress made in the Sustainable Development Goals (SDGs); there are no municipal subnational reports (ISNM). Tijuana is in the process of completing the First Municipal Subnational Report, and the learning curve could be replicated in other cities.

Table 1.
Project: municipal observatory of competencies for sustainable development goals (SDGs) Mexico 2030 agenda.

Tijuana (**Table 1**) with the support of the National Research Network in Social and Solidarity Economy (RIESS). The project is made up of more than 100 national and international researchers including the University of Guadalajara [11] and CIATEJ Research Center—CONAHCYT [12].

The research project has the support and methodological backing of the Executive Secretariat of the National Council of the 2030 Agenda for Sustainable Development of the Global Economic Intelligence Unit of the Ministry of Economy of the Government of Mexico as a strategy to contribute to the achievement of priority projects for the country, through collaboration with the Government and Academia.

2.1 Program for the professional development of teachers (PRODEP)

According to the Rules of Operation of the Program for the Professional Development of Teachers, issued by the Ministry of Public Education of the Government of Mexico in 2021, the guidelines are linked with the purpose of having a reference framework that allows identifying the actions that contribute to the fulfillment of the Sustainable Development Goals of the 2030 Agenda in the classroom through the educational programs. To this end, the PRODEP Program includes the 2030 Agenda for Sustainable Development as an Institutional Objective, contributing to action plan initiatives for people, planet, and prosperity. It also aims to strengthen universal peace within a larger concept of freedom. The importance of eradicating poverty in all its forms and dimensions, including extreme poverty, is recognized as the greatest challenge facing the world and is a prerequisite for sustainable development. The 2030 Agenda includes 17 SDGs; in the context of the education sector, it establishes SDG4, quality education, aimed at guaranteeing inclusive and equitable quality education, to promote lifelong learning opportunities for all, the great challenge of society on the planet [13].

The history of the PRODEP Program dates back to 1996, when the Program for the Professional Development of Teachers began at the university level for higher level, called PROMEP (Teacher Improvement Program), whose purpose was to invigorate the work of teachers working at the higher level. In normal education, it was later instituted, in 2009, within the PROFEN (Program to Strengthen Normal Schools), in which it was included as a new need in the field of knowledge generation in collegiate groups. Teachers would carry out research in addition to teaching, but reality shows scenarios different from those desired by several factors with the Priority Objective of revaluing teachers as fundamental agents of the educational process, with full respect for their rights, based on their professional development, continuous improvement, and vocation for service.

UNESCO states that the achievement of education for all was boosted in terms of coverage at the time. For the 2030 Agenda, education expands its scope beyond access, promoting learning achievements, different ways of accessing knowledge, safe and healthy educational environments, innovative methodologies, and the approach to lifelong education. In addition to educational achievement, quality education promotes skills for the promotion of peace, conflict resolution, mutual understanding, and sustainability; therefore, as far as young people are concerned, education must be related to their needs and interests, in which they incorporate significant learning and enable their adequate inclusion in the labor markets and the achievement of their projects of life [14].

2.2 Methodology QHS and DCS for sectoral development

Initiatives have been developed from academia, and the areas of postgraduate research, such as the Fifth Systemic Helix Methodologies (QHS) Methodology in Spanish, aimed at sectoral integration processes to analyze, evaluate, and promote projects where the different representatives of society interact, such as the government sector, education, companies, business chambers—unions, and specialized consultants [15]. In the case of the Sectoral Competency Dictionary Methodology (DCS Methodology in Spanish), this methodology strategy proposes the design of labor, professional, and research competencies for the design of Competency Standards [16] and thus respond effectively to the relevance and validity of knowledge skills, and attitudes necessary at each occupational level and position.

2.3 Municipal development agenda (MDA)

The Technological Institute of Tijuana of the TecNM, through the postgraduate program of Master in Administration as a Verification Unit, is the headquarters of the training work for the Public Servants of the State of Baja California. The training modules focus on delivery-reception of the Municipal Public Administration and Consultative Guide of Municipal Performance, which contemplates sustainable objectives according to the Global Agendas as is the case of the 2030 Goals of the UN [17]. The municipal development verification program is promoted by the National Institute for Federalism and Municipal Development (INAFED), which depends on the Ministry of the Interior (SEGOB) and is coordinated by the state governments to promote the improvement of Municipal Management. Currently, the five municipalities of the State of Baja California have implemented the INAFED program as a strategy for identifying and monitoring areas of opportunity generated through the verification processes developed by higher education institutions (HEIs) in Mexico [18].

In Baja California, the HEIs that develop the activity of social and academic responsibility of contributing with highly prepared professors for the application of the Instruments of the Municipal Development Agenda (ADM), now called the Municipal Development Guide (GDM), are the Technological Institute of Tijuana, through its Master's program in administration, and the Technological University of Tijuana. According to INAFED's national verification results, the municipalities that achieve the best results receive an Annual Recognition and Distinction at the Ministry of the Interior in Mexico City. Each year, the results generate opportunities for the development of action plans according to the gaps identified because of the surveys that are applied to the society as a complementary action to the annual verification of municipal performance and management, as a strategy to promote the strengthening and recognition of the talent management of public servants in Mexico.

2.4 Public and social economics research

Since the founding in May 2019 of CIRIEC Mexico (International Centre of Research and Information on the Public, Social, and Cooperative Economy), the Tijuana Institute of Technology of the TecNM has been part of the structure of the organization, occupying the Vice Presidency in the Northern region of Mexico. CIRIEC Mexico is a nonprofit association and national chapter linked to CIRIEC International, which is based in Belgium.

CIRIEC Mexico was born as a promise and a hope to contribute to another way of doing things in the economy. In the research and the very close accompaniment with the entities of the Social and Solidarity Economy (EESS) in their territory, under their cultural and socioeconomic contexts, with their resources (most of the time incredibly scarce), we hope that CIRIEC Mexico, little by little, achieves the cohesion and professional coherence required to face the social requirements of the countryside and the city [19].

2.5 TecNM scientific research projects

Strategies to influence Sustainable Development through Teaching and Research have generated initiatives to contribute to SDG 4: Quality Education. Observatory of the Integration of Engineering in the Economic Development Ecosystem of the Baja California Peninsula. The research focused on two areas of interest: the relevance of the study program for engineering careers at the TecNM in the region.

First, identification of the professional competencies of the engineers who intervene in the supply chain within the companies and manage local supply as a competitive strategy in the regional economic development ecosystem.

The second focus is the factors that intervene in entrepreneurship as MSMEs in the strategic services sector as a key strategy for the promotion of local supply as a professional development of engineers graduated from TecNM in the Baja California Peninsula [20]. Observatory of Sustainable Development in Graduate Study Programs in Baja California: the scientific research project focused on carrying out a systemic analysis in higher education institutions that incorporate the dimension of sustainability in their graduate level curricula in the State of Baja California through teaching, research, and university management.

To cement the variables to develop an Observatory of Sustainable Development of Graduate Study Plans in Baja California [21] as a tool for the analysis of the culture of sustainability and inclusion in the current university model, the project integrated three specific objectives:

1. To determine the level of culture of Sustainable Development in the Graduate Study Plans in Baja California.

2. To evaluate the implementation of the Sustainable Development Goals in the Graduate Study Plans in Baja California.

3. To design the perspective of the variables for an Observatory of Sustainable Development of Graduate Study Plans. The Project of the Observatory of Sustainable Development in Graduate Study Programs in Baja California is a continuation of the project Observatory of the Integration of Engineering in the Economic Development Ecosystem of the Baja California Peninsula.

The entry into force of the United Nations 2030 Agenda in 2015 makes it possible to promote sustainable development in its social, economic, and environmental dimensions. SDG 4 of the 2030 Agenda obliges the actors involved to carry out a series of actions in this area, guaranteeing inclusive and equitable quality education and promoting lifelong learning opportunities for all.

Among the benefits and impacts of the research project is the alignment with UNESCO's Medium-Term Strategy 2008–2013, which highlighted that development and economic prosperity require the capacity of states to teach their citizens and provide them with a culture and awareness about learning about sustainable development and the benefits for quality of life and integral development in society. It would provide local professional development for engineers who graduated from TecNM in the Baja California Peninsula.

2.6 TecNM's NODESS Tijuana project

The actions are derived from the TecNM Institutional Development Plan since 2019 to promote a NODESS Project (Nodes for the Development of Social and Solidarity Economy) of the City of Tijuana, integrated strategic allies; the Municipal Institute of Citizen Participation of the Municipality of Tijuana and the Confederation of Cooperatives of the Northern Border of Mexico, to promote the UN 2030 Goals, oriented to the thematic line of food sovereignty, with specific objectives to map the needs of the supply chain and areas of opportunity in social innovation. This is in accordance with the National Strategic Programs (PRONACES) of CONAHCYT (National Council of Humanities, Science and Technology), focusing efforts on the Supply Chain of the Dough and Tortilla Industry [22].

The project undertaken in 2019 as NODESS TIJUANA, crystallized, achieving the TecNM Tijuana, the national registry as the FIRST NODESS in the Northern Border Region of Mexico, occupying the leadership in the Border Municipalities of Baja California, Sonora, Chihuahua, Coahuila, Nuevo León, and Tamaulipas. The main objective of the NODESS TIJUANA Model presented at the national level by the TecNM Tijuana campus is the multiplier development of NODESS projects with a systemic approach (Fifth Systemic Helix) for the institutional link with the different sectors of society and with the aim of empowering the learning curve generated in the principles of the Social and Solidarity Economy.

TecNM Tijuana contributes to academic linkage and strategic research actions to the collaborative work with the Municipal Institute of Citizen Participation of the City of Tijuana as a Member of the Sectoral Subcommittees for the monitoring of the progress of objectives and goals of the Municipal Development Plan 2022–2024 and the Strategic Plan for Municipal Development 2022–2036, all under the focus of alignment of competencies from undergraduate and graduate educational programs to the National Strategic Programs of CONAHCYT, so that altogether, we can contribute to the challenges of the National Plan in each region of Mexico [23].

2.7 Commission for the promotion of the social and solidarity economy

Nowadays, communities in Mexico, whether urban or rural, must be thought of from a sustainability approach, this new approach to sustainable development involves three main spheres in perfect balance: economic, social, and environmental aspects. Social development cannot be generated without economic development and care for the environment, always bearing in mind that the goal of actions must be for the benefit of the development of people in an inclusive way and their communities.

The installation of the "Commission for the Promotion of the Social Economy" aims to identify the needs of the business exercise of the representatives of the Social and Solidarity Economy through the systematization of social innovations, aligned with the objectives of the national development plan to reduce the gaps of social inequality, as well as the instrumentalization of a social economy observatory of the goals for sustainable development of the 2030 Agenda, in the ecosystem of economic development of the City of Tijuana, to combat poverty and hunger. It requires the will to work in a coordinated manner the government, academia, social sector, representatives of civil society, as well as the citizens of Tijuana, with a systemic development approach [24].

2.8 TecNM training program in social and solidarity economy

At the national level, the initiative of the National Technological Institute of Mexico (TecNM) has been generated to offer a Training Program in Social and Solidarity Economy, which integrates diplomas, master's, and doctorates aimed at all people interested in the development, management, organization, and implementation of social and community development projects with a focus on Social and Solidarity Economy. The modality is face-to-face and mixed, and the doctorate is proposed to be virtual in an asynchronous way, thus seeking an offer of national and international scope based in TecNM Tijuana [25]. The general contents of the training in Social and Solidarity Economy (SSE) integrate aspects such as:

a. Fundamentals of SSE: introduction to economics, introduction to SSE, dimensions and features that identify SSE, purposes of SSE, and values.

b. Context of SSE: currents of the SSE, the SSE, and other alternative economies in Mexico: challenges and perspectives.

c. Legal Framework, Ecosystems, Innovation, and Entrepreneurship in SSE: legal framework that regulates the social sector of the economy and its economic activity in Mexico, types of cooperative societies, SSE ecosystems and innovation, and social entrepreneurship.

d. SSE Preliminary Projects and SSE Case Studies: elements of an SSE draft project, participatory tools for the development of an SSE draft project, and international and national experiences of SSE projects: case studies and presentation of an SSE draft project.

2.9 TecNM academic body ITTIJ-CA-12

The PRODEP Academic Body is a modality of conducting research with groups of research professors and researchers from one or more lines of research. Academic bodies are groups of full-time professors who share one or more lines of generation or application of knowledge (research or study) in disciplinary or multidisciplinary topics and a set of academic objectives and goals. At the Tijuana Institute of Technology of the TecNM, the first multidisciplinary team was formed to attend the topic of Sustainable Development, aligned with the UN's 2030 Goals.

The name of the Academic Body: "Sectorial Competitiveness, Social Innovation and Sustainable Development." It obtained its registration at the national level in Mexico in 2023, after a trajectory since 2019, code: ITTIJ-CA-12, founded in the Department of Economic and Administrative Sciences with Professors and Researchers with PRODEP Accreditation (Program for the Professional Development of Teachers) and National Distinction before CONAHCYT (National Council of Humanities, Science and Technology) of the Government of Mexico, as Members of the National System of Researchers. In the area of Economic and Administrative Sciences with Lines of Generation and Application of Knowledge: (1) sectoral competitiveness and (2) social innovation and sustainable development [26].

2.10 Voluntary report on progress on the SDGs in cities

Mexico will present its Fourth Voluntary National Report on the 2030 Agenda for Sustainable Development in 2024. The Executive Secretariat of the National Council of the 2030 Agenda supports efforts at the national level to develop subnational reports, and such is the case of the City of Tijuana through the processes of specialized training and orientation to shape the information of actions developed from the local level and thus add to the national results. The Municipal Institute of Citizen Participation (IMPAC) and the National Technological Institute of Mexico (TecNM) Tijuana Campus work in coordination with the NODESS Tijuana Program in the advancement of awareness of the UN 2030 goals, with workshops, conferences, and seminars at the local, state, national, and international levels of the good practices developed in the City of Tijuana toward the awareness of youth in primary, secondary, high school, and universities, as well as the different sectors of society.

For a greater context of the challenges that Mexico (United Mexican States, official name) must address the phases of awareness, empowerment, and specific actions in each municipality, Mexico has 32 Federal Entities (states) and 2475 municipalities, according to the National Institute of Statistics and Geography (INEGI), with a population of 131.1 million people and with a growth of 0.9 [27]. According to the UN's 2030 Agenda reports, Mexico has the following Voluntary Subnational Reports presented in **Table 2**, which includes state governments 8 out of 32 and municipal governments 5 out of 2475 [28].

Since 2015, the importance of the 2030 Agenda has been actively promoted, managing through different actions and tools planning at the local level, in cities, regions, state governments, and country governments, to implement actions in each SDG. According to the United Nations report of the Department of Economic and Social Affairs, **Table 3** presents the records of the countries that annually submit voluntary reports of a state, regions, or even at the city level [29].

The challenges are great, but the unity of all sectors of society can generate surprising actions. That is why, the key role of SDG 4, from the education sector, as an agent of change and leader, is to integrate, motivate, and articulate the different sectors of society; the reflection of the capabilities of society is its level of education.

Year	State government	Municipal government
2023	Puebla, State of Mexico	Tizayuca
2022	Oaxaca	
2021	Tabasco, Durango, Mexico City (Capital), State of Mexico	Merida, Guadalajara
2020	Yucatan	
2019	Mexico City (Capital), Oaxaca	

Table 2.
Mexico's 2030 agenda voluntary subnational reports.

Year	Participating countries	Number of countries
2023	Morocco, Australia, and Argentina. Belgium, Denmark, United States, Finland, Portugal, Uruguay, Brazil, Mexico, Malaysia, Japan, Netherlands	14
2022	Malaysia, Jordan, Netherlands, Spain, Brazil, Germany, Argentina, Portugal, United Kingdom, Uruguay, Bolivia, Paraguay, Belgium, Denmark, Italy, Australia, Guatemala, Tanzania, Mexico, Colombia, Finland, Japan, China	23
2021	Malaysia, Brazil, Norway, Finland, Mexico, Belgium, Peru, United States, Denmark, Japan, Spain, Argentina, South Africa, United Kingdom, Indonesia, Germany, South Korea, Norway, Canada, Sweden, China	21
2020	Spain, Germany, Finland, Belgium, Uruguay, United States, Brazil, Greece, Argentina, China, South Korea, Mexico, Mozambique, Albania	14
2019	Brazil, Spain, United States, United Kingdom, Finland, Mexico, Portugal, Argentina, Japan, Bolivia, South Korea, Philippines, Germany, Belgium	14
2018	Spain, Japan, Bolivia, United States, South Korea	5
2017	Philippines, Belgium, Australia, Spain, Brazil, Germany	6

Table 3.
Cities of the world with a voluntary local report of agenda 2030.

2.11 TecNM social and solidarity economy (SSE) research network

The TecNM Research Networks are the voluntary association of researchers or people interested in collaborating to address a problem of regional, national, or international magnitude from a multidimensional perspective and in an articulated way between national and international actors from academia, government, companies, and civil society. Networks can be understood as incubators of cooperation, where interactions, collaborations, and transfers between partners contribute to generating a multitude of outputs and results, both tangible and intangible [30].

Since 2023, TecNM Campus Tijuana has been the national leader of the research network in social and solidarity economy, with the purpose of developing research projects for three years of validity of accreditation as a nationwide network. The network is made up of committees for the training of human resources, professional practices, social service, certification of competencies, and development of initiatives to update the relevance of curricula and programs at different educational levels.

The focus of the research network is to support undergraduate and graduate students in projects that impact the different sectors of society, develop projects to analyze the capacities of supply, supply chain, productive ecosystems and productive vocations in the north, center and south of Mexico. The linkage activities promoted by the research groups and the Nodes for the Promotion of the Social and Solidarity Economy, known as NODESS, aim to promote different methodological strategies of intervention in organizations, for the analysis of the factors that encourage sectoral development, the application of the QHS Methodology; it integrates representatives of academia, government, companies, associations and consultants. The UN 2030 Agenda is a central point of the research network's work axes with an impact on the development and well-being of society, toward a sustainable approach, as well as the axes of the national development plan and the reference of local and state development plans. The vision of the academy is to seek dialogues and synergies between the different sectors of society, to know the needs and to develop solid solutions.

3. Results

Under the systemic analysis of information generated in Mexico, the last decade has been oriented to the contributions in favor of quality education for the awareness of sustainable development and the empowerment of the 17 Sustainable Development Goals (SDGs) of the UN 2030 Agenda. The statistics of the Ministry of Public Education in Mexico indicate that there is a gap between the goals and the indicators established to raise awareness among the population.

That is why, the Mexican regulations for the training and updating of full-time teachers with accreditation of the Program for the Professional Development of Teachers for the Higher Type (PRODEP) have generated actions to bring the information of the SDGs to the classrooms, but before transmitting the knowledge to the students, teachers must be trained at different educational levels to ensure that they are prepared to transmit the importance of each SDG in an effective and dynamic way. From the context of its evolution of the SDGs, Millennium Goals, it is for this reason that the national statistics on full-time teachers reflect a starting point of the need for greater efforts, programs that encourage the preparation of teachers and personnel who update educational programs to insert in each subject the necessary points toward critical thinking of Sustainable Development—concern for environmental conditions and the impact on future generations.

Table 4 presents the statistics of the universe of full-time professors. The statistic invites us to reflect that full-time professors are only 15% accredited in programs that include empowerment on the SDGs, but full-time professors in most educational institutions only represent 30% of total professors. As 70% of professors have a contractual and employment relationship of part-time or by subject hours, this generates great concern for achieving the mechanisms for training and updating teachers [31].

At TecNM 5 out of 10 engineers graduate nationwide, according to PRODEP statistics it has 8,286 full-time professors and 3,650 professors are accredited in PRODEP, a challenge of 66% of professors must prepare and be accredited with the Desirable Profile. The development of academic productivity mechanisms is a key to promoting that teachers are accredited with a desirable profile, for this skills and experience are required for research, project development and publication of articles and book chapters, as well as the training of human resources through theses. That is why **Table 5** presents the list of articles published by the Academic Body with registration ITTIJ-CA-12 of the Technological Institute of Tijuana of the TecNM with the lines of Research: a) Sectoral Competitiveness and b) Social Innovation and Sustainability

Mexico Universe		Accredited		TecNM Universe		Accredited	
276,000	100%	42,508	15%	8286	100%	3650	44%

Table 4.
Number of full-time professors with PRODEP accreditation.

Total articles	Citation	Index h	Index i10
114	198	7	5

Table 5.
Published articles on the QHS and DCS methodology.

Year	2016	2017	2018	2019	2020	2021
Full-time teacher	1750	2174	2930	3285	3478	3509
Academic bodies	559	682	682	817	848	868
% PRODEP	32	31	23	25	24	25

Table 6.
Evolution of the PRODEP indicators in TecNM.

Development, with a multidisciplinary team of researchers, with Collaboration Agreements at the national and international level [32].

The publications and citations obtained are products of research projects developed within the framework of national calls for scientific research projects for full-time professors and researchers and sectoral linkage activities. **Table 6** shows the relationship between the number of full-time professors of the TecNM and the composition of academic bodies for research.

The contractual situation of the professors limits the development of the desirable profile (PRODEP) and, with it, the comprehensive training to develop activities of academic management, linkage, research, and not only teaching. Part-time professors, or hourly professors, represent more than 70% of professors.

The Technological Institute of Tijuana received the Recognition of the Ministry of the Interior through the National Institute for Federalism and Municipal Development (INAFED) for the activities of Linkage and Advisory as an accredited VERIFICATION UNIT (UV) to develop the methodological application of the Municipal Development Agenda in the Municipalities of the State of Baja California as part of the objectives and social responsibility of linkage and applied research. The Ministry of the Interior, through the National Institute for Federalism and Municipal Development, INAFED, seeks to strengthen the capacities of municipal governments in the fulfillment of the SDGs. For this reason, on July 7, it will present a strategy so that, together, we achieve the objectives set out in the 2030 Agenda, considering the different realities, capacities, and levels of development of each municipal space, with full respect for its autonomy.

To achieve the goals of the SDGs, it is necessary for all of us, the three levels of government, the private sector, and civil society, to join and channel actions to transform our country and the world into a better place. **Figure 1** shows images of the activities to train public employees of the municipalities of the cities of Mexicali, Ensenada, Tecate, Playas de Rosarito, and Tijuana, Baja California, Mexico.

The National Institute for Federalism and Municipal Development (INAFED) of the Ministry of the Interior of the Government of Mexico states that the 2030 Agenda is an innovative development framework. Due to its scope, as well as the comprehensiveness of its policies, in compliance with its 17 Sustainable Development Goals (SDGs) and its 169 targets, it requires the broad involvement of all actors in all orders and levels of government. INAFED points out that "it is a commitment of the state" to implement the SDGs in all the municipalities of Mexico, but it requires the participation of the states, municipalities, the productive and social sectors, and citizens in general. Pointing out that "we are all actors responsible for its compliance and beneficiaries of its results." For this reason, it is a priority to promote the articulation of collaboration schemes and synergies between these actors. In the case of the National Technological Institute of Mexico, Tijuana Campus, it has become a Verification Unit of the Municipal Development Agenda, which contemplates the evaluation of actions aimed at the SDGs in the Cities. Under the traffic light evaluation scheme: (a) green refers to evidence of

Figure 1.
1,000,000 student project trained in SDGs UN 2030 goals.

sufficient compliance, (b) yellow there is evidence of attempts, partial evidence, but they reflect actions to initiate projects that are oriented toward the objectives of the SDG targets, and (c) red is the evaluation corresponding to the actions not developed and proposed in the municipal development plan; it is a reflection that no action was initiated; there is no minimum evidence of efforts to influence the SDGs [33].

Figure 2.
Social services activities with students for SDGs UN 2030 goals.

Figure 2 shows images of the evaluation processes carried out in each municipality of the State of Baja California, Mexico. Through professors and researchers who develop the verification processes of the Municipal Development Agenda, the principle of impartiality in the process of evaluation of municipal management is fulfilled. Once the evaluation processes have been carried out by the members of the Verification Unit of the Municipal Development Agenda (ADM), the results are sent to the national offices of INAFED in Mexico City. Once the results of the 2469 municipalities have been analyzed, a national ceremony will be held, and the best municipal compliance results will be announced. In 2019, the City of Tijuana had the best compliance result [34].

The Municipal Development Agenda (ADM) has evolved since 2022 and has changed its name to the Municipal Performance Advisory Guide (GDM). It is made up of eight modules that cover areas of responsibility of all municipal governments: Organization, Finance, Territorial Management, Public Services, Environment, Social Development, Economic Development, and Open Government. These are broken down into 31 themes and 115 indicators, of which 80 are management and establish the substantive documents for the municipal administration, and 30 are performance indicators that quantitatively measure the results achieved.

The participation of the municipalities is voluntary and will be formalized through an Act of the City Council and the registration application where the liaison is designated by the Municipal President. The new GDM contemplates mechanisms to carry out remote implementation, where the use of information technologies is essential when supporting the strengthening of municipalities, giving guidance so that higher education institutions can continue to collaborate with the same rigor as in person [35].

The GDM contemplates a specific item for the theme of education and aims to contribute to the strengthening of infrastructure, equipment, coverage, and provision of educational services in the municipality. Emphasizing three key aspects: education is a means to acquire, update, complete, and expand your knowledge, skills, abilities, and aptitudes that allow you to achieve your personal and professional development. Consequently, they contribute to their well-being and the transformation and improvement of the society they are a part of.

Basic education: it is made up of the initial, preschool, primary, and secondary levels. The services included in this type of education, among others, are (I) initial schooling and nonschooling; (II) general, indigenous, and community preschool; (III) general, indigenous, and community primary; (IV) secondary, including general, technical, community, or regional modalities authorized by the Secretariat; (V) secondary for workers, and; (VI) telesecondary. **Figure 3** presents the activities of the team of researchers that make up the Scientific Council of CIRIEC Mexico (International Center for Research and Information on the Public, Social, and Cooperative Economy).

The Board of Directors of CIRIEC Mexico was founded in 2018 in Acapulco, Guerrero, as a result of linkage and cooperation efforts between researchers from Mexico with representatives of CIRIEC Colombia, Brazil, Costa Rica, and Spain. CIRIEC Mexico was founded with four Vice Presidencies: TecNM Tijuana Institute of Technology; Indigenous Cooperative of Guerrero; Pascual Cooperative; and Autonomous University of Chapingo. Partner institutions were such as Universidad del Valle de México, Universidad Nacional Autónoma de México, Meritorious Autonomous University of Puebla. CIRIEC Mexico has been characterized by advancing in the analysis of the government's public policies, in the Social and Solidarity Economy, the promotion of cooperativism and sustainable development. There is

Figure 3.
Liaison activities between CIRIEC Mexico and TecNM Tijuana.

generation of intervention mechanisms in organizations for the analysis of productive chain ecosystems for the well-being of the most marginalized society. In this sense, various activities of an intellectual nature and linkage with the different sectors of society have been organized, seeking systemic actions of collaboration and management with the authorities of the federal, state, and local levels of government.

With the development of systemic initiatives generated through the identification of good practices in society, the competencies of specialized talent are strengthened, encouraging the application of the Fifth Systemic Helix Methodology (QHS in Spanish), for the integration of the different sectors of society and with it the mapping of productive vocations, supply chains, cementing the structure of regional development ecosystem models and a national benchmark [36].

3.1 TecNM SDG progress mapping report

The advances made by academia to influence quality education (SDG 4) have involved the development of scientific research projects to methodologically influence the analysis of the success factors or inhibitors of the lacing of the competencies of Sustainable Development. The project called *Observatory of Sustainable Development in Graduate Study Programs in Baja California* was promoted with the purpose of contributing to the lines of research of the postgraduate of the Master's Degree in Administration and Academic Body of the Technological Institute of Tijuana of the TecNM. In **Figure 4**, the presentation of results of the Research Project to the Society is presented.

The entry into force of the United Nations 2030 Agenda in 2015 has made it possible to promote sustainable development in its social, economic, and environmental dimensions. The agenda promotes the integration of the Education for Development (ED) and Education for Sustainability agendas. Goal 4: Quality Education of the 2030 Agenda obliges the actors involved to carry out a series of actions in this area, guaranteeing inclusive and equitable quality education and promoting lifelong learning opportunities for all. Among the benefits and impacts of the research project is the alignment with UNESCO's Medium-Term Strategy 2008–2013, which

Figure 4.
TecNM scientific projects with QHS methodology.

has highlighted that development and economic prosperity require the capacity of States to teach and provide their citizens with a culture and awareness about learning sustainable development and the benefits for quality of life and integral development in the society [37].

The Research Project *Observatory of Sustainable Development in Graduate Study Programs in Baja California* allowed a systemic analysis to be carried out on the problems raised in the scientific research project based on the limited information on the level of progress in the relevance and impact of the Sustainable Development Goals (SDGs) for the progress in the 2030 Goals of the UN (United Nations Organization) and the Programs National Strategic Programs (PRONACES) promoted by the National Council of Science and Technology (CONAHCYT) for the strengthening of sectoral sustainable development in the national territory, through the Study Plans of Graduate Programs in Mexico [38, 39].

The research project had three specific objectives:

1. Determine the level of culture of Sustainable Development in the Graduate Study Plans in Baja California,

2. Evaluate the implementation of the Sustainable Development Goals in the Graduate Study Plans in Baja California,

3. Design the prospective of the variables for an Observatory of Sustainable Development of Postgraduate Study Plans.

The results of the research project *Observatory of Sustainable Development in Graduate Study Programs in Baja California* propose the following approaches to the level of culture on the Sustainable Development Goals (SDGs) [40]. It is estimated on average that the level of culture is 53%, and the specific scope of SDGs is as follows:

- SDG 1: no poverty = 16% visible in graduate curricula in Baja California
- SDG 2: zero Hunger =16% visible in graduate curricula in Baja California
- SDG 3: good health and well-being = 83% visible in graduate curricula in Baja California
- SDG 4: quality education = 83% visible in graduate curricula in Baja California
- SDG 5: gender equality = 48% visible in graduate curricula in Baja California
- SDG 6: clean water and sanitation = 32% visible in graduate curricula in Baja California
- SDG 7: affordable and clean energy = 48% visible in graduate curricula in Baja California
- SDG 8: decent work and economic growth = 32% visible in graduate curricula in Baja California
- SDG 9: industry, innovation, and infrastructure = 48% visible in graduate curricula in Baja California
- SDG 10: reduced inequalities = 48% visible in graduate curricula in Baja California
- SDG 11: sustainable cities and communities = 64% visible in graduate curricula in Baja California
- SDG 12: responsible production and consumption = 80% visible in graduate curricula in Baja California
- SDG 13: climate action = 48% visible in graduate curricula in Baja California
- SDG 14: life below water = 32% visible in graduate curricula in Baja California
- SDG 15: life on land = 80% visible in graduate curricula in Baja California
- SDG 16: peace, justice, and strong institutions = 64% visible in graduate curricula in Baja California
- SDG 17: partnerships to achieve the goals = 80% visible in graduate curricula in Baja California

3.2 TecNM NODESS Tijuana project

The purpose of the formation of NODESS projects is to generate alliances between institutions of higher education, local government, and organizations of the social sector of the population to promote development, well-being, and build ecosystems. That is why, **Figure 5** presents the profile of NODESS Tijuana, which specializes in three strategic issues: (1) food sovereignty, (2) culture, and (3) education. With a metropolitan field of action and strategies for liaison and academic management, the goal was to train and raise awareness among more than 100,000 people throughout the State of Baja California, adding the efforts of professors, researchers, and representatives of the Society [41].

The training and awareness activities managed by NODESS Tijuana have been managed through the design of Social Service programs, which is a requirement for a degree in high school and higher education in Mexico. Social service refers to a voluntary activity that students do in a school or community. This activity is carried out to help others, raise awareness of education, and learn from life. Social service is carried out by some students who commit to dedicate their time and effort to doing something useful, allowing them to acquire experience, skills, and knowledge to be better people and better professionals. That is why, students are trained as instructors, after receiving specialized training in the training of people and the content of the 17 SDGs, to develop talks, workshops, and seminars with students of different educational levels. The students who have participated as instructors also develop complementary activities such as videos and activities on social networks to promote the understanding of each SDG and generating awareness in children. **Figure 6** shows the activities of workshops to train students, teachers, employees of business associations, trade unions, companies, and anyone interested in learning about the UN's 2030 Agenda [42].

As part of the results of the exercises in seminars, vodcasts, workshops, courses, and conferences on the SDGs with students from different educational levels, see **Figure 7**, surveys are carried out with the mentimeter tool, to develop word cloud, see **Figure 8**, to know what are the necessary actions to be developed by all to achieve progress in the implementation that affect the objectives and targets of each Sustainable Development Goal of the 2030 Agenda; the results are interesting. Words with greater emphasis are Solidarity, Commitment, Responsibility, Love,

Figure 5.
TecNM's NODESS Tijuana program.

Figure 6.
Dissemination of the SDGs of the UN 2030 goals.

Figure 7.
Project 100,000 students trained with NODESS Tijuana program.

Empathy, Respect, Innovation, Collaboration, Partnerships, Equality, Development, Teamwork, Leadership, Effort, Well-being, Goals, Recycling, Support, Knowledge, Learning, Volunteers, Discipline, Motivation, Awareness, Organization, Tolerance, Work, Equity, and Equality.

3.3 TecNM social economy and solidarity training program

The Tecnológico Nacional de México, as a Public Education institution aligned with the needs of the Federal Government, has designed a Continuing and

Figure 8.
Exercise word cloud in SDG seminars with students.

Figure 9.
National researchers on social economy of the TecNM, UdeG, and Ciatej.

Postgraduate Education program on Social and Solidarity Economy (SSE), which includes a special focus on sustainable development through the strengthening of the ecosystems of communities and the national territory [43].

The Social and Solidarity Economy Training Program (FESS) is integrated with the Development of Educational Programs for Diplomas, Master's, and Doctorates. **Figure 9** presents the list of strategic allies, professors, researchers, and specialists of the TecNM at the national level who have participated in the design of the educational programs [44].

3.3.1 General objective

Academic training in social and solidarity economy issues represents in the human resources of social and productive organizations the opportunity to strengthen talent management in skills to identify, manage, implement, and evaluate innovative projects with social responsibility and sustainability, in accordance with the demands of the regional, national, and international environment.

3.3.2 Specific objectives

Develop critical thinking in the local, regional, national, and international context with the contributions of the Social and Solidarity Economy from a sustainable perspective in ecosystems and supply chains.

Apply administrative, economic, and financial tools that contribute to the improvement of the management, operation, and development of Social and Solidarity Economy initiatives.

Generate initiatives with imagination, innovation, and intuition, which promote social and economic awareness for development, with strategies that meet the needs of communities.

Manage projects based on the principles and practices of the social, solidarity, and community economy with a collective social vision with sectoral competitiveness and sustainable development.

For the design of educational programs aimed at the Social and Solidarity Economy, as well as the efforts to comply with the UN 2030 Agenda, it is necessary to identify the state of the Art developed in the last decade in Mexico. **Table** 7 presents the actions carried out in different higher education institutions, as well as the initiatives of the law from the government to favor the development of the social economy, solidarity, and social innovation through cooperativism [45]. **Figure 10** shows the regional event organized by the University of Guadalajara, Social Leader, a training program for community leaders with initiatives in the SDGs, methodologically oriented projects to generate social innovation and well-being actions in the communities, and this program has collaborated through the Collaboration Network of Academic Bodies of the Technological Institute of Tijuana and CIATEJ—CONAHCYT Research Center [46].

The social and solidarity economy (SSE) is a set of initiatives based on a paradigm shift based on the collaborative work of people and the collective production of goods for self-consumption. To meet the needs of their members and their communities where they are developed, SSE seeks to generate relationships of solidarity and trust, community spirit and participation by strengthening the processes of productive integration, supply chains, value chain, sustainable consumption, distribution of savings and loans, and models of social innovation and sustainable development.

3.4 PhD project in social and solidarity economy at TecNM

The Tijuana Institute of Technology TecNM has been characterized by training and updating the management team of companies and organizations for more than 50 years in the City of Tijuana and in the region of Baja California, where outstanding professionals and researchers who have made important contributions to different sectors of society have graduated locally, nationally, and internationally. The TecNM Campus Tijuana offers an online and asynchronous Doctorate in Social and Solidarity Economy with Professional Guidance with a high academic level to strengthen professional competencies, providing the opportunity to update knowledge, learning and skills, as well as competencies for applied research. The Department of Economic and Administrative Sciences has maintained the traditional postgraduate Master's degree in Administration for more than 30 years. This new postgraduate degree has

	Actions developed in Mexico on SSE	Year
1	Social and Solidarity Economy Law	2012
2	Social and Solidarity Economy Law (Reform)	2019
3	Universidad IBEROAMERICANA	2013
4	INAES and Universidad IBEROAMERICANA convene the Academy of Social and Solidarity Economy	2015
5	PhD IN SOCIAL AND SOLIDARITY ECONOMY 1. Autonomous University of Chapingo 2. Meritorious Autonomous University of Puebla 3. Universidad Michoacana de San Nicolás de Hidalgo 4. Autonomous University of Aguascalientes 5. University of Guanajuato 6. University of Colima	2018
6	MASTER'S DEGREE IN SOCIAL AND SOLIDARITY ECONOMY Autonomous University of Guerrero (CONACYT)	2018
7	Presentation in the Chamber of Deputies in Mexico City CIRIEC Mexico Project (Founding Researchers) TecNM Tijuana (IES del Norte)	2018
8	Desarrollo del CIRIEC México (UAGro-Presidencia)	2019
9	CIRIEC Mexico (TecNM Tijuana–Vice-Presidency)	2020
10	Presidency of the Promotion of the Social Economy of the City of Tijuana—IMPAC—ODS	2021
11	Pre NODESS, NODESS, RED NODESS del TECNM	2022
12	Social Entrepreneurship Law Initiative	2023
13	Diploma and Master's Degree Project in Social and Solidarity Economy of the TecNM	2023
14	Project of the INTERNATIONAL NETWORK of Social and Solidarity Economy of the TecNM (RIESS)	2024
15	PhD in Social and Solidarity Economy of the TecNM	2024

Table 7.
TecNM training program in social and solidarity Economy.

Figure 10.
Award of the social leader program of the University of Guadalajara.

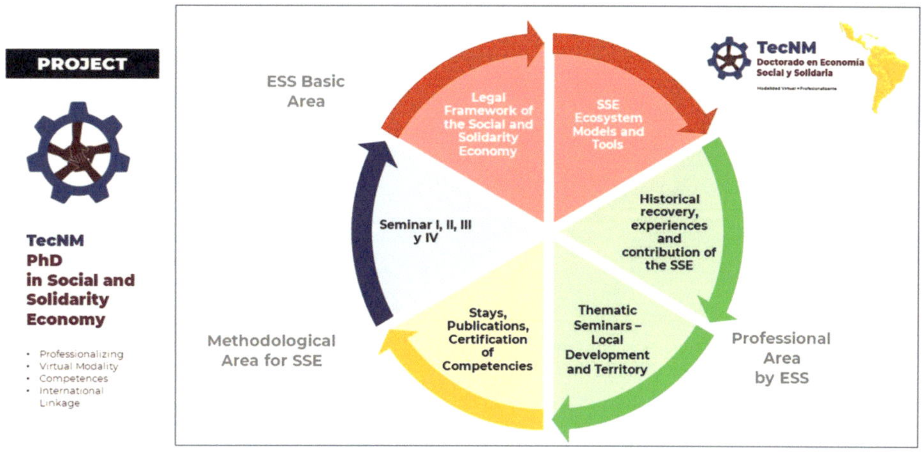

Figure 11.
Structure of the curriculum of the PhD in SSE at TecNM.

an impact both nationally and internationally. The structure of the curriculum of the Doctorate in Social and Solidarity Economy at TecNM is shown in **Figure 11** and is adapted to the demand for highly qualified professionals in an increasingly competitive world, in a cross-border region characterized by the economic dynamism of the cities of Tijuana, Baja California and San Diego, California [47].

The PhD in Social and Solidarity Economy generates an opportunity to meet the needs of the community of students and graduates of the sector of Higher Technological Education Institutions from different regions of the State of Baja California, nationally and even internationally. This is for a greater sectoral linkage under a sense of social responsibility and with the firm commitment of lines of knowledge generation oriented to the National Strategic Sectoral Programs (PRONACE). To overcome the frontiers of knowledge in the ecosystem of economic, social, and solidarity development, the field of action of research in sectoral competitiveness, sustainable development, and social innovation is promoted. **Figure 12** shows the institutions that are part of the National Research Network (RIESS) in Social and Solidarity Economy of the TecNM.

The Postgraduate Courses with a Professional Orientation train students for specialized professions, including stays of professionals and certification of competencies as a strategic part of the development of projects that mitigate, reduce, and improve the problematic situations that will arise as an applied research project by the Students of the PhD in Social and Solidarity Economy generating greater emphasis on practices and experience, for the strengthening of new knowledge, skills, and attitudes for the benefit of well-being and social progress.

The National Research Network on Social and Solidarity Economy (RIESS) of the TecNM has members from the Technological Higher Education Institutions, Technological Institutes and Public Universities, and Research Centers in the States of Baja California, Baja California Sur, Sonora, Chihuahua, Jalisco, State of Mexico, Querétaro, San Luis Potosí, Tamaulipas, Veracruz, Oaxaca, Tabasco, Campeche, Yucatán, and Quinta Roo. The RIESS Network divides the national territory into three: Zone 1, Northern Mexico; Zone 2, Central Mexico; and Zone 3, Southern Mexico. **Figure 13** shows the international links with researchers from countries such as El Salvador, Costa Rica, Panama, Colombia, Brazil, Argentina, and Chile [48].

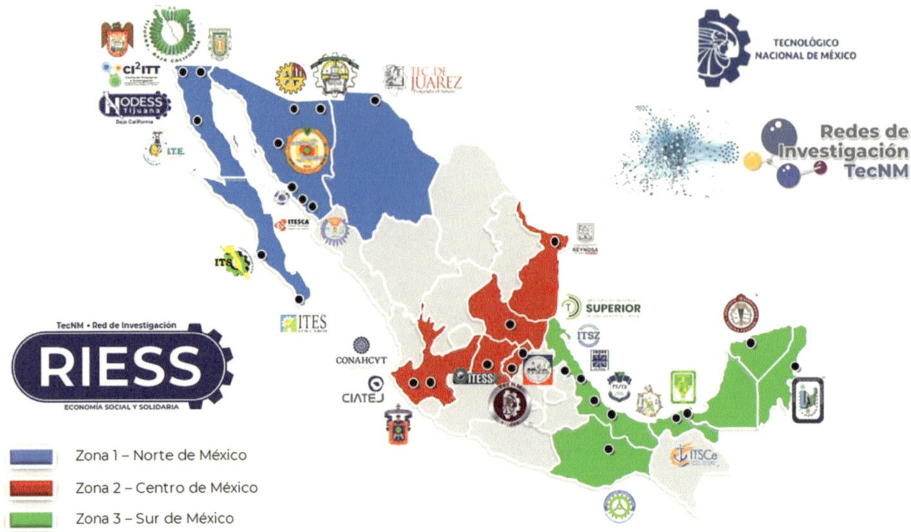

Figure 12.
TecNM social and solidarity economy (SSE) research network.

Figure 13.
International linkage of the social and solidarity economy network.

4. Conclusions and reflections

In the State of Baja California, Mexico, efforts have been made to integrate the Environmental Education Committee of regional scope, defining general agreements for periodic meetings to systematize the progress of the agreements defined in the State Environmental Education Committee; one of the key points of the Committee is to update the report on the progress of the UN 2030 Goals through the leadership of the Tijuana Institute of Technology, the National Network of Social and

Solidarity Economy (RIESS) and the Researchers of the Academic Body: "Sectorial Competitiveness, Social Innovation and Sustainable Development" ITTIJ-CA-12 of the TecNM Tijuana Campus [49].

The city of Tijuana managed to obtain the registration of the First NODESS in a border city in Northern Mexico in 2022, and the objective is to achieve inter-institutional work with the different educational institutions in the State of Baja California and in Mexico to identify actions in favor of the 17 Sustainable Development Goals (SDGs), To define the gaps in each SDG and thus develop joint projects and initiatives, it is worth mentioning that the Technological Institute of Tijuana is promoting through a Social Service Program with the goal of training and sensitizing 10,000 students of different educational levels in Tijuana and raising awareness among 100,000 people throughout the State of Baja California. The TecNM RIESS National Research Network aims to disseminate the importance of the 17 SDGs, PRONACES (National Strategic Programs of the Federal Government) and NODESS programs in other cities of the country through the sum of efforts of all sectors of society and inter-institutional linkage [50].

Acknowledgements

My deep gratitude for all the support and leadership of people from different institutions and public and private sector organizations. Especially my thanks and recognition go to Maestro Ramón Jiménez López, General Director of TecNM, Eng. José Guillermo Cárdenas López, Director of TecNM Tijuana, and Dr. Gaudencio Lucas Bravo, Academic Secretary of TecNM for the institutional support and trust to develop projects of national scope for the benefit of the large student community. Likewise, the Regional Leaders of the National Network for Investigation in Social and Solidarity Economy (RIESS Network), in the North of Mexico, Teachers Guadalupe Acuña Álvarez (SEP) and Carmen Adolfo Rivera Castillo (ITH), Dr. Blanca Esthela Zazueta Villavicencio (ITAP), in Central Mexico, Dr. Armando Alberto León López (CIIDET), in the South of Mexico, Dr. Jazmín Balderrabano Briones (ITUG). Sincerely to Dr. Rodolfo Martinez Gutierrez National Researcher SNI 2 CONAHCYT / TecNM Tijuana, Mexico.

Author details

Rodolfo Martinez Gutierrez[1*], Gaudencio Lucas Bravo[1],
Angel Ernesto Jimenez Bernardino[2] and José Daniel Padilla De la Rosa[3]

1 National Technological of Mexico (TecNM), Mexico

2 University of Guadalajara, Mexico

3 Center for Research and Assistance in Technology and Design of the State of Jalisco, A.C, CONAHCYT Research Center, Mexico

*Address all correspondence to: rodolfo.martinez@tectijuana.edu.mx

References

[1] UN Mexico. Sustainable development goals report. 2023: Special Edition. 2023. [Internet]. Available from: https://mexico.un.org/es/239254-informe-sobre-los-objetivos-de-desarrollo-sostenible-2023-edici%C3%B3n-especial [Accessed: December 12, 2023]

[2] Government of Mexico. SDG 4. Quality education. 2023. [Internet]. Available from: https://www.gob.mx/imjuve/articulos/ods-4-educacion-de-calidad [Accessed: December 12, 2023]

[3] Tijuana Institute of Technology. TecNM Tijuana campus achieves the first NODESS in the Northern Border of Mexico. 2023. [Internet]. Available from: https://www.tijuana.tecnm.mx/tecnm-campus-tijuana-logra-ser-el-primer-nodess-en-la-frontera-norte-de-mexico/ [Accessed: September 17, 2023]

[4] National Technological Institute of Mexico. Tijuana's TecNM pushes for the development of social economy in Baja California. 2020. [Internet]. Available from: https://www.tecnm.mx/?vista=noticia&id=304 [Accessed: September 17, 2023]

[5] Al-Raeei M. Scopus-based study of sustainability in the Syrian higher education focusing on the largest university. Open information Science. 2023;7(1):20220149. DOI: 10.1515/opis-2022-0149 [Accessed: February 06, 2024]

[6] Elnahla M, Wefki H. Sustainable development: A review of concepts, domains, technologies, and trends in smart cities. In: Negm AM, Rizk RY, Abdel-Kader RF, Ahmed A, editors. Engineering Solutions Toward Sustainable Development. IWBBIO 2023. Earth and Environmental Sciences Library. Cham: Springer; 2024. DOI: 10.1007/978-3-031-46491-1_26 [Accessed: February 06, 2024]

[7] Makarenko I, Plastun A, Petrushenko Y, Vorontsova A, Alwasiak S. SDG 4 and SDG 8 in the knowledge economy: A meta-analysis in the context of post-COVID-19 recovery. Knowledge and Performance Management. 2021;**5**(1):50-67. DOI: 10.21511/kpm.05(1).2021.05. Available from: https://www.mendeley.com/catalogue/de4b5fe8-5e72-360b-a3d7-ba80c90d4627/ [Accessed: February 06, 2024]

[8] Alfirević N, Malešević Perović L, Mihaljević KM. Productivity and impact of sustainable development goals (SDGs)-related academic research: A bibliometric analysis. Sustainability. 2023;**15**(9):7434. DOI: 10.3390/su15097434 [Accessed: February 06, 2024]

[9] De La Cruz Ramos RM, Yupanqui Villanueva CJ, Yupanqui Villanueva HR. Scientific production related to SDG 4: A global bibliometric analysis. Ciencia Latina Multidisciplinary Scientific Journal. 2023;**6**(6):13805-13823. DOI: 10.37811/cl_rcm.v6i6.4368 [Accessed: February 06, 2024]

[10] Ministry of Public Education. Rules of operation of the program for teacher professional development. 2022. [Internet]. Available from: https://dgesui.ses.sep.gob.mx/sites/default/files/2022-05/Reglas%20de%20operaci%C3%B3n%20PRODEP%202022.pdf [Accessed: October 01, 2023]

[11] University of Guadalajara. Academic activities and research. 2022. [Internet]. Available from: https://conectactica.udg.mx/2020/expertos.php [Accessed: February 06, 2024]

[12] Academic Chronicle. Formation of the NODESS-CIATEJ food sovereignty. The social solidarity economy is an alternative approach to the way in which economic activities are carried out. 2023. Available from: https://www.cronica.com.mx/academia/conformacion-nodess-ciatej-soberania-alimentaria.html [Accessed: June 29, 2023]

[13] Government of Mexico. In-Service Teacher Training Program (PRODEP) 2022-2026. [Internet]. Available from: https://www.mejoredu.gob.mx/images/programa-formacion-docente/docenteseb/programa-docentes-servicio-eb.pdf [Accessed: October 29, 2023]

[14] UNESCO. UNESCO and the sustainable development goals. 2021. [Internet] Available from: https://es.unesco.org/sdgs [Accessed: June 29, 2023]

[15] Martínez GR. Systemic fifth helix (QHS), a method for assessing the international competitiveness of the electronics sector in Baja California, Mexico. Administrative Investigation. 2012;**110**:34-48. Available from: https://www.redalyc.org/articulo.oa?id=456045338003 [Accessed: January 01, 2023]

[16] Martinez-Gutierrez R. Methodology of dictionaries of sector competences (DCS), to design standards of professional competences, research and labor. In: Nazir S, Ahram T, Karwowski W, editors. AHFE 2020. AISC. Vol. 1211. Cham. [Internet]: Springer; 2020. pp. 323-329. DOI: 10.1007/978-3-030-50896-8_46 [Accessed: January 01, 2023]

[17] Tijuana Institute of Technology. Master's Degree in administration receives recognition by SEGOB and government of BC as a municipal verification unit. 2019. [Internet] . Available from: https://www.tijuana.tecnm.mx/maestria-en-administracion-recibe-reconocimiento-por-segob-y-gobierno-de-bc-como-unidad-verificadora-de-municipios/ [Accessed: June 29, 2023]

[18] Ministry of the Interior. Municipal performance advisory guide. 2022. [Internet]. Available from: http://siglo.inafed.gob.mx/siguia/docs/2.%20Cuaderno%20de%20trabajo%20GDM%2022-24.pdf [Accessed: June 29, 2023]

[19] CIRIEC International. International center for research and documentation of the public, social and cooperative economy—Mexico. 2019. [Internet]. Available from: https://www.ciriec.uliege.be/en/our-network/national-sections/mexico/ [Accessed: June 29, 2023]

[20] Martinez-Gutierrez R, Ibarra-Estrada ME, Hurtado-Sanchez C, Carey-Raygoza CE, Chavez-Ceja B. Observatory for the Integration of engineering in the economic development ecosystem of the Baja California peninsula. In: Ahram T, Taiar R, editors. Human Interaction, Emerging Technologies and Future Systems V. IHIET 2021. Lecture Notes in Networks and Systems. Vol. 319. Cham. [Internet]: Springer; 2022. DOI: 10.1007/978-3-030-85540-6_162 [Accessed: June 29, 2023]

[21] Martinez-Gutierrez R, Marcela Solis-Quinteros M, Esther Ibarra-Estrada M, Ernesto J-BA. Observatory of Sustainable Development in Postgraduate Study Programs in Baja California. Sustainable Development. London, UK: IntechOpen; 2022. [Internet]. DOI: 10.5772/intechopen.104641 [Accessed: October 29, 2023]

[22] National Technological Institute of Mexico. TecNM Tijuana will install the first NODESS in the northern border region. 2022. [Internet]. Available from: https://www.tecnm.mx/?vista=noticia&id=2861 [Accessed: June 29, 2023]

[23] CONAHCYT. National strategic programs of the government of Mexico. 2019. [Internet]. Available from: https://conahcyt.mx/pronaces/ [Accessed: June 29, 2023]

[24] Tijuana Institute of Technology. Commission for the promotion of the social economy set up. 2020. [Internet]. Available from: https://www.tijuana.tecnm.mx/instalan-comision-de-impulso-a-la-economia-social/ [Accessed: June 29, 2023]

[25] TecNM—INAES. Social and solidarity economy nodes (NODESS). 2023. [Internet]. Available from: https://view.genial.ly/64d544b0f052d7001a7347ac/presentation-subtema-333-nodos-de-economia-social-y-solidaria-nodess [Accessed: June 29, 2023]

[26] Tijuana Institute of Technology. Presentation of a conference on public policy. 2022. [Internet]. Available from: https://www.tijuana.tecnm.mx/presentan-conferencia-sobre-politicas-publicas/ [Accessed: October 29, 2023]

[27] INEGI. National institute of statistics, geography and informatics. 2023. [Internet]. Available from: https://www.cuentame.inegi.org.mx/territorio/division/default.aspx#:~:text=Cada%20entidad%20federativa%20se%20constituye%20por%20municipios%2C%20de,de%20M%C3%A9xico%20se%20integra%20por%2016%20demarcaciones-%20territoriales [Accessed: June 29, 2023]

[28] Agenda 2030 Mexico. Guide to the development of voluntary subnational reports. 2023. [Internet]. Available from: https://www.gob.mx/cms/uploads/attachment/file/670811/Gui_a_para_elaborar_Revisiones_Voluntarias_Subnacionales.pdf [Accessed: February 06, 2024]

[29] United Nations. SDG localization and the voluntary local reviews. 2023. [Internet]. Available from: https://sdgs.un.org/topics/voluntary-local-reviews [Accessed: June 29, 2023]

[30] National Technological Institute of Mexico. TecNM research networks. 2021. [Internet]. Available from: http://redes-investigacion-tecnm.mx/ [Accessed: October 29, 2023]

[31] National Technological Institute of Mexico. Statistical PRODEP profile. 2021. [Internet]. Available from: https://prodep.dpii.tecnm.mx/estadistica/estadisticas2021.pdf [Accessed: October 29, 2023]

[32] Martinez Gutierrez R. Atena Editorial. Employability Observatory: Students and Graduates. 2022. [Internet]. Available from: https://atenaeditora.com.br/catalogo/ebook/observatorio-de-empleabilidad-estudiantes-y-graduados [Accessed: October 29, 2023]

[33] INAFED. Municipal performance advisory guide. 2022. [Internet]. Available from: https://www.gob.mx/cms/uploads/attachment/file/710671/2_Cuaderno_de_trabajo_GDM_22-24_iii__1_pdf [Accessed: October 29, 2023]

[34] Tijuana Institute of Technology. First day of training for public servants of BC developed by SEGOB in TNM-ITT. 2018. [Internet]. Available from: https://www.tijuana.tecnm.mx/primera-jornada-de-capacitacion-a-servidores-publicos-de-bc-desarrollada-por-segob-en-tnm-itt/ [Accessed: October 29, 2023]

[35] Tijuana Institute of Technology. For the second consecutive year, the TecNM Tijuana campus hosts the Baja California INAFED-SEGOB training day. 2019. [Internet]. Available from: https://www.tijuana.tecnm.mx/el-tecnm-campus-tijuana-por-segundo-ano-consecutivo-es-sede-de-la-jornada-de-capacitacion-de-baja-california/ [Accessed: October 29, 2023]

[36] Autonomous University of Guerrero. CIRIEC Mexico research networks. 2019. [Internet]. Available from: http://www.posgradoeinvestigacion.uagro.mx/investigacion/index.php/12-notas-investigacion/1004-ciriec-mexico

[37] Martinez-Gutierrez R, Esther Ibarra-Estrada M, Hurtado-Sanchez C, Galvan-Sanchez R, Ernesto J-BA. Perspective chapter: Prospective of sectoral competences for the design of dictionaries of professional competences and research. [Internet]. In: Higher Education—Reflections from the Field. Vol. 4. London, UK: IntechOpen; 2023. DOI: 10.5772/intechopen.109457 [Accessed: October 29, 2023]

[38] Tijuana Institute of Technology. The Tecnológico Nacional de México Tijuana Campus Was Once Again a Special Guest on the panel on competitiveness of the academic sector in Baja California. 2018. [Internet]. Available from: https://www.tijuana.tecnm.mx/el-tecnologico-nacional-de-mexico-campus-tijuana-de-nueva-forma-fue-invitado-especial-en-el-panel-de-competitividad-del-sector-academico-en-baja-california/ [Accessed: October 29, 2023]

[39] Tijuana Institute of Technology. Innovation center of the technological Institute of Tijuana. 2022. [Internet]. Available from: https://www.tijuana.tecnm.mx/centro-de-innovacion-del-instituto-tecnologico-de-tijuana/ [Accessed: October 29, 2023]

[40] Martinez-Gutierrez R, Ibarra-Estrada ME, Carey-Raygoza CE, Hurtado-Sanchez C, Chavez-Ceja B. Observatory of labor, professional and research competencies of the economic sectors in Baja California. In: Ahram T, Taiar R, editors. Human Interaction, Emerging Technologies and Future Systems V. IHIET 2021, Lecture Notes in Networks and Systems. Vol. 319. Cham. [Internet]: Springer; 2022. DOI: 10.1007/978-3-030-85540-6_164 [Accessed: October 29, 2023]

[41] CONAHACYT. Progress and results on SDGs and PRONACES. 2021. [Internet]. Available from: https://www.siicyt.gob.mx/index.php/transparencia/informes-conacyt/programa-institucional-2/4970-logros-pi-2021/file [Accessed: October 29, 2023]

[42] Tijuana Institute of Technology. Update of the Sep-Semarnat environmental education committee in Baja California. 2023. [Internet]. Available from: https://www.tijuana.tecnm.mx/actualizacion-del-comite-de-educacion-ambiental-sep-semarnat-en-baja-california/ [Accessed: September 16, 2023]

[43] Aguascalientes Institute of Technology. TecNM diploma in social and solidarity economy. 2023. [Internet]. Available from: https://ess.foprode.aguascalientes.tecnm.mx/ [Accessed: October 29, 2023]

[44] Ciudad Madero Institute of Technology. TecNM Master's degree in social and solidarity economy (MESS). 2023. [Internet]. Available from: https://www.cdmadero.tecnm.mx/index.php/oferta-educativa/maestrias/maestria-en-economia-social-y-solidaria [Accessed: October 29, 2023]

[45] CIIDET-TecNM. Workshops for the Master's degree and the diploma in social and solidarity economy of the TecNM. 2023. [Internet]. Available from: http://www.ciidet.tecnm.mx/ciidet/vistas/noticia.php?id=647f44842bb25d788803abb3 [Accessed: October 29, 2023]

[46] University of Guadalajara. Social leader program. 2023. [Internet]. Available from: https://cucea.udg.mx/es/noticia/14-jun-2023/

entrega-de-reconocimiento-del-proyecto-de-lider-social [Accessed: October 29, 2023]

[47] Tijuana City Hall. Installation of the commission for the promotion of the social economy of Tijuana. 2020. [Internet]. Available from: https://www.tijuana.gob.mx/webpanel/UMAIHipervinculos/Archivo/Hipervinculos/24-202061211528671-120201117.pdf [Accessed: October 29, 2023]

[48] ONDA UNED. Importance of academic linkage for sectoral development. 2023. [Internet]. Available from: https://www.youtube.com/watch?v=CPps2M0D9g8 [Accessed: October 29, 2023]

[49] MX Agenda. SDG workshop: Basic notions. 2023. [Internet]. Available from: https://www.youtube.com/watch?v=8YQgGqM71sg [Accessed: October 29, 2023]

[50] Universidad de Costa Rica. Vodcast 7 critical thinking. 2023. [Internet]. Available from: https://www.youtube.com/watch?v=7T8EXNnaMbI [Accessed: October 29, 2023]

Chapter 15

Perspective Chapter: Sustainability and Media Influence – The Role of Global Media in Creating a New Environmental Culture

Vesselina Valkanova and Nikolai Mihailov

Abstract

Media in today's world, particularly those that disseminate information through digital platforms, are one of the most influential sources of impact on people and society. The chapter is an attempt to review some basic points about the role of media—traditional and digital—in creating a new environmental culture within the context of the environment and environmental crisis. The chapter explores issues about the origin and development of environmental crisis as cultural phenomena, as well as some ideas about the moral assessment of media influence in informing, presenting and forming a position toward the problems of the environment. The concept of sustainability is changing both its content and its ground, becoming associated not only with the development of the natural and social environment, but also with attempts to build a new environmental culture. The role of the global media is to inform and motivate all members of society, especially young people for sustainability. More important conclusions are that media, ethics, and environmental knowledge must be developed in cooperation in order to explain and form specific moral consciousness, a new type of culture that does not allow indifference to the environment.

Keywords: media, culture, environment, media influence, digital communication

1. Introduction

The problems of the development of the environment, man's attitude to it, his interference in its functioning, and—unfortunately—its exhaustion and even ruin takes up an increasingly important place not only in the studies of biologists, ecologists, geographers, physicists but also in those of philosophers, sociologists, cultural scientists. And more recently, along with the digital change of media and the resulting global and decisive media influence on politics, worldview, culture and morality—also by media and media environment researchers. The concept of sustainability is changing both its content and its rationale, becoming associated not only with the development of the natural environment but also with attempts to build a new environmental culture, i.e., ways of thinking and behaving of contemporary societies and people in them that respect the principle of preserving and conserving natural diversity and impose norms of reverence for living and non-living nature, for ecosystems as sustainable natural systems that support and reproduce life on the planet, as well as for the environment created by humans—cities, towns, cultural heritage, etc. In this sense, education, public debate, and the media as an increasingly established social phenomenon of mass influence, play a major role both in preserving the heritage of previous generations as a material culture and in persuading society about the meaning of a sustainability policy. Moreover, in the contemporary social situation of digitalization and the technological revolution it has triggered—online platforms, social networks, and media pervasiveness—the latter is becoming a global tool for social impact. The digital change of society's media architecture crosses national borders and creates a global community, a global audience, which is daily subjected to media influence. For example, one of today's major environmental issues—climate change—is widely discussed and commented on in the media. The basic idea—a developing view that culture and environment are inseparable and arise from/within each other [1, 2]—can be complemented by the observation that global impact is needed to reinforce the norms and values of this culture, and this is what modern global media is capable of. One of the main objectives of our research is to investigate the role of global media influence on audiences successfully and with good consequences in order to promote the ideas of sustainability as development, anticipation, and trends of building a new environmental culture of respect and care for both natural and man-made eco-social systems. In order theoretically and methodologically to ground the process of influence of the media on the new environmental culture, we refer to ideas, studies, and authors who work in the field of ecology, environmental science, culture, and media and their impact on the formation of new values and norms of human attitude toward the natural and cultural environment. Thus, the methodological basis of our research are scientific publications of scientists who study not only ecological problems but also investigate their connection with the sociocultural activity of people, part of which is the media as an institution and as public communication. As well as those that explore the value dimensions of these issues.

2. The concept of sustainability

The idea of such an attitude toward the natural environment and its resources, which allows for the preservation of its diversity and protects it from destruction, has existed for a long time in human culture. There are various documents and prescriptions in which we can read on how to limit the use of natural resources to a reasonable extent and to enable the natural environment to renew itself as a natural ecosystem without losing any of its diversity. This also applies to the culture of the human social environment, with the idea that it should not be built at the expense of the natural environment, let alone destroy it. Rachel Carson's book *Silent Spring* (1962) and her formulation of "the obligation to endure" is an important theoretical step in this direction. "Only within the moment of time represented by the present century [20th century – V.V.; N.M.] has one species – man – acquired significant power to alter the nature of his world" [3, 4]. The idea of this obligation according to R. Carson is based on the fact that we develop deadly pesticides much faster than natural processes of adaptation [3, 5]. Sustainable development concerns not only the treatment of the natural environment, but also that which man builds for himself and which must meet conditions that do not threaten or endanger his residence in it. In both aspects, the concept of sustainability has been questioned due to the numerous and frequent environmental, social, or health-epidemiological crises widely reported and interpreted by the media.

Although various scholars and researchers have drawn attention to the possibility of development without depletion of natural resources [2], the concept of sustainability has gained wide publicity and started to be used in everyday discourse notably actively since the so-called Brundtland report [6], which in a special section called The Concept of Sustainable Development describes the different requirements and positions to ensure the maximum satisfaction of human needs in harmony with man and nature. In a broader sense, this concept has great significance to human culture, to the way we think about and perceive the world, and of course to the way we act or make sense of the natural world within the framework of cultural heritage.

2.1 Environmental culture: culture and environment

The problems of the human relation to the environment—physical, natural, social, cultural—that surround people and in which they socialize and realize themselves as individuals have long been the focus of sociologists, anthropologists, philosophers and social ecologists. In today's world, media, especially "new media"—disposed on digital online platforms—have become an integral part of the environment of human habitation and communication. This is the way they enter into what we define as culture. Regarding the interaction of culture as a specific human identification and presence in the world, which differs from the natural (a place of habitation unaffected by human's intervention) environment—there are different positions among the authors dealing with this problem. For example, the well-known Norwegian philosopher Arne Naess writes that there is a process of increasing ecological unsustainability, which is global in scope. The point is that it is hard to change the process into one of "decreasing unsustainability". We need to shift our point of view in ways of thinking from that of ecology to that of environment, according to Naess. "The ... term [environment] suggests that we have to do with something outside humanity, something we regrettably are dependent upon" ([4], p. 143). It is man's living in the world in a way, connected with nature, and its "rules" create sustainability, thinks Prof. Naess. According to his observations from his stay in Himalayas the specific and based on traditions of Sherpa culture in Nepal define strict rules regarding how to make use of trees and other raw materials for heating. Sherpa people live very high in the mountains—up to 4000 meters in the Himalayas with really cold and long winters. So, the vital need for heating is clear, but only dead trees were used ([4], p. 144). Prof. Naess argues that there are also ethical dilemmas, i.e., questions about values and choices concerning the natural environment. Traditional ethics is anthropocentric, i.e., it takes only human beings as being capable of moral choices and therefore subject to the moral responsibility of agency as beings possessing reason and therefore, the possibility of free choice. Environmental culture, according to Prof. Naess, creates another ethical system—that of the treatment of natural objects—plants, animals, ecosystems—to which man must also have responsibility. The ethics and practice in relation to forests and vegetation, as a rule, had been local in the sense of protection of natural reserves and materials, not in general ([4], p. 144). Professor Ness argues that this type of ethics—i.e., rules of human behavior toward nature—is at least local at present and does not have general acceptance or global reach and is therefore only inherent in some local cultures. According to Prof. Naess, delectable ecological ethics with moral practice in many non-industrial countries are valid and dominant in primarily local areas ([4], p. 144).

The idea about the environment is linked to the one about the nature. Nature and culture are interconnected. In different cultures nature means something that exists, we do not have uniquely vision of it animate or inanimate, and this carries a strong meaning, writes the Norwegian philosopher." Nature exists as a material and conceptual reality... different human groups understand and define nature according to their cultural traditions..." ([4], p. 42). The notion of nature and the way to interact with it is therefore not universal and may be shaped by a cultural context through which individuals or societies view its elements—animate or inanimate. It is through culture that man forms his relationship to nature, either as a source of raw materials and resources or as an important part of the environment he inhabits, rather than perceiving it directly. And the preservation of the sustainability of this environment also means the preservation and conservation of the human presence in the world—its culture, achievements, progress. In other words, the attitude toward nature, the awareness of its inseparable presence and impact on people, depends entirely on them. Different cultures around the world have different ways of perceiving their environment and its dimensions. Examining the Andean vision of nature, the Spanish University of Santiago de Compostela lecturer German Vargas Callejas describes it as an ecosystem where humans, flora, fauna but also supernatural or spiritual beings constitute a whole (social and spiritual) which attaches purpose and meaning to everyday life, the things we find in it and the actions we take. "A global and holistic reality where human beings represent only one more element in the whole" ([5], p. 43). Similar ideas about the environmental worldview of different cultures, which includes not only natural but also supernatural elements, can be observed in different countries around the world. The main understanding that connects them is that it is this wholeness that creates stability, becomes a guarantee of it and "meets the needs of the present without compromising the ability of future generations", which is also the accepted definition of sustainability. However, there is another type of culture for which Prof. Arne Ness, we think, correctly notes that the small minority of people around the world (about 500 million) are responsible for most of the decline and even for the degradation of conditions for maintenance of life on Earth and allow and support patterns of culture that favor irresponsible reproduction ([4], p. 148).

2.2 Media and media influence: global media

Media in today's world, especially those that disseminate information through digital platforms, are one of the most influential sources of impact on people and society, including in shaping their political, moral, economic, or worldview culture [7]. There are also different theories dealing with media influence, which can be described with the term media effects. One of the first studies of this effect, also known as magic bullet theory, claimed that "ideas from the media were in a direct causal relationship to behavior...the media could inject ideas into someone the way liquids are injected thorough a needle" ([8], p. 295). Marshall McLuhan also developed influential ideas about media impact with his concept of the crucial importance of communication and the technological nature of media in shaping the culture of societies. The Canadian thinker accepts that the visual sense is capable of creating the forms of space and time. These forms are continuous, uniform, and connected. That is why Euclidean space is the prerogative of visual and literate man, writes McLuhan.

With the appearance of electric circuity and the immediate move of information a change occurs. "The Euclidean space recedes and the non- Euclidean geometries emerge.... The printed word created the Public. The Public consists of separate individuals each with his own point of view. Electric circuity.... creates the mass... [It] consists of individuals profoundly involved in one another" ([9], p. 112).

One of the most influential books to date that explores the specifics, mechanisms, and trends of media influence is *personal influence: the part played by people in the flow of mass communications*, by Elihu Katz and Paul F. Lazarsfeld. The authors thoroughly analyze the nature of media, the phenomenon of public opinion, and the one of mass communication. They consider media of communication as "a new kind of unifying force – a simple kind of nervous system – reaching out to every eye and ear, in a society characterized by an amorphous social organization and a paucity of interpersonal relations" ([10], p. 16). Both researchers with one mind emphasize the role of the media in contemporary society not only to inform but also to build common perceptions and attitudes toward important events, facts or processes in politics, worldviews, history, and most of all—toward the general social environment. The media, which convey information, sense, meanings, become that mechanism that connects people in the contemporary technologized and global societies and create a common notion, evaluation, attitude toward something that is not possible for individual or immediate observation by individuals. In other words, they build culture. In fact, many things in the world are inaccessible to direct empirical observation. So people must to rely on each other for making sense of and rationalize things ([10], p. 55).

In a more general theoretical perspective, journalism today is not always associated with the notion of broadcasting—the classical understanding of the dissemination of audio and video content, but also with online impact, multimedia or interactive implementation, or direct digital contacts with the audience. Professional communication in this always important and sometimes risky profession is going through significant changes that can be called fundamental. "Much of this change is experiential (in this case – from experience): media audiences are *experiencing* media convergence in the situations they find themselves in while connecting to network infrastructures" ([11], p. 6).

Media influence will continue to be the subject of various studies and even measurements. Indeed, as media change, and especially their technological nature—analog, electronic, networked, digital—so does the scope and significance of their influence and impact. The more the reach of the media expands by changing the way it is disseminated, the more influence it shows in society, and the more decisively it has the potential to change its culture. The emergence of what we will call global media has been driven by the technological Internet revolution and digital technologies, changing both the way media is used and the scope of their impact. But perhaps most important in this process is the emergence of a new culture virtually across the entire world—the culture of digital information and of technological media influence, culminating today in the emergence of AI in such systems as ChatGPT. This decisive change imposed by the development of technology raises an important question for journalists—about what important concepts mean in their professional work and whether a source based on artificial intelligence can answer questions concerning important moral, political, or even personal choices. "All this will raise questions about censorship, objectivity, and the nature of truth" [12].

The emergence of global media is associated with the ever-expanding influence of the Internet and digital communication technologies, as well as the emergence of increasingly sophisticated and increasingly compact devices for offering and perceiving information and media content. It is also part of the process of globalization—a large-scale transformation of the social, political, economic, cultural, and communication order, which affects the entire world. The way in which information, with the help of digital technologies, is distributed globally and there are no obstacles (except political or ideological) to reach every corner of the world is called by some authors the information highway ([13], p. 41). This creates conditions for global media influence and forms a new social culture of attitude and perception of media content. Including through the media, to perceive and realize events and processes affecting the future of the whole world, not only at the local level. In this regard, the problems and processes that affect the global environment of humanity play a major role. Among them, one of the first places is the one related to sustainability and development.

2.3 New environmental culture: the role of media

The problems related to the concept of nature (natural environment) and people's attitude toward it are finding more and more space in the media all over the world. Some researchers even talk about environmental journalism, i.e. for such a genre and specialization in professional journalism that engages in work in the field of creating sustainability in the environment—current and future state, current and pressing issues, policy decisions related to it, etc. Important issues in this regard are those about the state of the climate and biological diversity and, for example, related ideas about the European Green Deal and the subsequent European Climate Pact, as well as the EU biodiversity strategy for 2030, which affect not only one separate country. Climate change is a phenomenon that transcends national boundaries. It is pointed out very often as a symbol of a "world risk society". We can observe climatic changes all over the world, on all continents. Climate change also shows social, economic and political impacts around the globe ([14], p. 39). In addition to all this, people are responsible for something on which their present, and even more so, future depends. In this sense, any particularization and prioritization of private interests in the field of the natural environment carries an extremely serious risk. "When the British colonized Australia, they considered it 'no man's land' (terra nullius), refusing to recognize the validity of any claims to ownership of this land by the local population. In the same way, people living today dispose of the resources of the planet Earth, treating the future as 'no one's time' (tempus nullius), as if this time will not be inhabited by people of future generations" ([15], p. 419). People's general concern for the climate is becoming one of the new and important moral norms and values of the culture. And it is no surprise that "climate change has received a considerable amount of media coverage globally since the mid-2000s" ([14], p. 40). On the other hand, though, the authors note that "*media coverage* [of climate change – V.V., N.M.] *is strongly event driven and episodic, and it focuses on similar events across countries*" ([14], p. 44). This inconsistency of coverage of environmental issues affecting all people can be observed in other respects, in coverage of environmental risks, for example, or in the media presentation of scientific articles and studies dealing with the state of the climate, natural resources, biodiversity, and the future of generations living in a polluted environment with depleted resources. The conclusions of Painter and Schaefer are based on the idea of paying attention to exploring the difference in media coverage of climate change (and other environmental issues) in different countries, as well as the implications of the emergence of digital media (according to the authors, traditional media are in crisis in many countries) in reporting on the global nature of climate change—"climate change has become a global media issue...in many countries, a "societal turn" seems to have occurred in which coverage turned away from the scientific aspects of climate change toward the political and social ones" ([14], p. 52). This is also one of the important characteristics of media influence: making society aware of the social problems that it faces and that require a common solution, providing facts, but also insisting on a clear journalistic position—social, political, moral—through which to point the way to solving various contradictions and social risks.

According to a worldwide study by the Reuters Institute for Journalism from 2020, online news sites of the biggest and most influential media organizations in the world are the second most popular news source across all markets [16]. In the survey cited respondents mention specialized publications covering climate issues (13%). Also they pay attention to alternative sources of information about the topics such as social media and blogs (9%). By comparison analog media such as printed newspapers and radio are pointed out as less important as a source of information and news than dialog with friends and family. Research points out that on average, 7% of respondents across all media markets expressed opinion that they do not pay attention to climate change news [16]. The research also shows a difference in the use of sources and media influence according to the age of the respondents. Young people tend to use digital online platforms much more, including because of the possibility of being followers of famous figures there—actors, athletes, activists, intellectuals, or they consider them more independent of political and economic dependencies and in this sense—for "more transparent". The ubiquity of online media platforms enhances interest and engagement in the communication process through their networked nature. This, on the one hand, predisposes the users of these platforms to a greater engagement with environmental issues. The ability to share content on social networks and platforms related to environmental issues is an undeniable new hallmark of environmental culture. On the other hand, this is undoubtedly an example of the affirmation of this environmental culture, of a new, active attitude toward the natural environment, reflection on and sharing of information about its problems related to care for its sustainable development. In the research, we find some examples. In the United states, those who individuals who accept that climate change is not serious threat are as likely to share news about climate change online (by social media or email) as those who are notably disturbed about climate change. By contrast, in Chile, where the researchers record the highest degree of care for climate change and its consequences for the environment, most of the online sharing comes from those with the highest levels of concern. This is the picture that researchers observe in most other countries [16].

The power and the influence of the media in the contemporary society, as well as the professionals working in them—journalists, editors, managers, etc., to educate in the spirit of environmental culture and achieve sustainability is very significant. We must note that sustainability does not refer to the surrounding natural environment (nature) only, nor only to the human and social one, but is related to the education of specific values and ideals as part of the new culture. This undoubtedly includes such activities as influencing and informing through the media of new educational, informational, and political initiatives that support the affirmation of the new environmental culture. An important observation is that a large number of articles and publications on the topic are mainly concerned with the economic and business aspects of the risks of overconsumption. As some authors indicate the economic dimension seem over—dominant in the way the sustainability is used and discussed. Also very often there is predominant emphasis on economic growth and consumerism when we deliberate human development, to the detriment of its other forms and manifestations ([17], p. 6). The interest in the heritage of the ancestors, the commitment to preserve it for the future and for generations to come is also part of what could be called sustainability, because without it such concepts as identity, memory, cultural heritage, political continuity, linguistic and moral culture. Heritage can be material or intangible assets, cultural artifacts that are part of one's memory and identity (in our country—Bulgaria, the statue of The Madara Rider or Madara Horseman is famous, a monument left from the Early Middle Ages) or ideas that identify the spirit and culture of a given nation. "Sustainability means including the living community in the heritage process and taking into account the cultural vitality of population...and finding ways to ensure...that the past contributes quality of life..." ([17], p. 13).

3. Conclusion

The role of the media in the building and development of the culture of the modern technological world will increasingly grow. This process is determined both by the maximizing sensitivity of societies to the problems of the environment, by the general concern for sustainable development in all respects, and by the need to anticipate and, if possible, avoid crisis situations that threaten the natural and cultural environment, which we live. For the modern world, all types of crisis—environmental, techno genic, and health, political, economic—seem to have become something natural, even according to some—almost inevitable. The need for a new environmental awareness and a new culture with values aimed at stopping the depletion of natural resources and preserving cultural heritage and wealth is obvious. In this regard, in conclusion, we can share the idea that the social and technological power of the global media, specially trained journalists with actual materials and columns, and with today's decisive media influence regarding the formation of values and worldview positions should be directed to the preservation and protection of the environment, biodiversity, and cultural diversity. Along with this, more in-depth work and even more research related to the influence of the media on the achievement of sustainability and the formation of a new environmental culture is needed. Similar research is needed both for the purpose of studying the mechanisms and most effective ways of impact, especially through digital media platforms, and for determining the most important topics and positions that need public discussion and analysis. They will help to explain why, in different countries, the attitude toward the environment—natural and that of culture—is not unambiguous—sometimes positive, but in places—with mistrust and whether this phenomenon is a consequence of an inadequate media environment or other factors.

Acknowledgements

This article/research was funded by the Bulgarian National Science Fund, contract number № КП-06-Н65/9 from 12.12.2022, within the project "Transformation and convergent models of journalism in the process of digitalization and in digital society in a communicative context: a dispositive analysis."

DOI: http://dx.doi.org/10.5772/intechopen.1005581

Conflict of interest

The authors declare no conflict of interest.

Author details

Vesselina Valkanova and Nikolai Mihailov*
Faculty of Journalism and Mass Communication, Sofia University, Bulgaria

*Address all correspondence to: nikolajkm@uni-sofia.bg

References

[1] Zandvliet D. Culture and environment: Weaving new connections. In: Zandvilet D, editor. Culture and Environment: Weaving New Connections. Leiden, Boston: Brill Sense; 2019. pp. 1-15

[2] Mihailov N, Sakelarieva L. Environmental alarmism: The club of Rome and its critics. Studia Ecologiae et Bioethicae. 2016;**4**:129-145

[3] Carson R. The obligation to endure. In: Herndl C, editor. Sustainability. A Reader for Writers. Oxford: Oxford University Press; 2014. pp. 4-12

[4] Naess A. Culture and environment. Humboldt Journal of Social Relations. 1994;**20**(1):143-149

[5] Callejas GD. Paradigms in the relationship between human beings and nature in the Andes, 2019. In: Culture and Environment: Weaving New Connections. Leiden, the Netherlands: Brill; 2019. pp. 41-59

[6] Sustainable development is development that meets the needs of the present without compromising the ability of future generations to meet their own needs. World Commission on Environment and Development, 1987. Report of the World Commission on Environment and Development: Our Common Future. Norway: Brundtland Commission; 2019

[7] Valkanova V, Mihailov N. Digital media and dynamics of contemporary public sphere: Towards a theoretical framework. Philosophy. 2023;**32**(3):284-292

[8] Biagi S. Media Impact. Boston, MA: California State University; 1990

[9] McLuhan M. In: Moos MM, editor. Environment to Anti- Environment: Media Research: Technology, Art, Communication. Amsterdam: G+B Arts; 1997. pp. 110-121

[10] Katz E, Lazarsfeld P. Personal Influence. Glencoe, IL: Free Press; 1955

[11] Dwyer T. Media Convergence. McGraw-Hill; 2010

[12] The battle for search. The Economist. 2023;**446**(9333):7

[13] Campbell R, Martin C, Fabos B. Media and Culture. St. Martin's: Bedford; 2000

[14] Painter J, Schaefer M. Global Similarities and Persistent Differences: A Survey of Comparative Studies on Climate Change and Communication, Climate Change and the Media. Vol. 2. New York: Peter Lang Publishing; 2018. pp. 39-59

[15] Stavru S. Climate narrative(s): Ethical metamorphoses. Sociological Problems. 2022;**2**:415-432. (In Bulgarian)

[16] Simge A. How People Access News about Climate Change [Internet]. 2020. Available from: https://www.digitalnewsreport.org/survey/2020/how-people-access-news-about-climate-change/

[17] Auclair E, Fairclough G. Living between past and future. In: Theory and Practice in Heritage and Sustainability. New York: Routledge; 2015. pp. 1-22

Chapter 16

Living Culturally: The Pillar of Sustainable Development of the Multi-Ethnic Country of Indonesia

Retno Hanggarani Ninin, Hammad Zahid Muharram and Noer Fauzi Rachman

Abstract

Indonesia as a multi-ethnic country faces challenge to embrace many standards for development as every ethnic community adopts their own cultural values. There were cases by which government, on behalf of equalizing standard, create national policy and regulation to develop communities by the way of national standard. It was at that time that cultural values were considered by the government as a factor inhibiting development and wellness. Recently, critical awareness has emerged at the national level regarding the climate crisis, resulting in the emergence of a new paradigm in development. Modernization that leaves a large carbon footprint is no longer the only approach in development. A cultural way of life that is naturally more environmentally friendly at this point is interpreted as more ideal development because it is oriented towards sustainable prosperity. Various examples of culture-based life that represent a sustainable way of life will be presented.

Keywords: Indonesia, climate awareness, community development, cultural value, sustainability

1. Introduction

Indonesia in this decade has carried out massive development in various fields. Development in infrastructure and human are an important part of Indonesia's development, as Indonesia is transforming from developing to developed country. Because modernity is considered a characteristic of developed countries, physical developments such as housing and infrastructure are planned and standardized based on a modern perspective. Modern standards for buildings and infrastructure include brick, steel, and cement, among others. Anything beyond that, such as from wood or bamboo, is considered a building that does not meet the standards. Apart from being applied to aspects of modernity, standardization is also applied to other aspects such as health and welfare. In terms of health standards, housing must be exposed to sunlight and have access to fresh air, as well as basic needs such as the availability of toilets and clean water. For welfare, house construction is focused on building quality for comfort and the ratio of occupants to building area.

The implementation of housing quality standardization is aimed at the good of humanity, namely so that people can adapt to the progress of civilization and maintain the quality of their health and welfare. Unfortunately, under the umbrella of modernization, the increasing focus on human interests diverts human attention to environmental interests. In fact, not only are they distracted from thinking about environmental interests, they even consider people, groups, or communities who preserve the natural environment where they live as unmodern, backward, and traditional. Often their culture, which prioritizes nature over modernization, is considered an obstacle to development.

Living on the archipelago consisting of more than 13,400 islands, the Indonesian nation counts almost 276 million population [1] comprising over 300 ethnic groups. According to the 2010 Population Census, Indonesia has 1340 tribes with categories of ethnic names, sub-tribe names, and sub-names of sub-tribes [2]. Those various ethnic groups live in 34 provinces. Indonesia is the most populous Muslim country in the world, but has six official religions, including Islam, Christianity, Catholicism, Hinduism, Buddhism, and Confucianism, and also acknowledges indigenous faith. The proportion of religious adherents in Muslim is 87.2%, Christian 6.9%, Catholic 2.9%, Hinduist 1.7%, Buddhist 0.7%, and Confucius 0.05% [3].

Bhinekka Tunggal Ika (Unity in Diversity) is the Indonesian national's motto and refers to the true sense of unity (Indonesianness) among the multiculturality of the society [4]. The existence of *bahasa Indonesia* as a language of unity is an external representation of the value of respecting differences. Indonesia has around 750 regional languages [5] that are used when people communicate within their ethnic groups, but it is easy for them to communicate across ethnic groups because every Indonesian citizen can speak *bahasa Indonesia*.

Blessed with a tropical climate and because of its geographical location, Indonesia has an extremely high level of biodiversity and endemicity [6]. As of May 2020, access of 4,147,875 hectares (out of targeted 12.7 million hectares) of social forestry (Adat Forests) has been distributed to 6620 forest farmer groups. The access has enabled the creation of 7311 social forestry business groups across country, which are successfully increasing the productivity and value of on-farm and off-farm commodities and of village small-scale industries. Adat Forests are defined as forests located within territories over which Adat communities hold traditional rights (Adat). Access to these forests will enable indigenous peoples to fulfill most of their living needs, not only using forests as a source of food, but also raw materials for making houses and other facilities.

2. Settlement culture

Communities that have local wisdom to cultivate the use of natural materials in building settlements are spread across various traditional communities in Indonesia. They build their traditional houses from natural materials without going through the fabrication process. The natural ingredients used are those available in the surrounding area. For example, the use of bamboo (Bambusoideae) to build houses, barriers between land, river crossing bridges, and rafts for water transportation. Indonesia's geographical location allows bamboo to grow in many areas. Data on Indonesia.go.id shows that although there is no valid data regarding the area of bamboo land in Indonesia, it is estimated that the area covered by bamboo reaches more than 1 million ha. Indonesia is a country that has around 176 types of bamboo, out of around 1439 types of bamboo known in the world, and around 50% of them are endemic. As an endemic plant, the availability of bamboo is relatively close to communities in various regions in Indonesia. Thus, in the value change and supply chain aspects, using bamboo for buildings and facilities will leave a much smaller carbon footprint compared to construction using fabricated materials. Moreover, it is stated in the portal that the total biomass that bamboo plants can store is around 87.35 tons per ha, so bamboo has an important ecological role in facing environmental threats and the negative impacts of climate change [7].

Bamboo is an excellent alternative material for construction substitutes such as iron, concrete, or bricks for several reasons, including the fact that the bamboo plant itself is a significant CO_2 (carbon dioxide) absorber during its growth, bamboo as an organic material can be recycled, and the manufacturing process uses less energy and water than other industries used in wood, metal, and concrete construction [8]. In "modern society", bamboo is not frequently used as a permanent building material, but it is in villages and indigenous communities. Bamboo plant development conserves soil, water, and biodiversity while also collecting CO_2. Thus, bamboo can reduce water and energy consumption, boost CO_2 absorption (the conversion of CO_2 into organic matter), and be used to create houses and structures [9]. Ecological, economic, and social advantages of the usage of raw materials derived from bamboo plants can help local people develop sustainably. Based on this sustainability value, instead of conceptualizing it as unmodern, it should instead be claimed as modernization, because of its alignment with nature and sustainability (**Figure 1**).

Traditional indigenous communities certainly do not have formal scientific knowledge about bamboo, but they have cross-generational knowledge from their

Figure 1.
Traditional settlement made of bamboo in Temanggung, Central Java, Indonesia (photo by Hammad Zahid Muharram).

ancestors who respect nature and its sustainability [10]. Unfortunately, the collective values that constitute local wisdom regarding the culture of building housing and public facilities are faced with defining the quality of housing in favor of the use of fabricated materials, thus positioning their culture as "low quality" or "unfit" or "not meeting standards". Thus, community groups who try to maintain local wisdom are often marginalized at the national policy level as obstacles to development and modernization. Therefore, community groups who try to maintain local wisdom are often marginalized at the national policy level and are considered to be an obstacle to development and modernization so that they are considered worthy of being "regulated" with national regulations which often ignore the uniqueness of indigenous communities [11]. This pressure was the trigger for the struggle of indigenous communities to obtain collective rights to manage their communities and customary lands [12].

In fact, the idealism of top-down based development is an attempt to colonize the knowledge and awareness of traditional communities, so that they realize that they are wrong and therefore need to adapt to the national or universal standards [13]. Through this colonialization of thought, indigenous people's life space buildings and infrastructure are treated as tourism artifacts, while for residential purposes and living facilities, communities must change their residential culture to follow new standards that are framed as high quality, decent, and in accordance with health and welfare standards.

In line with the transition of national leadership in Indonesia and the increasing ecological awareness of society, development in Indonesia is undergoing a transformation that provides space for the culture and customs of the community to grow and develop [14]. This is an opportunity to decolonize, so that traditional culture in society is not considered an obstacle to development, but instead culture must be part of development itself. Because the culture of society and many indigenous communities in Indonesia is traditional and very close to nature, a paradigm that gives space to culture will be able to increase the ecological value and sustainability of development. Because the culture of society and many indigenous communities in Indonesia is traditional and very close to nature, a paradigm that gives space to culture will be able to increase the ecological value and sustainability of development. The idea of prioritizing indigenous technology is in line with Vernacular Architecture [15], namely native and unique building construction that is produced without the need for imported components or processes.

In dealing with the difficulties of climate change and environmental sustainability, comparing traditional and modern houses is critical. **Table 1** compares the carbon chains of traditional and modern communities, spanning the complete product life cycle, from raw material manufacture to construction, usage, and demolition or recycling.

Table 1 shows that in terms of production, bamboo produces less carbon emissions (low) compared to the carbon emissions produced by brick and concrete (high). This is because as a building material, bamboo is obtained through harvesting bamboo plants. To produce bamboo, the production process is planting seeds while the regeneration of the plant is through bamboo shoots. One mature bamboo clump can produce 10–20 stems each year [16] and grows quickly, reaching 30–90 cm per day [17]. Through this natural production process, bamboo not only produces low carbon emissions, and it can even bind carbon during the production process, namely when it is growing as a bamboo plant. In contrast, brick and concrete are building materials produced by factories. It is stated that the cement industry in Indonesia emits 0.833 tons

	Bamboo		**Bricks and concrete**	
	Carbon emissions	**Environmental impact**	**Carbon emissions**	**Environmental impact**
Production	*Low*, as bamboo can grow quickly and be replanted.	*Minimal*, as bamboo grows naturally and requires few inputs.	*High*, as cement production involves burning limestone and fossil fuels.	*High*, as limestone and clay mining can damage ecosystems.
Processing/ construction	*Low*, as bamboo production requires less energy	*Low*, as bamboo can be obtained at low cost and requires minimal processing	*Medium to high*, as bricks and cement tend to be heavy and require large amounts of energy during construction.	*High*, as brick and cement production has left a large carbon footprint.
Usage/ maintenance	*Low*, as bamboo is lightweight and requires little energy during construction	*Low*, as bamboo is strong and durable	*Depends on* repair and maintenance needs.	*Depends on* the use of chemical materials and products in maintenance.
Disassembly/ replacement	*Low*, as bamboo can be recycled or decomposed naturally.	*Minimal*, as bamboo leaves no toxic waste.	*High*, as demolition of brick and cement buildings can generate toxic waste.	*High*, as construction waste can contaminate soil and water.

Table 1.
Carbon emission comparison of traditional and modern buildings.

of CO_2 for every ton of concrete [18]. In terms of environmental impact, because bamboo is a product of nature and is used directly, there is no production waste as produced by factories, so bamboo has a much lower environmental impact than the impact of brick, cement, and concrete production.

Compared to the construction aspect, bamboo uses less energy because all the supporting components are also natural from bamboo, such as pegs and ties, so there is no need for factory-made components. There is also no need for heavy equipment like constructing a house using brick, cement, and concrete, so bamboo house construction produces a very low carbon footprint compared to brick houses. As an illustration, building a small to medium house in Indonesia emits 22,179 kg of CO_2, equivalent to 22 tons of carbon [19]. Through these figures, it can be estimated the impact of damage to the environment, for the construction of many modern houses in just one community.

In terms of maintenance, bamboo is quite durable, generally able to last between 15 and 40 years [20]. Meanwhile, modern properties require maintenance at least once every five years. It is estimated that carbon emissions from house maintenance will reach around 46 tons within 25 years [19]. In the process of replacing buildings with waste, when bamboo is no longer used it will produce organic waste, while the remains of damaged concrete structures will pollute the soil. Thus, it can be said that as a building material bamboo is more sustainability-oriented.

Bamboo trees, with their high nitrogen consumption, can assist minimize water pollution by absorbing excess nutrients from agriculture, industrial, livestock, and sewage treatment. Bamboo clumps may absorb up to 12 tons of CO_2 per hectare, making them an effective fresh air recharger [20]. In addition to being an ecologically beneficial building material, bamboo is currently regarded as a low-cost alternative material. Bamboo has a great ability to absorb shocks, making it ideal for home construction materials that can withstand high winds and seismic pressures such as earthquakes [20].

Bamboo is becoming increasingly popular as a building material in the tourism industry. Bamboo resorts are growing as an appealing option, not only to fulfill the needs of modern travelers, but also to reduce negative environmental repercussions. Bamboo is a plentiful native resource in Indonesia, grows quickly, and requires little processing effort. Bamboo resorts decrease reliance on traditional building materials like concrete and steel, which have a bigger environmental impact.

The production and construction process of bamboo resorts has a lower carbon footprint compared to conventional construction. Bamboo grows fast and can be obtained at low cost, reducing carbon emissions associated with the production and transportation of building materials. In addition, the lightweight construction requires less energy during the building process.

Bamboo resorts, as an eco-friendly tourism alternative, can serve as an inspiration for tourist sites around the world. By continuing to develop and implement sustainability principles, Indonesia can preserve its natural beauty while also contributing positively to global sustainability and inspiring future environmental action.

The development described above only involves infrastructure development and the use of natural bamboo materials. However, in reality cultural involvement in development is much broader than just "infrastructure development" or just "bamboo use". In terms of development, if we talk about "development in all fields", then what is meant by development is not just the provision of housing, but also includes the provision of clothing and food.

3. Clothing culture

In terms of development, when we talk about "development in all fields", what is meant by development is not just the provision of housing, but also includes the provision of clothing and food. When the provision of clothing is aimed at fulfilling desires, not needs, then discussing clothing from a functional perspective, namely as body covering, is certainly not enough. The clothing industry includes fabric production, designing models, selecting designers, and tailors, whose goal is not to make clothes that fulfill functional aspects, but rather to make them based on "what people want". It is at this stage that providing clothing in a modern format will become a human activity that leaves a massive carbon footprint, as occurs in the modernization of settlements.

In contrast to providing clothing in a modern format, indigenous communities in Indonesia still maintain their culture in providing clothing. For example, the traditional clothing of the people of Nusa Tenggara is called *tenun ikat* [21]. *Tenun ikat* is made from natural materials that are taken and processed manually, woven with hand-operated wooden tools (**Figure 2**), and colored with natural dyes such as tree bark or flowers [22]. The plants needed to make cloth and to provide color grew around them. Another example is the Javanese, Madurese, Sundanese, and Betawi communities on the islands of Java and Madura. Their traditional clothes, called batik, are not only made and colored with natural materials that grow around them,

DOI: http://dx.doi.org/10.5772/intechopen.1005077

Figure 2.
Process of making tenun ikat in Kupang, Nusa Tenggara Timur, Indonesia (photo by Hammad Zahid Muharram).

and they do not even wash them using soap but a type of stone called *lerak* [23]. To smooth clothes, they do not use an iron, but hang them by placing a weight on the edge of the cloth so that the cloth is pulled down. The indigenous Papuan people on Papua Island and the indigenous Dayak people on Kalimantan Island have their own culture in terms of settlement. In Papua, there are even bags called noken which are made from part of a kind of trees [24]. All of them use natural ingredients and traditional processes that save carbon footprint. Maintaining these cultures instead of "modernizing" them is a choice that will contribute to the sustainability of natural resources and the climate. The condition is not to colonize indigenous peoples by framing their culture as backward and not modern so that it must be replaced with another culture that is "more in line with civilization".

4. Food culture

Indonesia's geographical location makes it a country that has a climate that supports abundant food availability throughout the year, on land, rivers, and in the sea. With such a climate there is no need for food preservation technology because food sources are available in abundance throughout the year. However, traditional food preservation technology develops uniquely in each region. This need is not related to climatic conditions, but rather because the abundant harvest exceeds needs, so the food needs to be preserved so that it can be stored longer for later consumption.

The most popular food preservation is meat preservation. The main reason is because hunting produces large animals such as buffalo or cows which produce enough meat to last one day's food, whereas in the past there were no refrigerators to store food. An example of preserving meat in the indigenous communities of Nusa Tenggara is by burning it using wood over low heat and for a long time, resulting in ready-to-eat dried meat called "se'i" [25] and can be stored for quite a long time without needing a refrigerator. Meanwhile, in traditional Malay communities, meat is preserved by cooking it into a dish called "rendang" [26] which allows the food to be stored for a long period of time at room temperature. The difference between "se'i" and "rendang" is that "se'i" developed in areas with minimal spice plants so it takes the form of marinated dried meat, while "rendang" comes from areas rich in spices so it takes the form of meat cooked using lots of spices.

As a country whose sea area is wider than land, Indonesia also has traditional fish preservation technology which is popular in almost all coastal communities throughout Indonesia. The fish product is called "ikan asin" (salted fish). Salting fish is done by drying the caught sea fish in the sun in an open area for days or even several weeks depending on the quality of sunlight exposure in each area. Through drying, sea fish dries out, leaving behind dried salted fish flesh containing sea salt. "Se'i", "rendang", and "salted fish" are three examples of foods in Indonesia that are often stored for a long time without needing to be refrigerated. The traditional food preservation industry is very diverse according to the local culture of the community, but they all have one thing in common, namely that they do not require preservatives and fabrication processes that consume large amounts of energy due to the use of machines, electricity, and fuel.

5. Village revitalization and circular economy

In early 2013, a bamboo bicycle model named spedagi was developed, using Petung bamboo (*Dendrocalamus asper*) as the main material, one of the strongest and most abundantly available bamboo species in rural Indonesia. This large-diameter, thick-walled bamboo is not only strong but also allows for the manufacture of bicycle frame rods of uniform size. The construction of bamboo slats "usuk bamboo", the roof frame of a rural house, is a source of inspiration to increase the stiffness of bamboo culms. The combination of oval-section slats and metal lugs resulted in a frame design that is not only beautiful and distinctive, but also strong and comfortable to use.

The Spedagi bamboo bike is produced at Studio Magno in Kandangan Village, Temanggung, Central Java [27]. Magno's artisans have been trained to produce products with high-quality craftmanship. This capital is required in the manufacture of the Spedagi bamboo bike frame. The Spedagi bamboo bicycle is made from local Petung bamboo that has come of age. The bamboo is cut and split to size, then cured and dried using an oven. Next, the bamboo is shaped into bamboo sticks of a certain size and assembled with metal joints (lugs) produced by local craftsmen. The final process is finishing and coating to protect the bamboo from water or other materials that could damage it. It takes at least 60 hours of labor to produce 1 Spedagi bamboo bike frame.

Spedagi bamboo bicycle is not just a means of transportation made from sustainable materials (bamboo), Spedagi is a trigger, an icon, and a symbol of a social movement that focuses on the issue of village revitalization [28]. The Spedagi bamboo bike is a metaphor of village revitalization itself. Made from bamboo, which has been perceived as an ancient, cheap, weak, and easily damaged material, through serious efforts can be made into a product that is not only beautiful, functional, and comfortable, but also has a high selling value (**Figure 3**).

Bamboo bicycles are not only a means of transportation but also a means of social transformation. A well-designed bicycle, done with high craftmanship and an important part of a social movement, makes Spedagi different from other bicycles. Bamboo bicycles have national influence, having been used by the Indonesian president in diplomacy with the Australian prime minister and delivered to country delegations during Indonesia's G20 presidency.

Figure 3.
Spedagi bamboo bike (photo by Hammad Zahid Muharram).

One of the Spedagi movement's social projects, which has inspired hundreds of similar events in Indonesia, is *Papringan* market (*pring* means bamboo in Javanese, and then, *Papringan* is more or less "about bamboo"), a traditional market in a bamboo forest that blends food and art with traditional wisdom on nature and environmental care [29]. *Papringan* market was founded and operated by a group of young people from *Ngadiprono Hamlet* who are members of *Komunitas Mata Air*, a community dedicated to environmental conservation.

Papringan market was previously a neglected area that was often used for garbage disposal. The market occupies an area of 2500 square meters owned by several villagers. There are at least twenty vendors at the location. The handicrafts, vegetables, and traditional food and drinks sold at *Papringan* market have gone through a selection process, so their quality is guaranteed (**Figure 4**).

Customers must exchange money for *pring* money or bamboo money. Visitors are also prohibited from using plastic wrap on the goods and meals they purchase. For this reason, the management provides woven bamboo baskets and box-shaped containers. Bamboo bikes and *Papringan* market are examples of how cultural wealth and modernization may be used to improve environmental stewardship. From there, a circular economy is formed that can enable communities to fulfill each other's needs [29]. In the face of climate change issues, sustainable village revitalization necessitates collaboration among the community, government, and other stakeholders. Communities can achieve development that preserves cultural heritage while also reducing their carbon footprint by using local understanding, implementing sustainable technologies, and raising public awareness. This is a difficult sustainability task, but as Stephenson [30] believes, culture has the capacity to do it.

6. Indonesian government commitment

The Indonesian government currently has various sustainable development programs which are an external expression of the government's commitment to involving

Figure 4.
Papringan market (photo by Hammad Zahid Muharram).

culture as part of development. One of these programs is the Indonesia FOLU Net Sink 2030 program, namely an effort to reduce the greenhouse effect through the development of agroforestry, especially bamboo agroforestry [7]. This source states that bamboo forests can be a means of protection from severe environmental damage when natural disasters such as earthquakes and tropical storms occur. It is further said that based on an agroforestry perspective, bamboo populations can also restore critical land. In critical land in East Nusa Tenggara, Indonesia, 2.5 million bamboo stalks have been planted on an area of 1300 ha. The Indonesian government, through the Ministry of Environment and Forestry, is targeting bamboo planting on an area of 300 thousand ha in 74,000 villages in various regions in Indonesia. This program shows the Indonesian government's commitment to involving local culture as part of development.

The government program involving bamboo is just one of a number of programs involving indigenous communities to manage development in a way that is consistent with their traditional customs and culture. Another program is what is called social forestry. If bambooization specifically focuses on bamboo trees, then social forestry focuses on "forests". In 2016, the Social Forestry program was launched through a regulation by the Minister of Environment and Forestry [31]. Prior to that year, management of Indonesia's forest areas was dominated by *Hak pengusahaan Hutan* (Forest Concession Rights), a policy to permit to use forests for economic purposes. Even though in 1978 the community-based forest resource management approach, namely Forest for People, was declared at the 8th World Forestry Congress in Jakarta and Indonesia hosted the event, decades later the issue and agenda of "forests for the people in Indonesia" was still not be a priority [32]. The social forestry program is actually the result of a long process of struggle by indigenous communities to regain the power to manage forests in their territories, which were previously nationalized, recognized as state property, and handed over their management to entrepreneurs. Most of the entrepreneurs who receive permits for forest management rights come from outside the managed forest area. It is not surprising that they do not feel ownership of the forest and are not motivated to preserve the forest. For them, forests for which management permits have been granted are economic resources.

Instead of managing forests for the benefit of society or the state, entrepreneurs focus more on utilizing forests to maximize economic profits, and this is done with a modern approach so that forest use reaches massive levels, as does its impact on damage. Once again, modernization is an approach that is not in favor of sustainability. Forestry entrepreneurs have been proven to leave forests in a damaged state after receiving massive benefits. Most uses involve logging to use the wood or felling to use the land for other economic purposes such as mining. This modernization approach is very different from the traditional approach of indigenous peoples who view the forest as a "home" where they live and "depend on their livelihood", so that their culture in terms of forest use also aims to protect the forest [10]. In traditional community management, trees are not cut down but maintained. An example is a program called foster trees [33], which is sharing responsibility for caring for trees in the forest so that the trees are maintained until they grow tall and widen in diameter. Another program is the development of ecotourism [34], namely making local resources and wisdom a tourist destination, which allows people to live through "forest use" without having to cut down trees. This program also opens access to the outside world to recognize the ways or culture of people in various forest village communities (communities of people who live in, on the edge or around the forest) in managing and protecting forests (**Figure 5**).

Figure 5.
Example of ecotourism. Source: [35].

Through this and other similar programs, forest village communities with their traditional living culture have proven that their way of life has the potential to absorb far more carbon than modern approaches to forest management. Therefore, it can be said that living with traditional culture is not only caring for the forest but also caring for the earth. Communities with traditional lifestyles are groups that from an ecological perspective are sustainable communities and whose lives are connected to nature [36]. The Indonesian government's commitment to social forestry has now been proven by the establishment of around 90 customary forests in Indonesia with a forest area of more than 75,000 hectares, and nearly 200 locations are in progress with a forest area of almost 3 million hectares [37]. From a conservation perspective, the Indonesian government has recorded 552 conservation areas covering an area of 27.14 million hectares located in 6381 villages throughout Indonesia [38]. Wiratno, the Director General of Natural Resources and Ecosystem Conservation, Ministry of Environment and Forestry of the Republic of Indonesia, reported that the scope of conservation includes ecosystems in high mountain (snowy) rain forests, low mountain rain forests, kerangas forests, swamp forests, peat forests, karst, savannas, mangroves, coastal forests, sandbanks, seagrass beds, coral reef ecosystems, and freshwater lake ecosystems. All conservation is managed by the community through the "Culture as a Base for Nature Conservation" program.

7. Conclusion

Indonesia, as a country with many tribes and many cultures, proves that living a traditional way of life is the best choice from a sustainability perspective. Even though indigenous communities throughout Indonesia are aware of this and are also supported by environmental scientists and environmental activists, efforts to realize "culturally friendly" development have still not reached the ideal level. This business is still connected with community groups who have economic and political interests in development. Based on experience and history in Indonesia, efforts to fight for cultural interests in development need to involve a number of stakeholders, especially indigenous peoples, cultural and environmental scientists and practitioners, as well as journalists. This is where psychology comes into play, as many sustainability challenges are interconnected in human behavior [39].

Indigenous communities are the main actors whose rights to forests and customary territories need to be "reclaimed". The term "reclaim" has a complex meaning, namely not just reclaiming rights to manage forests and customary territories, but reclaiming a way of life that supports nature, giving nature the right to grow and develop, not exploiting it to fulfill human desires (not needs). There is a need for decolonization of knowledge, so that the pro-environmental behavior of indigenous peoples in terms of providing housing, public facilities, food management, providing clothing, etc., is not considered ancient, backward, and not modern. This behavior needs to be assessed as sustainability-oriented behavior so that it is worthy of implementation and promotion.

Because they are the main actors, indigenous communities must initiate this effort. They need to fight to have their "voices" considered in national development policies and plans. It is at this stage that indigenous communities need to receive assistance so that they can fight for it using the language of the laws and regulations that apply in Indonesia. That is the role of environmental scientists and practitioners.

Environmental scientists and practitioners are the groups best able to provide evidence that people's traditional way of life is an alternative that has high sustainability value. Naturally, their behavior is sustainable behavior. Therefore, this behavior needs to be facilitated so that it can be realized. This behavior and lifestyle cannot be realized if their interactions with forests or nature are intervened by political or legal decisions that hand over the rights to manage forests and nature in their area to parties outside the community. Based on experience in Indonesia, the number of groups of environmental scientists and practitioners providing assistance and joint efforts with indigenous community groups is still small compared to needs.

Journalists are parties who play a role in disseminating information about indigenous peoples' movements, in order to attract the attention of the general public to provide the necessary support. Data obtained through Google searches shows that there is still very little news about environmental issues, including the movement of indigenous communities to take over forest management rights and bambooization programs. Community search data also shows that not many people are looking for information about environmental issues and indigenous communities. This situation is reciprocal. Because there are not many people who are interested in environmental issues, in the journalist's view, reporting about them does not have "news value". On the other hand, because it is considered to have no news value, not many journalists report it in the mass media, so there is not much information about it, and as a result, the public is not adequately exposed to these issues. Thus, even though the movement towards sustainability has been launched in many groups of society, it still requires a long struggle by every element of society who has cultural and environmental awareness, so that this movement becomes a massive, popular movement and has a broad impact. The three stakeholder components, namely indigenous peoples, environmental scientists, and activists, as well as Indonesian journalists, have succeeded in achieving certain achievements in the past and present in Indonesia, and there is hope and optimism that by continuing to maintain similar efforts, these achievements will continue to increase in future years. Although individual and collective efforts have been made and a number of successes have been achieved, research shows that the transformation of sustainability thinking at the individual level is not enough [36]. How sustainability spreads from the individual level to local, regional, and global communities is still homework that needs to be done.

Author details

Retno Hanggarani Ninin*, Hammad Zahid Muharram and Noer Fauzi Rachman
Universitas Padjadjaran, Indonesia

*Address all correspondence to: rhninin@unpad.ac.id

References

[1] Direktorat Jenderal Kependudukan dan Pencatatan Sipil. Kemendagri Serahkan Jumlah Penduduk per Kecamatan ke KPU untuk Pemilu 2024. Diambil kembali dari Kemendagri.go.id; 2022. Available from: https://dukcapil.kemendagri.go.id/blog/read/kemendagri-serahkan-jumlah-penduduk-per-kecamatan-ke-kpu-untuk-pemilu-2024

[2] Indonesia.go.id. Suku Bangsa. Diambil kembali dari INDONESIA.GO.ID. Portal Informasi Indonesia; 2017. Available from: https://indonesia.go.id/profil/suku-bangsa/kebudayaan/suku-bangsa

[3] Indonesia.go.id. Agama. Diambil kembali dari Indonesia.go.id Portal Informasi Indonesia; 2023. Available from: https://indonesia.go.id/profil/agama

[4] Pursika IN. Kajian Analitik terhadap Semboyan "Bhinneka Tunggal Ika". Jurnal Pendidikan dan Pengajaran. 2009;**42**(1):15-20

[5] Haryanto. Merdeka Belajar untuk Revitalisasi Bahasa daerah yang Terancam. Diambil kembali dari Sekretariat Kabinet Republik Indonesia; 6 April 2023. Available from: https://setkab.go.id/merdeka-belajar-untuk-revitalisasi-bahasa-daerah-yang-terancam/

[6] Ministry of Environment and Forestry Republic of Indonesia. The State of Indonesia's Forests 2020: Executive Summary. Jakarta: Ministry of Environment and Forestry, Republic of Indonesia; 2020

[7] Setiawan A. Bambu sebagai Mitigasi Perubahan Iklim. Indonesia: Portal Informasi; 2023. Available from: https://indonesia.go.id/kategori/keanekaragaman-hayati/6927/bambu-sebagai-mitigasi-perubahan-iklim?lang=1

[8] Trianingsih L, Hidayah R. Studi perbandingan efektivitas material bambu dan batu bata sebagai konstruksi dinding. INERSIA. 2014;**10**(1):44-52. DOI: 10.21831/inersia.v10i1.4428

[9] Devi AS, Singh KS. Carbon storage and sequestration potential in aboveground biomass of bamboos in north East India. Scientific Reports. 2021;**11**(837):1-8. DOI: 10.1038/s41598-020-80887-w

[10] Tha Samdhana Institute; Molecula Indonesia. Final Report: Perception toward the "Masyarakat Adat". Bogor: The Samdhana Institute; 2022

[11] Rachman NF, Masalam H. The trajectory of indigeneity politcs against land disposession in Indonesia. Sriwijaya Law Review. 2017;**1**(1):122-142. Available from: http://journal.fh.unsri.ac.id/index.php/sriwijayalawreview/article/view/13

[12] Afiff SA, Rachman NF. Institutional activism: Seeking customary forest rights recognition from within the Indonesian state. The Asia Pacific Journal of Anthropology. 2019;**20**(5):453-470. DOI: 10.1080/14442213.2019.1670245

[13] Rachman NF, Siscawati M. Forestry law, Masyarakat Adat, and struggles for inclusive citizenship in Indonesia. In: Antons DC, editor. Routledge Handbook of Asian Law. London, New York: Routledge; 2016

[14] Kementerian Lingkungan Hidup dan Kehutanan Republik Indonesia. Peraturan Menteri No. P.34/MENLHK/SETJEN/KUM.1/5/2017 tentang

Pengakuan dan Perlindungan Kearifan Lokal dalam Pengelolaan Sumber Daya Alam dan Lingkungan Hidup. Jakarta: Kementerian Lingkungan Hidup dan Kehutanan; 2017

[15] Brown R, Maudlin D. Concepts of Vernacular Architecture. Diambil Kembali Dari Academia. Edu; 2011. Available from: https://www.academia.edu/16402794/Concepts_of_Vernacular_Architecture

[16] Jannah M, Baharuddin, Taskirawati I. Potensi dan Pemanfaatan Tanaman Bambu pada Lahan Masyarakat di Desa Kading Kabupaten Barru; potency and utilization of community bamboo plantation in the Kading Village of Barru District. Jurnal Perennial. 2019;**15**(2):87-92. Available from: https://journal.unhas.ac.id/index.php/perennial

[17] Raka ID, Wiswasta IA, Budiasa I. Pelestarian Tanaman Bambu sebagai Upaya Rehabilitasi Lahan dan Konservasi Tanah di Daerah Sekitar Mata Air pada Lahan Marginal di Bali Timur. Agrimeta, Jurnal Pertanian Berbasis Keseimbangan Ekosistem. 2011;**1**(1):1-11

[18] Atmaja IG, Haryono E. Analisis Landscape Capacity Industri Semen di Pulau Jawa. Yogyakarta: Universitas Gadjah Mada; 2014

[19] Aria Z, Purwanto L. Analisis Emisi Karbon Rumah Tipe-45 di Kota Palangkaraya dengan Single-Subject Experimental. Jurnal Arsitektur Komposisi. 2021;**14**(2):93-101

[20] Artiningsih NK. Pemanfaatan Bambu pada Konstruksi Bangunan Berdampak Positip bagi Lingkungan. METANA, Media Komunikasi Rekayasa Proses dan Teknologi Tepat Guna. 2012;**8**(01):1-9

[21] Redaksi PI. Budaya dan Sejarah Kain Tenun Ikat khas NTT. Diambil kembali dari Pariwisata Indonesia. t.thn. Available from: https://pariwisataindonesia.id/budaya-dan-sejarah/kain-tenun-ikat-khas-ntt/

[22] Salma IR, Syabana DK, Satria Y, Cristianto R. Diversifikasi Produk Tenun Ikat Nusa Tenggara Timur Dengan Paduan Teknik Tenun Dan Teknik Batik. Dinamika Kerajinan dan Batik. 2018;**35**(2):85-94. Available from: https://media.neliti.com/media/publications/272154-diversifikasi-desain-produk-tenun-ikat-n-3aecb6da.pdf

[23] Laela E, Isnaini REY, Sayogo R. Efektivitas Sabun Alami terhadap Warna Batik. Dinamika Kerajinan dan Batik. 2018;**35**(2):119-124. Available from: https://media.neliti.com/media/publications/272160-none-6129eb5f.pdf

[24] Kementerian Pendidikan dan Kebudayaan Republik Indonesia. Noken. Diambil kembali dari Warisan Busaya Takbenda Indonesia; 2013. Available from: https://warisanbudaya.kemdikbud.go.id/?newdetail&detailTetap=77

[25] Wisnubroto K. Se'i, Cita Rasa Daging Asap dari Timor. Diambil kembali dari INDONESIA.GO.ID. Portal Informasi Indonesia; 2021. Available from: https://indonesia.go.id/kategori/kuliner/2572/se-i-cita-rasa-daging-asap-dari-timor?lang=1

[26] Arsyad MZ. Jejak Jalur Rempah dalam Sepotong Rendang Daging. Diambil kembali dari Jalur Rempah; 2023. Available from: https://jalurrempah.kemdikbud.go.id/artikel/jejak-jalur-rempah-dalam-sepotong-rendang-daging

[27] Spedagi. Spedagi. Diambil kembali dari spedagi. 2012. Available from: https://www.spedagi.com/bamboo-bike

[28] Spedagi. Coolaboration. In: Diambil kembali dari The 3rd International Conference on Village Revitalization. 2018. Available from: https://icvr.spedagi.org/en/home-2/

[29] Suyatna H. Dampak Sosial Ekonomi Pasar Papringan di Dusun Ngadiprono. Diambil kembali dari Universitas Gadjah Mada; 2019. Available from: https://etd.repository.ugm.ac.id/penelitian/detail/178101

[30] Stephenson J. Culture and Sustainability: Exploring Stability and Transformation with the Cultures Framework. Dunedin, New Zealand: Palgrave Macmillan; 2023. DOI: 10.1007/978-3-031-25515-1

[31] Kementerian Lingkungan Hidup dan Kehutanan. Peraturan Menteri Lingkungan Hidup dan Kehutanan No. 33 tahun 2016. Jakarta: Kementerian Lingkungan Hidup dan Kehutanan; 2016

[32] Sanudin, Awang SA. Evaluasi Kehutanan Sosial: Tantangan Generasi 3. Yogyakarta: Kementerian Lingkungan Hidup dan Kehutanan: Direktorat Jenderal Konservasi Sumber Daya Alam dan Ekosistem; 2019

[33] Mirza IN, Irwanto RR. Evaluasi reforestasi di kawasan konservasi Taman Buru Gunung Masigit Kareumbi, Sumedang. Prosiding Seminar Nasional Masyarakat Biodiversitas Indonesia. 2015;**1**(7):1615-1621

[34] Hendriyani IG. Siaran Pers: Menparekraf Sebut Penyelenggaraan "Indonesia Ecotourism Summit" Tingkatkan Kualitas PariwisataBerkelanjutan. Diambil kembali dari Kemenparekraf.go.id; 2023. Available from: https://www.kemenparekraf.go.id/hasil-pencarian/siaran-pers-menparekraf-sebut-penyelenggaraan-indonesia-ecotourism-summit-tingkatkan-kualitas-pariwisata-berkelanjutan

[35] Chusnul C. Ekologi Adat Komunitas Ammatoa. Yogyakarta: Center for Religious and Cross-cultural Studies (CRCS); 2021

[36] Thierman UB, Sheate WR. The way forward in mindfulness and sustainability: A critical review and research agenda. Journal of Cognitive Enhancement. 2020;**5**:118-139. DOI: 10.1007/s4146-5-020-00180-6

[37] Supriyanto B. Rencana Aksi Kolaborasi Kegiatan Kampus Merdeka Dengan Perhutanan Sosial. Jakarta: Kementerian Lingkungan Hidup dan Kehutanan; 2022

[38] Wiratno. Kebudayaan sebagai Basis Konservasi Alam. Jakarta, Indonesia: Ministry of Environtment and Forestry Republic of Indonesia (Unpublished document). 2018

[39] Riemer M, Schweizer-Ries P. Psychology and sustainability science: Complexity, normativity, and transdisciplinary in meeting sustainability challenges. Umweltpsychologie. 2012;**16**(1):143-165

Chapter 17

Fostering a Cultural Melting Pot: The UAE's Approach to Multicultural Integration and Heritage Preservation

Basak Ozoral

Abstract

In our globalized world, the concepts of cultural assimilation and multicultural integration have gained significant attention, especially in regions experiencing a surge in cultural diversity. The coexistence and interaction of diverse cultures are essential for fostering social harmony and well-being. This phenomenon is especially evident in areas where rigid local and religious cultures coexist, often sparking discussions on identity and belonging. The United Arab Emirates (UAE) serves as a remarkable example of a nation navigating cultural change through policies that preserve its local heritage while seamlessly integrating diverse elements within a harmonious multicultural framework. This study explores how the UAE has successfully embraced globalization while safeguarding its values, culture, and identity. Despite some criticisms and challenges, the UAE's achievements in this regard deserve recognition. Additionally, we'll discuss Ibn Khaldun's theories on the transformations that occur when Bedouin societies transition to urban life, illustrating them with examples from the UAE.

Keywords: United Arab Emirates, cultural heritage in the UAE, globalization, multicultural integration, identity

1. Introduction

As a result of the dramatic economic development in the United Arab Emirates in recent years, the spread of economic liberalization and freedom of thought is observed. Consequently, Emiratis often find themselves entangled in an identity debate between modernity and tradition. Some religious leaders, politicians, and older citizens accuse the younger generation of adopting "Western," "non-Arab," or "non-Islamic" values. However, portraying the Arab world as radical, inflexible, and regressive, as the West often does, does not accurately reflect the identity and lifestyles of the Arabs living in today's United Arab Emirates ([1], p. 74).

Samuel Huntington's thesis on the "clash of civilizations" suggests a conflict between Western and Islamic values, and this view still finds acceptance. However,

Figure 1.
The meeting of modern and traditional. Foto by Mohamed Reshad, https://unsplash.com/photos/camel-on-white-car-during-daytime-MsHMkIQXAao.

as demonstrated by the example of the United Arab Emirates, many Emirati Arabs can reconcile their beliefs, modern values, cosmopolitan identities, and respect for individual freedoms with their commitment to their families and communities, and they may not perceive contradictions in their identities.

The methodological framework underpinning this article's exploration of the United Arab Emirates' culture is characterized by an exhaustive and meticulous review of pertinent literature. This discerning examination encompasses a myriad of scholarly books, articles, and comprehensive reports that pertain to the subject matter. This rigorous review is essential to facilitate the synthesis of pre-existing knowledge, enabling a nuanced analysis of multifaceted cultural dimensions, historical trajectories, and contemporary influences embedded within the rich tapestry of the UAE's cultural fabric. Moreover, this endeavor is bolstered by an exploration of both primary and secondary sources, spanning governmental publications, scholarly research, and ethnographic studies. The amalgamation of these diverse sources promises to yield a holistic and discerning comprehension of the intricate cultural landscape prevailing within the United Arab Emirates (**Figure 1**).

2. Development of the UAE as a new hub in the Middle East

The United Arab Emirates (UAE) is a federation that was established on December 2, 1971, comprising seven Emirates, formerly known as the Trucial States. The most prominent Emirates among these are Abu Dhabi and Dubai, with the other five being Ajman, Fujairah, Ras Al Khaimah, Sharjah, and Umm Al-Quwain. Abu Dhabi's commencement of oil exports in 1962 marked a significant moment in the embryonic stages of the formation and development of the UAE [2]. Rapid infrastructure development and modernization within the Emirates have led to visible transformations in the lifestyle of the local population. The preservation of cultural heritage is a focal point of the UAE Vision 2021 plan, emphasizing the commitment to safeguarding the nation's identity and fostering a sense of belonging among its citizens. The UAE achieved significant recognition by ranking 16th out of 138 economies in the World Competitiveness Index for 2016–2017, showcasing its prominent position in the global competitive arena [2].

This achievement is remarkable in a multicultural setting where over 85% of the population comes from foreign countries, and more than 200 nationalities peacefully coexist while preserving their cultural values ([3], p. 56). The values and heritage that form the foundation of Emirati identity hold significant importance within this multicultural context. The United Arab Emirates, serving as a geographic and cultural space that hosts people from various corners of the world, showcases the intricate fabric of cultural diversity. By successfully serving as a converging point for both native and international populations, the country creates an environment where numerous cultural traditions coexist harmoniously. At the core of the United Arab Emirates' multifaceted, multicultural integration strategy lies a steadfast commitment to fostering an environment where diverse cultures converge, guided by principles of mutual respect and deep intercultural understanding. This convergence is not merely symbolic but also makes concrete contributions to the country's social and economic development.

Significantly, the commitment of the United Arab Emirates extends beyond the promotion of multicultural integration; instead, it presents itself as a meticulous guardian of its highly valued cultural heritage. The country undertakes a range of initiatives, including the establishment of museums, archaeological excavations, and cultural events, all aimed at preserving its rich history and cultural fabric for future generations. In doing so, the United Arab Emirates not only strengthens its own cultural identity but also ensures the enduring preservation and continuation of its esteemed cultural heritage.

The United Arab Emirates has adopted an approach can be called as "Multicultural Integration and Heritage Conservation: The United Arab Emirates' Perspective ([4], pp. 41–44)." This approach exemplifies the importance of embracing cultural diversity and, more significantly, fostering social prosperity and advancement. While initially implemented in the UAE, this approach also has the potential to serve as a model for fostering harmonious cultural cooperation and safeguarding cultural treasures in various regions and countries.

3. Cultural roots of the UAE

The culture of the United Arab Emirates carries the characteristics of Bedouin Arab culture, but throughout history, the local population, which was originally a small tribal community, changed with the influence of foreign nationals who migrated to the region in the mid-20th century ([5], pp. 50-55). Therefore, Emirati culture offers a richness where Arab, Islamic, and Persian cultures intertwine, and it also carries influences from East African and South Asian cultures. Islam has a significant impact on local architecture, music, clothing, cuisine, and lifestyle ([5], pp. 50-55).

The city of Al Ain, located in the Emirate of Abu Dhabi, is part of the United Arab Emirates, which consists of seven separate Emirates and has been recognized as a UNESCO World Heritage site. The Emirate of Sharjah was declared the "Cultural

Capital of the Arab World" by UNESCO in 1998 and was selected as the "Capital of Islamic Culture" by the Organization of Islamic Cooperation (OIC) in 2014 (Culture - The Official Portal of UAE Government). This international recognition is considered an indicator of the cultural richness and importance of the United Arab Emirates, and in this context, the UAE plays a leading role in the preservation of cultural heritage and the promotion of cultural interaction, serving as a potential example to other countries and regions.

To understand the extraordinary journey that has brought this vibrant Arab country to its current status, it is helpful to take a brief look at its socio-economic history. The famous historian Ibn Khaldun's saying, "geography is destiny," resonates significantly in this context. The geographical area where the United Arab Emirates is located has definitively shaped its destiny. Indeed, this arid and hot region has historically stood as one of the most arid and hot regions in the world, burdened with extremely challenging living conditions. The scarcity of water resources made agricultural ventures impossible, leading to consistently low population density. Despite the presence of luxury hotels, shopping centers, and skyscrapers today, half a century ago, the areas they now adorn were vast desert expanses left to the wilderness.

The native population of Emirates can trace its origins back to Bedouin tribes. These tribes, facing challenges like scarcity of water and food due to the arid desert climate of the Arabian Peninsula, were referred to as Bedouins. Most Bedouin tribes maintained a nomadic way of life, but some engaged in activities such as fishing along the eastern coast or cultivating grapes, wheat, and dates. Limited-scale trade also took place in these settlements, with craftsmen like tailors, leatherworkers, and blacksmiths present ([6], p. 26; [7], pp. 22–23). Nomadic tribes provided protection from attacks for settled communities in exchange for essential supplies, forming the basis of this mutual exchange, which was vital for survival. Caravan trade was also common among nomadic tribes as these caravans facilitated the transportation of goods like spices, leather, precious stones, silk, and tea from India to various regions of Anatolia, Africa, and Asia. To survive the harsh climate and geographical conditions of the desert, mutual support and cooperation among nomadic tribes were necessary, leading to the tight maintenance of familial and tribal relationships. Although not written, an oral culture was significant, and agreements were made verbally, always honored as societal status was of great importance, and no one wanted to tarnish their reputation. Bedouin tribes lived with a patriarchal culture where men were expected to be more warrior-like, protective, and brave, as demanded by the harsh climate and geography. Additionally, Bedouins possessed qualities considered important virtues, such as hospitality, generosity, and kindness. Like other nomadic communities, Bedouins could not establish more advanced cultural structures, thus leading to a simpler and more primitive lifestyle ([8], p. 23).

In Bedouin tribes, blood ties were the most binding, and members of the same tribe had mutual responsibilities toward each other. In today's United Arab Emirates, traces of the traditional organizational structure of Bedouin society can still be observed in various aspects of social, political, and cultural life. At the pinnacle of this social and economic organization is the Sheikh, responsible for organizing the tribe. To achieve this, the Sheikh oversees the economic needs of families, ensures internal harmony, resolves disputes, administers justice when necessary, and provides protection to those in need. Regular meetings called Mejlis are held to discuss matters of concern to the community, where the Sheikh listens to the opinions of important members of the tribe and makes decisions. These meetings also include influential traders who have continued the tradition of finalizing trade agreements and decisions to the present day.

The above description indeed reflects how Ibn Khaldun depicted the lifestyle and characteristics of Bedouin (nomadic) communities in his work "Muqaddimah." In "Muqaddimah," Ibn Khaldun describes the qualities of the Bedouins as follows: "The Bedouins are those who live apart from society, lead a wild life exclusively in the desert, and in the process become wild and uncultured, distancing themselves from all forms of protection and settling within fortified places and walls. As a result, they take it upon themselves to defend, not entrusting this responsibility to others. They do not rely on others in this regard. Bedouins constantly carry weapons and remain vigilant against potential threats along their journey. They seldom rest or sleep, even when they sit or prepare their mounts. They avoid sleep, even at night. They are cautious and alert in the face of any noise or disturbance. They can enter vast, uninhabited, and desolate deserts and plains with confidence and fearlessness, without needing assistance from others. They have developed patience and courage as inherent qualities. Except for rare occasions when someone calls for help or when there is an individual pursuing them with hostile intent, they rely on their patience and courage." ([9], p. 323).

Ibn Khaldun's observations emphasize the impact of economic conditions and social changes on the development and character of human societies, highlighting the fluidity of human nature and culture in response to changing circumstances. As explained by Ibn Khaldun, these changes in the path of urbanization, coinciding with the discovery of oil resources in the Arabian Peninsula, marked a historical turning point. This transformation, deeply underscored by the introduction of oil as a new and significant economic resource, played an effective role in accelerating the transition from a once-nomadic Bedouin society to a more settled way of life. ([9], p. 627) Ibn Khaldun emphasizes the idea that the development and dissemination of knowledge are closely associated with urbanized and culturally advanced societies, as opposed to nomadic or less developed communities.

Indeed, the progress of civilization in the Arabian Peninsula revealed itself as a tightly interconnected, complex process with the transition from nomadic life to settled communities. The discovery of oil resources not only represented a profound economic shift but also served as a catalyst for the establishment of essential infrastructure, educational institutions, and technological advancements necessary for the development of a sophisticated and urbanized society. From such a perspective, we can understand how crucial oil has been in reshaping the path of societal progress within the region. In summary, Ibn Khaldun's insights offer a valuable perspective on understanding the interaction between economic transformation, knowledge development, and the evolution of society. The urbanization transition driven by the emergence of oil as a significant economic resource demonstrated how a change in economic foundations could ultimately contribute to profound societal changes, ultimately adding to the intricate fabric of civilization in the Arabian Peninsula ([10], p. 4).

In the 1930s, the discovery of oil in Bahrain in 1932 led to a radical transformation in the political and economic landscape of the Arabian Peninsula. Dubai emerged as a significant port, particularly in the early 20th century, after Sheikh Maktoum bin Hashar abolished taxes and controls in 1904. This attracted Indian and Iranian merchants and led to the growth of trade ([11], p. 71). The United Arab Emirates underwent a gradual transformation in turning its oil and gas reserves into a source of income, with significant changes starting in the 1960s. Rapid development began, especially with the increase in oil prices in 1973. Within 25 years, the United Arab Emirates became one of the countries with the highest living standards. Today, one-fifth of the population living in the United Arab Emirates are millionaires, and education and healthcare services are provided free of charge. As of 2012, the United

Arab Emirates is one of the world's richest countries in terms of per capita income (The World Bank [12], p. 72). Aslında geleneksel yaşam tarzı, petrol keşfi ve eski Trucial Devletleri'nin 1971'de bağımsızlığına kavuşmasının hızlı değişiklikleri sonucunda derin bir dönüşüm geçirdi ve Birleşik Arap Emirlikleri'nin kurulmasına yol açtı ([13], p. 77).

The significant economic development brought about by oil reserves also brought about a societal change in terms of social and cultural dimensions. Life in the United Arab Emirates underwent a major transformation with the beginning of modernization of traditional structures. Its competitive economy attracted people from all around the world to work, live, and explore in this region. This has an impact on traditional values, lifestyles, and cultural structures. The cultural and lifestyle changes experienced by Emiratis will be discussed below by looking for past and present practices.

4. The cultural characteristics of the United Arab Emirates in the past and today

4.1 Family, women, and social order

The family institution was the most important institution in Emirati society and continues to hold a central position, serving as the foundation of its cultural values and social structure. Despite the increasing prevalence of nuclear families in contemporary Emirati life, the tradition of extended family gatherings remains deep-rooted in the nation's cultural fabric. Special occasions, especially on Fridays and various holiday celebrations, continue to serve as moments when extended families come together, sharing communal meals, and providing opportunities to strengthen the bonds that connect them.

In Emirati families, the role of the mother holds a respected position and carries the title "Umm," signifying her vital position within the household and her significant contributions to family life. This title emphasizes respect and reverence for mothers, highlighting their central role in nurturing and sustaining family bonds ([2], p. 10).

Polygamous households are traditionally and legally still a valid arrangement. They combine elements of both the nuclear and extended family structures, with spouses living separately, sometimes with their own families or in-laws. Husbands divide their time among these households as needed. Islamic principles allow a man to have up to three wives, with the condition of treating them fairly. However, practical differences due to interpretations and sources can impact family well-being. Multiple marriages are no longer accepted by educated women, as they choose to establish their own lives when faced with this situation, given their economic independence. However, it is not realistic to assume that this reflects the general population, as the number of such confident and empowered women is not very high." ([14], p. 455; [15], pp. 75–92).

In accordance with the patriarchal cultural norms prevalent throughout the Arabian Peninsula, sons are granted higher respect than daughters ([16], p. 14). It is not difficult to conclude such a result concerning symbolic differences marking the birth of sons, such as more elaborate celebrations following the birth of sons and the ritual slaughter of two rams upon the birth of a son. These symbolic distinctions marking the birth of sons are characteristic features of practices commonly observed in other parts of the world where patriarchal family structures dominate.

Families adhering firmly to rigid interpretations of these gender norms may invest significant effort in monitoring their daughters' activities and movements while granting more leniency in supervision to their sons. Consequently, sons often enjoy the freedom to do what they desire and go where they please, without needing to explain their whereabouts or activities to their families.

This tolerance toward young sons is not an uncommon phenomenon in the context of the United Arab Emirates and has raised concerns among school social workers and educators, particularly regarding discipline and academic performance [17].

When Islamic values and traditional structures come together, gender roles in family and social life are also clearly defined. According to the religious understanding, a woman's place within the family is the home, and her responsibility is to take care of the family members. Economically, it is the responsibility of men to provide for the family and earn money. In the past, women had to independently manage various tasks, including childcare, household chores, water collection, food gathering, weaving, sewing family clothes, and raising animals.

Regarding the changing role of women, Emirati women today have more opportunities compared to their mothers. The older generation generally focused on household responsibilities, but these tasks are now more equally shared. Younger women are becoming more visible in society, as they can receive education and attain a status in line with their aspirations ([18], p. 260).

The government provides positive support for increasing women's participation in different fields and industries (Sheikh Zaid participated in the Gulf Women's Conference in Abu Dhabi in 1984, where his wife, Sheikha Fatima, spoke about the necessity of women working). Young female students have positive thoughts about the future and can indeed find suitable opportunities upon graduation. Although they consider balancing their professional lives with family life and continuing their careers when they have children, the majority still do not contemplate working outside the home ([18], p. 270).

In both individual and societal contexts, women are grappling with the contradictions and unrest created by deep-rooted traditions and customs. In the United Arab Emirates, educated women who have gained economic independence following the increase in higher education and career opportunities find themselves caught between two different worlds. Their professional identity as active members of society clashes with the traditional values that confine them, limiting their visibility in society and restricting their ability to travel without a male companion. This creates a peculiar dilemma where a female teacher who educates hundreds of young minds and prepares them for the future, a female doctor who saves lives in critical surgeries, cannot enjoy the freedom to travel wherever they desire [19].

Additionally, due to the opportunities brought about by high economic power, Emirati women hire domestic helpers for household services and childcare, thus reducing a significant portion of their domestic responsibilities. The roles of women in modern United Arab Emirates have evolved, with some traditional roles disappearing alongside greater opportunities.

In fact, the persistence of these cultural structures and values reflects the resistance of Emirati traditions in the face of modernization and globalization, emphasizing the importance of preserving the family structure. Government policies even point to the continued significance of certain old traditions in the rapidly changing social context for the preservation of Emirati identity. These cultural elements not only provide continuity and a sense of identity but also strengthen a strong sense of community and unity among Emirati society, thereby enhancing the country's social fabric.

These remarkable developments have a relatively new origin and have created a significant generational difference in shaping formative experiences and life expectations between previous and subsequent generations within the family. Often, the previous generation tends to idealize the past a bit and has praised the Bedouin way of life, actually emphasizing society's commitment to common moral values, as "good-hearted" and "well-behaved" individuals from that era. Despite the challenges, people believed they had closer family ties and neighborly support.

4.2 Elderly and youth in the United Arab Emirates

The population of the United Arab Emirates (UAE) is predominantly young, with a median age of 32.6 years as of 2020. In contrast, Germany had a median age of 45.7 years, and Japan's was 48.4 during the same year. Given these age demographics, the current focus of public policy in the UAE is on addressing the needs of its youth and the working-age population below 65, making up 98.7% of the total population. Notably, in the UAE, older individuals are highly respected ([20], p. 2).

It's not surprising that, due to Emiratis' strong commitment to family values, local customs, and reverence for the elderly, roughly 80% of the surveyed population agrees that society has an obligation to care for the elderly. Moreover, over 75% of respondents believe that older individuals contribute significantly to society through their knowledge, experiences, and voluntary contributions to their communities ([20], p. 2). In the UAE, a majority of the population (62%) personally knows someone who receives support or aged care in their own home. However, only 32% are familiar with someone residing in a residential aged care facility or hospital.

While it can be argued that old age still enhances family and societal status in the United Arab Emirates, and even today, elderly individuals continue to live with their families or under their close care and protection, this situation is changing. The country's shifting demographics, changes in the patterns of disease and dependency, older peoples' expectations, and shifting family structure will create pressure on the demand for aged care in the future. Actually, research demonstrates that there are misconceptions among Emirati families about having public nursing homes. For people in this society placing their older parents into nursing homes is synonymous with disobedience; to them, it is a sign of dishonor and embarrassment. Changing cultural concepts must be taken to the higher levels of policymakers so they may develop continuous public awareness campaigns of such matters ([21], p. 11).

In the UAE, 40% of the population is under the age of ten, raising concerns about potential political instability due to a significant youth population. Emirati men often wear traditional clothing, while teenage boys frequently combine traditional attire with baseball caps. Emirati girls blend Western and Arab fashion by wearing traditional black dresses paired with Western-style clothing underneath.

Dubai Mall, along with several other upscale shopping malls in the UAE, is a prominent shopping destination that houses branches of Western retail establishments. The presence of these stores is gradually bridging the gap between wearing a hijab and wearing fashionable attire from well-known Western brands like Zara, H&M, Mango, and DKNY. DKNY, for example, introduced an Islamic Ramadan collection for the summer and fall of 2014, designed with urban women in mind. This collection includes a range of vibrant trench coats, complemented by trousers, khakis, and stylish hijab options ([1], p. 84).

Citizens of the United Arab Emirates (UAE) typically live with extended families and have strong social bonds with their relatives. Non-citizens, on the other hand, usually live in nuclear families. They exercise more parental control over their children compared to Western parents. Many families in the UAE can afford foreign domestic helpers, who play a valuable role in child-rearing despite their low wages ([22], pp. 130–131). Therefore, in recent years, caregivers' cultures and languages have had a significant influence on children's upbringing. This situation raises concerns that it may lead to a more comfortable acceptance of Western norms and foreign languages in the younger generation, potentially distancing them from traditional values.

In the United Arab Emirates, girls, especially concerning pre-marital sexual relations prohibited by religion and tradition, face stricter social constraints compared to boys. Religion holds significant importance in the lives of the youth in the UAE and influences their actions in various ways. Most Muslim girls in the region adopt practices such as covering their heads in public spaces and abstaining from alcohol, while some view these rules as personal preferences and minor transgressions. Others adhere more strictly to their beliefs, utilizing conservative religious aspects to reaffirm their cultural and religious identities. Many young Muslim men regularly attend mosque prayers, and both boys and girls generally perform daily prayers, reflecting the significance of Islam in their lives and their respect for their religion and culture ([22], p. 134).

4.3 Education

In the United Arab Emirates (UAE), literacy has significantly increased over the past 30 years, with a notably higher literacy rate among women. Schools and universities in the UAE are often segregated by gender, with separate campuses for male and female students.

Until the 1950s in the UAE, educational institutions were known as "kuttab," where only young boys and girls attended to learn the Quran, memorize it, and receive some basic mathematics instruction. Most girls would attend these schools before reaching adolescence, preparing for marriage and later household duties ([23], p. 51). However, in contemporary times, female students have the opportunity to attend education at all levels, including universities, in coeducational schools or schools and campuses exclusively for females ([24], p. 313). While many female students pursue education in the social sciences and arts, only a few opt for fields like engineering, medicine, or law ([23], pp. 55–56).

In Dubai, the Dubai Education Council (DEC) is dedicated to achieving global educational standards, emphasizing international accreditation and quality assurance while preserving cultural and social values. Their mission and strategy include:

1. Developing global benchmarked educational standards.

2. Encouraging schools to contribute to societal development.

3. Enhancing and supporting quality education through initiatives.

4. Promoting continuous improvement and standardization in education.

5. Providing support to the Ministry of Education and relevant stakeholders.

6. Recommending proposals to the Dubai Executive Council to fulfill its vision for education reform.

Education reform in the UAE focuses on higher standards, professionalism, interactive learning, and integrating English into subjects like math and science. ADEC, DEC, and MOE are driving this reform while preserving cultural identity. Since its inception, ADEC has significantly expanded education access, with around 650,000 students in 1,256 schools in 2006–2007, with 60% in public schools ([25], pp. 16–19).

Despite substantial developments in the UAE's education system, the government acknowledges that education is a long-term investment requiring continual updates in policies, infrastructure, and human resources to prepare graduates for the modern job market and contribute to the country's development. This awareness led to the establishment of Education 2020, a five-year plan initiated by the UAE Ministry of Education (MOE) with the goal of improving teaching methodologies and student learning outcomes. Vision 2020 is not a rigid vision but rather underscores the importance of continuous enhancement in response to evolving conditions within the educational system and society. It serves as a strategic guide for the UAE's educational progress, with periodic assessments conducted every five years to gauge advancements and chart future courses of action. Vision 2020 is grounded in four primary pillars: Islamic heritage, national heritage, the nation's constitution, and its current developmental priorities. These four pillars provide the groundwork for planning and decision-making, aiding in the evaluation of the country's internal and external landscapes, taking into account social, economic, strategic requirements, and necessary skills. Vision 2020 aspires to transform education by integrating it with various facets of life, ultimately boosting individual self-realization and aligning with the nation's demographic, strategic, economic, and societal needs ([24], p. 315).

4.4 Cuisine and food culture in the United Arab Emirates

The culinary traditions of the United Arab Emirates (UAE) reflect a rich history and a blend of cultural influences. Historically, the local population in the region has adhered to a simple yet health-conscious dietary tradition, characterized by the utilization of locally sourced ingredients. Goat and camel meat, dates, and milk have long been staples of the Emirati diet, offering not only sustenance but also a connection to the arid and challenging environment of the desert ([26], pp. 2–3).

However, the UAE's rapid modernization and globalization have significantly impacted its food culture. Today, the UAE stands as a global hub, attracting individuals from all corners of the world. As a result, the culinary landscape of the UAE has evolved to encompass a diverse array of international cuisines. Chinese, Japanese, American, and Indian cuisines, among others, have found a prominent place in the hearts and palates of the UAE's multicultural population.

One prominent aspect of this culinary transformation is the proliferation of international fast-food chains throughout the country. These establishments have gained immense popularity among the people, offering a convenient and quick dining experience. While traditional Emirati dishes remain cherished and are often enjoyed by families at home, the emergence of fast food has changed dining habits and preferences, particularly among the younger generation.

This shift toward dining out in restaurants, meeting friends, and spending leisure time in shopping malls is emblematic of the changing social fabric of the UAE. However, it is essential to acknowledge that this trend has also brought about a decline in the traditional Arab practice of hosting guests at home for meals. The act of serving homemade dishes to family and friends has been an integral part of Arab hospitality for generations, fostering a sense of togetherness and community. The shift toward restaurant dining, while convenient, has, to some extent, led to the diminishing of these cherished traditions ([26], p. 3).

Furthermore, the adoption of unhealthy eating habits, often associated with the consumption of fast food and processed foods, has contributed to a growing health concern in the UAE: obesity. Obesity rates have been on the rise in recent years, posing significant challenges to public health. This underscores the importance of promoting balanced and nutritious dietary choices among the population to mitigate the adverse effects of these changing food preferences.

Amidst these shifts in culinary habits, one enduring symbol of Emirati hospitality is Arabic coffee, or 'Gahwa.' This aromatic and traditional beverage continues to hold a special place in Emirati culture. It is often served to guests as a symbol of welcome and goodwill. However, even this tradition has seen an evolution with the introduction of large coffee chains like Starbucks, which have brought the American-style coffee-drinking culture to the UAE.

In summary, the influence of globalization is unmistakably apparent in the culinary landscape of the United Arab Emirates. While traditional Emirati dishes and hospitality practices endure, they now coexist with a dynamic array of international flavors and dining habits. These changes in food culture not only reflect the UAE's position as a global crossroads but also have far-reaching effects on family dynamics and societal life, as well as public health challenges, such as the rising prevalence of obesity.

4.5 Architecture

Traditional Emirati homes are known as structures that represent the pre-independence era, designed to accommodate extended families, reflecting a rich cultural and architectural heritage. These historic homes used various sustainable construction materials readily available in the region. In particular, structures in the region included tents and huts made from palm leaves, while more complex residences near the coast incorporated materials such as coral stone, mud, and locally sourced mangrove wood. One distinctive architectural feature of these traditional Emirati homes was the presence of the "barjeel," a wind tower. These wind towers were designed for both functional and environmental purposes, cleverly capturing prevailing winds and providing natural ventilation and cooling to interior spaces ([2], p. 9). This local architectural feature played a significant role in mitigating the harsh desert climate and offered inhabitants a more comfortable living environment. Moreover, the barjeel was an integral part of the architectural identity of these homes and was essential in shaping the cultural and social fabric of Emirati society.

However, with the rapid urbanization and modernization in the United Arab Emirates, traditional barjeels have unfortunately faded into obscurity, almost disappearing from the contemporary architectural landscape. A shift toward modern construction techniques and technologies, coupled with a demand for larger and more Western-style homes, has contributed to this decline. Nevertheless, there is a growing recognition of the importance of preserving Emirati heritage and cultural identity. As a result, there has been renewed interest in the characteristic wind tower design. Modern architects and urban planners have started incorporating elements inspired by the traditional barjeel into contemporary architectural projects. This revival not only shows respect for the region's cultural heritage but also serves as a source of pride and identity for the Emirati people. Consequently, the architectural evolution from traditional Emirati homes to contemporary urban structures reflects not only the changing landscape of the Emirates but also underscores the enduring significance of cultural heritage. The legacy of the barjeel and its impact on modern architecture emphasizes the importance of preserving and celebrating the rich cultural tapestry of the United Arab Emirates, even in the face of rapid modernization. However, now homes are designed for nuclear families rather than extended ones, so large houses have become less popular, and apartment living is on the rise. Furniture and home design have become very similar to what is seen in Western countries ([2], p. 9). Floor seating has given way to tables. Despite traditional influences, the impact of globalization is felt in architecture and home design, and traces of significant societal change can be seen in architecture as well.

4.6 Vacation

However, due to the significant increase in per capita income over the years in the United Arab Emirates, the dynamics of vacation habits have undergone a significant transformation. A noteworthy development in this regard has been the globalization of vacation destinations. This transformation can be attributed to various factors such as increased financial resources, improved connections, and an increasingly cosmopolitan outlook among Emirati citizens ([17], pp. 578–580).

One of the key catalysts for this change has been the emergence and growth of local airlines within the United Arab Emirates. Specifically, the United Arab Emirates is home to several leading airlines like Emirates, Etihad Airways, Flydubai, and Air Arabia, which have greatly contributed to making international travel more accessible to local residents. These airlines have effectively bridged geographical distances, making destinations worldwide just a single flight away from the United Arab Emirates.

This expansion in the aviation industry has not only assisted Emiratis in exploring different cultures and landscapes but has also played a role in promoting global connectivity and cultural exchange. The availability of direct flights to numerous international destinations has allowed Emirati families to embark on enriching journeys worldwide, exposing them to a multitude of cultural experiences and perspectives.

Before gaining independence, family vacations were mostly limited to local destinations, often involving visits to nearby regions and occasional trips to neighboring countries like those in the Gulf Cooperation Council (GCC) or countries in the Indian Subcontinent. These trips usually served the purpose of visiting distant relatives or shopping. The desert landscape, defined by its unique charm and serenity, continued to be a preferred destination for weekend getaways among Emirati families ([13], pp. 80–82).

As a result, the evolution of Emiratis' vacation preferences from local and regional destinations to a more global outlook reflects broader socio-economic developments in the United Arab Emirates. The increase in per capita income, coupled with the establishment of leading local airlines, has opened up new horizons that were previously unthinkable for Emirati families. This change not only enriches their personal lives but also contributes to the country's relationship with the global community.

4.7 Sports

The concept of nations dividing the world is widely recognized, but it's essential to acknowledge that it is not an inherent or natural state. The process through which individuals come to identify with a nation is a multifaceted question with diverse answers. National identity can be influenced by both top-down, state-driven initiatives and bottom-up, grassroots practices, often intertwining with one another. This phenomenon is particularly evident when examining heritage sports in the United Arab Emirates (UAE).

In recent times, Gulf states, including the UAE, have actively pursued hosting significant international sporting events, particularly in elite sports such as tennis, cycling, sailing, golf, and equestrian sports, as a means to enhance their global prestige and project an image of opulence and luxury. For instance, the construction of Abu Dhabi's Yas Marina Circuit, costing $1.3 billion, following Bahrain's Formula One Grand Prix in 2004, exemplifies this strategic approach ([27], p. 12). These events serve as platforms to showcase the UAE as a cosmopolitan, affluent, and modern destination while simultaneously promoting urban development and positioning cities like Dubai and Abu Dhabi as sporting hubs for elite tourists. These initiatives align with the nationalistic objective of countering Orientalist stereotypes about the Middle East in the Western world and actively challenging negative perceptions through promotional materials featuring state-of-the-art sports venues and infrastructure. The primary target audience for such events is international, with the aim of reshaping perceptions of the UAE and its people [28].

In contrast to elite sports, state leaders also promote heritage sports that reflect local Emirati culture and traditions. The promotion of local culture in heritage sports is carefully curated and aestheticized. While heritage sports operate in a different sphere from elite sports, they share the common goal of enhancing a positive Emirati national identity that combines elements of modernity with tradition.

When addressing international audiences, elite sports question negative perceptions about the Middle East, while cultural heritage sports represent a more localized aspect of Emirati culture and national traditions. Cultural heritage sports are carefully organized to reflect a modern yet tradition-based image of Emirati national identity. Additionally, the state actively involves Emirati citizens as primary participants and spectators by supporting festivals, competitions, and various events, thereby promoting the growth and institutionalization of these sports. Examples include the Cultural Programs and Heritage Festivals Committee in Abu Dhabi, which organizes events promoting cultural heritage sports such as camel racing, dog racing, and falconry. "Falconry, in particular, stands out as a prominent cultural heritage sport in the UAE" ([29], p. 12).

Sports heritage is a process that defines the symbols of the past and imbues the past with meaning through direct participation or admiration by people. Promoting sports heritage in the UAE is supported by both the government and its citizens. These sports celebrate traditional and modern values by elevating Emirati culture. Whether performed to pique the interest of tourists or as a hidden passion, sports heritage strengthens the Emirati identity and contributes to internalizing the international geopolitical vision.

To preserve their cultural heritage, Emiratis continue traditions like camel racing that enhance social cohesion and a sense of belonging. Camel racing also contributes to shaping the state's ideological and political identity.

5. Effects of multiculturalism

The federal system in the United Arab Emirates allows for the implementation of local traditions and governance in different Emirates. While federal rules apply to fundamental matters like currency and foreign policy, each emirate has its own laws and codes of conduct, making it difficult to define the state and society in a singular way.

Dubai is the second-largest emirate in the United Arab Emirates, accommodating approximately 14% of the local Arab population. While it derives only 7% of its income from oil revenues, the remaining 93% comes from the non-oil service sector. Dubai serves as a significant trade hub in the Middle East, notably through the Jebel Ali Free Zone, contributing a substantial portion of the United Arab Emirates' non-oil gross domestic product (GDP) and accounting for 60% of the total. The rest of the country is actively working to expand its infrastructure and offer free trade agreements to Western and Asian nations in an effort to diversify its economy. Under the leadership of Sheikh Mohammad bin Rashid Al Maktoum, Dubai has successfully realized its vision of becoming a lucrative global trade center.

Dubai stands as an emirate where over 200 nationalities coexist in peace, thanks to its high income, comfortable living conditions, educational opportunities, healthcare services, and a tolerant societal structure. In the 2014 United Nations Development Index, the United Arab Emirates ranked 40th out of 187 countries worldwide ([30], p. 3). However, the influx of immigrants to Dubai is not solely due to its high standards.

The percentages of the migrant population in the UAE by country can be seen below (**Table 1**).

Various factors contribute to the influx of foreign labor in the region. One primary reason is the low local population. Since 1971, the government's main objective has been to create a robust economy independent of natural resources, such as oil and gas. Consequently, significant investments have been made in tourism, construction, transportation, and finance sectors since the 1980s ([7], p. 107). However, to achieve the goal of economic diversification, which has been identified as a priority, there is a shortage of the required workforce within the country, necessitating the influx of labor from abroad ([7], p. 106).

"Foreign workers, who make up over 90% of the country's workforce, come from countries such as India, the Philippines, Bangladesh, and Pakistan. The majority of these workers are male, with 58% coming from South Asia, 17% from Asia, and 8.5% from Western expatriates" ([3], p. 53). While the number of female workers from Bangladesh, Indonesia, and the Philippines has increased in recent years, women generally work in domestic chores, childcare, and cleaning jobs, while men tend to work in the construction sector and low-profile service jobs. Additionally, the region attracts a highly educated workforce, often hailing from the United Kingdom, the

Country	Percentage of population
India	39.20
Pakistan	9.40
Bangladesh	9.50
Philippines	6.10
Iran	4.76
Egypt	10.20
Other South Asians	2.30
All other countries	12.80
Other countries	17.94

Source: [31]

Table 1.
UAE population by nationality 2024.

United States, Australia, Canada, and Western Europe, working in industries such as petroleum, education, finance, and investment.

In the Emirates, foreign curriculum schools have been established for the children of these foreign workers, offering music, foreign language, sports, art courses, and international restaurants. Various groups come together at shows and festivals, celebrating special days and holidays from different faiths ([4], pp. 42–43).

The establishment of the Ministry of Tolerance in the UAE in 2016 aimed to declare tolerance as a virtue and an integral part of Islamic culture. With over 200 nationalities living peacefully and successfully in the UAE, the UAE society has unquestionably become an example of tolerance and inclusivity. Now, the Federal Government is willing to promote and embrace tolerance and understanding as the core values of the society ([4], p. 42).

In 2016, the establishment of the Ministry of Tolerance in the United Arab Emirates (UAE) was undertaken with the aim of emphasizing tolerance as a virtue and an integral part of Islamic culture. Tolerance is observed at individual, institutional, and national levels. With over 200 nationalities peacefully coexisting and thriving in the UAE, the Emirati society has unquestionably set an example of being a tolerant and inclusive country. The Federal Government is now committed to promoting acceptance and understanding as fundamental values of society ("Tolerance – UAE Government Official Portal," n.d.).

The UAE government, while preserving Islamic culture and Bedouin heritage, approaches different cultures with respect and tolerance, reducing tensions by facilitating easy access to authorities. Islamic culture is not rigorously enforced due to globalization, and the government places economic globalization ahead of religious beliefs. In this context, the Emirates rapidly adapts to the changing future while preserving the state and societal identity.

However, as a consequence of multiculturalism, English has prominently emerged as the most widely used common language. In the United Arab Emirates (UAE), the dominance of the English language in private-sector activities has posed a threat to the marginalization of Arabic. Furthermore, many young Emirati citizens fluently speak Hindi due to the presence of Asian caregivers. According to a survey conducted in 1999, nearly half of the respondents preferred English as their medium of education, while 30% opted for a mixture of English and Arabic, and only 23% favored Arabic-only education. This situation contradicts the government's policies that promote nationalism and traditional values [32].

Despite the existence of multiple identities and affiliations, policies, media, and education have fostered an increasingly prevalent Emirati identity in the UAE. Many citizens primarily identify as Emirati, as opposed to Arab or other localized identities.

6. The impact of globalization on culture in Western societies

Globalization has had a profound and multifaceted impact on the cultures of Western societies. Here are some key aspects to consider:

1. Cultural Exchange and Diversity: Globalization has facilitated greater cultural exchange, as Western societies interact with people from all over the world. This has led to increased cultural diversity within Western countries, with the influx of immigrants and the adoption of various cultural elements from different regions.

2. Spread of Western Values and Consumerism: Western values, such as individualism, democracy, and human rights, have been promoted and embraced globally through media, international organizations, and economic ties. Similarly, consumerism, driven by Western advertising and brands, has become a global phenomenon.

3. Impact on Traditional Values: Globalization has brought both positive and negative consequences for traditional values in Western societies. On one hand, it has encouraged greater tolerance and acceptance of diverse perspectives. On the other hand, it has led to concerns about the erosion of traditional cultural values and practices.

4. Cultural Homogenization and Hybridization: Western popular culture, including music, films, fashion, and fast food, has spread globally, leading to a degree of cultural homogenization. However, it has also resulted in cultural hybridization, where elements from different cultures blend to create new forms of expression.

5. Technology and Communication: Advances in technology, particularly the internet and social media, have revolutionized communication and information exchange. This has allowed for the rapid dissemination of cultural products and ideas, further blurring geographical and cultural boundaries.

6. Challenges to National Identity: Globalization has sometimes challenged traditional notions of national identity in Western societies. Increased immigration, international business, and global connectivity have led to discussions about what it means to be a "citizen" of a particular country in an increasingly interconnected world.

7. Cultural Hegemony: Some critics argue that globalization, particularly in media and entertainment, can lead to cultural hegemony, where Western cultural values dominate and marginalize indigenous cultures ([33], pp. 22–24; [34], pp. 173–175).

In summary, globalization has had a profound impact on Western cultures, fostering cultural exchange, influencing traditional values, and reshaping the way people perceive and interact with the world. It has both enriched and challenged cultural identities, and its effects continue to evolve as globalization continues to shape our interconnected world.

7. Conclusion

Under the influence of globalization, the Emiratis' adoption of a holistic identity and their rejection of simplistic labels like West versus East, modern versus traditional, and reason versus religion in favor of their place in modern life are significant and noteworthy developments. They seek to respect human rights, embrace liberalism, show tolerance toward different values, and pursue modernization without sacrificing values such as family, honor, or tradition. In contemporary United Arab Emirates, efforts to reduce disadvantages that may arise from issues related to class, gender, ethnicity, or sect are highlighted, with the goal of providing equal opportunities for every citizen without compromising these values.

The United Arab Emirates has undergone significant economic and social changes in the past 30 years, and traditional family structures and the status of women have changed dramatically. Altering the roles of women in both family and society faces resistance in conservative and traditional settings. Nowadays women can pursue education, careers, and increased visibility in society without compromising their roles as wives and mothers. Their success in the business world, politics, and the preservation of their traditional identities serves as a testament to the possibility of this balance. This dynamic is particularly valuable for women navigating the tension between traditional values and modernization, offering them a meaningful guide. It's noteworthy that women, while upholding traditional roles, also leverage opportunities available in the modern world for their benefit.

The active involvement of women in the economy and their increased productivity are vital for a strong, diversified economy, independent of natural resources. Although women's rights in the UAE have made progress compared to global standards, they still face limited freedom and autonomy. Despite being well-educated and successful in business, many Gulf women are expected to conform to traditional social roles and defer to male family members' decisions, leading to significant societal conflicts.

Within the context of this investigation, it becomes evident that while the paramount dimension of societal transformation pertains to the reshaping of familial structures and the redefinition of women's roles, there exists a concurrent phenomenon where traditional societal frameworks undergo a process of integration with contemporary Western paradigms across various domains, including architecture, sports, entertainment, political culture, and education. This convergence between historical conventions and modern Western models, as discerned within the sociocultural milieu, signifies a multifaceted evolution.

Not only the changing roles of women in family and society, but also in various aspects of culture in the United Arab Emirates, such as education, architecture, travel and leisure, dining, and even the impact on the elderly and youth in the context of globalization and modernity, when examined, the resulting conclusion is as follows.

Drawing inspiration from the insights of the renowned scholar Ibn Khaldun, this confluence of societal dynamics is underscored by the concomitant processes of economic development, urbanization, and engagement with diverse cultural perspectives. It becomes apparent that the Emirates, once characterized by a nomadic Bedouin ethos, is experiencing a profound shift toward more sedentary values and structures.

The paradigmatic exemplar of this transformative trajectory is encapsulated within the United Arab Emirates (UAE), wherein a compelling illustration emerges of the inexorable interplay between globalization and the ramifications of economic advancement. This nexus invariably engenders liberalization and a discernible transformation of societal paradigms, even within traditionally conservative and religiously oriented societies.

Noteworthy is the observation that the populace appears to embrace these transformations, thereby implying a societal consensus on the beneficial outcomes of this ongoing process of liberalization and modernization. It is conceivable that this trajectory, hitherto manifested predominantly within the cultural sphere, may progressively extend its reach into the realm of political discourse in the foreseeable future, thereby fostering a more profound and holistic societal metamorphosis.

The study suggests that the impact of globalization on the culture of a highly traditional and religiously conservative country like the United Arab Emirates may not be significantly different from its impact on Western countries, as outlined in the previous section.

Author details

Basak Ozoral
Istanbul Commerce University, Political Science and International Studies, Istanbul, Turkey

*Address all correspondence to: ozoralbasak@gmail.com

References

[1] Momani B. Arab Dawn Book: Identity, Arab Youth and the Demographic Dividend They Will Bring. Toronto: University of Toronto Press; 2015

[2] Kotsi F, Balakrishnan MS. Cultural Reflections from the United Arab Emirates (UAE). Dubai, UAE: Zayed University; 2017

[3] Watanabe A. The complexities of social inclusion and exclusion among intermarried filipino migrantworkers in the United Arab Emirates. In: Lian KF, Hosoda N, Ishii M, editors. International Labour Migration in the Middle East and Asia. Vol. 8. Springer; 2019. pp. 55-73. DOI: 10.1007/978-981-13-6899-8_4

[4] Baycar H. Promoting multiculturalism and tolerance: Expanding the meaning of "unity through diversity" in the United Arab Emirates. Digest of Middle East Studies. 2023;**32**(1):40-59

[5] Morton MQ. Keepers of the Golden Shore: A History of the United Arab Emirates, London, UK: Reaktion Books; 2016

[6] Onley J. The Arabian Frontier of the British Raj: Merchants, Rulers, and the British in the Nineteent Century Gulf. Oxford; New York: Oxford University Press; 2007

[7] Davidson C. Dubai: The Vulnerability of Success. New York: Columbia University Press; 2008a

[8] Davidson CM. Dubai, the Vulnerabitiy of Success. London, UK: HURS; 2008b

[9] Khaldun I. In: Uludag S, editor. Mukaddime, Cilt I-II. Istanbul: Dergah Publishing; 2007

[10] Ozoral B. Bedevi tend to Burj Khalifa: An analysis of the progress of Dubai from the view of Ibn Khaldun. Journal of Ibn Khaldun Studies. 2018;**3**(1):1-12

[11] Gabriel EF. The Dubai Handbook. Ahrensburg: Institute for Applied Economic Geography; 1987

[12] The World Bank. World Development Indicators. 2013. Available from: https://databankfiles.worldbank.org/public/ddpext_download/WDI-2013-ebook.pdf

[13] Schvaneveldt PL, Kerpelman J, Schvaneveldt JD. Generational and cultural changes in family life in the United Arab Emirates: A comparison of mothers and daughters. Journal of Comparative Family Studies. 2005a;**36**(1):77-91

[14] Al-Krenawi A, Graham JR, Slonim-Nevo V. Mental health aspects of Arab- Israeli adolescents from polygamous versus monogamous families. The Journal of Social Psychiatry. 2002;**142**(4):446-460

[15] Al-Krenawi A, Graham JR. Principles of social work practice in the Muslim Arab World. Arab Studies Quarterly. 2003;**25**(4):75

[16] Jawad HA. The Right of Women in Islam. Basingstoke: Macmillan Press; 1998

[17] Crabtree SA. Culture, gender and the influence of social change amongst Emirati families in the United Arab Emirates. Journal of Comparative Family Studies. 2007;**38**(4):575-588

[18] Ramazani N. Arab Women in the Gulf. MiddleEast Journal. 1985;**39**(2):258-276

[19] Samira I, Islam S. Arab women in Science, Technology, Engineering and Mathematics Fields, The Way Forward. World Journal of Education. 2017.

Available from: http://www.sciedupress.com/journal/index.php/wje/artie/view/12619/0

[20] Ridge N. What the UAE Population Thinks of Aging and Aged Care Shahed Almarabta, STRATEGIC REPORT No. 05. *Sheikh Saud Bin saqr Al Qasımı foundation of Policy Research*. 2021. Available from: https://f.hubspotusercontent10.net/hubfs/5081768/Aged%20Care%20Report%20.pdf?hsCtaTracking=754c1c40-a16d-4288-8519-7ae42a1f615c%7C39210993-569a-4e51-884e-b627c8f61913

[21] Al-Ali A. Aging in the UAE and services available for the elderly: Structured interviews with experts in the field Dubai School of Government Policy Brief, no 34. 2013. Available from: https://mbrsgcdn.azureedge.net/cmsstorage/mbrsg/files/91/91a79863-e776-4b22-ae2c-9a7449b99f80.pdf

[22] Monshipouri M. Muslims in Global Politics, Identities, Interests, and Human Rights. Pennsylvania: University of Pennsylvania Press; 2009

[23] Soffan L. The Women of the United Arab Emirates. London: Routledge; 1980

[24] O, Sullivan K. Education reform in the UAE – bringing private schools into the fold. Journal of Teaching and Education. 2015, April;**04**(01):311-320

[25] Alia A, Shaheen S, Haikal R, Nakhoul C, Baddar H, Aly E. The Development of Education in the UAE. Presented to Abu Dhabi University-Abu Dhabi, UAE. 2018. Available from: https://www.academia.edu/36115381/The_Development_of_Education_in_the_UAE

[26] Martin E. Nation building and social change in the United Arab Emirates through the invention of Emirati cuisine. The Sociological Review. 2021;**3**(69):538-559

[27] Mosly A. Sports Diplomacy in the GCC. Gulf Research Center; 2022. Available from: https://www.grc.net/documents/62c5512d17feaSportsDiplomacyintheGCCNew.pdf

[28] Koch N. Sports and heritage in the UAE. LSE Middle East Blog. 2018. Available from: blogs.lse.ac.uk /mec/2018/12/15/sports-and-heritage-in-the-uae/

[29] Valek N. Abu Dhabi Sport Tourism: from zero to hero. In: Conference: Sustainable Development of Sport Tourism. Portorož, Slovenia; 2015. pp. 11-21. Available from: https://www.researchgate.net/publication/277775617_Abu_Dhabi_Sport_Tourism_from_zero_to_hero

[30] Malit F, Youha A. Labor Migration in the United Arab Emirates: Challenges and Responses. Washington: Migration Policy Institute; 2013. Available from: http://www.migrationinformation.org/Feature/display.cfm?ID=965

[31] Global Media Insight. 2024. Available from: https://www.globalmediainsight.com/blog/uae-population-statistics/#:~:text=The%20total%20expat%20population%20in,approximately%2089%25%20of%20the%20population.

[32] Al-Issa A, Dahan L. Language and identity construction in the United Arab Emirates: Challenges faced in a globalized World. In: Bassiouney R, Walters K, editors. The Routledge Handbook of Arabic and Identity. London: Routledge; 2020

[33] Held D, McGrew A, Goldblatt M, Perraton J. Global Transformations: Politics, Economics and Culture. Politics at the Edge. London: Palgrave Macmillan; 2000. pp. 14-28

[34] Martell L. The third wave in globalization theory. International Studies Review. 2007;**9**(2):173-175

Chapter 18

Measuring the Poverty of Herder Households in Mongolia

Ganchimeg Gombodorj and Károly Pető

Abstract

Mongolia faces difficulties in eradicating rural poverty, the same as other developing countries in the World. The selection of appropriate measurement is an essential prior step with regard to the complexity of poverty types and specifications. The most widely used poverty measuring methods are Income/cost Per Capita (IPC) per day and multidimensional poverty index (MPI). In the case of nomadic herder households in Mongolia, a livestock number is used as a proxy indicator to measure livelihood and poverty. Considering that, a herder household with less than 200 animals is considered poor, in general principle. This study uses a survey of 350 herder households, divided into three livelihood groups based on the number of livestock following the World Bank methodology; three groups are namely poor, lower wealthy and wealthy herder households. The livelihood of households was measured using IPC and MPI-3 methods, and the study results suggest that the MPI-3 method is not entirely appropriate to measure the herder household's poverty and needs to adjust and add some indicators. According to their difference, part-time and periodic herders should be identified separately from full-time herders regarding measuring and studying livelihood and poverty levels.

Keywords: poverty, herder household, Mongolia, livestock number, MPI-3, IPC

1. Introduction

The United Nations Development Programme (UNDP) calls for action to end poverty, protect the planet, and ensure that by 2030 all people enjoy peace and prosperity. The United Nations' 17 Sustainable Development Goals (SDGs) start with no poverty, followed by zero hunger, good health and well-being, quality education, reduced inequalities, clean water and sanitation and continue with other goals—most of the SDGs are connected to the household's livelihood and well-being.

Poverty is poor people's inability to generate income and opportunities for everyday life. In the effort of 25 years, extreme global poverty was declining continuously. However, the fight to eradicate poverty has suffered its most serious blow. This setback is mainly due to significant challenges—COVID-19, conflicts, and climate change. In 2019, 660 million people lived in extreme poverty as they lived on less than $1.90 per day [1]. By all measures, poverty today is predominantly rural. Seventy percent of the impoverished lives in rural areas; 60% work in agriculture [2]. The 2023 release of the MPI presents estimates for 110 developing countries, including

Mongolia. According to this research, about 1.1 billion people live in multidimensional poverty, with a population covering 6.1 billion [3].

Mongolia is the second largest landlocked country, bordered by Russia and China, with an area of 1.5 million square kilometres of territory that accounts for the 18th biggest country in the world with a population of 3.4 million. Mongolia's current gross domestic product is about 13.7 billion USD, and its GDP per capita is USD 4566.14 [4]. According to the World Bank view, Mongolia has significantly reduced poverty in the past three decades. However, macroeconomic volatility and slower growth have led to stagnation and poverty reduction in recent years [5].

Mongolians are among the most educated people in Central Asia. The national average basic education rate (9 years) is 97.1%. In the 2017–2018 school year, 1.03 million students study in 2393 educational institutions nationwide [6].

Mongolia, one of the few countries where pastoralism is a source of livelihood, has a distinctive antique culture of nomadic herding [6]. The country has been home to a remarkably resilient, land-based, nomadic pastoral life and cultural heritage. The mining boom since the 2000s has put intense pressure on pastureland and water sources, which are the basis of Mongolian nomadic herders' livelihoods. Globally, mining causes encroachment on lands used and inhabited by nomadic communities. It results in forced displacement, loss of livelihoods, migratory routes and community cohesion [7].

Herders hold large tracts of land, are custodians of intangible cultural heritage, and significantly contribute to preserving customs and culture. Due to the impact of globalisation, preserving cultural heritage and identity is challenging for smaller nations. Mongolia's modernisation process has posed a challenge to traditional nomadic lifestyles. The younger generation continues to seek urban and technology-dependent lifestyles [8]. Mongolia is one of the most sparsely populated countries. As of 2019, the average population per 1 km^2 in rural areas is 2.1 people, while 327.6 people in the capital city [9]. Half of the total population lives in the capital city. Therefore, improving the herders' livelihood and reducing rural poverty will support and preserve Mongolia's heritage.

Agriculture has historically been a vital and traditional sector of the Mongolian economy and will continue to be so in the foreseeable future. The livestock sector is an important sector that accounts for 85% of agriculture. Almost 30% of Mongolia's population is engaged in the livestock sector, which owns 64 million livestock heads. Herders provide the urban population with milk and meat and serve as the raw material suppliers for industries that produce wool, cashmere, leather, and meat processing. There is a saying, 'Give what you have'. Mongolia has a long-standing tradition of donating animals and animal products, as evidenced by their contributions during significant events such as World War II, the Korean War, and the COVID-19 pandemic. During World War II, they generously provided livestock, meat, and animal products to the Soviet Union. Similarly, during the Korean War, Mongolia supported North Korea by donating 10,000 horses. More recently, during the COVID-19 pandemic, 30,000 sheep were donated to China [10].

The livestock sector is growing and contributing significantly to the Mongolian economy. However, the sector is vulnerable to natural disasters and problems of environmental degradation, animal health, rural poverty, productivity, migration, successors, health and education [11]. Herders are in charge of their means of subsistence, the productivity of their livestock, inputs and outputs and natural risk. Due to the numerous risks that Mongolia experiences, especially flooding, drought, winds, storms, earthquakes and forest fires [10], herders' livelihoods are highly dependent on natural threats. The last big *dzud*[1] happened in 2010 and 2019 and enormously impacted livestock numbers and household livelihoods, and herders lost 15 million animals [12–15]. Pasture and water are common pool resources in rural Mongolia. Thus, herders openly access these common pool resources. They apply seasonally moving and rotational grazing management of pastures as a standard pasture management tool against overgrazing and degradation of pastureland. A household is considered the fundamental decision-making unit regarding production and consumption [16, 17]. Rural households' majority income is earned from animal-origin products. With the effort of Government programmes, favourable weather, increased livestock product demand and prices, better market access, government subsidies, and public transfers, herders' livelihoods improved in the last decade. The rural poverty level of 58.4% in 2010 was reduced to 30.5% in 2020 [18]. Poverty drives herders into more minor sustainable resource utilisation. A share of herders' households with less than 200 animals is 55% of the total number of herder households [4], while their herd size accounts for only 12% of the total number of livestock [19]. The expenses of herders with 101–200 animals were nearly equal to their revenue but only enough for subsistence; they were left with no money [20, 21]. They are not worth spending on long-distance moves for pasturing livestock. Investment in breeds and innovation is almost impossible in this group of households.

The large number of livestock and unequal distribution of herd size lead to environmental deterioration. Specifically, rangeland degradation ranges from 9 to 90% [22] and 38% of already degraded pasture will likely take more than 10 years to restore [14, 22]. Poor herder households contribute more to pasture degradation [23]. Thus, when a herder household selects various livelihood strategies or changes existing strategies, it can significantly influence pasture quality [24, 25] and the herder's well-being [26]. These changes may further affect a pastoral region's sustainable development [27]. Pastoralism is one of the low-input livestock production systems that is environmentally friendly and has minimal effect on the climate, biodiversity, and water resources [28].

Reducing poverty is a practical way to improve people's quality of life [29]. In the case of herding households, there is no regular income recording system, free consumption of meat, milk, and dairy products from their farms, no records are kept, which makes it difficult to measure their poverty and livelihood. Also, the income from animal husbandry is seasonal, and the price fluctuates a lot; the location is far from the central settlement, and the income from natural resources and tourism is not registered, which became the basis for measuring poverty and livelihood based on the number of livestock. The nomadic herder's lifestyle poses challenges for data collection and they left behind the household socio-economic survey. To fill this gap, the nomadic household pilot survey was carried out in 2019 using an innovative sampling approach, Random Geographic Cluster Sampling [30]. However, it is in an initial step and needs deep study; most reports and studies advised that only a household with up to 200 animals is at a subsistence level [20, 21].

Income/cost Per Capita per day and Multidimensional Poverty Index I, II and III are widely used internationally [3, 31]. Therefore, this study was conducted to compare the method of measuring poverty by the number of animals with the results determined by the IPC and MPI methods.

[1] Dzud-Very harsh winter with snow and cold.

2. Materials and methods

2.1 Poverty measurement methods and indicators

Correct measurement of poverty is essential for poverty elimination. Several methods for measuring poverty are accepted and widely used. Multidimensional poverty measurement is proposed by the Alkire-Foster (AF) method, referred to as the global Multidimensional Poverty Index (MPI). The MPI is the product of the headcount or incidence of multidimensional poverty and the intensity of multidimensional poverty [31]. The MPI measures three dimensions: health, education and living standards [32]. The health includes nutrition and child mortality indicators. The second dimension of MPI—Education consists of years of schooling and school attendance. The World Bank measures also use these two indicators for education [33]. MPI—Standard of living dimension has six indicators: electricity, sanitation, water, floor, cooking fuel, and assets. However, the World Bank applies only electricity, sanitation and water. The MPI methodology does not use a monetary measure of poverty, while the World Bank uses the monetary poverty measure. An extreme or international poverty line refers to daily consumption/income being less than US$ 1.90 per person. World Bank updated the new extreme poverty line of $2.15 per person per day in 2022, which replaces the $1.90 poverty line in 2018 [34]. Our survey was conducted in 2017; thus, we used $1.90.

The MPI has three sub-weighting structures: MPI-1, MPI-2 and MPI-3.

- MPI-1 weights for global multidimensional poverty and uses AF weightings and 33% + threshold: health (1/3), education (1/3) and standard of living (1/3).

- MPI-2 is multidimensional poverty, and it has equal weightings for 10 indicators: Nutrition (1/10), Child mortality (1/10), schooling (1/10), Attendance (1/10), electricity (1/10), Sanitation (1/10), Water (1/10), Floor (1/10), Cooking fuel (1/10) and Assets (1/10).

- MPI-3 is a non-health multidimensional poverty. It has equal weightings, and the 50% + threshold is poor [35]. Each indicator is equally weighted within its dimension. A deprivation score of 1/3 (one-third of the weighted indicators) distinguishes the multidimensionally poor and nonpoor. If the deprivation score is 1/3 or greater, the household (and everyone in it) is classified as multidimensionally poor. Individuals with a deprivation score greater than or equal to 1/5 but less than 1/3 are classified as vulnerable to multidimensional poverty. Finally, individuals with a deprivation score greater than or equal to 1/2 live in severe multidimensional poverty. The MPI ranges from 0 to 1, and higher values imply higher multidimensional poverty and show multidimensional poverty composition [31].

The number of livestock was used to measure nomadic herders' poverty in rural Mongolia. By classifying the World Bank study in 2009, the following five wealth groups were defined [36], and it is widely used. Hence:

i. Wealthy households with more than 500 head of livestock (in extensive systems) or smaller herds if in intensive or semi-intensive production such as dairy cattle, providing full-time employment for household members, good access to pastures and inputs, good winter preparation (hay and fodder production) and ability to move long distance if required.

ii. Full-time herders are growing herds of between 200 and 500 animals. Households are typically from middle or lower wealth levels. While herding can support the household, they are vulnerable to *dzud* and restricted from increasing herd numbers by access to capital and pastures.

iii. Full-time herders with fewer than 200 animals (and frequently less than 100) are poor households, highly vulnerable to *dzud,* often dependent on state support, and for whom herding may not be a preferred livelihood but who have no alternative sources of income.

Indicators	Variable definitions and coding
	Education variables
Years of Schooling	No household member (age ten or older) has completed five years of schooling.
School attendance	Any school-aged child in the household is not attending school up to class 9. Suppose a household has such a child, coded 1.
Electricity	A household with no electricity (wind, solar panel, generator, or central gridlines) should be coded 1.
Water	The household does not have access to safe drinking water, or safe water is more than a 30-minute walk round trip or lacks access to 'limited-standard' drinking water. Access to drinking water for four seasons. The question is, 'Does your household have access to drinking water?' If the respondent answered yes to any of the following answers: (1) protected hand well; (2) mechanised well; (3) unprotected hand well? Those answers were coded 1. If the respondent selected other answers (river, spring, snow, ice and others), they were coded 0.
Housing	AF and MPI use indicators of the floor. We replaced the floor with the number of housing. Due to nomadic life and the continued moving, 98.7% of herders live in *Ger*[1] houses. In addition to the *ger*, it is common for herders to have winter quarters, barns and containers in the centre of the *bagh*[2], *soum* and *aimag*s to store their belongings. Therefore, in order to determine housing deprivations, a household is coded as poor if it does not have any type of residential or non-residential housing in addition to their living *ger*. Suppose a household has only one ger, coded as 1.
Cooking fuel	Mongolian nomadic and extensive herding households use dung, wood, or charcoal. Therefore, all answers are coded 1.
Sanitation	Mongolian rural households have no standard sanitation facilities. They use a bio-, green- and environmentally friendly sanitation. Building standard sanitation facilities while living a nomadic life is also very expensive. Then, all herding households answered that they had no facilities, so they were all coded as 1.
Assets	The household does not own more than one of the following: radio, TV, telephone, bike, motorbike or refrigerator and does not own a car or truck. Suppose a household owns two assets, which should be coded 0, nonpoor.

Source: Adapted from Refs. [33, 37] and authors.[1]*The ger is a portable home ideally suited to Mongolia's climate and nomadic herders' lifestyle: the round wooden framework (known as khana) and the felt cover, the central support columns as uni, the smoke hole is toono. Eighty-eight separate wooden poles, each around 1.5 m, are used for the ger frame, with two central columns supporting the entire structure.*
[2]*The bagh is the smallest level of the administrative unit in Mongolia. Mongolia has 1615 bagh.*

Table 1.
Mongolian herder households' multidimensional poverty index-3 (MPI-3) indicators and definitions.

iv. Periodic herders, usually with fewer than 200 animals, come in and out of the livestock sector, depending on their economic circumstances. Often, herding is used as a safety net or fallback option. This group includes recent migrants to urban areas for whom the move may not be permanent.

v. Part-time herders with less than 200 animals may or may not own supplement income from livestock with other sources, including 'ninja[2]' mining, farming, and seasonal migration to urban centres for temporary employment.

This research contributes to the Akire-Foster methodology but only chooses the MPI-3, comparing the three wealth groups of herder's households in Mongolia.

Within the preferred method, the selected groups are as follows:

- Herder households with up to 200 livestock;
- Herder households with 200–500 livestock;
- Herder households with more than 500 livestock.

The herding households have no appropriate data on nutrition and child mortality due to lack of time and other conditions. Therefore, we chose to apply the MPI-3 methodology, which excludes the health dimension. All responses to the questionnaire were recoded as 1 or 0, with 1 indicating poor and 0 indicating nonpoor. **Table 1** shows that the indicators were adjusted according to the characteristics of nomadic herder households based on the Alkire Foster and the World Bank methods, where eight indicators of education and standard of living dimensions were used.

2.2 Survey sampling and a study area

A total of 350 herder households' data was collected in 2017, covering 15 *soums*[3] from 9 *aimags*[4] of Mongolia. Using the Neiman equal distribution method, the total sample size determined the number of units surveyed from the selected *soums* 350:

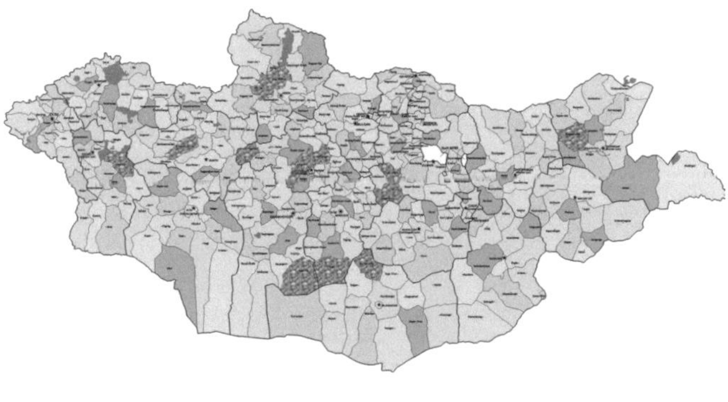

Figure 1.
Herder households survey sampling from the selected soums in Mongolia. Source: Authors' illustration.

[2] Ninja mining-Small scale mining people without permission.

[3] *Soum* is the second level of the administrative unit of Mongolia. Mongolia has 330 *soums*.

[4] *Aimag* is the third level of the administrative unit of Mongolia. Mongolia is divided into 21 *aimags*.

15 = 23. As a result, 23 herder households from each selected soum participated. The herders were chosen from each *soum,* representing the ten different livestock groups (**Figure 1**).

3. Results and discussion

In Mongolia, the poverty level of urban and rural households has been consistently different. According to the poverty profile survey, rural poverty in Mongolia has generally decreased since 2010. Like in other developing countries, rural poverty in Mongolia has always been higher than urban poverty (**Figure 2**).

3.1 Results of poverty measurement based on the number of livestock

Table 2 shows the results of calculating the poverty of herder households based on the number of livestock using a survey of 350 herder households (see the Appendix: **Table A1**). These three livelihood groups will further be used as a reference to define poverty by applying the MPI-3 and IPC methods.

By grouping the number of livestock, 29% of all households have less than 200 livestock in the poor category. This figure is nearly the same as Mongolia's rural poverty level; our study sample could represent the herder's typical characteristics.

Figure 2.
Mongolia poverty headcount. Source: Author illustration based on data of Ref. [38].

Indicators	Household with less than 200 livestock—Poor household	Household with 200–500 livestock-Lower wealthy household	Household with 500 more livestock—Wealthy household
Number of households	102	132	116
Minimum livestock number	6	200	518
Maximum livestock number	198	492	3059
Average livestock number	108.5	340.0	854.0
Percentage	29%	37.7%	33.4%

Source: Authors' calculation.

Table 2.
Descriptive statistics of selected herder households and grouping of wealth classification.

On the other hand, wealthy households account for 33.4% of 350 herder households. The herd distribution or inequality between herder's households is relatively high compared to households with 6 livestock and 3059 livestock.

3.2 Results of poverty measurement on multidimensional poverty Index-3

The Multidimensional Poverty Index-3 are calculated through the following three stages:

1. Calculating indexes of the different wealth group

2. Estimating the mean value of each livelihood group

3. Calculating of Multidimensional Poverty Index-3.

In the first stage, survey responses were coded by eight indicators explained in **Table 1** within the three livelihood groups, i.e., (i) poor households, (ii) lower wealthy households and (iii) wealthy households. For instance, herder household response was 'no assets running with electricity', coded by 1 and 'one and more assets running with electricity', coded by 0 or nonpoor, etc.

In the second stage, the mean value of each of the eight indicators was estimated in three groups, i.e., poor 102 households, lower wealthy 132 households and wealthy 116 households. The estimated average value of each group is shown in **Table 3**.

Cooking fuel and sanitation are crucial indicators for herder households in Mongolia. There are no differences between wealth groups regarding cooking fuel and sanitation. All herder households use dung, wood and charcoal depending on the availability and regional differences. Herders living in the western part of Mongolia have more access to wood, while herders in the eastern part use dung and coal due to the limited availability of firewood. Herder households have not or impossible to use modern hygiene and sanitation tools.

The indicator close to the highest poverty level or above 50% of the threshold is housing, followed by access to drinking water in all three livelihood groups.

Poverty dimension	Poverty indicators	Mean		
		Poor household	Lower wealthy household	Wealthy household
Education	Years of Schooling	0.04	0.01	0.02
	School attendance	0.09	0.13	0.14
Standard of living	Electricity	0.03	0.02	0.02
	Water	0.46	0.39	0.42
	Housing	0.58	0.59	0.52
	Assets	0.20	0.12	0.01
	Cooking fuel	1	1	1
	Sanitation	1	1	1

Source: Authors' calculation.

Table 3.
Results of 1st stage of MPI: Mean for the poverty indicators of herder households' groups.

Internationally, housing conditions, quality, and years of operation are the indicators of measuring poverty. In the case of herder households, they use traditional dwellings named *ger* with 4–5 *khanas* all year round (see footnote 3). According to the study of changes of herders' move in 2019, 46% of herders households move 3–4 times annually, 30.1% of herders households move more than 4 times annually but 5.8% of herders households are not move all year around. The number of livestock is strongly correlated with a the number of herder households' annual moves [39]. Therefore, herder's household needs to have a house suitable for moving easily several times a year. Within three livelihood groups of herder households, the housing code was more than 0.5, which means most of them to have more than one housing asset for living, as storage or some housing is in *soum* or *aimag* centre for their school child living.

Water supply has ranked after housing. The survey questioned what are the drinking water supply sources during winter, spring, summer and autumn. Response polls: 1-Protected hand well, 2-Engineering well, 3-Unprotected hand well, 4-Protected nature springs, 5-Open waters from river and lake, 6-Snow and ice water, 90-Others with specification. The case of respondents selected 1–3, then it considered these herder households are secure with drinking water sources. All other responses were considered as difficulties for drinking water supply and coded by 1. By specification of nomadic herders' living it is not possible to stay close to engineering well and central water system in all year around. Therefore, nearly half of the herder households in our survey lacked a drinking water supply.

Interestingly, lower-wealth households' housing and school attendance indicators are poorer than those of poor herders. That can be explained by the frequency of annual moves as explained above. Thus, herders with less than 200 animals mainly stay close to *soum* or *bagh* centre all year round, which has better social and economic services.

In terms of housing, wealthy herder households can have more than one house and storage building. For the poor herders, a household with a few livestock is likely to be a periodic or part-time herder household. This is primarily because these households frequently have additional income as defined by the World Bank.

The poverty indicator of school attendance is higher in lower wealthy and wealthy households because of the lack of workforce. In recent years, as the number of livestock has increased, herders have taken measures such as hiring assistant herders. Lower-wealth households interviewed said that despite the lack of workforce and, moreover, they have a lack of financial resources to hire someone. Therefore, their children dropped out because of the parent's need for assistance.

As of 2021, 5.9% of Mongolian herders were assistant herders [40]. The number of wage-based labour contracts has increased in Mongolia, but neither clientelism nor kin-based cooperation has been replaced [41]. As defined by Mearn, the tradition of helping one's family and working together remains preserved and typical. Kinship support constantly helps poor herder families in difficult times, when they lack food and have health issues [12]. There is a propensity for an expanding number of herders to proactively consider becoming assistant herders as the number of livestock increases. The 'Regulation for the Selection of National Best Herding Households' was authorised under the Government of Mongolia's Resolution No.05, dated 6 January 2021. This rule states that to eliminate poverty, the best herder award must establish full- or part-time jobs for young herders. As this happens, an increase in the number of paid assistant herders will be expected in the upcoming years [11].

One advice to the Government of Mongolia is to consider dropping out reasons, school starting years and school locations, especially for six-year-olds. Average survey respondents live 19.4 km from the bagh centre and 23.9 km from the soum centre. In addition, herders are moving from one place to another year-round, making it challenging to manage their school-age children. It is typical for herders to send their kids to school dormitories, relatives' or grandparents' homes, or live with one of their parents.

The education level is an important measure to mitigate poverty as usual. Eliminating poverty must include educating the head of the herder's household and its members. However, 62.8% of the heads of herder households were educated for 6–12 years. Still, the rate of uneducated poor herders is as high as 9.8%, four times higher than that of wealthy herders.

In the third step, the MPI-3 index is determined. As explained in Section 2.1, MPI-3 is measured by the equally weighted eight indicators; thus, one indicator equals 0.125 coefficient. **Table 4** shows the MPI-3 values estimated by using the converted indicators. Conversion is made for each indicator in **Table 3** multiplied by the 0.125 (**Table 4**). The average and sum value of MPI-3, education and living standards were calculated in order to compare other similar studies.

Regarding education (25%) and standard of living dimensions (75%), the two lower groups are the same, and the wealthy group has a difference of 0.01 and 0.02, respectively.

The poverty index of poor and lower wealthy households is similar in value with regards to the MPI-3 estimation. A deprivation score of one-third of the weighted indicators distinguishes between the multidimensionally poor and nonpoor. If the deprivation score is 1/3 or greater, the household (and everyone in it) is classified as multidimensionally poor. Following this criterion, all three livelihood groups' herders were considered multidimensionally poor. However, the rural poverty level is high in Mongolia, and severe multidimensional poverty (greater than or equal to 1/2) was not the case in our

Poverty dimension	Poverty indicators	Coefficient	MPI-3 index			
			Poor household	Lower wealthy household	Wealthy household	Average MPI-3
Education	Years of Schooling	0.125	0.005	0.001	0.002	0.003
	School attendance	0.125	0.011	0.016	0.017	0.015
Sum of education		0.25	0.016	0.016	0.017	0.018
Standard of living	Electricity	0.125	0.004	0.002	0.002	0.003
	Water	0.125	0.058	0.048	0.053	0.053
	Housing	0.125	0.072	0.074	0.065	0.070
	Assets	0.125	0.025	0.015	0.001	0.014
	Cooking fuel	0.125	0.125	0.125	0.125	0.125
	Sanitation	0.125	0.125	0.125	0.125	0.125
A sum of the standard of living		0.75	0.408	0.408	0.389	0.389
	Total	1	0.424	0.406	0.390	0.407

Source: Authors' calculation.

Table 4.
Results of the multidimensional poverty Index-3 (MPI-3).

study. A possible explanation for this is that all herder households have assets and other means for basic needs such as housing (*ger*), food, drinking water, and energy resources.

The results obtained here have been compared with the most recent MPI for Mongolia using a survey in 2018. According to this study, 7.3% of the population in Mongolia is multidimensionally poor, while an additional 15.5% is classified as vulnerable to multidimensional poverty [31]. The study is part of a study covering 110 developing countries released in 2023 using the MPI-1 method, which includes health, education and standard of living. According to our study, the MPI-3 average index is 0.41 for herder households, indicating that rural poverty is much higher than the national average. Also, it is necessary to change the indicators used in the study, such as sanitation and floor, with other appropriate indicators, reflecting the characteristics of herder households and further research.

3.3 Results of poverty measurement on monetary poverty

According to the international scholars' methodology, households' wealth and poverty are measured by Cost/Income Per Capita per day in US dollars. Therefore, the IPC indicator estimation was included compared to other studies.

Although herder households have income from 17 possible sources of income, they differ in terms of size, income frequency and amount per household member. **Table 5** shows the calculation results comparing the cash income per member of the herding households per day in US dollars (exchange rate MNT 2632.2 = USD 1).

Household size characteristics matter of poverty. The higher the proportion of young people among household members, the greater the likelihood of poverty. The number of family members in South Asia, Latin America, and the Caribbean is 3–5 in Sub-Saharan African countries with six or more members. More than 55% of households are in poverty, regardless of the MPI criteria [37]. Household size is significant when calculating poverty using the IPC method. If the family size is large, then the per capita income is low.

As seen in **Table 5**, the average income per capita per day of the poor and lower wealthy is pretty similar; the median is 1.2 and 2.2, which is very close to the poverty line. The minimum IPC of each wealth group is far below the poverty line of 1.9 US$ or 2.15 US$.

Descriptive statistics	Poor household		Lower wealthy household		Wealthy household	
	LN	IPC	LN	IPC	LN	IPC
Mean	109.8	2.30	342.6	2.72	1026.9	5.17
Median	108.5	1.20	340	2.20	854	3.96
Std.Dev	54.5	2.86	82.78	2.17	531.5	3.88
Range	192	13.47	292	11.48	2541	23.13
Minimum	6	0.00	200	0.29	518	0.39
Maximum	198	13.47	492	11.77	3059	23.52
Count	102		132		116	
Confidence level (95%)	10.70		14.25		97.75	

Source: Authors' calculation.

Table 5.
Herder households' income per capita per day in USD by wealth groups.

The median number of livestock in poor households is 108.5, and the mean is 110. The IPC median of poor households with up to 200 animals almost coincides with the international poverty line of US$1.9. However, we can see in **Table 5** that the max IPC of poor households is several times higher than that of lower wealthy and wealthy households. This result indicates that these households with a few livestock have higher sources of income other than livestock income. There may be another explanation for this. According to the classification of the World Bank, herders are divided into five groups. However, those categories are not well distinguished in the study of rural households and herder households. Part-time and periodic herders participate in the full-time herders survey, which is typical. Therefore, studying part-time and periodic herders separately in future poverty research is necessary. Part-time and periodic households with livestock (named 'households with livestock' in official statistics of Mongolia) often have other livelihood strategies, such as living in soum centres, school teachers and unemployed people. By the end of 2021, the percentage of full-time herder households will account for 76.6% of total households with livestock.

According to researchers on the poverty of rural households, increased income diversification is one of the ways out of poverty. However, income diversification is low in all three wealth groups. A total of 33.1% of herder households have non-livestock income sources, i.e., state pensions and benefits, wages, handicraft income and natural resource mining. Fruit picking, logging, dung picking, onion picking and precious stone picking income are not registered and taxed. In addition, some households have income from planting and selling various crops, business activities and artificial mining, and one household earns income from tourism-related services.

Non-livestock income accounts for more than 90% of the income of poor and low-income households, while it accounts for less than 1% of the total income of the wealthiest households. This may be because pastoralism requires human resources and does not have many strategic options [28]. In addition, it is difficult and a burden for herders to do other types of business simultaneously. Regarding the composition of non-livestock income, pension income is the highest proportion in middle-income households and in general, pension income is contributed to all three groups of households.

The social insurance coverage rate of herders has gradually increased yearly, from only 6% in 2000 to 35.2% by the end of 2021 [40]. Although the participation rate is rising, it is much lower than the national average. Besides, the amount of pension received is a minimum level. Paying and receiving higher amounts is one possible way to increase the livelihood of poor herder households, improve well-being and promote more sustainable use of natural resources.

4. Conclusions

It can be concluded that when determining the livelihood of herders by grouping them by the number of livestock, there is a difference between the groups. Still, when evaluating the livelihood by the MDI method, the differences and boundaries between the livelihood groups of herders are minimal.

MDI method is not perfectly appropriate for measuring the poverty of Mongolian herder households. Therefore, updating the measurement indicators by specification of herder households in Mongolia is necessary. For instance, essential equipment and tools for livestock herding and animal husbandry, owned or contracted hay-making areas, and pasture user agreements are vital indicators.

The number of livestock is a critical indicator of wealth and assets, at least in the case of herders households in Mongolia. However, our study results warn that considering the number of livestock is insufficient to classify the poverty of herders' households in Mongolia. It is confirmed that using the number of livestock to measure the livelihood of herder households ignores the other incomes of herder households. Hidden and unregistered income contributes to the high poverty level.

Conflict of interest

The authors declare no conflict of interest.

A. Appendix

Questions/variables	Mean	Median	Minimum	Maximum	Count
Number of livestock	502	352	6	3059	350
Number of livestock by sheep unit	786	536.5	10.8	4527	350
Family size	4	4	1	11	350
Education level of family head	3	3	1	6	350
Education level of wife	3	3	1	6	345
Drop out reason	59	90	3	99	111
What type of following houses do you have?	2	2	1	4	350
How mane khanas have in your current ger?	7	5	3	80	350
What is your household's major winter water source?	4	3	1	90	350
What is your household's major spring water source?	3	3.00	1.00	6.00	350
What is your household's major summer water source?	4	4	1	90	350
What is your household's major autumn water source??	4	3	1	90	350
Do you households have an electricity means? If yes, what type?	1	1	1	1	13
Number of household's appliances use electricity?	2	2	0	6	350
What television provider use in your households?	2	1	1	90	305
Do your households have additional income rather than livestock? 1-yes, 2-no	2	2	1	2	350
What is your household's annual expenditure?	574,939	24,666	446,301	47,500	350

Questions/variables	Mean	Median	Minimum	Maximum	Count
What is your household's annual income?	12,017,078	531,845	9,315,001	—	350
Annual income per capita?	3,280,238	167,696	2,408,786	—	350
Additional income?	534,290	118,553	—	—	259
Part time job income?	766,667	218,581	600,000	500,000	3
Income from state pension and welfare?	1,702,451	143,276	720,000	48	210
Income from crop production?	100,000	—	100,000	100,000	1
Income from forage sale?	1,000,000	—	1,000,000	1,000,000	1
Income from potato and vegetable growing?	644,000	240,179	500,000	70,000	5
Income from meat sales?	4,122,848	280,567	2,500,000	—	287
Income from hide and skin sales?	298,353	20,450	200,000	—	304
Income from animal fibre (cashmere, wool etc.)?	4,736,683	291,503	3,185,000	75,000	340
Income from milk and dairy products sales?	1,132,932	155,273	600,000	30,000	88
Income from natural resources (fruits, timber, dung, onion, stone so on)?	425,000	209,662	400,000	75,000	3
Income from tourism services?		—	250,000	250,000	1
Income from hand made products?	1,704,737	143,164	720,000	720	210
Other income?	300,000	—	300,000	300,000	1

Table A1.
Descriptive statistics of primary data: Monetary values are in Mongolian tugrik.

Author details

Ganchimeg Gombodorj[1*] and Károly Pető[2]

1 Mongolian University of Life Sciences, Ulaanbaatar, Mongolia

2 Institute of Rural Development, Regional Economy and Tourism Management, Faculty of Economics and Business, University of Debrecen, Debrecen, Hungary

*Address all correspondence to: ganchimeg.g@muls.edu.mn

References

[1] World Bank. Long COVID:World Bank East Asia and Pacific Economic Update 2021 (October). Vol. 22. Washington DC: World Bank Group; 2021

[2] UNDP. Ending Poverty by 2030: UNDP's Perspective and Role. 2016

[3] UNDP. Global Multidimensional Poverty Index (MPI). United Nations Development Programme. 2023

[4] NSO. Mongolian Statistical information service [Internet]. 2023. Available from: https:/www.1212.mn/ [Accessed: February 6, 2024]

[5] World bank. Overview: The World Bank in Mongolia. 2023. Available from: https://www.worldbank.org/en/country/mongolia/overview [Accessed: February 8, 2024]

[6] UNESCO. Mongolia education policy review: Towards a Lifelong Learning. 2020

[7] Dovchin U, Dovchin S. The discourse of the Anthropocene and posthumanism: Mining-induced loss of traditional land and the Mongolian nomadic herders. Ethnicities. 2023. p. 8

[8] Lkhaajav B. How Mongolia's nomadic culture is navigating modernization. Diplomat Media INC. 2023. p. 1

[9] National statistics office of Mongolia. Монгол улсын хүн ам (Mongolian population). Ulaanbaatar; 2019

[10] Campi A. Mongolia and the coronavirus outbreak – Rethinking strategies for regional integration. Mongolian Journal of International Affairs. 2020;**21**(Dec):13-49

[11] Gombodorj G. Poverty, Sustainability and Livelihood Strategies of Herder Households in Mongolia. Library of University of Debrecen; 2023. Available from: https://hdl.handle.net/2437/359271

[12] Mearns R. Sustaining livelihoods on Mongolia's pastoral commons: Insights from a participatory poverty assessment. Development and Change. 2004;**35**(1):107-139

[13] Oniki S, Dagys K. Recovery from a winter disaster in Töv Province of Mongolia. Journal of Arid Environments. 2017;**139**:49-57

[14] Engler JO, Wesche K, Kaczensky P, Dhakal P, Chuluunkhuyag O, von Wehrden H. Biophysical variability and politico-economic singularity: Responses of livestock numbers in south Mongolian nomadic pastoralism. Ecological Economics. 2021;**187**(Mar 2020):107073. DOI: 10.1016/j.ecolecon.2021.107073

[15] Khan TS, Gottschalk J. Investigating the transmission channels behind Dutch disease effects: Lessons from Mongolia using a CGE model. In: Policy Research Working Paper of Worldbank. Washington DC: World Bank, Washington, DC; 2017. Available from: https://openknowledge.worldbank.org/handle/10986/28360

[16] Ellis F. Rural Livelihood Diversity in Developing Countries: Evidence and Policy Implications. London: Overseas Dev Institute; 1999. p. 40

[17] Zhao Y, Yan Y, Liu Q, Li FY. How willing are herders to participate in carbon sequestration and mitigation? Inner Mongolian grassland case. Sustainability. 2018;**10**(2808):7

[18] NSO and WB. Mongolia Poverty Update 2020. Ulaanbaatar: NSO; 2021

[19] FAO and MoFALI. Action plan for the Mongolian agenda for sustainable livestock. Global agenda for sustainable livestock. 2018

[20] NSO and WB. Mongolia Poverty Update 2018. Ulaanbaatar: NSO; 2020

[21] JICA and MoFA. Mongolia The study for improvement plan of Livestock farming system in Rural Area [Internet]. 2006. Available from: https://openjicareport.jica.go.jp/pdf/11821865_01.pdf

[22] Densambuu B, Sainnemekh S, Bestelmeyer B, Budbaatar U. National Report on the Rangeland Health of Mongolia: Second Assessment. Ulaanbaatar: SDC; 2018

[23] Tumur E, WJM H, Heerink N, Agipar B. Critical factors enabling sustainable rangeland management in Mongolia. China Economic Review. 2020;**60**:101237

[24] Li A, Wu J, Zhang X, Xue J, Liu Z, Han X, et al. China's new rural "separating three property rights" land reform results in grassland degradation: Evidence from Inner Mongolia. Land Use Policy. 2018;(71, December 2017)**27**(1):170-182

[25] Alexander S, Aronson J, Whaley O, Lamb D. The relationship between ecological restoration and the ecosystem services concept. Ecology and Society. 2016;**21**(1):5

[26] Dong X, Dai G, Ulgiati S, Na R, Zhang X, Kang M, et al. On the relationship between economic development, environmental integrity and well-being: The point of view of herdsmen in northern China grassland. PLoS One. 2015;**10**(9):1-23

[27] Singh PK, Hiremath BN. Sustainable livelihood security index in a developing country: A tool for development planning. Dierkes J, editor. Ecological Indicators. 1st editio. 2010;**10**(2):442-451

[28] Houzer E, Scoones I. Are Livestock Always Bad for the Planet ? Rethinking the Protein Transition and Climate Change Debate. Brighton: PASTRES; 2021 80 p

[29] Werhane P, Newton L, Wolfe R. Alleviating Poverty through Profitable Partnerships: Globalisation, Markets, and Economic Well Being. 2nd ed. London, Brighton: Routledge; 2020. DOI: 10.4324/9780429329043

[30] Hiragaikuko M, Uochi I, Gabriela RAD. Counting the uncounted – How the Mongolian nomadic survey is leaving no one behind [Internet]. World Bank. 2020. Available from: https://blogs.worldbank.org/en/opendata/counting-uncounted-how-mongolian-nomadic-survey-leaving-no-one-behind [Cited 2024 May 13]

[31] UNDP. Briefing note for countries on the 2023 multidimensional poverty index: Mongolia. UNDP. 2023. p. 1

[32] Jiao W. Analyzing multidimensional measures of poverty and their influences in China's Qinba Mountains. Chinese Journal of Population, Resources and Environment. 2020;**18**(3):214-221

[33] World Bank. Piecing Together the Poverty Puzzle. Piecing Together the Poverty Puzzle. 2018

[34] World Bank. Fact Sheet: An Adjustment to Global Poverty Lines. 2022. Available from: https://www.worldbank.org/en/news/factsheet/2022/05/02/fact-sheet-an-adjustment-to-global-poverty-lines [Accessed: December 1, 2023]

[35] Robles Aguilar G, Sumner A. Who are the world's poor? A new profile of global multidimensional poverty. World Development. 2020;**126**:104716. Available from: https://www.sciencedirect.com/science/article/pii/S0305750X19303651

[36] Worldbank. Livestock Sector Study, Mongolia Volume 1 – Syntesis Report. 2009

[37] Aguilar GR, Sumner A. TATT. Who are the world's poor? A new profile of global multidimensional poverty [Internet]. NV-1 O. Washington, DC: Center for Global Development; 2019. (Working paper / Center for Global Development 499 (January 2019)). 1973. Available from: https://www.cgdev.org/publication/who-are-worlds-poor-new-profile-global-multidimensional-poverty

[38] National Statistics Office, World bank. Mongolia 2020 Poverty Report: A Decade of Progress and Stagnation in Poverty Reduction. Washington DC and Ulaanbaatar: NSO; 2022

[39] Punsantsogvoo TDK, Gankhuyag U, Ts O. Малын тооны бүлэглэлтээс үүдэн малчдын нүүдэлд гарч буй өөрчлөлт (Changes in herders seasonal moving depend on livestock number). 2019. Available from: https://www.researchgate.net/publication/335060854_MALYN_TOONY_BLEGLELTEES_DEN_MALCDYN_NDELD_GARC_BUJ_RCLLT

[40] Mongolian Statistical Information Service. Agriculture sector 2021 [Internet]. Ulaanbaatar; 2022. Available from: https://www.1212.mn/en

[41] Murphy DJ. From kin to contract: Labor, work and the production of authority in rural Mongolia. Journal of Peasant Studies. 2015;**42**(2):397-424

Chapter 19

Blueprints for Sustainability: An In-Depth Study of Construction

Khululekani Ntakana and Aidi Ahmi

Abstract

This study systematically explores sustainable construction, a crucial intersection of environmental stewardship and the built environment. It aims to map the scholarly landscape, analyzing publications from 1996 to 2024 sourced from Scopus, using robust bibliometric methods including keyword co-occurrence, trend analysis, and multiple correspondence analyses. This approach identifies key thematic developments over time, such as material innovation, environmental impact assessment, and emerging technologies, highlighting an expanding trajectory in research evidenced by increasing publications and citations. Significant research contributions with high citation metrics are pinpointed, emphasizing their influence in advancing sustainable methodologies. However, the study's scope is constrained by reliance on a single database, potentially limiting the breadth of global research coverage. Nonetheless, it offers valuable insights for academia, industry, and policymakers, suggesting future interdisciplinary inquiries and regional research explorations. This research distinguishes itself through its comprehensive temporal and thematic mapping of sustainable construction, using advanced bibliometric analyses to inform stakeholders about the field's dynamic contours and future research directions crucial for sustainable development in construction.

Keywords: sustainable construction, bibliometric analysis, thematic evolution, scholarly impact, publication trends, future research directions, sustainable practices

1. Introduction

The detrimental effects of the construction industry on the environment are still being discussed. Almost every publication that has been released on sustainable construction starts out by emphasizing this problem. The use of land and virgin land such as forests and wetlands, which frequently results in the loss of biodiversity and soils, is one of the construction industry's widely acknowledged significant and irreversible effects on the environment. Other effects include the extensive use of natural resources, many of which are nonrenewable; the pollution of air and water reserves; the consumption of water and energy; the generation of waste; and the noise generated by construction activities [1]. Therefore, there is concern over how to enhance construction methods to lessen their negative effects on the environment [2]. Sustainable construction is the alternative. To meet the requirements for sustainable construction, current building designs, engineering processes, manufacturing technologies, and construction techniques must be modified [3, 4].

Understanding sustainable construction is advanced by looking at how the idea of sustainable development has evolved [5]. Sustainable development refers to initiatives that aim to meet social demands while being cautious to reduce any potential harm to the environment. Sustainable development, according to Lacasse [4], is the use of physical and natural resources to allow individuals and communities to achieve their immediate social, economic, and cultural requirements without jeopardizing the ability of future generations to do the same. The concept of sustainability has very few scholarly papers that link it back to African roots. According to Hill and Bowen [5], early human civilizations like the South African "Bushmen" likely had an innate understanding of sustainability. Though it may sound insulting, the term "Bushmen" was perhaps appropriate at the time. They contended that Africans understood the value of using nature's resources sustainably and had firsthand knowledge of how dependent humans are on the planet's life support systems for their existence. Four "pillars" comprise the principles of sustainable construction: social, economic, biophysical, and technological. These pillars are accompanied by a set of general, process-oriented principles that can be used as a checklist in real-world situations [5].

The fields of sustainability, sustainable development, and sustainable construction have seen the publication of several documents. This bibliometric analysis therefore aims to provide a comprehensive analysis of the existing body of literature in the domain of sustainable construction, offering insights into the evolving trends, key contributors, and influential research themes. Thus, the following research questions (RQ) have been designed:

- RQ1: What are the temporal publication trends in sustainable construction research from its inception until the present, and how do these trends reflect the evolving focus and priorities within the field?

- RQ2: How do authorship patterns and contributions across different countries and institutions shape the global research landscape in sustainable construction, and what does this reveal about international and institutional research collaborations?

- RQ3: What are the characteristics of the most influential publications within the sustainable construction domain, and how do they contribute to the field's theoretical and practical advancements?

- RQ4: Which thematic clusters and keywords are most prevalent in sustainable construction literature, and what does this indicate about the central research themes and the interconnectivity of topics within the field?

- RQ5: How have the themes in sustainable construction research evolved over time, and what emerging trends are signaled by recent keyword usage and citation patterns?

The subsequent sections of this chapter are structured as follows: Section two provides an in-depth review of existing literature on sustainable construction, elucidating the historical development and underlying theoretical foundations. Section three outlines the adopted bibliometric methodology, discussing the tools and techniques employed for the analysis. Section four presents the findings, showcasing trends and patterns identified through the bibliometric analysis. Section five engages in comprehensive discussion, offering nuanced insights into the implications of the findings within the broader context of sustainable construction research. Section six examines the contributions, limitations, and potential avenues for future research, contributing to the ongoing discourse on policy development, economic implications, and environmental sustainability within sustainable construction studies.

2. Literature review

This section presents an overview of the current research on sustainable construction and the studies that employed bibliometric analysis in the domain of sustainable construction.

2.1 Empirical studies on sustainable construction

The field of sustainable construction has been the subject of numerous studies. Scholars have examined a wide range of topics related to sustainable construction, including sustainability assessment instruments, tools for use, technology utilized, and building techniques and materials. Subsequently, in order to create a sustainable construction standard, Zimmermann et al. [6] carried out a study with the goal of looking at sustainable construction benchmarks. These benchmarks aim to specify the conditions that structures and buildings must satisfy in order to support sustainable development in Switzerland. The assessment recorded the allowable building impacts with respect to energy consumption and pollution loads during the building's construction, maintenance, and operation. Furthermore, Plessis [7] created a strategic framework for sustainable construction in developing countries due to the significant development that is required in the built environment and its physical infrastructure in these nations. Arguing that it is much more urgent to make sustainable interventions now, while the built environments are being produced, rather than trying to change things after the fact, despite the fact that there are many obstacles in the way of the implementation of sustainable building technologies and practices.

Scientists have accused the construction industry of being the root of numerous environmental issues. According to Ding [2], this is because the project assessment stage, when environmental considerations are most effectively integrated, has received little to no attention when it comes to the significance of choosing ecologically friendly designs. Ding thus looked at the evolution, function, and constraints of environmental building evaluation techniques that are applied to determine the sustainability of buildings in developing nations. This review established that existing methods for assessing the environment do not conveniently and sufficiently consider environmental effects in a single instrument, which makes them ineffective for aiding in the evaluation of sustainable development as a whole.

Pitt et al. [8] developed the sustainability measuring standards and examined the drivers and barriers of sustainable construction practices. This analysis found that building codes and financial incentives/penalties were the two main drivers, whereas affordability was a major barrier to sustainable building. Abidin looked into Malaysian developers' awareness and application of the sustainable construction concept in order to bolster the findings of the later study. It was shown by this review that only large developers knew about and could apply sustainable techniques to their projects. Due to their lack of knowledge about the concept and worries about cost, many developers were still hesitant and uneasy [9, 10]. Similar to this, Djokoto et al. [11] looked at Ghana's construction industry's barriers to sustainable construction. According to this review, the primary barriers to sustainable construction include the following: a lack of public knowledge, a lack of government backing, higher starting costs, a lack of strategy to encourage sustainable development, and a lack of demand for sustainable structures. Similarly, Alsanad [12] looked into Kuwait's awareness of, and actions toward, sustainable construction, as well as its impediments. It is interesting to note that the outcomes of this review match those of the earlier research. It has been determined that there is a circle of responsibility for the challenges facing sustainable construction. The owner's viewpoint on the application of sustainable construction was investigated by Gan et al. [13]. The research revealed that the primary barriers preventing owners from implementing sustainable construction practices are financial viability, awareness, legislation, and regulation.

While other scholars are concerned about cost as a barrier to the application of sustainable construction, Robichaud and Anantatmula [14] investigated green project management practices for sustainable construction. They contended that adjustments must be made to conventional project management procedures and practices in order for project managers to produce sustainable construction in line with clients' budgetary expectations. According to this analysis, green project management techniques can significantly increase the value of a sustainable building project and still complete it within reasonable budgetary limitations.

A few academics concentrated on the advancement of environmentally friendly building materials and technologies. In a review of the materials and technologies to be utilized in sustainable construction, Lacasse [4] found that existing building designs, engineering processes, manufacturing technologies, and construction methods must be modified to meet the demands of sustainable construction. In Hong Kong's congested urban surroundings, Jaillon and Poon [1] examined the sustainable construction elements of using prefabs. They looked at the triple-bottom line of employing prefabs in high-rise buildings and contrasted it with traditional construction techniques. According to this analysis, employing prefabs has more triple-bottom-line advantages for sustainable construction than traditional building techniques. Furthermore, the utilization of solid waste from industry and agriculture to manufacture "waste-create bricks" was evaluated by Raut et al. [15]. They contended that recycling these waste materials appears to be an affordable option for designing green buildings in addition to being a workable solution to pollution issues. Similar research was done in 2016 by Kariyawasam and Jayasinghe [16] on the use of cement-stabilized rammed earth as a sustainable building material. They maintained that when a cradle to grave life cycle approach is considered, earth can serve as an ideal building material with the potential to lower the carbon footprint. Panda et al. [17] carried out an experimental study to examine the application of concrete with in-process "printed" reinforcement or concrete with 3D printable fiber reinforcement for sustainable building. Hager et al. [18] questioned if 3D printing of buildings and building components would be a part of sustainable construction in the future. They maintained that 3D printing technology may be viewed as an environmentally friendly derivative that offers practically limitless opportunities for the realization of geometric complexity. The creation of new, environmentally friendly materials with the potential to guarantee qualities like strength, thermal insulation, or transparency is one benefit of this technology. Furthermore, in order to produce sustainable building materials, Tang et al. [6] examined the latest developments in recycling solid waste from construction and municipal sources. Recycling these waste materials into precursor, aggregate, additive, reinforcing fiber, or filler materials for geopolymer composite products is the goal.

The adoption of sustainable techniques will eventually become a source of competitive advantage for contractors as the building industry requires it. A study by Tan et al. [19] looked at sustainable construction methods and the connection between contractors' commercial competitiveness and sustainability performance. This study provides a framework for implementing sustainable building practices to increase contractors' competitiveness, which will aid contractors in developing their sustainable strategies for meeting a changing competitive climate. Rajendran and Gambatese [20] looked at the development and validation of a sustainable construction health and safety rating system. This system aimed at providing an opportunity to rate projects based on the importance given to construction workers' safety and health and the degree of implementation of safety and health elements. In contrast, integrated delivery systems for sustainable building were studied by Barrett et al. [3]. This is the result of their realization that there is a growing consensus regarding the necessity for proper strategies and actions to improve the sustainability of the built environment and building activities.

2.2 Previous studies on bibliometric analysis and sustainable construction

Scholars have employed bibliometric techniques in the past several years, which are more closely related to the approach applied in this study. However, there is no record of any bibliometric analysis study conducted in the domain of sustainable construction pre-2020 in the Scopus database. The first recorded study was authored by Det Udomsap and Hallinger [21], which aimed at documenting and synthesizing research trends in the domain of sustainable construction over a period of 25 years (1994–2018). This study used bibliometric methods to review the document volume, growth trajectory, geographic distribution of scholars, scholars who have emerged as thought leaders, intellectual structure, and "hot topics" in sustainable construction research. This review documented four dominant lines of inquiry, namely, alternative materials for sustainable construction, sustainable construction management, recycling and waste reduction, and social sustainability in construction management. In contrast to the broad perspective adopted by the previous authors, 2 years later, Wang and Guo [19] used bibliometric methods to review sustainable development through the perspective of Construction 4.0. Also known as the Fourth Industrial Revolution in the construction industry, Construction 4.0 is a concept that is fundamentally altering the industry, production value chains, and corporate structures by connecting intelligent production processes with embedded system production technologies. The authors conducted a content analysis of the impacts of Construction 4.0 on sustainable development based on triple-bottom-line dimensions. They also identified and summarized the critical success factors of Construction 4.0 toward sustainable development.

According to the review's findings, Construction 4.0 will eventually gain traction as a quick and affordable solution for problems relating to sustainable development. In exploring the potential of Construction 4.0 for sustainable development, the most researched and used technologies were found to include BIM, 3DCP, machine learning, digital twins, IoT, VR, and blockchain. Further to the earlier reviews, Maier [22] conducted a bibliometric review on the perspective of using wood waste, especially the one that resulted from construction and demolition activities, to produce sustainable bioenergy. This review documented that although there is not much research on

Author & Year	Objective of the Study	Attributes Examined	Total Documents Examined	Data Source and Coverage	Research Gap
[21]	To document and synthesize research trends in the domain of sustainable construction over a period of 25 years (1994–2018)	• Citation analysis, • Author co-citation analysis, • Keyword co-occurrence analysis.	2877	Scopus (1994–2018)	Analysis did not focus on a specific sustainable construction method. Rather adopted a broader perspective in this domain.
[19]	To conduct content analysis of Construction 4.0 impacts on sustainable development based on triple-bottom-line dimensions. To identify and summarize the critical success factors of Construction 4.0 toward Sustainable Development.	• Publication trend, • The most prolific journals, • Countries, • Institutions, • Researchers, and • Keywords analysis.	229	Scopus and WoS (2014–2022)	Analysis was based on construction 4.0 and sustainable development.
[22]	To investigate the perspective of using wood waste, especially the one that resulted from the construction and demolition activities, to produce sustainable bioenergy.	• Number of publications, • Most studied topics, • Keyword co-occurrence, • Trend topic plot	4934	WoS (2012–2021)	Analysis was based on the use of wood waste from construction and demolition to produce sustainable bioenergy.
[23]	To explore the challenges that the construction sector faces in developing countries in integrating sustainability practices into infrastructure delivery.	• Number of publications, • Citation patterns, • Content analysis, • Emerging trends, • Most productive authors, • Collaboration among authors, • Most productive journals, • Most active countries, • Publications with highest impact.	28	Scopus (2011–2021)	Analysis was based on the implications in the implementation of sustainable construction practices.

Author & Year	Objective of the Study	Attributes Examined	Total Documents Examined	Data Source and Coverage	Research Gap
[24, 25]	To map the existing literature in sustainable construction practices research field.	• Growth trajectory of publications, • Most productive and influential authors, • Collaboration among authors, • Geographical distribution of publications, and • Emerging knowledge areas.	94	Scopus (2002–2021)	Analysis was based on the implications on construction productivity.
[26]	To investigate the contribution of three major technologies that are considered under the umbrella of Construction 4.0, namely, Building Information Modeling (BIM), advanced manufacturing (including 3D printing, modular construction, offsite construction, and prefabrication), and Digital Twins.	• Map the currently discussed technologies. • Group these technologies into themes	556	WoS (2005–2022)	Analysis was based on construction 4.0 and sustainable construction.

Source: Developed by the authors.

Table 1.
Summary of previous studies.

the topic of using wood waste from the construction industry, it has the potential to become a major topic in the future if we consider the main benefits, particularly those connected to the incorporation of carbon dioxide in the wood elements of construction, and the numerous ways in which the resultant waste can be reused.

While there are numerous methods that the construction industry may implement to lessen its adverse environmental effects, some academics believe that these practices come at a cost. Tjebane et al. [23] conducted a bibliometric review on the challenges that come with the implementation of sustainable construction practices in developing countries. This review focused on the integration of sustainability practices into infrastructure delivery. However, they only reviewed 28 documents and it is not clear what cost implications are in relation to sustainable construction practices. Although time and effort are identified as the two most important factors in developing nations, the exact impact of these factors on sustainable construction techniques is not precisely measured. Finally, the review notes that developing nations lack both building standards and adequate ecological regulatory frameworks with regard to environmental sustainability. Furthermore, Adebowale and Agumba [24, 25] also employed bibliometric methods to understand the implications on construction productivity when/if sustainable construction practices are used. The review analyzed the growth trajectory of publications, productive and influential authors, author collaboration, geographical distribution, and emerging knowledge areas in sustainable construction. It identified eight emerging areas: lean construction, procurement, energy, recycling, construction ecology, life cycle costing, construction waste, and sustainable designs.

Contrary to several of the above reviews, the current review did not focus on a specific sustainable construction method (i.e., Construction 4.0, wood waste) or implications in the implementation of sustainable construction practices. Rather, a broad perspective on "sustainable construction" was adopted. While the bibliometric methods employed in this review most closely resemble the method used by Det Udomsap and Hallinger [21], our approach was to extend and fill in gaps in trends reported in these prior reviews. **Table 1** provides an overview of prior bibliometric analysis research in the field of sustainable construction.

3. Methods

3.1 Search strategy

The search strategy for identifying literature on sustainable construction is meticulously depicted in **Figure 1**, which serves as a flow diagram outlining the methodological approach underpinning the research. This strategy is critical to the method section as it elucidates the systematic procedures employed to retrieve the most relevant scholarly articles for analysis. The flow diagram indicates that the database Scopus was utilized as the primary source for literature retrieval. Scopus is a large multidisciplinary database containing citations and abstracts from industry journals, patent records, books, peer-reviewed articles, and conference publications and provides tools for analyzing, tracking and visualizing the search outcome [29]. It is renowned for its comprehensive coverage and is widely recognized for its rigor in indexing peer-reviewed literature across various disciplines, including science, technology, medicine, social sciences, and arts and humanities [30]. It can provide an extensive scientific research perspective from all over the globe and is presently considered as one of the major relevant information sources in the research community [29, 31]. In fact, all peer-reviewed articles published in Scopus database come from well-known and leading academic publishers such as Emerald, Elsevier, Springer, Inderscience, and Taylor and Francis Group [32, 33]. Studies such as [34] have conducted comparisons between search engines and discovered that Scopus and Web of Science are the two most efficient and objective databases for literature search. However, all published bibliometric review documents have used only one database and mostly Scopus due to its abovementioned features.

To refine the search and ensure the specificity of the results, the search field was confined to article titles, capturing the essence of research focused explicitly on sustainable construction. The time frame for the search was left open, signifying an all-encompassing retrieval strategy that aimed to include the entire corpus of available literature on the topic without temporal limitations. This approach enables a holistic overview of the field's development over time. The language of the articles was not restricted, allowing for the inclusion of research published in

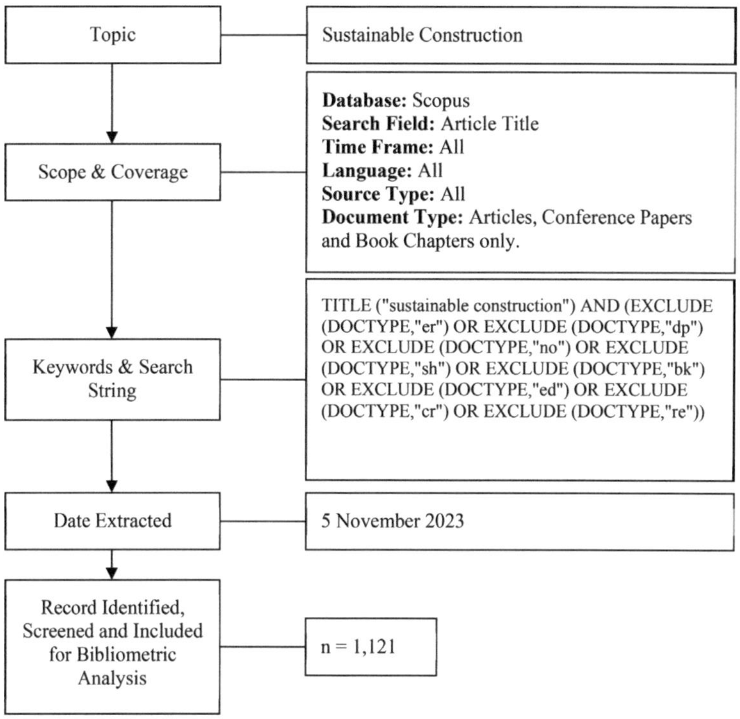

Figure 1.
Flow diagram of the search strategy. Source: Zakaria et al. [27], Moher et al. [28].

all languages, thereby enriching the dataset with global perspectives. The document types were narrowed to articles, conference papers, and book chapters only, thereby focusing on textually rich and substantive forms of scholarly communication that are most likely to contribute to the theoretical and practical advancements in the field.

A detailed search string was crafted, incorporating the term "sustainable construction" and employing Boolean operators to exclude less relevant document types. This strategic formulation ensured a targeted yet exhaustive exploration of the database, which is vital for a robust bibliometric analysis. The date of the data extraction is explicitly noted as November 5, 2023, providing a clear temporal context for the search and ensuring reproducibility of the research process. The records identified and screened for inclusion numbered 1121, indicating a substantial body of literature for subsequent bibliometric analysis. This flow diagram, inspired by the work of Zakaria et al. [27] and Moher et al. [28], embodies a rigorous, transparent, and reproducible search strategy that is essential for the integrity of bibliometric studies. It serves not only as a blueprint for the present analysis but also as a guide for future research endeavors aiming to synthesize knowledge and assess trends within the domain of sustainable construction.

3.2 Data cleaning, harmonization, and tools

The bibliometric analysis necessitates a rigorous process of data cleaning and harmonization to ensure accuracy and reliability. In this study, the sophisticated functionalities of OpenRefine and biblioMagika [35] were harnessed to systematically cleanse and organize the data extracted from Scopus. The initial phase involved downloading the Scopus dataset as.csv files and meticulously sifting through key columns—author names, affiliations, and keywords—employing advanced clustering functions to detect and correct inconsistencies. After the initial automated cleaning, a thorough manual review was undertaken to verify the rectified keywords and consolidate cells with multiple entries. This step was crucial to re-establish data integrity post-segmentation. Subsequently, the dataset, now pristine and consistent, was reverted to its original format for in-depth analysis.

The study then advanced to the analytical phase, employing biblioMagika [35] to further enhance the bibliometric assessment, particularly by standardizing data related to authors, affiliations, and countries. OpenRefine complemented this process by fine-tuning keyword data [36]. Visual representations of the findings were generated using VOSviewer [5] and Biblioshiny [37], enabling the visualization of the research's conceptual networks. This integrative approach, utilizing multiple tools, facilitated a comprehensive examination of the sustainable construction research domain.

4. Results

This section presents the outcomes of the comprehensive analysis conducted on the landscape of sustainable construction studies, addressing the research questions outlined in the background section. The primary objective is to provide a robust exploration and elucidation of the sustainable construction research domain, offering valuable insights and knowledge for practitioners, policymakers, academics, and other key stakeholders.

4.1 Documents profiles

An overview of the documents retrieved from Scopus and examined for this investigation is shown in **Table 2** below. The table indicates that the first Scopus publication in sustainable construction occurred in 1996. Thus, data recorded between 1996 and 2024 was used in this investigation. A total of 15,656 citations were documented in 1121 publications from 3525 authors throughout this time span, yielding an h-index of 58.

In addition, **Table 3** below shows the type of sources from which the data used in this inquiry were extracted. As shown in the table, most of the data (53,08%) were journal articles, followed by conference proceedings (31.94%), book series (8.47%), books (3.57%), and trade journals (2.94%).

A further analysis in terms of languages used in the data was conducted. **Table 4** below shows the number of publications in different languages. The table indicates that most data (98.57%) were published in English. Some documents used other languages, such as German, Spanish, Portuguese, Italian, French, Greek, and Polish.

The final factor in determining a current trend, after linguistic patterns, is the subject area, which indicates which subject areas have recognized sustainable construction. Based on the subject area, this analysis groups the published documents shown in **Table 5**. The distribution of the literature on sustainable construction generally shows that it is found in a variety of subjects, primarily the social sciences, energy, environmental sciences, engineering, and materials sciences.

Main Information	Data
Publication Years	1996–2024
Total Publications	1121
Citable Year	28
Number of Contributing Authors	3525
Number of Cited Papers	773
Total Citations	15,656
Citation per Paper	13.97
Citation per Cited Paper	20.25
Citation per Year	559.14
Citation per Author	4.44
Author per Paper	3.14
Citation sum within h-Core	12,965
h-index	58
g-index	100
m-index	2.07

Table 2.
Main information.

Source Type	Total Publications (TP)	Percentage (%)
Journal	595	53.08%
Conference Proceeding	358	31.94%
Book Series	95	8.47%
Book	40	3.57%
Trade Journal	33	2.94%
Total	**1121**	**100.00**

Table 3.
Source type.

Language	Total Publication	Percentage
English	1105	98.57%
German	13	1.16%
Spanish	4	0.36%
Portuguese	3	0.27%
Italian	2	0.18%
French	1	0.09%
Greek	1	0.09%
Polish	1	0.09%

Table 4.
Languages.

Subject Area	Total Publications	Percentage
Engineering	753	67.17%
Environmental Science	288	25.69%
Energy	189	16.86%
Social Sciences	189	16.86%
Materials Science	178	15.88%
Business, Management, and Accounting	148	13.20%
Computer Science	123	10.97%
Earth and Planetary Sciences	70	6.24%
Physics and Astronomy	34	3.03%
Arts and Humanities	29	2.59%
Chemical Engineering	25	2.23%
Economics, Econometrics, and Finance	24	2.14%
Decision Sciences	23	2.05%
Chemistry	15	1.34%
Mathematics	15	1.34%
Agricultural and Biological Sciences	14	1.25%
Medicine	11	0.98%
Biochemistry, Genetics, and Molecular Biology	6	0.54%
Multidisciplinary	3	0.27%
Immunology and Microbiology	2	0.18%
Pharmacology, Toxicology, and Pharmaceutics	1	0.09%
Psychology	1	0.09%
Undefined	1	0.09%

Table 5.
Subject area.

4.2 Publication trends

Table 6 and the accompanying **Figure 2** articulate the temporal trends in sustainable construction research, revealing an ascending trajectory in both the number of publications and citations over the years. This bibliometric data, encompassing the period from 1996 to 2024, provides a longitudinal perspective on the field's scholarly activity and its reception within the academic community. From the nascent beginnings in 1996 with just two publications, there has been a notable escalation in output, reaching a peak in 2023 with 161 publications. The trend suggests a burgeoning interest and expanding research activity in sustainable construction. The table also denotes the total citations (TC), peaking significantly in 2020, which reflects the growing influence and recognition of the research contributions in this domain. The average citations per publication (C/P) and per cited publication (C/CP) offer additional insight into the impact of the research, with the highest C/P observed in the years 1997 and the highest C/CP in 2008.

Year	TP	NCA	NCP	TC	C/P	C/CP	*h*	*g*	*m*
1996	2	4	2	84	42.00	42.00	2	2	0.07
1997	3	5	2	397	132.33	198.50	2	3	0.07
1998	3	5	0	0	0.00	0.00	0	0	0.00
1999	2	4	2	27	13.50	13.50	2	2	0.08
2000	6	12	6	229	38.17	38.17	6	6	0.25
2001	10	10	6	213	21.30	35.50	3	10	0.13
2002	5	6	3	132	26.40	44.00	3	5	0.14
2003	4	9	2	20	5.00	10.00	2	4	0.10
2004	4	5	2	47	11.75	23.50	1	4	0.05
2005	11	21	8	286	26.00	35.75	5	11	0.26
2006	11	29	6	127	11.55	21.17	4	11	0.22
2007	27	53	13	293	10.85	22.54	5	17	0.29
2008	24	53	15	1307	54.46	87.13	8	24	0.50
2009	17	46	13	478	28.12	36.77	7	17	0.47
2010	33	79	18	619	18.76	34.39	8	24	0.57
2011	32	72	23	1309	40.91	56.91	12	32	0.92
2012	35	92	26	448	12.80	17.23	10	21	0.83
2013	41	115	34	332	8.10	9.76	9	17	0.82
2014	37	119	27	544	14.70	20.15	12	23	1.20
2015	47	128	35	866	18.43	24.74	14	29	1.56
2016	47	134	38	1280	27.23	33.68	16	35	2.00
2017	53	136	47	1112	20.98	23.66	16	32	2.29
2018	73	230	61	1301	17.82	21.33	19	35	3.17
2019	86	296	68	739	8.59	10.87	14	25	2.80
2020	114	388	93	1532	13.44	16.47	22	33	5.50
2021	107	370	83	1135	10.61	13.67	20	29	6.67
2022	125	457	77	547	4.38	7.10	11	18	5.50
2023	161	644	63	252	1.57	4.00	9	11	9.00
2024	1	3	0	0	0.00	0.00	0	0	—
Total	1121	3525	773	15,656	13.97	20.25	58	100	2.071

Notes: TP = total number of publications; NCA = Number of contributing authors; NCP = number of cited publications; TC = total citations; C/P = average citations per publication; C/CP = average citations per cited publication; h = h-index; g = g-index; m = m-index.

Table 6.
Publication by year.

Bibliometric indices such as the h-index, g-index, and m-index serve to further characterize the publication trends. The h-index, which indicates both productivity and citation impact, shows a steady increase, reaching its zenith in 2021, suggesting that the field has produced a body of work that has been extensively cited and

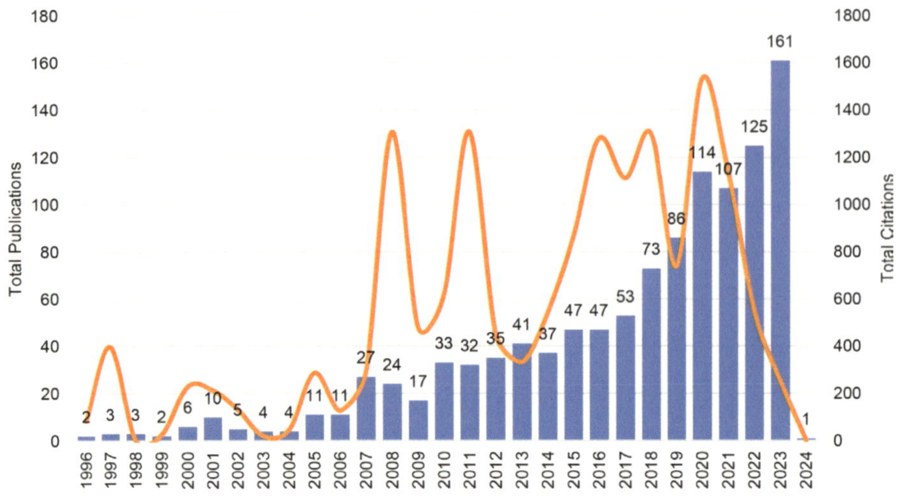

Figure 2.
Total publications and citations by year.

continues to influence new research. The g-index corroborates this growth pattern, emphasizing that the most cited articles have received substantial attention, thus reinforcing the field's academic significance. The m-index, indicative of the rate at which publications are cited, presents an upward trend, signifying rapid recognition and citation uptake of the research output in recent years. Collectively, the publication trend data manifest the dynamic evolution of sustainable construction research, underscoring a field in expansion, with increasing academic contributions that resonate with the wider scholarly and practical community. This expansion reflects the escalating importance placed on sustainability within the construction industry and the corresponding scholarly efforts to address the challenges it presents.

4.3 Publications by authors

Table 7 provides a systematic account of the most prolific authors contributing to the field of sustainable construction, delineating their scholarly influence *via* a suite of bibliometric indicators. This table, which focuses on authors with at least five publications, reveals not only the volume of research output but also the impact and recognition within the academic community as evidenced by citation metrics. At the forefront is Oke, Ayodeji Emmanuel from the Federal University of Technology in Nigeria, with an impressive 28 publications and a total of 321 citations. This indicates a significant contribution to the body of literature, with a substantial number of these works resonating within the scholarly community, as demonstrated by a high h-index of 11. Closely following is Aigbavboa, Clinton from the University of Johannesburg, South Africa, whose 22 publications have amassed 340 citations, reflecting a considerable impact as well, with an h-index of 8. The array of authors from diverse geographical and institutional backgrounds underscores the global nature of sustainable construction research. For instance, Ahmed Farouk Kineber from Prince Sattam Bin Abdulaziz University, Saudi Arabia, shows a strong citation per publication ratio, suggesting that his work is not only voluminous but also influential. Similarly, authors from universities in the United Arab Emirates, Brazil, the United States, Lithuania, Australia, and the United Kingdom

Author's Name	Current Affiliation	Country	TP	NCP	TC	C/P	C/CP	h	g	m
Oke, Ayodeji Emmanuel	Federal University of Technology	Nigeria	28	25	321	11.46	12.84	11	17	1.57
Aigbavboa, Clinton	University of Johannesburg	South Africa	22	17	340	15.45	20.00	8	18	0.89
Kineber, Ahmed Farouk	Prince Sattam Bin Abdulaziz University	Saudi Arabia	9	7	116	12.89	16.57	5	9	1.67
Olofinnade, Oluwarotimi	Covenant University	Nigeria	8	4	51	6.38	12.75	3	7	0.60
El-Sayegh, Sameh	American University of Sharjah	United Arab Emirates	7	5	125	17.86	25.00	4	7	1.33
Mymrin, Vsevolod	Universidade Tecnológica Federal do Paraná	Brazil	6	5	29	4.83	5.80	3	5	0.60
Aliu, John	University of Georgia	United States	6	5	35	5.83	7.00	2	5	0.50
Zavadskas, Edmundas Kazimieras	Vilniaus Gedimino Technikos Universitetas	Lithuania	5	5	246	49.20	49.20	4	5	0.29
Zuo, Jian	University of Adelaide	Australia	5	5	325	65.00	65.00	5	5	0.42
Rydin, Yvonne	University College London	United Kingdom	5	4	46	9.20	11.50	3	5	0.18
Shen, Li-yin	Hangzhou City University	China	5	4	289	57.80	72.25	3	5	0.20
Shafiq, Nasir	Universiti Teknologi PETRONAS	Malaysia	5	5	45	9.00	9.00	3	5	0.75
Radziejowska, Aleksandra	AGH University of Krakow	Poland	5	2	3	0.60	1.50	1	1	0.11
Khan, Kaffayatullah	King Faisal University	Saudi Arabia	5	5	43	8.60	8.60	4	5	0.80
Kingsley, Chukwuma	Federal University of Technology	Nigeria	5	5	102	20.40	20.40	5	5	1.67
Ismail, Zulhabri	Universiti Teknologi MARA	Malaysia	5	5	48	9.60	9.60	4	5	0.31
Joshua, Opeyemi	Covenant University	Nigeria	5	4	36	7.20	9.00	3	5	0.43
Deifalla, Ahmed Farouk	Future University	Egypt	5	5	23	4.60	4.60	3	4	1.50
Avanci, Monica A.	Universidade Tecnológica Federal do Paraná	Brazil	5	5	29	5.80	5.80	3	5	0.60

Notes: TP = total number of publications; NCP = number of cited publications; TC = total citations; C/P = average citations per publication; C/CP = average citations per cited publication; h = h-index; g = g-index; m = m-index.

Table 7.
Most productive authors with a minimum of five publications.

exhibit high citation metrics, indicative of the far-reaching impact and international recognition of their research.

The bibliometric indices, such as the h-index, g-index, and m-index, serve to further qualify the research output. The h-index reflects the number of publications each author has that have been cited at least h times, providing a measure of productivity coupled with impact [34]. The g-index offers an additional layer, acknowledging authors whose works have been highly cited, and the m-index adjusts the h-index to account for the number of years the author has been active, offering a rate of impactful productivity. These metrics not only highlight individual authors' contributions but also potentially reflect the research focus and academic vigor of their respective institutions. The data thus provide a multidimensional understanding of authorial impact in the realm of sustainable construction, highlighting leading voices and key contributors whose work is shaping the evolution of the field.

4.4 Publications by institutions

In the domain of sustainable construction studies, institutional contributions play a pivotal role in advancing research and disseminating knowledge. **Table 8** presents a comprehensive enumeration of the most productive institutions that have published a minimum of seven scholarly articles. This quantitative evaluation is based on several bibliometric indicators including the total number of publications (TP), total citations (TC), number of cited publications (NCP), average citations per publication (C/P), average citations per cited publication (C/CP), and indices such as the h-index, g-index, and m-index, which collectively offer a nuanced view of an institution's scientific influence.

The University of Johannesburg in South Africa leads in terms of publication count, indicating a significant concentration of research activity within the region. However, when considering the impact through citation metrics, the Hong Kong Polytechnic University stands out, with a notably high number of average citations per publication. This illustrates not only the volume of research output but also the degree to which the academic community values the research.

The indices presented in the table offer further insights into the qualitative impact of the research. The h-index and g-index suggest that universities such as the University of Johannesburg, the Federal University of Technology Akure, and Universiti Teknologi PETRONAS have a substantial number of publications that have been frequently cited, reflecting a consistent recognition of their research contributions. The m-index, which adjusts the h-index for the academic age of an institution, provides additional context on the speed and extent to which these publications have been cited, with the Federal University of Technology Akure showing a particularly high m-index, indicative of swift and widespread citation uptake. This data underscores the diverse and global nature of research in sustainable construction, highlighting not only the prolific academic output but also the far-reaching impact and engagement that these institutions have within the broader research community.

4.5 Publications by countries

The bibliometric assessment presented in **Table 9** and **Figure 3** elucidates the global dissemination and impact of research in sustainable construction studies,

Institution	Country	TP	TC	NCP	C/P	C/CP	*h*	*g*	*m*
University of Johannesburg	South Africa	30	362	21	12.07	17.24	9	19	1.00
Federal University of Technology Akure	Nigeria	24	218	22	9.08	9.91	8	14	1.33
Universiti Teknologi PETRONAS	Malaysia	19	169	15	8.89	11.27	8	13	0.73
Covenant University	Nigeria	18	115	13	6.39	8.85	5	10	0.50
Universiti Teknologi Malaysia	Malaysia	17	248	13	14.59	19.08	7	15	0.50
Hong Kong Polytechnic University	China	15	813	13	54.20	62.54	9	15	0.50
Universiti Teknologi MARA	Malaysia	15	110	9	7.33	12.22	6	10	0.46
University of Hong Kong	China	13	253	10	19.46	25.30	7	13	0.35
Universiti Sains Malaysia	Malaysia	13	443	9	34.08	49.22	6	13	0.43
University of Moratuwa	Sri Lanka	12	218	8	18.17	27.25	4	12	0.50
American University of Sharjah	United Arab Emirates	11	129	7	11.73	18.43	4	11	0.33
University College London	England	9	101	6	11.22	16.83	3	9	0.13
Vilnius Gediminas Technical University	Lithuania	9	351	9	39.00	39.00	7	9	0.50
Delft University of Technology	Netherlands	9	174	6	19.33	29.00	5	9	0.23
Loughborough University	England	9	247	9	27.44	27.44	8	9	0.38
RMIT University	Australia	9	112	6	12.44	18.67	4	9	0.22
Technical University of Košice	Slovakia	8	37	4	4.63	9.25	3	6	0.27
Visvesvaraya National Institute of Technology	India	8	353	6	44.13	58.83	4	8	0.31
University of Minho	Portugal	8	74	5	9.25	14.80	3	8	0.18
University of New South Wales	Australia	8	163	8	20.38	20.38	5	8	0.36
Chongqing University	China	7	223	5	31.86	44.60	3	7	0.23

Institution	Country	TP	TC	NCP	C/P	C/CP	*h*	*g*	*m*
King Faisal University	Saudi Arabia	7	102	7	14.57	14.57	5	7	0.63
University of Technology	Iraq	7	961	6	137.29	160.17	4	7	0.25
Prince Sattam Bin Abdulaziz University	Saudi Arabia	7	87	5	12.43	17.40	3	7	1.00

Notes: TP = total number of publications; NCP = number of cited publications; TC = total citations; C/P = average citations per publication; C/CP = average citations per cited publication; h = h-index; g = g-index; m = m-index.

Table 8.
Most productive institutions with a minimum of seven publications.

as measured by the scholarly output of various countries. The analysis employs a suite of bibliometric indicators to gauge the scientific contributions of each country, which collectively offer a multifaceted view of research influence and reach. India emerges as the most prolific contributor with 140 publications, followed by the United Kingdom, United States, Malaysia, and China, indicating a vibrant and diverse geographic distribution of research activity within the domain. However, the United Kingdom and the United States lead in citation metrics, suggesting that while India has a substantial output, the research from the United Kingdom and United States exerts greater influence on the academic community, as reflected in higher citation counts.

The h-index, which measures both the productivity and citation impact of the publications, shows the United Kingdom and United States are leading, with indices of 21 and 18, respectively, suggesting that a substantial number of publications from these countries have been cited frequently. The g-index, which gives more weight to highly cited articles, reinforces this pattern, indicating that the United States and United Kingdom have produced influential research that has garnered substantial attention. The m-index, a measure of the scientific "maturity" or speed at which a country's research is cited, further nuances this landscape. Malaysia stands out in this regard, with an m-index of 1.29, indicating rapid recognition and citation of its research contributions in the field.

4.6 Publications by source titles

Table 10 delineates the scholarly journals and conference series that serve as the most active venues for the dissemination of research in sustainable construction. This table offers an in-depth look at the publishing landscape by listing source titles that have a minimum of eight publications within the field, along with a suite of bibliometric indicators that reflect the reach and influence of the research published within them. "Sustainability (Switzerland)" emerges as the most prolific source title, with 76 publications to its name and a substantial citation count of 953, signifying its central role in shaping discourse in sustainable construction. This is further underscored by an h-index of 18, indicating that a significant number of articles in this journal have been frequently cited, reflecting both the quality and the impact of the research published.

Country	TP	TC	NCP	C/P	C/CP	*h*	*g*	*m*
India	140	1083	80	7.74	13.54	16	32	0.89
United Kingdom	108	1609	83	14.90	19.39	21	40	0.78
United States	100	1781	76	17.81	23.43	18	42	0.69
Malaysia	98	1321	72	13.48	18.35	18	36	1.29
China	88	1169	62	13.28	18.85	19	34	0.79
Australia	70	2200	56	31.43	39.29	21	46	1.11
Nigeria	55	429	46	7.80	9.33	12	20	1.20
South Africa	50	1135	38	22.70	29.87	14	33	0.50
Saudi Arabia	40	467	33	11.68	14.15	12	21	0.75
Hong Kong	36	1145	27	31.81	42.41	14	33	0.70
Indonesia	32	100	18	3.13	5.56	5	10	0.56
Italy	31	289	21	9.32	13.76	8	17	0.42
Germany	30	102	16	3.40	6.38	7	10	0.37
Brazil	25	283	17	11.32	16.65	8	16	0.33
Egypt	24	240	21	10.00	11.43	9	15	0.56
Portugal	21	234	14	11.14	16.71	8	15	0.47
Poland	21	478	13	22.76	36.77	7	21	0.37
Pakistan	19	156	16	8.21	9.75	6	12	1.00
United Arab Emirates	19	241	15	12.68	16.07	7	15	0.58
Canada	18	98	10	5.44	9.80	5	9	0.19
Turkey	17	198	12	11.65	16.50	5	14	0.42
Spain	17	168	14	9.88	12.00	7	12	0.44
Iran	16	181	14	11.31	12.93	8	13	0.67
Ghana	15	247	13	16.47	19.00	7	15	0.70
South Korea	14	157	12	11.21	13.08	7	12	0.30
Netherlands	14	230	10	16.43	23.00	7	14	0.32
Sri Lanka	13	237	8	18.23	29.63	4	13	0.50
France	12	128	7	10.67	18.29	5	11	0.31
Greece	12	51	3	4.25	17.00	3	7	0.14
Russian Federation	12	77	8	6.42	9.63	5	8	0.50
New Zealand	12	290	8	24.17	36.25	6	12	0.35
Switzerland	11	255	8	23.18	31.88	5	11	0.26
Lithuania	10	365	10	36.50	36.50	8	10	0.57
Belgium	10	128	9	12.80	14.22	3	10	0.13

Notes: TP = total number of publications; NCP = number of cited publications; TC = total citations; C/P = average citations per publication; C/CP = average citations per cited publication; h = h-index; g = g-index; m = m-index.

Table 9.
Top countries with a minimum of 10 publications.

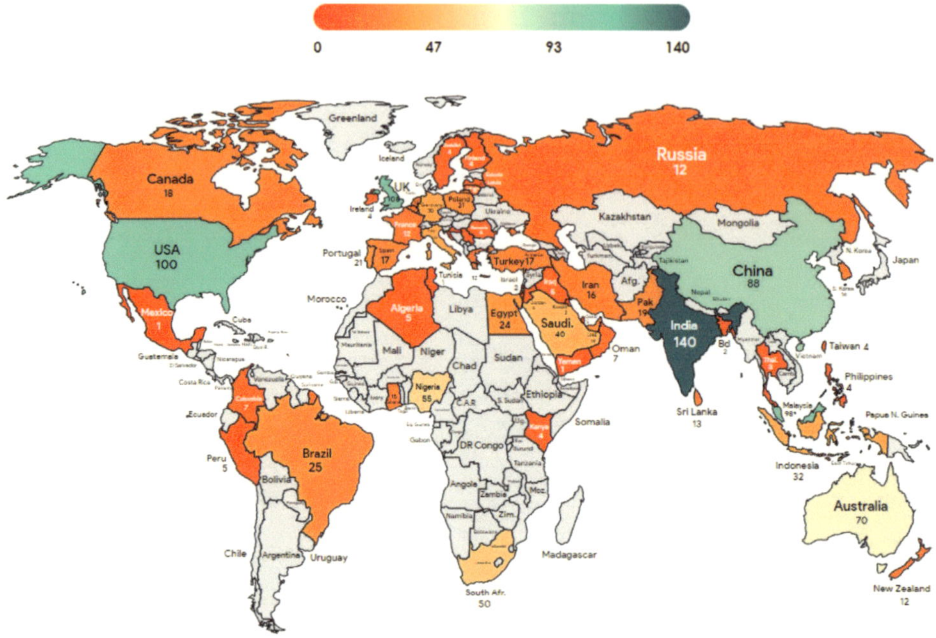

Figure 3.
Scientific country production of publications in sustainable construction studies.

Other noteworthy source titles include "Lecture Notes in Civil Engineering" and "IOP Conference Series: Earth and Environmental Science," which, while having a lower citation count, signify important platforms for emerging research and early-stage dissemination of innovative ideas in the field. The "Journal of Cleaner Production" and "Construction and Building Materials" stand out not only for their publication count but also for their high citation metrics, which include average citations per publication (C/P) and per cited publication (C/CP). These metrics, alongside robust h-indices, indicate that these journals are pivotal in hosting research that significantly advances the field. The table also reveals the dynamism within the field, showcasing a variety of platforms that cater to diverse research interests and methodological approaches, from the materials sciences to economic and policy considerations in sustainable construction. The inclusion of multiple conference series highlights the importance of academic conferences as forums for the exchange of cutting-edge research and collaboration among scholars.

4.7 Highly cited documents

Table 11 encapsulates a bibliometric analysis of the top 20 highly cited articles within the domain of sustainable construction. This analysis not only highlights the academic impact of these seminal works but also reflects the thematic directions that have garnered significant attention in the field. The table ranks the articles by the total number of citations they have received, further contextualized by the average citations per year, offering a temporal perspective on their sustained relevance. At the apex of this list is Ding's [2] pivotal work on the role of environmental assessment tools in sustainable construction, published in the Journal of Environmental Management. This article has resonated profoundly within the academic community,

Source Title	TP	NCA	NCP	TC	C/P	C/CP	*h*	*g*	*m*
Sustainability (Switzerland)	76	323	61	953	12.54	15.62	18	27	2.57
Lecture Notes in Civil Engineering	39	118	13	21	0.54	1.62	2	2	0.25
IOP Conference Series: Earth and Environmental Science	32	111	16	69	2.16	4.31	5	7	0.45
Journal of Cleaner Production	26	110	26	1167	44.88	44.88	18	26	1.80
Construction and Building Materials	26	100	22	1037	39.88	47.14	15	26	1.15
IOP Conference Series: Materials Science and Engineering	23	68	17	79	3.43	4.65	5	7	0.71
Materials Today: Proceedings	20	64	18	191	9.55	10.61	8	13	2.00
International Journal of Construction Management	16	63	15	208	13.00	13.87	9	14	0.69
Buildings	15	65	13	229	15.27	17.62	7	15	0.64
Construction Management and Economics	15	41	15	1231	82.07	82.07	11	15	0.41
Procedia Engineering	15	39	15	682	45.47	45.47	10	15	0.91
Advanced Materials Research	14	38	11	61	4.36	5.55	4	7	0.31
AIP Conference Proceedings	13	45	6	13	1.00	2.17	2	3	0.25
World Construction Symposium	9	25	3	7	0.78	2.33	2	2	0.40
Sustainable Construction Materials and Technologies	9	23	5	32	3.56	6.40	3	5	0.27
Smart and Sustainable Built Environment	9	34	8	115	12.78	14.38	5	9	1.25
Journal of Building Engineering	8	30	8	123	15.38	15.38	5	8	1.25
Proceedings of International Structural Engineering and Construction	8	26	2	2	0.25	1.00	1	1	0.13
Portugal SB 2007 - Sustainable Construction, Materials and Practices: Challenge of the Industry for the New Millennium	8	17	4	12	1.50	3.00	2	3	0.12
Construction Innovation	8	28	7	278	34.75	39.71	4	8	0.27
E3S Web of Conferences	8	20	2	2	0.25	1.00	1	1	0.20

Notes: TP = total number of publications; NCA = Number of contributing authors, NCP = number of cited publications; TC = total citations; C/P = average citations per publication; C/CP = average citations per cited publication; h = h-index; g = g-index; m = m-index.

Table 10.
Most active source titles with a minimum of 8 publications.

No.	Authors	Title	Source	Cites	Cites per Year
1	Ding [2]	Sustainable construction-The role of environmental assessment tools	Journal of Environmental Management	802	50.13
2	Robichaud and Anantatmula [14]	Greening project management practices for sustainable construction	Journal of Management in Engineering	414	31.85
3	Hill and Bowen [5]	Sustainable construction: Principles and a framework for attainment	Construction Management and Economics	395	14.63
4	Panda et al. [17]	Anisotropic mechanical performance of 3D printed fiber reinforced sustainable construction material	Materials Letters	386	55.14
5	Hager et al. [18]	3D Printing of Buildings and Building Components as the Future of Sustainable Construction?	Procedia Engineering	314	39.25
6	Raut et al. [15]	Development of sustainable construction material using industrial and agricultural solid waste: A review of waste-create bricks	Construction and Building Materials	309	23.77
7	Abidin [9, 10]	Investigating the awareness and application of sustainable construction concept by Malaysian developers	Habitat International	285	20.36
8	Jaillon and Poon [1]	Sustainable construction aspects of using prefabrication in dense urban environment: A Hong Kong case study	Construction Management and Economics	283	17.69
9	Tan et al. [19]	Sustainable construction practice and contractors' competitiveness: A preliminary study	Habitat International	224	17.23
10	Plessis [7]	A strategic framework for sustainable construction in developing countries	Construction Management and Economics	214	12.59
11	Pitt et al. [8]	Toward sustainable construction: Promotion and best practices	Construction Innovation	199	13.27
12	Rohracher, [23]	Managing the technological transition to sustainable construction of buildings: A socio-technical perspective	Technology Analysis and Strategic Management	183	7.96
13	Zimmermann et al. [6]	Benchmarks for sustainable construction: A contribution to develop a standard	Energy and Buildings	178	9.37
14	Gan et al. [13]	Why sustainable construction? Why not? An owner's perspective	Habitat International	158	17.56

No.	Authors	Title	Source	Cites	Cites per Year
15	Tang et al. [6]	Advanced progress in recycling municipal and construction solid wastes for manufacturing sustainable construction materials	Resources, Conservation and Recycling: X	132	33.00
16	Djokoto et al. [11]	Barriers to sustainable construction in the Ghanaian construction industry: Consultants perspectives	Journal of Sustainable Development	132	13.20
17	Alsanad [12]	Awareness, Drivers, Actions, and Barriers of Sustainable Construction in Kuwait	Procedia Engineering	122	13.56
18	Rajendran and Gambatese [20]	Development and initial validation of sustainable construction safety and health rating system	Journal of Construction Engineering and Management	119	7.93
19	Colangelo et al. [38]	Life Cycle Assessment (LCA) of different kinds of concrete containing waste for sustainable construction	Buildings	117	19.50
20	Kariyawasam and Jayasinghe, [16]	Cement stabilized rammed earth as a sustainable construction material	Construction and Building Materials	117	14.63

Table 11.
Top 20 highly cited articles.

as evidenced by its 802 citations, averaging 50.13 citations per year, indicating its foundational influence on the research that followed.

Robichaud and Anantatmula's [14] examination of greening project management practices and Hill & Bowen's [5] treatise on sustainable construction principles are also notably influential, with citations reflecting their enduring impact over the years. These works have served as cornerstones in defining and advancing the concepts and practices of sustainable construction. The inclusion of Panda et al.'s [17] research on 3D printed fiber-reinforced materials highlight the innovative technological advances in the field. Concurrently, Hager et al.'s [18] and Raut et al.'s [15] contributions underscore the growing interest in 3D printing technologies and the development of sustainable materials from industrial and agricultural waste. Zainul Abidin's [9, 10] investigation into Malaysian developers' awareness and application of sustainable construction concepts and Jaillon and Poon's [1] case study on prefabrication in Hong Kong's urban environment emphasize the geographical diversity and practical applications of sustainable practices in construction.

The table further includes works that delve into the competitive advantages of sustainable construction, barriers to its adoption in different regional contexts, and frameworks for its promotion. For instance, Du Plessis [7] offers a strategic framework for sustainable construction in developing countries, reflecting a nuanced understanding of the challenges and opportunities unique to these contexts.

Each entry within this bibliometric compilation presents a significant contribution to the literature, collectively charting the intellectual evolution of sustainable construction. The articles serve as benchmarks for current research and offer a repository of knowledge for academia and industry, guiding future studies and informing sustainable practices. This assemblage of highly cited works therefore not only narrates the history of sustainable construction research but also sets the agenda for its future trajectory.

4.8 Conceptual structure of the literature

The conceptual structure of sustainable construction literature embodies a complex network of interrelated themes and evolving paradigms. This segment aims to meticulously unravel the intricate fabric of this scholarly domain, applying rigorous bibliometric methodologies to elucidate the thematic cores and peripheries that characterize the field. The examination is set against a backdrop of multifaceted research interests, ranging from material innovation to policy formulation, each thread contributing to the rich tapestry of sustainable construction discourse. Through the analytic lenses of co-occurrence, trend analysis, and multidimensional mapping, this section will provide a structured synthesis of the literature, reflecting the depth and breadth of academic endeavors that define and propel the study of sustainability within the construction industry. The ensuing analysis is designed not only to catalog the existing body of knowledge but also to identify emergent trends and potential lacunae, thereby offering a scaffold upon which future research can robustly build.

4.8.1 Thematic clusters in sustainable construction literature

The bibliometric analysis presented in the findings underpins a comprehensive examination of thematic clusters in the sustainable construction literature, highlighting the intricate relationships between various research themes and concepts within this field. This analysis, facilitated by VOSviewer, a tool renowned for its capability to visualize bibliometric networks, allows for exploring co-occurrences of author keywords in the literature (Jan [39]). The methodology and findings from this study provide a structured insight into the prevailing research themes, their interconnections, and their evolution over time, contributing significantly to the body of knowledge in sustainable construction.

In the context of bibliometric studies, co-occurrence analysis involves examining how often pairs of items, such as keywords, appear together in a set of documents [40]. VOSviewer employs this analysis to map and visualize the relationships and clusters among items based on their co-occurrences. This approach offers an intuitive understanding of how research themes are interconnected, indicating areas of concentrated study and potential gaps in the literature. The application of VOSviewer's co-occurrence analysis to author keywords reveals the thematic structure and dynamics within the sustainable construction domain, highlighting the collaboration and focus areas in the research community.

The network visualization in **Figure 4** was generated using the Full counting method, with a minimum threshold of five occurrences for keyword inclusion out of 2339 identified keywords in sustainable construction literature. This method ensures that only the most relevant and frequently discussed topics are included, enhancing the visualization's clarity and focus by assessing keyword significance without fractionalization.

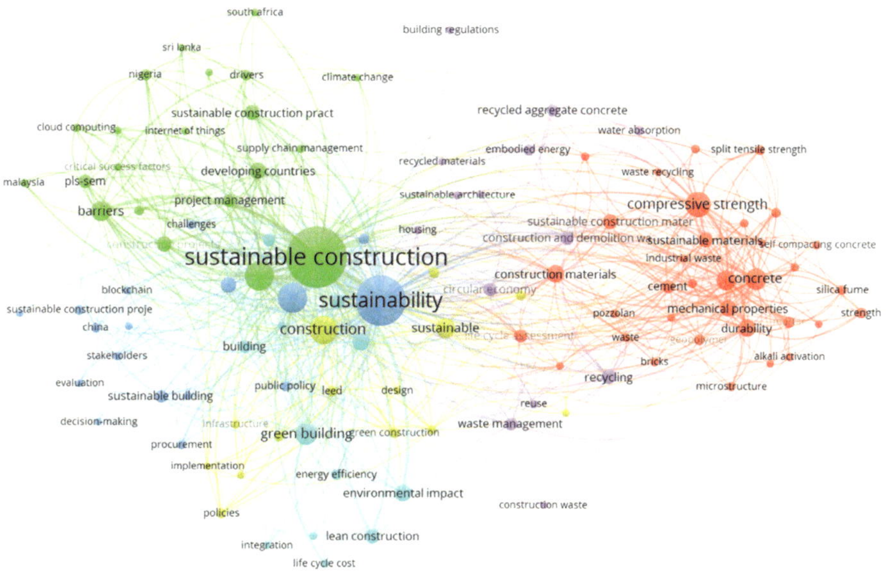

Figure 4.
Network visualization of the author's keywords https://app.vosviewer.com/?json=https://drive.google.com/uc?id=1EMLXGY0T7KPT6On6ldDGfXuWaBk4hDwU.

The analysis delineated six primary themes within the sustainable construction literature, as evidenced by the clustering of keywords. These themes encompass a broad range of topics, each contributing unique insights into the field of sustainable construction:

- **Materials and Technologies**: This theme includes keywords such as "aggregate," "alkali activation," "bamboo," and "cement," reflecting the research focus on sustainable materials and innovative construction technologies. The exploration of alternative materials like bamboo, fly ash, and geopolymer concrete signifies the pursuit of reducing environmental impact while enhancing material properties and efficiency.

- **Sustainability and Environmental Impact**: Keywords such as "sustainability," "sustainable construction," "environmental impact," and "energy efficiency" highlight the core objective of sustainable construction research: to mitigate environmental impacts through efficient use of resources and adoption of sustainable practices. This theme underscores the interdisciplinary nature of sustainable construction, integrating principles of environmental science, engineering, and architecture.

- **Management and Practices**: Encompassing "construction management," "project management," "risk management," and "supply chain management," this theme addresses the operational aspects of implementing sustainable construction practices. It reflects the growing emphasis on optimizing project delivery, enhancing productivity, and adopting best practices for sustainability.

- **Policy and Regulation**: With keywords like "building regulations," "policies," and "public policy," this theme emphasizes the role of legislative frameworks and policies in shaping sustainable construction practices. It highlights the importance of aligning construction activities with regulatory standards to promote sustainability.

- **Performance and Evaluation**: Keywords such as "compressive strength," "life cycle assessment," and "energy efficiency" pertain to the assessment of building materials and structures in terms of their sustainability performance. This theme focuses on developing and applying metrics and methodologies for evaluating the environmental and operational performance of construction materials and projects.

- **Innovation and Future Directions**: Featuring "blockchain," "building information modelling (BIM)," and "internet of things (IoT)," this theme illustrates the forward-looking research into incorporating advanced technologies into sustainable construction. It explores how digitalization and technological innovations can enhance efficiency, transparency, and sustainability in construction projects.

This bibliometric analysis enhances the body of knowledge in sustainable construction by providing a structured overview of the research landscape, identifying key themes, and highlighting interdisciplinary connections. By visualizing thematic clusters and analyzing keyword co-occurrence, it uncovers potential research avenues and serves as a road map for future studies, emphasizing the need for integrated approaches from materials science, environmental studies, technology, and management to address sustainability challenges in construction.

4.8.2 Evolution of author's keywords

The overlay visualization map in **Figure 5** uses a rainbow color scheme to illustrate the temporal evolution of research themes in sustainable construction based on author keyword co-occurrences. This technique enhances bibliometric analysis

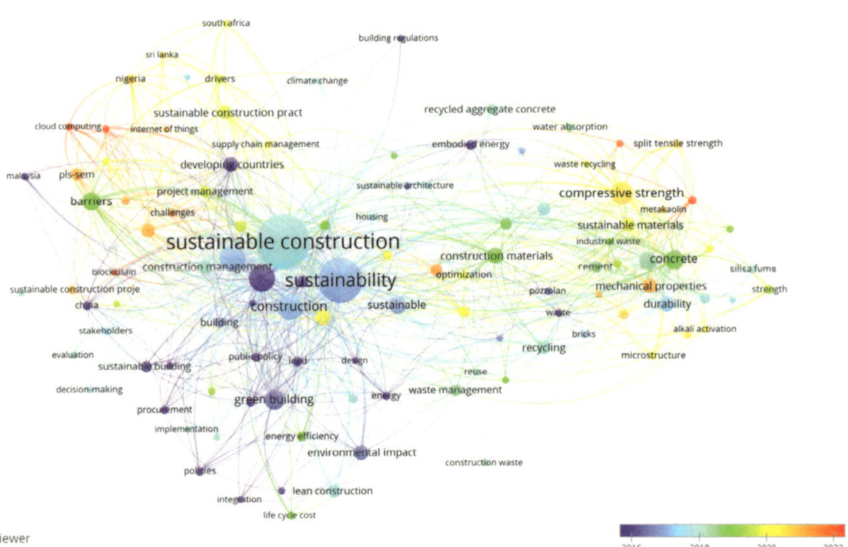

Figure 5.
Overlay visualization of the author's keywords.

by showcasing interconnectivity and thematic clusters, providing a dynamic perspective on the chronological development, shifting research focus, and emerging trends in the field.

Overlay visualization maps extend the utility of co-occurrence analysis by incorporating an additional dimension of data—time, in this case—onto the existing network of keywords [40]. In **Figure 5**, the rainbow color scheme signifies the research timeline, with colors transitioning from one end of the spectrum to the other (typically, from red, representing earlier works, to violet, indicating more recent publications). This color-coded strategy allows for the immediate visual identification of the temporal distribution of research efforts across different themes, providing insights into the progression and current frontiers of the field.

The rainbow color scheme in the overlay visualization map provides a nuanced view of temporal shifts in sustainable construction research. Keywords shifting from red to violet signify the transition from foundational concepts to recent advancements and emerging sustainability concerns, illustrating the field's trajectory from established practices to novel technologies.

1. **Emergence of New Materials and Technologies**: Keywords in the violet to blue range, such as "geopolymer concrete," "building information modelling (BIM)," and "internet of things (IoT)," reflect the recent surge in exploring advanced materials and digital technologies. This shift underscores the field's progression toward integrating cutting-edge solutions to enhance sustainability and efficiency in construction.

2. **Sustained Focus on Sustainability and Environmental Impact**: The presence of keywords across the entire color spectrum related to "sustainability," "environmental impact," and "energy efficiency" indicates a consistent and enduring emphasis on reducing the environmental footprint of construction activities. The evolution of these topics, from foundational to current research, underscores the ongoing refinement and expansion of sustainability concepts in construction.

3. **Shifts in Management Practices and Policy**: Keywords transitioning from warmer to cooler tones, such as "construction management," "supply chain management," and "public policy," illustrate the evolving understanding of how management practices and regulatory frameworks influence sustainable construction. The shift toward more recent research in these areas signals a growing recognition of the complexity of implementing sustainable practices within existing systems.

4. **Future Directions and Innovations**: The overlay visualization highlights the dynamic nature of the field, with recent keywords in cooler tones pointing toward emerging areas of interest, such as "circular economy," "carbon capture," and "smart cities." These trends indicate a forward-looking approach to addressing sustainability challenges through innovative concepts and interdisciplinary integration.

The overlay visualization map in **Figure 5** significantly contributes to sustainable construction literature by offering a comprehensive, temporal perspective on the field's evolution, clarifying current research, and forecasting future priorities. This dynamic tool guides researchers, policymakers, and practitioners toward important emerging areas, enriching the discourse on sustainable construction and encouraging ongoing adaptation and advancement to meet evolving sustainability challenges.

4.8.3 Trend topics

The analysis of trend topics within the sustainable construction literature, as depicted in **Figure 6**, is based on the frequency and temporal distribution of author keywords over a set period, utilizing bibliometric tools such as Biblioshiny. This analysis enables the identification of the most salient topics and their trajectories within the field. Through the frequency of keywords and their median occurrence year, the evolution of research interest and the potential future direction of the field can be discerned.

The data presented provides a longitudinal view of the focus areas within sustainable construction research from the early 2000s. It is evident from the data that certain themes have persisted over time, while others have emerged more recently, reflecting shifts in the field's dynamics and responses to global sustainability challenges.

1. **Persistent Themes**: Keywords such as "public policy" and "sustainable building" that emerged around 2006 and had their median occurrence around 2010 to 2013 demonstrate a sustained interest over an extended period. This indicates a continued recognition of the fundamental role that policy and construction practices play in achieving sustainability goals.

2. **Evolving Themes**: The median occurrence of keywords such as "sustainable development" and "environment" shifted toward the latter half of the 2010s, suggesting an increased focus on the broader implications of construction activities. This evolution signifies a deepening of the sustainability discourse, acknowledging the complex interplay between construction practices and environmental stewardship.

3. **Emerging Themes**: Recent keywords with median occurrences post-2020, such as "compressive strength," "building information modelling (BIM)," "circular economy," and "blockchain," reflect developing areas of interest that align with the latest technological advancements and conceptual shifts toward a more holistic understanding of sustainability. This trend highlights an inclination toward innovation and the adoption of new methodologies to enhance sustainability in construction.

4. **Future Trajectories**: The presence of keywords like "circular economy," "blockchain," and "structural equation modelling" with median occurrences in 2023 suggests that future research may pivot toward these areas. It indicates a trajectory where the construction industry may increasingly embrace circular economy principles, leverage blockchain for transparency and efficiency, and utilize advanced statistical methods like structural equation modeling to understand the interrelations among various sustainability factors better.

The trend topics analysis complements the previous thematic cluster analysis by adding a temporal dimension that showcases the evolution of research focus areas.

Figure 6.
Trend topics based on the author's keywords.

While the thematic cluster analysis reveals the landscape of sustainable construction research at a structural level, the trend topics analysis elucidates how this landscape changes over time. The two analyses provide a comprehensive picture of the field's progression, offering insights into the current state and potential future research directions. This figure serves as a map guiding researchers, policymakers, and industry professionals through the evolving terrain of sustainable construction research. It is a visual chronicle of the field's past, a mirror reflecting its present, and a compass pointing toward future research horizons. By understanding these trends, stakeholders can better align their practices with the cutting edge of sustainability, ensuring that the built environment contributes positively to society and the ecosystem.

4.8.4 Word cloud analysis

Figure 7, showcasing a word cloud visualization of the top 50 author's keywords, provides a visually engaging representation of the frequency and prominence of terms within the sustainable construction literature. The word cloud is a graphical depiction where the size of each term is proportional to its frequency or importance in the dataset, offering an immediate sense of which concepts are most dominant within the corpus of analyzed documents [41].

The word cloud analysis indicates "sustainable construction" as the most frequent keyword, with a remarkable frequency of 271. This term's prominence underscores the central focus of the field on developing construction practices that are not only efficient and economical but also environmentally conscious and socially responsible. "Sustainability" follows with a frequency of 186, reinforcing the overarching theme of the research domain. It signifies the multidimensional aspect of sustainability encompassing economic, environmental, and social factors, which are integral to

Figure 7.
Word cloud visualization of the top 50 author's keywords.

the discourse in sustainable construction. Other keywords like "sustainable development," "construction industry," "construction," and "compressive strength" indicate the breadth of focus areas within the field. They represent the diverse aspects of research, from industry-wide considerations to specific material properties. Further to that, these keywords from the word cloud are categorized below:

1. **Material Innovation**: Keywords such as "concrete," "fly ash," "recycled aggregate concrete," and "geopolymer" highlight an active area of research focused on innovative materials that offer sustainable alternatives to traditional construction materials. The prominence of these terms suggests a significant research interest in finding solutions to reduce the carbon footprint and enhance the recyclability of construction materials.

2. **Technological Advancements**: The term "building information modelling" (BIM), with a notable frequency, represents the digital transformation of the construction industry, emphasizing the importance of information management and technology in enhancing the sustainability of construction projects.

3. **Methodological Approaches**: "Analytical hierarchy process" and "multi-criteria decision making" indicate the utilization of sophisticated decision-making frameworks in the field. These methodological tools are crucial for evaluating and prioritizing various aspects of sustainability in construction projects.

4. **Environmental Concerns**: The presence of terms like "environmental impact," "life cycle assessment," and "circular economy" reflects an acute awareness within the research community of the environmental considerations that must be integrated into sustainable construction practices.

5. **Global and Local Perspectives**: The mention of specific locations such as "developing countries," "Nigeria," and "China" reveals a geographical dimension to the research, acknowledging that sustainable construction challenges and solutions may vary significantly across different regions.

This word cloud visualization captures the essence of the sustainable construction literature and provides a snapshot of the key terms that are currently shaping research in the field. This representation aids in the comprehension of the vast body of literature, providing a quick reference point for the most influential concepts and potentially guiding future research directions. As a result, it serves as both a reflective and a prospective tool, offering insights into the current state of research and suggesting pathways for future inquiry.

4.8.5 Thematic map

Figure 8 presents a thematic map of the author's keywords, which is a strategic tool in bibliometric analysis, particularly useful for evaluating the structure and evolution of academic fields. This map is generated from the data provided by Biblioshiny and is instrumental in classifying the identified keywords into four distinct quadrants based on two dimensions: centrality and density. Centrality reflects a keyword's strength of the association with other keywords in the network, while density represents the internal strength of the cluster to which the keyword belongs [29].

The thematic map categorizes keywords into four quadrants:

1. **Motor Themes** (High Centrality, High Density): These are well-developed and pivotal themes within the field that are strongly related to other research areas. In the context of sustainable construction, "sustainable construction," "sustainability," and "sustainable development" are identified as motor themes, which indicates that they are central, well-established areas of research that have a strong interconnection with other topics.

2. **Niche Themes** (Low Centrality, High Density): These themes are highly specialized and developed but relatively isolated from other research areas. Keywords such as "optimisation," "bamboo," and "industrial waste" fall into this quadrant, suggesting they are specific, in-depth areas of study within sustainable construction that have not yet permeated broadly across the field.

3. **Emerging or Declining Themes** (Low Centrality, Low Density): These are either new areas of interest that have not yet formed a cohesive body of research or older themes that are losing prominence. "Microstructure" is an example of such a theme, which may indicate an emerging interest that could develop further or a once-important topic that is waning in relevance.

4. **Basic Themes** (High Centrality, Low Density): These are fundamental to the field's development and have potential for growth, but the research is not yet well-developed. "Compressive strength," "concrete," and "fly ash" are seen as basic themes, which are foundational to the construction research area and have the potential to develop into more cohesive themes.

The thematic map's illustration provides profound insights into the research landscape of sustainable construction, particularly delineating the central and peripheral topics that shape the field's scholarly discourse. The Motor Themes quadrant, wherein the terms "sustainable construction," "sustainability," and "sustainable development" are prominently placed, reflects the comprehensive and foundational nature of these concepts. Their position within the quadrant is indicative of their extensive influence and integration across various subdomains of sustainable construction research, signifying their role as cornerstone themes that not only have spawned extensive

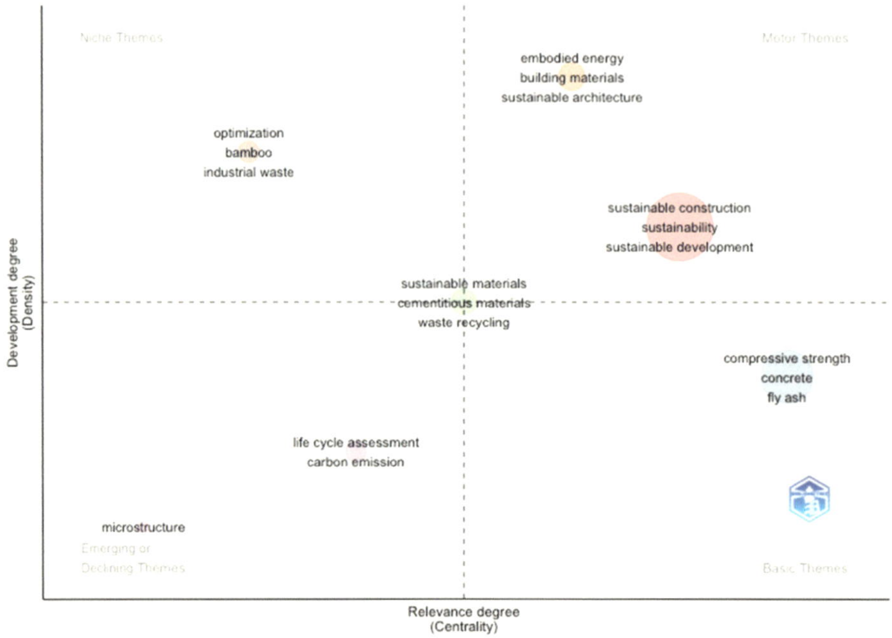

Figure 8.
Thematic map of the author's keywords.

research streams but also have established strong linkages with a multitude of other research areas. Conversely, the Niche Themes quadrant contains specialized topics such as "optimisation" and "bamboo," which, despite their considerable research depth and the sophistication of the scholarly work they represent, appear to have a delimited impact on the field at large. This suggests that these areas, while being of significant academic interest and depth, have not yet permeated the broader sustainable construction research community to the extent that they transform or significantly influence prevailing research paradigms.

In the Basic Themes quadrant, the inclusion of "compressive strength," "concrete," and "fly ash" suggests these topics are fundamental to the field's ontology. These themes, by virtue of their essential nature in the practice and theory of sustainable construction, harbor the potential to further coalesce and mature into more cohesive research clusters. As the discipline evolves, it is anticipated that these foundational topics will increasingly interlink, thereby amplifying their centrality and density within the research network. Lastly, the thematic map denotes a dynamic research terrain through the Emerging or Declining Themes quadrant. This segment of the map is particularly telling of the field's fluidity, capturing themes that are at a pivotal juncture—like "microstructure"—which may either ascend to prominence due to their increasing relevance or wane as they become peripheral to the field's central concerns. This dynamic interplay underscores the evolving nature of sustainable construction research, where themes continually adapt and transform in response to the field's internal developments and external societal demands.

The thematic map offers a nuanced perspective on the structure of the sustainable construction field, providing researchers with a visual guide to the relative development and interconnectivity of various research themes. It aids in identifying which areas have the most influence on the field, which are highly specialized, which are nascent or declining, and which are fundamental but underdeveloped. This understanding is crucial for researchers, funding bodies, and policymakers to make informed decisions about where to allocate resources, what areas to prioritize, and which topics might require more interdisciplinary collaboration to enhance their development and centrality within the field.

4.8.6 Co-word analysis

Figure 9 presents a visual depiction of a co-word analysis using multiple correspondence analysis (MCA), a methodological approach often employed in bibliometrics to uncover and interpret the complex patterns of relationships among terms within a body of literature. This form of analysis is particularly adept at revealing the multidimensional structure of the data by positioning the keywords within a conceptual space delineated by two principal dimensions.

The analysis is generated from a dataset wherein each keyword is plotted based on its coordinates along two dimensions, which are mathematical abstractions derived from the patterns of keyword co-occurrences. The positioning of keywords in this multidimensional space allows for the interpretation of the closeness and distance between terms, where proximity suggests a higher degree of association or co-occurrence within the literature, and distance implies lesser relation or independence. Keywords such as "sustainable construction," "sustainability," "sustainable development," and "construction industry" share proximal spaces, suggesting these terms are frequently discussed in conjunction within the literature, thereby forming a thematic cluster indicative of a significant research stream in the field of sustainable

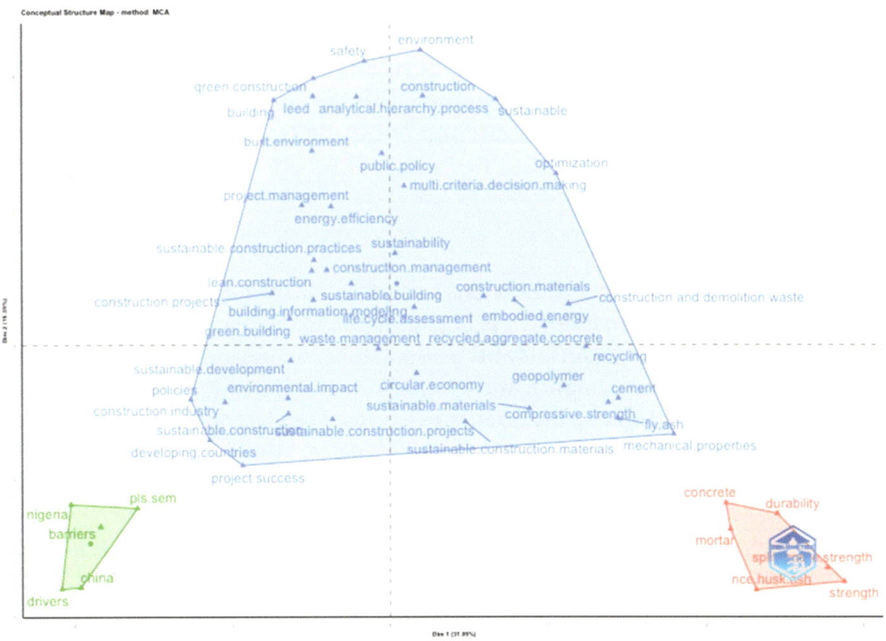

Figure 9.
Co-word analysis of the author's keywords using multiple corresponding analysis.

construction. The central clustering of these terms within the first quadrant signifies a robust, interconnected body of research that is likely driving much of the discourse in this domain.

Conversely, terms like "concrete," "durability," "mechanical properties," and "fly ash" appear further apart from the central cluster, particularly aligned along the second dimension, indicating these are distinct but related constructs within the field. Their positioning suggests that while they are relevant to the discourse on sustainable construction, they may represent a more technically focused strand of research, possibly related to the material science aspects of sustainability in construction. Notably, certain terms such as "barriers," "pls.sem," "nigeria," and "drivers" are located peripherally in the third cluster, potentially signifying niche or emerging areas within the sustainable construction research that are not as densely connected to the central themes. Their relative isolation from the core cluster may reflect the novelty of these topics within the field or may indicate areas of research that are currently less integrated into the mainstream discourse but may hold potential for future development.

Multiple correspondence analysis, as demonstrated in **Figure 9**, thus provides a nuanced and data-driven way to visualize and understand the landscape of sustainable construction research. By mapping out the relational dynamics of keywords, MCA enables researchers to identify not only the dominant themes within the literature but also the extent to which these themes are interrelated, thereby offering a comprehensive overview of the field's intellectual structure. This methodological approach enriches the bibliometric toolkit by offering a sophisticated means to navigate the complex terrain of academic research, facilitating the identification of core, peripheral, and potentially emerging domains of inquiry.

5. Discussion

The bibliometric analysis presented in the results section offers a comprehensive landscape of sustainable construction studies. This discussion section interprets these findings, considering their implications for practice and suggesting directions for future research.

5.1 Implications for practice

The significant increase in publications and citations over the years suggests that the sustainable construction industry is becoming increasingly pivotal. The consistent thematic emphasis on materials and technologies, management and practices, policy and regulation, performance and evaluation, and innovation indicates that the industry should continue to invest in these areas. For instance, the persistent focus on materials like bamboo and geopolymer concrete underscores the necessity for continued development and implementation of alternative, sustainable construction materials. Moreover, the importance of management practices in sustainable construction highlighted in the literature calls for enhanced training and knowledge dissemination among construction managers [14]. The focus on policies and regulations indicates that practitioners should be well-versed in the latest legislative changes and understand their implications for everyday practice. Additionally, the assessment of building materials' performance suggests a need for industry-wide standards to evaluate sustainability metrics effectively.

DOI: http://dx.doi.org/10.5772/intechopen.1005619

5.2 Recommendations for future research

The field of sustainable construction stands at a crossroads of technological innovation and environmental stewardship, with research playing a crucial role in steering the industry toward a greener future. As the bibliometric analysis indicates a surge in academic interest, it is imperative that future research endeavors are channeled to address both contemporary challenges and anticipate emerging trends.

Longitudinal Analysis of Evolving Themes: Future research should use longitudinal studies to trace the evolution of sustainable construction themes, identify patterns, and predict future trends. These studies could provide valuable insights into integrating emergent technologies like blockchain for resource tracking and circular economy principles for material reuse and recycling, potentially revolutionizing the construction industry.

Interdisciplinary Approaches to Sustainability: Sustainable construction requires a collaborative approach, integrating disciplines like materials science, environmental science, technology, and management. Future research should promote interdisciplinary studies to develop holistic solutions, such as creating low-carbon materials and ensuring their effective integration into construction projects.

Innovative Methodologies and Technologies: Future research should explore how technologies like Building Information Modeling (BIM), augmented reality, and 3D concrete printing can enhance sustainability in construction by improving precision, reducing waste, and enabling prefabrication. Additionally, studies should examine the life cycle impacts of construction projects to integrate sustainability from design through to end-of-life.

Global and Local Sustainability Practices: Research must adopt a global approach while acting locally to address the diverse challenges of sustainable construction across different geographical and cultural contexts. Comparative studies should analyze regional sustainability practices to understand how local factors influence success and identify adaptable best practices for various contexts.

Impact Assessment of Sustainable Practices: Future research should develop robust metrics and methodologies to assess the environmental, social, and economic impacts of sustainable construction practices quantitatively and qualitatively. This includes creating comprehensive life cycle assessment models that account for both the environmental impacts and the social and economic benefits.

The research community must stay vigilant and proactive as the construction industry evolves, exploring ways to make sustainability both practical and economically viable. Findings from this research will shape policies, guide industry practices, and educate future generations of construction professionals.

6. Conclusion

The conclusion of a scholarly paper serves to encapsulate the study's findings and implications, offering a cohesive synthesis and proposing directions for future scholarly and practical pursuits. This paper's comprehensive bibliometric analysis of sustainable construction literature from 1996 to 2024 presents a nuanced understanding of the field's evolution, key contributors, and thematic concentrations. The analysis has demonstrated a substantial growth in sustainable construction research, underscored by an ascending trajectory of publications and citations. It revealed that the field is expanding not only in terms of quantity but also in the depth and influence of its academic contributions, as evidenced by robust bibliometric indices such as the h-index, g-index, and m-index. These indices attest to a body of literature that is both prolific and profoundly impactful, resonating across the academic community and informing practice.

The research further identified a global dispersion of contributions, with a notable predominance of English-language publications, reflecting the international reach and accessibility of the discourse on sustainable construction. The diversity of source types—from journals to conference proceedings—highlights the multifaceted platforms through which sustainable construction research is disseminated. A thematic exploration illuminated six primary clusters within the literature: materials and technologies, sustainability and environmental impact, management and practices, policy and regulation, performance and evaluation, and innovation and future directions. These clusters capture the breadth of sustainable construction research, encompassing technical, environmental, managerial, policy, and innovative aspects. The study's conceptual mapping and co-word analysis have delineated the prevailing research themes and have pointed out potential gaps and emerging trends. The multidimensional nature of sustainable construction research has been vividly captured through the utilization of advanced bibliometric tools, providing a clear visualization of the field's thematic structure and evolution.

The findings have significant implications for construction industry practices, suggesting a need for integrated approaches that consider environmental, technological, and managerial dimensions of sustainability. The analysis underscores the importance of innovative materials and construction methods that can minimize environmental impact and enhance efficiency. Policymakers and practitioners are encouraged to draw upon this research to guide strategic decisions and invest in areas identified as having high relevance and potential impact. The emphasis on policy and regulation within the literature suggests that legislative frameworks must evolve to support the adoption of sustainable practices and technologies.

The paper has highlighted several areas ripe for future research, including the necessity for longitudinal studies that can track and predict the evolution of sustainability themes, and interdisciplinary collaborations that can address sustainability challenges from multiple angles. In conclusion, this bibliometric analysis has provided a comprehensive and systematic exploration of sustainable construction research. The study reaffirms the dynamic and expansive nature of the field, signaling a continuing trajectory of growth and influence. The insights garnered here serve not only as a testament to the field's academic maturity but also as a beacon for future scholarly endeavors and industry practices. As the world increasingly prioritizes sustainability, the role of research in sustainable construction becomes ever more pivotal, steering us toward a future where the built environment harmonizes with ecological imperatives, economic demands, and social equity.

Disclosure statement

The authors reported no potential conflict of interest.

Declaration of generative AI And AI-assisted technologies

While preparing this work, the authors used ChatGPT to do the paraphrasing. After using this tool/service, the authors reviewed and edited the content as needed and took full responsibility for the content of the publication.

DOI: http://dx.doi.org/10.5772/intechopen.1005619

Author details

Khululekani Ntakana[1*] and Aidi Ahmi[2]

1 Department of Construction Management and Quantity Surveying, University of Johannesburg, South Africa

2 Tunku Puteri Intan Safinaz School of Accountancy, Universiti Utara Malaysia, Malaysia

*Address all correspondence to: kntakana@uj.ac.za

References

[1] Kariyawasam KKGKD, Jayasinghe C. Cement stabilized rammed earth as a sustainable construction material. Construction and Building Materials. 2016;**105**:519-527. DOI: 10.1016/j.conbuildmat.2015.12.189

[2] Gan X et al. Why sustainable construction? Why not? An owner's perspective. Habitat International. 2015;**47**:61-68. DOI: 10.1016/j.habitatint.2015.01.005

[3] Det Udomsap A, Hallinger P. A bibliometric review of research on sustainable construction, 1994-2018. Journal of Cleaner Production. 2020;**254**(120073). DOI: 10.1016/j.jclepro.2020.120073

[4] Panda B, Paul C, Tan MJ. Anisotropic mechanical performance of 3D printed fiber reinforced sustainable construction material. Materials Letters. 2017;**209**:146-149. DOI: 10.1016/j.matlet.2017.07.123

[5] Jan van Eck N, Waltman L. Manual for VOSviewer Version 1.6.17. Universiteit Leiden; 2021

[6] Zimmermann M, Althaus HJ, Haas A. Benchmarks for sustainable construction: A contribution to develop a standard. Energy and Buildings. 2005;**37**(11 SPEC. ISS):1147-1157. DOI: 10.1016/j.enbuild.2005.06.017

[7] Robichaud LB, Anantatmula VS. Greening project management practices for sustainable construction. Journal of Management in Engineering. 2011;**27**(1):48-57. DOI: 10.1061/(asce)me.1943-5479.0000030

[8] Raut SP, Ralegaonkar RV, Mandavgane SA. Development of sustainable construction material using industrial and agricultural solid waste: A review of waste-create bricks. Construction and Building Materials. 2011;**25**(10):4037-4042. DOI: 10.1016/j.conbuildmat.2011.04.038

[9] Abidin NZ. Investigating the awareness and application of sustainable construction concept by Malaysian developers. Habitat International. 2010a;**34**(4):421-426. DOI: 10.1016/j.habitatint.2009.11.011

[10] Ammar A et al. Synergies between construction 4.0 technologies and sustainable construction: A bibliometric analysis. In: Proceedings of the Creative Construction e-Conference. 2022. pp. 365-370. DOI: 10.3311/ccc2022-046

[11] Hager I, Golonka A, Putanowicz R. 3D printing of buildings and building components as the future of sustainable construction? Procedia Engineering. 2016;**151**:292-299. DOI: 10.1016/j.proeng.2016.07.357

[12] Chadegani AA et al. A comparison between two main academic literature collections: Web of science and scopus databases. Asian Social Science. 2013;**9**(5):18-26. DOI: 10.5539/ass.v9n5p18

[13] Jaillon L, Poon CS. Sustainable construction aspects of using prefabrication in dense urban environment: A Hong Kong case study. Construction Management and Economics. 2008;**26**(9):953-966. DOI: 10.1080/01446190802259043

[14] Tang Z et al. Advanced progress in recycling municipal and construction solid wastes for manufacturing sustainable construction materials.

Resources, Conservation and Recycling: X. 2020;**6**(March):100036. DOI: 10.1016/j.rcrx.2020.100036

[15] Tan Y, Shen L, Yao H. Sustainable construction practice and contractors' competitiveness: A preliminary study. Habitat International. 2011;**35**(2):225-230. DOI: 10.1016/j.habitatint.2010.09.008

[16] Maier D. The use of wood waste from construction and demolition to produce sustainable bioenergy—A bibliometric review of the literature. International Journal of Energy Research. 2022;**46**(9):11640-11658. DOI: 10.1002/er.8021

[17] Rajendran S, Gambatese JA. Development and initial validation of sustainable construction safety and health rating system. Journal of Construction Engineering and Management. 2009;**135**(10):1067-1075. DOI: 10.1061/(asce)0733-9364(2009)135:10(1067)

[18] Jan van Eck N, Waltman L. VOSviewer Manual Version 1.6.15. Universiteit Leiden; 2020

[19] Wang K, Guo F. Towards sustainable development through the perspective of construction 4.0: Systematic literature review and bibliometric analysis. Buildings. 2022;**12**(10). DOI: 10.3390/buildings12101708

[20] Rohracher H. Managing the technological transition to sustainable construction of buildings: A socio-technical perspective. Technology Analysis and Strategic Management. 2001;**13**(1):137-150. DOI: 10.1080/09537320120040491

[21] Fayad AAS et al. Towards concise reporting through integrated reporting: A bibliometric review. Meditari Accountancy Research. 2023;**32**(3):832-856. DOI: 10.1108/MEDAR-10-2021-1470

[22] Pitt M et al. Towards sustainable construction: Promotion and best practices. Construction Innovation. 2009;**9**(2):201-224. DOI: 10.1108/14714170910950830

[23] Tjebane MM et al. Challenges for the implementation of sustainable construction practices in developing countries: A bibliometric review. In: Skatulla S, Beushausen H, editors. Proceedings of the 19th International Conference on Computing in Civil and Building Engineering (ICCCBE). Switzerland: Springer; 2022. pp. 109-123

[24] Adebowale OJ, Agumba JN. A bibliometric analysis of sustainable construction practices—Implication on construction productivity. In: Haupt TC et al., editors. Proceedings of the 15th Built Environment Conference. Switzerland: Springer; 2022a. pp. 181-194

[25] Barrett PS, Sexton MG, Green L. Integrated delivery systems for sustainable construction. Building Research and Information. 1999;**27**(6):397-404. DOI: 10.1080/096132199369237

[26] Colangelo F et al. Life cycle assessment (LCA) of different kinds of concrete containing waste for sustainable construction. Buildings. 2018;**8**(5). DOI: 10.3390/buildings8050070

[27] Zakaria R, Ahmi A, Ahmad AH, Othman Z, Azman KF, Ab Aziz CB, et al. Visualising and mapping a decade of literature on honey research: A bibliometric analysis from 2011 to 2020. Journal of Apicultural Research. 2021;**60**(3):359-368. DOI: 10.1080/00218839.2021.1898789

[28] Moher D, Liberati A, Tetzlaff J, Altman DG, Prisma Group. Preferred reporting items for systematic reviews and meta-analyses: The PRISMA statement. PLoS Medicine. 2009;**6**(7):e1000097

[29] Hill RC, Bowen PA. Sustainable construction: Principles and a framework for attainment. Construction Management and Economics. 1997;**15**(3):223-239. DOI: 10.1080/014461997372971

[30] Punj N, Ahmi A, Tanwar A, Abdul Rahim S. Mapping the field of green manufacturing: A bibliometric review of the literature and research frontiers. Journal of Cleaner Production. 2023;**423**(138729):1-22. DOI: 10.1016/j.jclepro.2023.138729

[31] Plessis C. Du. 'A strategic framework for sustainable construction in developing countries'. Construction Management and Economics. 2007;**25**(1): 67-76. DOI: 10.1080/01446190600601313

[32] Abdul Rahman NA, Ahmi A, Jraisat L, Upadhyay A. Examining the trend of humanitarian supply chain studies: Pre, during and post COVID-19 pandemic. Journal of Humanitarian Logistics and Supply Chain Management. 2022a;**12**(4):594-617. DOI: 10.1108/JHLSCM-01-2022-0012

[33] Alsanad S. Awareness, drivers, actions, and barriers of sustainable construction in Kuwait. Procedia Engineering. 2015;**118**:969-983. DOI: 10.1016/j.proeng.2015.08.538

[34] Ding GKC. Sustainable construction-the role of environmental assessment tools. Journal of Environmental Management. 2008;**86**(3):451-464. DOI: 10.1016/j.jenvman.2006.12.025

[35] Ahmi A. biblioMagika. 2023. Available from: https://aidi-ahmi.com/index.php/bibliomagika

[36] Ahmi A. OpenRefine: An approachable tool for cleaning and harmonizing bibliographical data. In: 11th International Conference on Applied Science and Technology 2022 (11th ICAST 2022) in Istanbul, Turkey. AIP Conference Proceedings. Vol. 2827. AIP Publishing; 2023b. DOI: 10.1063/5.0164724

[37] Aria M, Cuccurullo C. Bibliometrix: An R-tool for comprehensive science mapping analysis. Journal of Informetrics. 2017;**11**(4):959-975

[38] Djokoto SD, Dadzie J, Ohemeng-Ababio E. Barriers to sustainable construction in the ghanaian construction industry: Consultants perspectives. Journal of Sustainable Development. 2014;**7**(1):134-143. DOI: 10.5539/jsd.v7n1p134

[39] Katuu S. Converting qualitative data into quantitative survey instruments: A detailed guide. In: Stacey A, editor. Proceedings of the 18th European Conference on Research Methodology for Business and Management Studies. ACPIL: Jhannesburg, South Africa; 2019. pp. 201-210. Available from: www.academic-bookshop.com/

[40] Lacasse MA. Materials and technology for sustainable construction. Building Research and Information. 1999;**27**(6):405-408. DOI: 10.1080/096132199369246

[41] Mansour AZ et al. Discovering the global landscape of fraud detection studies: A bibliometric review. Journal of Financial Crime. 2022;**29**(2):701-720. DOI: 10.1108/JFC-03-2021-0052

Chapter 20

The Energy Transition as a Portal to Exploring the Justice Dimensions of Global Sustainability

Lorelei Hanson

Abstract

In this chapter I examine a sociotechnical system deemed central to achieving global sustainability, renewable energy. Through a focus on Canadian policy, I explore carbon reduction through fuel substitution and electrification, common pathways being promoted to advance an energy transition. By combining a transformative climate justice framework with an examination of embodied energy injustices, I centre an interrogation of power and injustice to examine the existing and magnified impacts of low-carbon pathways, particularly on peoples and communities that are marginalized. I argue that the problems of injustice encountered across the energy transition are endemic to achieving sustainability overall, and that we need to ground our ambitious commitments to address climate change and global sustainability in considerations of justice and injustice if we are to make headway on addressing these crises.

Keywords: energy transition, global sustainability, transformative climate justice, embodied energy injustice, renewables, equity, power

1. Introduction

Late 2023, The United Nations University (UNU) and the Global Systems Institute (GSI) with the University of Exeter, each publicly released reports that identify how multiple life-system Earth processes are perilously close to risk tipping points. "A risk tipping point is the moment at which a given socioecological system is no longer able to buffer risks and provide its expected function, after which risk of catastrophic impacts to these systems increases substantially" ([1], p. 4). The UNU identifies 25 interconnected tipping points across Earth systems, five of which are already close to being crossed due to existing levels of climate change warming and three more threatened to occur in the 2030s as the global average temperature exceeds 1.5° Celsius. The GSI [2] emphasizes how existing governance systems institutions and decision-making processes are inadequately responding to the level of threat posed by these negative tipping points. They argue, "[t]he current approach of linear incremental change favoured by many decision makers is no longer an option" ([2], p. 3) and instead they need to adapt to facilitate transformational change. Similarly, the independent scientists who authored the 2023 UN Global Sustainable Development Report [3], insist that global sustainability will not be achieved through fragmented and incremental change, and business-as usual approaches. "The only way forward is to transform how we think, live, produce and consume in order to achieve a new equilibrium that balances resilience, security and well-being, and does so in harmony with nature" (p104).

Justice and equity are key to achieving global sustainability. As Inger Andersen, Executive Director of the United Nations Environment Programme, succinctly states: "Justice is an essential part of the environmental discussion" [4]. Such awareness informed the Brundtland Commission [5], which understood sustainable development as being "development that meets the needs of the present without compromising the ability of future generations to meet their own needs." Likewise, justice and equity have been integrated into the UN's Sustainable Development Goals by grounding them in three principles: a human rights-based approach, leave no one behind, and gender equality and women's empowerment [6].

Yet, in the translation and implementation, sustainability has been attenuated, obfuscated, and subverted. Too often understandings of sustainability, be they conceptual or translated into initiatives, strategies and actions, treat the larger sociopolitical power dynamics as incidental contingencies external to the challenges facing much of humanity, in order to allow business-as-usual to proceed. In the process, justice, equity and fairness are consigned as optional supplements that are narrowly defined or as matters to be considered after the fact, or by someone else, rather than being a guiding framework for advancing sustainability. While there is no way for a single initiative or activity to address the full range of sustainability challenges, routine considerations of justice and injustice are paramount to fundamentally addressing the multiple socioecological crises facing humanity. Technical solutions play a role in transitions, but too often are framed as 'the solution'; yet achieving sustainability requires transformative approaches that address the root causes of the problems, and enact new ways of doing, relating and being.

To illustrate this pattern of conceptual oversight as it pertains to sustainability, I'll interrogate some of the chains of injustice associated with common energy transition pathways, supply-side decarbonization policies focused on fuel substitution and electrification. While often necessary components of energy transition, these initiatives are far from sufficient to address the complexities of how energy is deeply integrated within modern sociotechnical extractive systems, not only nationally, but globally. Such attenuation not only is largely ineffective in achieving global sustainability and transition, but it frequently propagates injustices by exacerbating the already uneven and unequal burdens associated with climate change without adequately attending to the differential responsibilities for action [7, 8].

I undertake an integrative assessment of the local and global implications of energy transition policy choices by engaging with a transformative climate justice framework combined with an examination of embodied energy injustice. Combining these two frames of analysis assists in centring an interrogation of power and injustice in the critical analysis of low-carbon pathways. Questions of justice are relevant to most discussions related to the solutions and strategies used to address the impacts of climate change: from who gets what, to where the infrastructure will be located, for whom the impacts of both climate change and responses to it are most severely experienced, to whose knowledge counts and who gets to decide in these debates, and what are the sources, allocation, and prioritization of funding for mitigation and adaptation [9]. Making power a central focus in critically examining energy transition discourse, policy and actions directs attention to "the social and institutional relations and inequalities that both produce climate change and profoundly shape responses to it" ([10], p. 1). High-level climate and energy transition discourse tends to frame climate change as largely a scientific problem with a technocratic solution, and in so doing obscure both the highly political nature of these debates and the highly uneven impacts and patterns of exchange [10, 11].

To outline some of these situated justice effects, I examine the Canadian government's energy transition policies and programs. Key Government of Canada documents I utilized include: the Emission Reductions Plan (ERP) and supporting materials released by Environment and Climate Change Canada, the lead department for advancing the ERP, supporting documentation related to the ERP provided by the Canada Energy Regulator, the performance audit of the implementation of the ERP as compiled by the Auditor General of Canada, and related energy data compiled by Statistics Canada and Natural Resources Canada.

2. Transformative climate justice

In its most basic framing, the concept of climate justice has been utilized to highlight how climate change is having the most severe impacts on those who are the least responsible for the greenhouse gas emissions (GHGs) that destabilize the planet, and how these same individuals and communities are the most excluded from the decision-making processes about how to respond to climate change and its effects [10]. In short, climate change has uneven and unequal burdens [8]. Vulnerabilities to the impacts of climate change both reflect and exacerbate structural injustices driven by race, class, caste, ethnicity, gender, ableism and speciesism. The policies and programs adopted for climate change mitigation and adaptation can generate economic benefits and improved well-being but when not firmly rooted in principles of justice across the energy chain they often magnify injustices [12]. People and other beings in many parts of the world are already struggling with loss of livelihood, displacement, environmental degradation, pollution, natural resource depletion, food insecurity and habitat loss, and these circumstances are intensified by climate change. For those facing severe disadvantages, these are not future scenarios that may eventualize, but rather existing realities that are well past tolerable levels and beyond their capacities to adapt [13].

There are four pillars central to transformative climate justice: distributive, procedural, recognition and intergenerational. Distributive justice concerns issues related to who benefits and who loses, or who is disproportionately burdened because of projects, programs and policies meant to both mitigate and adapt to climate change. In a globally carbon-constrained context, distributive justice also raises issues about who is most responsible (current versus historical) and entitlement (whose needs are most pressing, and who decides who can emit how much?) [10]. Procedural justice focuses attention on the inclusion and exclusion of individuals and groups in processes for making decisions, and in ensuring that these processes are fair, accountable and transparent [10]. They are four key elements integrated within procedural justice [14]: "(1) access to information; (2) access to and meaningful participation in decision-making; (3) lack of bias on the part of the decision-makers; and (4) access to legal processes for achieving redress" (p437). Issues of procedural justice often arise in place-based and natural resource projects, and raise questions about free, prior and informed consent related to industrial development and resource allocation decisions [10]. Recognition justice is related to distributive and procedural justice, but it fills a gap that arises when the unequal power relations between different actors is not considered [15]. Recognition justice attempts to address this oversight by focusing on the differences between Indigenous, subaltern and other marginalized groups and the dominant society, and how these differences are embedded within and manifest in sociocultural and institutional historical and contemporary actions and processes. Grounded in human dignity, recognition justice draws attention to understanding differences alongside protection of equal rights for all by extending consideration for the cultural, social, economic, and political marginalization and discrimination some communities have had to and continue to endure [10]. An example is the recognition of the carbon debts associated with the colonial and racialized ways in which the exploitation of fossil fuel and mineral resources has caused myriad social, cultural and economic injustices for Indigenous communities, and often resulted in dispossession that undermines their land relations central to their culture and identity. Finally, intergenerational justice highlights the exacerbated and long-term injustices created by climate change and holds current decision-makers and polluters accountable for the impacts of their decisions and actions on the well-being of future generations. Future generations will not have the same opportunities to satisfy their own needs because of the decisions being made today and hence, will be confronted with increased risks and burdens due to escalating GHG levels [16].

Climate justice as both a concept and praxis has evolved from an examination of the impacts of climate change on disadvantaged communities to also consider the justice dimensions of energy transition to a post-carbon economy. Energy transition signifies a shift in an energy system from one dominant source of energy or set of energy resources to another [17, 18]. In popular discourse, the term 'just transition' signifies employment losses in fossil-fuel industries and the economic and social consequences of such labour disruptions on communities, regions and businesses supporting them, and decline in services and amenities reliant on these legacy industries [19]. While the discourse of energy transition is often framed as a solution to the social problems or worries of the present, these processes nevertheless produce winners and losers. Potential benefits such as new employment opportunities, involvement in decision-making processes, and access to advanced, low-carbon and efficient technologies are unevenly spread across populations and socio-economic groups [17]. Energy transitions also typically result in increased energy insecurity as the cost of energy rises, at least in the short- to medium-term, which disproportionately impacts rural, racialised, low-income, elderly and renting populations, as well as communities in the Global South. Hence, failures to adequately engage with the configurations of justice and injustice in energy transition can "lead to aggravated poverty, entrenched gender [and racial] bias[es], and non-participation as outcomes or by-products of 'blinkered' decision-making" ([18], p. 67).

Increasingly scholars and activists concerned about the justice dimensions of energy transition are exploring and applying the concept of energy justice. Sovacool and Dworkin [14] define energy justice as "a global energy system that fairly disseminates both the benefits and costs of energy services, and one that has representative and impartial energy decision-making" (p436). They elaborate that energy justice involves three key elements: costs or how the hazards are imposed on others; benefits and access, and; procedures for decision making or concerns with due process and representation. Sovacool and Dworkin's definition of energy justice draws attention to both local and global impacts, structures and processes, and inspection of how energy production has increasingly expanded the 'commodity frontiers' across borders, particularly in the form of industrialization and extractive activities in middle and low-income countries. Yet amongst policy makers, scholars and activists, energy justice is often distilled down to what some call the '3 A's': availability, accessibility and affordability [18]. As a part of this, energy justice advocates tend to conflate harms with injustices, not considering the differences between environmental harms experienced by affluent communities versus the injustices imposed on impoverished, historically marginalized communities [20].

Pursuit of a just and democratic energy transition requires consideration of all four pillars of justice (distributional, procedural, recognition and intergenerational) in alignment with a focus on "energy availability and access, affordability, due process, accountability and transparency and inter- and intra-generational equity" ([17], p. 570). Such analyses necessitate investigation of the contextual and intersectional axes of oppression and differentiation that operate across the full energy system assemblage [8]. The transition pathways promoted by national and international policies are largely embedded within concentrated patterns of existing economic and political power and can be expected therefore to systematically produce highly uneven and inequitable consequences within and across societies, as well as magnify existing injustices [10]. Hence, efforts to mitigate and adapt to the impacts from climate change will necessarily not only benefit some groups over others but constrain the pathways available, leading to maladaptation, and even violence, for others. As Newell and Adow [11] explain, "the technologies of control furnished in the colonial era, such as direct extraction and dispossession of land and livelihoods, are now getting replayed through climate colonialism in the form of green grabs and reducing emissions from deforestation and degradation (REDD) forest projects" (p33).

Understandings of environmental and climate justice are seen by many Indigenous and racialized scholars and activists to be bound to Western commitments to modernity and colonial exploitation, favoring particular sources of knowledge, justice and ways of being in the world [8, 21]. For example, distributive equity implies nature is a distributable good – resource, commodity, or externality – which is antithetical to many Indigenous cosmologies which centre around relational modes of being, living in relation to other humans and nonhumans in non-dominating and non-exploitive ways [7, 22]. Therefore, transposing Western concepts of justice, energy assemblages and transition frameworks to the global South, and to Indigenous and vulnerable communities across the global North, runs the risk not only of being maladaptive but of imposing additional burdens on human and more than human communities [7]. There are multiple understandings of what constitutes energy access and need, and numerous ways to configure the sociotechnical elements for energy provision and wellbeing with differential impacts on various populations and communities [23]. Consequently, valuing experiential and lived knowledge situated in specific places should be central in energy transition decision-making. Transformative climate justice makes the case for a plurality of climate knowledge and appropriate accommodation of this reality. It emphasizes dramatic shifts in decision-making processes that allow for a broader range of actors and voices, and different forms of knowledge and wisdom shaping energy transition and climate adaptation priorities [10, 24].

Energy transitions are technologically, socially and politically constituted; they "are composed of technologies, infrastructures supporting their use, specific business and financing models, social practices, and policy instruments and regulations" ([15], p. 1193). To interrogate "the social and institutional relations and inequalities which produce climate change and profoundly shape responses to it" ([10], p. 1) requires consideration of the full energy 'assemblages' of incumbent and emerging energy systems [23]. An energy assemblage encompasses "three interrelated dimensions: (1) the tangible elements of the energy system, which include technology, infrastructure, market, production equipment, consumption patterns and distribution chains; (2) actors and their conduct, which comprise new strategies and investment patterns, as well as changing coalitions and capabilities of actors; and (3) socio-technical regimes that contain formal regulations and policies, institutions as well as mindset and belief systems, discourse and views about normality and social practices" ([25], p. 46). Energy assemblages are neither static or deterministic, as they are "endowed with the capacity to act in different ways, depending on their configuration – agencies and arrangements are not separate" ([23], p. 167). Assemblage thinking thus draws attention to the plurality of energy system possibilities and multi-scalar operations of power that underlie them.

Combining transformative climate justice with an embodied energy injustice analysis allows for an interrogation of injustice and power rooted within energy assemblages and proposed energy transition pathways. "Transformative climate justice focuses on the need to disrupt dominant power relations and shift decision-making processes that lock in and reproduce injustice" [26]. To this end, transformative climate justice directs attention to 1) the need for "a stronger integration of social justice with the causes of and responses to climate change"; ([10], p. 2), and 2) linking climate change adaptation and mitigation by examining justice needs and causes related to climate change as they are being enacted across various scales and domains [10]. Those who advocate for transformative climate justice support a reduction in atmospheric carbon levels, but they also understand climate change as a symptom of a much larger crisis associated with modernity, colonialism and capitalism. An embodied energy injustice analysis helps reveal the hidden and distant injustices across the full energy assemblage and outline the chains of energy injustice. Consideration of embodied energy injustices provides a framework to situate and understand place-based injustices and link them to the broader dynamics of global capitalism [20].

3. Canada's emissions reduction plan

Canada is a nation richly endowed with diverse and vast energy resources compared to most other countries in the world. In 2021, Canada produced 4% of the global energy supply and ranked sixth in primary energy production [27]. Based on 2022 data, Canada was also sixth in global per capita electricity production, with Iceland, Norway and Bahrain being the top three producers. The sources for energy consumed in Canada include gas (31.0%), oil (30.2%), hydropower, (26.4%), nuclear (5.5%), coal (2.7%) and renewables (3.6%). Globally, the sources of energy consumed include oil (31.6%), coal, (26.7%), gas (23.5%), hydropower (6.7%), nuclear (4.0%), and renewables (6.8%) [28]. Fossil fuels are the source for 63.9% of the energy consumed in Canada but constitute only 17.4% of the sources for electricity, which compares globally to 81.8% for all energy consumed and 61.3% for electricity sources. As a comparison, in Sweden 26.3% of all energy consumed and 1.8% of the electricity consumed has a fossil fuel source [28].

Canada was a signatory to the Paris Agreement in 2015, an international commitment to reduce global greenhouse gases to keep the global average temperature increase below 2 degrees Celsius and preferably below 1.5 degrees Celsius compared to pre-industrial temperatures. In alignment with this international commitment, in 2021 the Government of Canada committed to reduce national GHG emissions to 40–45% below 2005 levels by 2030. That same year they also passed the Canadian Net-Zero Emissions Accountability Act to enshrine in legislation Canada's commitment to achieve net-zero greenhouse gas emissions by 2050. The Act also focuses on ensuring that the federal government's efforts are transparent and accountable by requiring the Minister of Environment and Climate Change to set emission targets and publish emission reduction plans and progress reports. The Act also requires the Commissioner of the Environment and Sustainable Development to report on the federal government's success in implementing actions to meet their GHG emission reduction targets [29]. To fulfill these commitments, in March 2022 the Minister of Environment and Climate Change released Canada's 2030 ERP, which serves as the cornerstone of the federal government's strategy to reduce national GHG emissions and which they declare to be "an ambitious and achievable roadmap" [30]. According to the Canada Energy Regulator [31], "Canada's energy transition is occurring in many different areas of the economy and is influenced by various technologies, policies, and market trends."

The ERP [32] states that the Government of Canada is taking bold action in its energy transition. In the Plan's overview, the federal government outlines its approach to achieving their climate goals by:

- helping to reduce energy costs for homes and buildings;
- empowering communities to tackle action by expanding the Low Carbon Economy Fund to incentive actions by the provinces and territories, municipalities, universities, colleges, schools, hospitals, businesses, not-for-profit organizations, communities and organizations;
- making it easier for Canadians to switch to electric vehicles;
- driving down carbon pollution from the oil and gas sector;
- powering the economy with renewable electricity to electrify more activities from heating and cooling buildings to various industrial processes;
- helping industries develop and adopt clean technology in their journey to net-zero emissions;
- investing in nature and natural climate solutions;
- supporting farmers as partners in building a clean, prosperous future; and
- maintaining Canada's approach for pricing pollution.

The Canadian government's strategy for achieving net-zero by 2050 is advanced through a host of federal laws, regulations and programs including: pricing carbon pollution on fossil fuel end use and industrial facilities; an oil and gas emissions cap and methane reduction regulations; a tax credit for investment in carbon capture and storage technology; a low carbon technologies investment fund to support the development and adoption of such technologies; clean electricity regulations aimed at decarbonizing electricity by phasing out carbon-emitting electricity generation and increasing non-emitting sources such as renewables, biofuels, and nuclear; sales mandates and incentives for electrification, increasingly stringent emission standards and fuel-switching in the transportation sector; higher efficiency regulations that apply to building design and energy technologies; reduction regulations to phase out of hydrofluorocarbon use; fertilizer emission reduction targets; and methane emission regulations for large industrial emitters and municipal landfills [33]. The federal government made some efforts to identify and accommodate groups that might be disproportionately burdened by measures in the plan. Consequently, they provide targeted funding to support both clean energy projects in the North and Indigenous communities, and public transportation in rural communities, and allow some exemptions from carbon pricing and targeted support for groups such as farmers, fishers and rural residents [29]. Decarbonization and supply-side energy management are the primary focus of the plan, but there are incentives targeted to energy efficiency, mostly related to retrofitting existing buildings. Given its emphasis on mitigation, the ERP largely reinforces the conventional mitigation-adaptation dichotomy, as demonstrated by the absence of any mention of adaptation in the Net-Zero Emissions Accountability Act [32]. The ERP is a complex policy that aligns with a projection model in which policies and measures incentivize fuel switching, primarily to electricity, greater use of biofuels and hydrogen and adoption of zero emissions vehicles to create 50.6% of total GHG emission reductions while allowing for national economic growth [32].

In outlining the major components of the ERP, my intention is not to provide a detailed evaluation of it, but rather to use it to illustrate how supply-side technological solutions and market mechanisms are emphasized as the appropriate tools to usher in an energy transition, with little consideration of the chains of global-local injustices that underlie particular energy and climate policy choices.

4. Adopting clean energy technologies to address climate change impacts

It will be challenging to achieve the goals of Canada's 2030 ERP for a variety of reasons. Canada is the second largest country in the world with a population of just over 40 million [34]. Overall, Canada has a relatively low-emitting electricity grid with 82.6% coming from low GHG emission sources such as renewables and nuclear, but Alberta, where much of Canada's oil, natural gas and bitumen deposits are located, as well as its neighbor Saskatchewan, still rely extensively on fossil fuels for electricity generation [31]. This is not a consequence of inadequate renewable resource endowments as both these provinces have some of the greatest potential to expand wind and solar energy production. The intransigence to a shift away from traditional energy sources is political. The transportation sector also presents significant challenges as most Canadians are driving more and larger vehicles, Canada's light duty vehicles produce the highest CO_2 emissions per kilometer in the world, and the EV market is novel and small [35].

Despite a rapid build out of renewable energy capacity and increased electric vehicle sales globally, Canada continues to lag behind many other nations in adopting clean energy technologies. In 2022, globally electric vehicle sales exceeded 10 million, constituting 14% of all new vehicle sales worldwide [36], whereas electric vehicle registrations in Canada were 3% [37]. Global renewable energy capacity rose by 13% in 2022 to 340 gigawatts [36] with record high investments of $1.3 trillion, a 70% increase from 2019 [38]. In Canada in 2022 there was a 10.5% increase in renewable energy investments and an installed capacity of renewables of 19 GW [39]. Canada's contribution to world renewable electricity production is 2% compared to China at 17% and Nigeria at 6% [27].

More significantly than its slow adoption of green technology, what impedes Canada's energy transformation is it carbon intensive economy, and its continued support for and economic reliance on mining activities, including the extraction of oil, gas and bitumen. Despite its promises to reduce national GHG emissions 40–45% below 2005 levels by 2030 and achieve net-zero greenhouse gas emissions by 2050, Canada's fossil fuel production continues to grow; in 2022, primary energy production increased in Canada by 3.9% [40]. Fossil fuels accounted for over 64% of per capita primary energy consumption in Canada in 2022, compared to solar, wind and other renewables at 3.6%, and nuclear and hydropower at 32% [41]. Similarly in a global context, though increases in energy demand are shifting to renewables and nuclear, in aggregate, "government plans and projections would lead to an increase in global coal production until 2030, and in global oil and gas production until at least 2050" [42]. Under the current measures in the ERP, Canada's oil production (crude oil plus natural gas liquids) will increase by 25% over 2022 levels by 2035 and remain roughly constant until 2050 [33]. Consequently, even if all the emission reduction measures in Canada's ERP are fully realized, the reductions will only reach 35% below the 2005 level, missing the federal government's 40–45% goal [29]. Despite claims made by federal government officials, such as the Canada Energy Regulator who declared that Canada's energy transition is occurring in many different parts of the economy, there is no transition yet. Renewables are expanding the choice of energy sources and accommodating some of increase in increased energy demand, but they are still not replacing fossil fuels [43].

Additionally, while energy consumption is falling in many high-income countries, not so in Canada. In 2002, energy consumption rose in Canada by 4.2% [40]. Alongside Norway, Iceland, the United States and countries in the Middle East like Oman, Saudia Arabia and Qatar, Canada has one of the highest levels of energy consumption per capita in the world; the average Canadian consumes more than 100 times as much the average person living in some lower-income countries [44]. This in part explains why Canada, as well as the United States, are the only G7 countries that have increased their GHG emissions since signing the Paris Agreement [45]. Canada being a dominant emitter of GHGs is a not only a current reality but a historical trend; a study of cumulative emissions from 1850 to 2021 reveals that Canada ranks tenth for the cumulative amount of CO_2 emitted by countries, and second for cumulative per capita emissions of CO_2 [46]. Hence, even using the most superficial definition of the term, the Canadian government is neither participating in nor concertedly enabling an energy transition. Further, they are not assuming adequate responsibility for their share of the climate change crisis thereby exacerbating distributive and intergeneration injustices. Instead, the continued growth in Canada's emissions continues to contribute disproportionately to global changes such as sea level rise, extreme weather events and temperature increases that are not only impacting people distant from its shores, but contributing to record wildfires and extreme temperatures that are affecting many in Canada, most acutely those who are Indigenous, marginalized, have existing health problems and live in remote communities.

Even if Canadian government could more quickly deploy renewables this does not side-step the multitude of injustices associated with relying almost exclusively on technologies and existing markets as solutions to climate change and its impacts. The reliance on green technologies and markets alone inadequately addresses distributional justice issues such as energy poverty and energy insecurity and how the deployment of renewable energy can exacerbate these conditions. According to Carley and Konisky [17], there is ample evidence that that energy costs are a significant financial burden for households across many high-income nations, particularly for those who are rural low-income, elderly, non-white and renting populations. Low-income and non-white households are more likely to live in energy inefficient dwellings, have inefficient appliances and poor structural building conditions, all of which results in more pronounced impacts from climate change than that experienced by other households. They note that these same populations also tend to have limited access to low-carbon and energy efficiency technology programs featuring rebates and incentives for low-emission and electric vehicles, residential solar panels, community solar, smart meters, efficient appliances and LED lightbulbs; it's mostly high-income households who are the beneficiaries of such green incentives.

The energy system changes in Canada's most populous province, Ontario, serves as an example of the possible inequitable burdens associated with renewable energy roll out in Canada. Between 2003 and 2014 Ontario phased out coal, introduced a demand-side management program, and encouraged less carbon-emitting electricity through introduction of suite of renewable energy programs and legislation, and smart grid development. From 2006 to 2016 electricity rates increased on average 19% [47]. The Province of Ontario introduced 40 initiatives to address energy poverty that temporarily lowered electricity bills, offered rebates, electricity bill credits, tax relief and emergency assistance aligned with costs of energy, inefficient housing or low income, but failed to acknowledge and respond to those who suffer from the cumulative costs of energy, inefficient housing *and* low incomes [47]. In other words, due to initiatives adopted to increase renewable electricity on the provincial grid, the burden for the most disadvantaged people in Ontario was magnified, thereby exacerbating existing injustices. It is important to note that energy is largely a provincial responsibility but as the federal government is trying to decarbonize energy grids across the country, they have a role in collaborating with and assisting assist provincial governments to transition their energy systems. The federal government is collaborating with its provincial counterparts to some extent on energy matters, such as dedicating $458.5 million for low interest loans and grants to housing providers to make affordable (i.e., low-income) housing more energy efficient [48]. However, only 3% of Canada's population live in low-income housing but 7.4% of the population live below the poverty line [49]. Also, Canada presently faces housing deficit and affordability challenges across most of the country, and most Canadians are experiencing higher-than normal cost of living increases (e.g., in 2022 there was a national 6.8% inflation rate) [50]. Hence, existing federal funding associated with the EPR dedicated to offsetting increased energy prices associated with renewable energy roll outs appears inadequate for those already facing economic disadvantages and will likely magnify the hardships they face.

Yet, for many Indigenous communities, renewable energy holds prospects for not only decarbonization but self-determination. The history of renewable energy developments in Canada, particularly large-scale hydro developments, are characterized by conflict and disregard for Indigenous ways of life, and largely are examples of neo-colonialism, extractive economics and dispossession [51]. Nevertheless, over the last couple of decades many Indigenous communities are taking advantage of federal and provincial renewable energy incentives and programs to install solar panels and wind turbines on their traditional lands and buildings to achieve a host of goals: take action on climate change by replacing diesel generators and other fossil-fuel based electricity sources with less carbon polluting sources; provide employment opportunities, particularly for youth; create an economic opportunity that helps financially support the community; address energy poverty in their communities; and build energy self-reliance to advance Indigenous sovereignty and self-determination [52, 53].

An embodied energy injustice perspective directs one to examine injustices across the full energy assemblage, and thereby account for burdens associated not only with consumption but also production in both proximate and distant contexts [20]. An emphasis on the wide-spread production and deployment of renewable energy technologies as key to energy transition means "[m]inerals and metals will play a key role in the transition to a significantly lower carbon future" ([54], p. xvi). In a 2017 study, the World Bank illustrated how technologies assumed to populate the clean energy shift - such as wind, solar, hydrogen and electricity systems - are more material intensive in their composition than current traditional fossil-fuel-based energy supply systems. While recycling and reuse will have a role in meeting future mineral demand, nevertheless as demand for renewable technologies increases so does demand for mining. Transition minerals, also termed critical minerals, are necessary for modern technologies such as renewable electricity, batteries and electric vehicles [55]. According to the International Energy Agency [56], the energy transition to renewables will dramatically increase demand for critical minerals by 4 to 6 times in 2040, compared to 2020. Currently mining of critical minerals is highly concentrated in several specific geographic regions and dominated by a few major companies who control a significant portion of the global production and trade of specific critical minerals. "For example, more than 70% of platinum is mined in South Africa; 70% of cobalt is mined in the Democratic Republic of Congo; more than 60% of natural graphite in China; almost 50% of nickel in Indonesia; almost 50% of lithium in Australia and almost 50% of dysprosium in China" ([57], p. 37). Mineral processing is even more concentrated, with China being the dominant player (100%) for minerals such as natural graphite and dysprosium (a rare earth mineral) and manganese (over 90%) [57]. While critical mineral reserves are more evenly distributed than production, low-income nations currently account for much of the global production of materials needed to fulfill national renewable projections, and their share in reserves is even greater. For example, Bolivia has 21 million tonnes of lithium reserves, more than any other country [57].

Given that most critical minerals do not have extraction and refining process supply chains well established yet, and critical mineral deposits are dispersed more evenly across the planet than traditional mineral and metal production, governments and industry organizations speak of the energy transition as offering future opportunities to low-income nations to diversify the mining sector [57]. Yet, the geological presence of resources alone is insufficient to undertake mining projects and effectively integrate them into the global critical mineral supply chain; it requires considerable upfront capital and expertise to locate, extract, process, transport and distribute these minerals. The global mining industry is highly concentrated, dominated by a few large multinational corporations and state owned/controlled enterprises, "yielding small and often oligopolistic markets" ([57], p. 15). Some of the largest mining companies are vertically integrated, operating across numerous stages of the commodity value chain and across multiple countries, controlling a significant portion of global production and trade in minerals and metals [57]. For example, at present most global mineral exploration is undertaken by countries that are members of the OECD, such as Australia, Canada and the United States [57]. Indeed, Canada is a mining superpower. With extensive domestic production and companies present in 96 foreign countries in 2021 [58], Canada produced:

> *more than 60 minerals and metals...ranking in the top five countries in the global production of 15 minerals and metals, including several critical minerals essential to new technologies, such as cobalt, copper, precious metals, nickel, and uranium. It has the potential to expand in lithium, magnesium and rare earth minerals [59].*

Canadian mining assets abroad totalled $188 billion in 2020 [60].

Mining has a complex legacy in Canada, both internationally and domestically. Mineral production has stimulated non-Indigenous exploration and settlement, investment in infrastructure such as roads, railways and power production, and development of secondary industries, businesses and services including health care and education. But mining operations have also left behind social, economic, and environmental ruin, including a toxic legacy of tailings ponds and waste rock dumps, ecological disturbance and loss, food and water contamination that affect nearby communities, and social and economic dislocation, particularly for Indigenous communities [61]. Mining is responsible for 10% of global GHGs [57], is the fourth largest driver of deforestation and negatively impacts biodiversity at multiple scales [62]. For many Indigenous communities throughout Canada, the advent of mining on their traditional lands has been "synonymous with the key moments of their colonial history, when outside forces brought major social and economic changes while compromising local relationships to the land-based economy" ([63], p. 266). "From an industry and government perspective, there is a prevailing sentiment that times have changed" and they have repaired socio-ecological relations; they have become sustainable stewards of natural resources "and the historical exclusion of Indigenous communities from the benefits of the mineral economy can be dismissed as simply a relic of the bad old days" ([63], p. 266). Certainly, Indigenous peoples in Canada have gained influence over regional decision-making that affects their territories [63]. Aboriginal rights are protected by the Constitution and governments must "consult and accommodate Indigenous people wherever a government agency contemplates conduct that may impact Aboriginal rights" [53]. Also, in 2021 the Canadian government agreed to abide by the United Nations Declaration on the Rights of Indigenous Peoples, which "requires Canada to involve Indigenous participation in all matters that concern them" and "mandates obtaining their free, prior and informed consent on these matters" [64]. Yet, such accountability and respect for Indigenous communities is not fully realized across Canada and does not necessarily apply to mining activities outside the boundaries of Canada. For example, a damning case has been made against OceanaGold Corporation, a gold mining and exploration company incorporated in British Columbia and with offices in Vancouver, Canada and Brisbane Australia. According to Mining Watch Canada [65], OceaniaGold have committed widespread human rights and environmental abuses, and several of their mining operations threaten endangered species, conservation lands, watersheds, and Indigenous communities. They are not alone. As International Human Rights lawyer, James Yap, explains:

> *The activity of ripping minerals and other resources from the Earth has three notable features: first, it is a highly complex and labour-intensive endeavour. Second, it causes massive disruption to the Earth and those living on it. Finally, it is capable of generating enormous wealth for those who stand to profit... When mining takes place in Canada, serious disputes and controversies definitely arise - ...mining has also caused a lot of human rights and environmental harm in Canada — and ... Indigenous peoples have borne the brunt of this. But in the relative sense, Canada*

> *does have strong laws and institutions that are reasonably effective at preventing at least the very worst kinds of abuses. ...[But] Canadian mining companies currently operate in all corners of the world. Regrettably, they've acquired a particularly bad reputation globally for causing serious human rights abuses [66].*

Given these hegemonic imperialist power relations, without a focus on justice the degree to which transition metals will enable low-income countries to address the inequities and violence related to the mining operations of high-income nations within their borders and fairly compete in the global mining economy seems limited.

Domestically, critical mineral deposits are located across Canada, with concentrated reserves in a few regions. One of the largest potential reserves is situated in northwest Ontario with a potential yield of $60 billion [67]. Known as the 'Ring of Fire', this crescent-shaped deposit of minerals spans almost 5000 square kilometers and is the traditional territory of more than a dozen Indigenous First Nations. Most of the mining claims fall within the Hudson Bay Lowlands, an area of carbon-rich peatlands that are an important global carbon sink and part of the world's second largest complex of wetlands. The vast peatlands of the Hudson Bay Lowlands support extensive river systems, provide critical habitat for wildlife, including endangered species like caribou and wolverines, and are an important cultural landscape and food security source for Indigenous peoples [68]. For many of the First Nations in this region, the subsistence economy accounts for a large proportion of their communities' livelihood as residents still rely heavily on wild meat like caribou, moose, Canada goose and lesser snow goose [67]. This is a remote landscape with no roads or transmission lines, and only accessible by ice roads in the winter, or by air [69]. The Premier of Ontario and other proponents argue that the minerals in the Hudson Bay Lowlands are key in the push to lower GHG emissions, but environmentalist and some First Nations argue that the costs of disturbing these peatlands outweighs any of the benefits [69]. If all the peatlands in the Ring of Fire were disturbed an estimated 130 to 250 megatonnes of carbon could be released [68].

Mining for transition minerals also can contribute to intergenerational and recognition injustices, including the displacement and dispossession of Indigenous communities from their traditional lands. Bolivia, for example, with its wealth of lithium reserves has a weak regulatory regime that results in a spotty human rights track record. Despite children under 14 not being technically allowed to work, the law is rarely enforced, and many children are employed in mining activities [70]. Further, while the 2009 Bolivian constitution includes comprehensive guarantees of Indigenous peoples' rights to land titling and free, prior, and informed consent on development projects, Indigenous peoples frequently face barriers in exercising those rights [71]. Globally, an estimated 54% of energy transition mineral deposits are located on or near Indigenous peoples' lands [57]. In the Hudson Bay Lowlands, there are more than 28,000 mining claims held by 14 companies and no consensus amongst the First Nations about these prospective developments [72]. Marten Falls and Webequie First Nations are leading an environmental assessment on proposed roads into the region, viewing them as access for their community members to health, education and employment services [72]. In contrast, three of the affected First Nations have issued a moratorium against projects proposed for the Ring of Fire region [64] and four First Nations - Grassy Narrows, Big Trout Lake, Wapekeka and Neskantaga - have signed a mutual cooperation agreement with the intent of stopping unwanted encroachments on their traditional territories, which cover about 60,000 square kilometers of land [73]. They are opposed to mining and prospecting on their lands because it puts their way of life under threat and assert that the Ontario provincial government has pushed forward on development in the region without free, prior and informed consent [73].

Whether in support of or opposed to energy transition through renewable energy, there is largely consensus amongst most Indigenous and settler communities of the need to be properly consulted about energy transition developments that impact their lives. Procedural justice is a cornerstone of transformative climate justice and highlights the fairness of, and equitable opportunities in energy decision-making processes [17]. Studies demonstrate that public participation in environmental decision making can provide important knowledge, inform policy and other solutions, and lead local residents to perceive the result of decisions more positively [74]. Nevertheless, "public participation and deliberation around questions of energy policy have traditionally been very weak" ([10], p. 5). Not only are there inadequate opportunities for participation in many energy decision- making processes, but there is a lack of complete information and access to scientific and legal expertise. At a deeper level, the fairness of these processes can reflect deeper procedural inequities between rich and poor nations, or rich and poor communities, and raises recognition justice issues about what constitutes an appropriate response to the energy challenges confronting societies.

In accordance with the Canadian Net-Zero Emissions Accountability Act, the Minister of Environment and Climate Change is required to provide opportunities for comment on the ERP. The federal government did this, to a limited extent. A survey portal was launched December 10, 2021 that remained open until January 21, 2022, and submissions by e-mail were also welcomed. The on-line portal received 15,788 responses and 14,380 e-mail submissions were received from the 38.25 million people living in Canada – 0.08% of the population [32]. The submission from the Assembly of First Nations (AFN) noted the very tight timelines for input and how this constrained their ability to conduct a full assessment of the proposed plan. Further, they stressed that the federal government needs to engage directly with First Nations in more meaningful conversations for subsequent versions or updates of the ERP.

Powatomi scholar and activist Kyle Whyte [21] responds to the urgency rhetoric that often accompanies discussions and strategies focused on avoiding the dangers related to climate change, and questions whether this should prescribe the approach taken to address this socio-ecological challenge. Whyte explains, that in "contemporary indigenous intellectual traditions and political activism, relationships of mutual responsibilities, infused with appropriate qualities like consent and reciprocity, are often referred to as kin relationships." He argues that while people in the Indigenous climate movement resolutely agree on the need for swift action on climate change, establishing kin relationships cannot be sacrificed for the sake of urgent actions without committing further injustices. Sweeping urgent actions led by settler societies to lower GHGs in the absence of "first establish[ing] and repair[ing] the qualities of consent, trust, accountability and reciprocity" will only exacerbate climate vulnerability and exploitation, particularly for Indigenous peoples. While qualities like consent and reciprocity are critical to the cross-sectoral collaborations required to lower carbon footprints, "these are precisely the kinds of qualities of relationships that take time to nurture and develop". Yet, not taking the time to build such justice-oriented kin relationships will perpetuate recognition injustices and undermine an effective collective response to climate change.

5. Conclusion

Global sustainability requires safeguarding and restoring the Earth's life support systems. Organizations like the UNU [1] and GSI [2], as well as many independent scientists [3] are raising alarms about how we are undermining global sustainability as many of our planetary life support systems are reaching major tipping points. They stress how existing governance systems, institutions and decision-making processes are inadequately responding to these grave threats by using incremental change approaches. Instead, significant transformation at both local and global scales and structural changes within dominant production and consumption patterns is required.

The Canadian government's ERP is not facilitating transformational change; like most government and corporate energy transition policies it is a at best introducing a few incremental changes that allow business-as-usual to proceed. Largely dependent upon supply-side technocratic solutions such as renewable energy and electric vehicles, it promotes economic growth while reinforcing and exacerbating the "entrenchment of patterns of socially and ecologically uneven exchange rather than resolving these in a globally just manner" ([22], p. 183). As the Assembly of First Nations point out, energy transition and sustainability challenges that place growth as a key objective should "presume a continuation of the structurally inequitable and racist systems that have led us to this compounding environmental crisis in the first place" ([28], p. 144). Further, a focus on building out renewable energy systems and growing the economy does nothing to decrease consumption of resources and the waste this produces, and hence in the absence of other socio-political changes, does nothing to remake the exploitive relationships with other peoples and ecosystems ([75], p. 85). As respected British environmentalist Jonathon Porritt [76] explains, "We're trapped in an energy-intensive way of life, driven by frenetic consumerism, causing massive damage to both people and the planet, and putting at risk the very future of humankind."

Transformative sustainability policies at all levels cannot be delivered in isolation from the pursuit of justice. Social scientists and Indigenous scholars remind us that these transformative actions must be informed by recognition of "the structurally inequitable and racist systems that have led us to this compounding environmental crisis in the first place" ([32], p. 144). It requires challenging our assumptions about authority and expertise, "and opening up debates to plural knowledge and experiences' ([9], p. 7) as a part of considering how to redistribute power and resources.

There is an opportunity to chart a different course away from the collapse of life supporting Earth systems, but if it is to be just and equitable, we need to give concerted attention to the drivers of existing and new socioecological injustices. A transformative climate justice perspective directs attention to the drivers of injustice across spatial scales and how these can be contested and prevented in advancing sustainability transformations [10]. It helps highlight the institutional structures and policy processes that produce and maintain inequalities between countries and regions, but also identify how national-level sustainability priorities and policy processes can exacerbate inequities and vulnerabilities within countries that have been shaped by long histories of encounter with colonialism, imperialism, warfare, and other forms of extractivism and violence [7]. Analysis of embodied energy injustices provides a framework to situate and understand how power operates in creating these place-based socioecological injustices and links these injustices to the broader dynamics of global capitalism.

Conflict of interest

The author declares no conflict of interest.

Author details

Lorelei Hanson
Athabasca University, Athabasca, Canada

*Address all correspondence to: lorelei.hanson@athabascau.ca

References

[1] Eberle C, O'Connor J, Narvaez L, Benavides MM, Sebesvari Z. Interconnected Disaster Risks: Risk Tipping Points. Bonn: United Nations University – Institute for Environment and Human Security; 2023. DOI: 10.53324/WTWN2495

[2] Lenton TM, Armstrong McKay DI, Loriani S, Abrams JF, Lade SJ, Donges JF, et al. Global Tipping Points – Summary Report 2023. 2023. Available from: https://global-tipping-points.org

[3] Independent group of scientists appointed by the Secretary-General. Global Sustainable Development Report 2023: Times of Crisis, Times of Change: Science for Accelerating Transformations to Sustainable Development. United Nations; 2023. Available from: https://sdgs.un.org/sites/default/files/2023-09/FINAL%20GSDR%202023-Digital%20-110923_1.pdf

[4] United Nations. Climate Action. Available from: https://www.un.org/en/climatechange/inger-andersen-climate-justice

[5] World Commission on Environment and Development. Our common future. United Nations; 1987. 247 p. Available from: https://www.are.admin.ch/are/en/home/media/publications/sustainable-development/brundtland-report.html

[6] United Nations. Universal Values. What Does the 2030 Agenda Say about Universal Values? 2024. Available from: https://unsdg.un.org/2030-agenda/universal-values

[7] Álvarez L, Coolsæt B. Decolonizing environmental justice studies: A Latin American perspective. Capitalism Nature Socialism. 2020;**31**(2):50-69. DOI: 10.1080/10455752.2018.1558272

[8] Sultana F. Critical climate justice. Geographical Journal. 2022;**188**(1):118-124. DOI: 10.1111/geoj.12417

[9] Huff A, Naess LO. Introduction: Reframing climate and environmental justice. IDS Bulletin. 2022;**53**(4):1-12. DOI: 10.19088/1968-2022.135

[10] Newell P, Srivastava S, Naess LO, Torres Contrevas G, Price R. Toward transformative climate justice: An emerging research agenda. Wiley Interdisciplinary Reviews: Climate Change. 2021;**12**(6):1-17. DOI: 10.1002/wcc.733

[11] Newell P, Adow M. Cutting the supply of climate injustice. IDS Bulletin. 2022;**53**(4):31-45

[12] Schlosberg D, Collins LB. From environmental to climate justice: Climate change and the discourse of environmental justice. Wiley Interdisciplinary Reviews: Climate Change. 2014;**5**(3):359-374. DOI: 10.1002/wcc.275

[13] Srivastava S, Bose S, Parthasarathy D, Mehta L. Climate justice for whom? Understanding the vernaculars of climate action and justice in marginal environments of India. IDS Bulletin. 2022;**53**(4):101-123

[14] Sovacool BK, Dworkin MH. Energy justice: Conceptual insights and practical applications. Applied Energy. 2015;**142**:435-444

[15] Romero-Lankao P, Rosner N, Brandtner C, Rea C, Mejia-Montero A, Pilo F, et al. A framework to Centre justice in energy transition innovations. Nature Energy. 2023;**8**:1192-1198. DOI: 10.1038/s41560-023-01351-3

[16] Foundation for the Rights of Future Generations. About us. Available from: https://www.intergenerationaljustice.org

[17] Carley S, Konisky DM. The justice and energy implications of clean energy transition. Nature Energy. 2020;**5**:569-577. DOI: 10.1038/s41560-020-0641-6

[18] Jenkins K, Sovacool BK, McCauley D. Humanizing sociotechnical transitions through energy justice: An ethical framework for global transformative change. Energy Policy. 2018;**117**:66-74. DOI: 10.1016/j.enpol.2018.02.036

[19] Lachaz A. A War of Words between Alberta and Ottawa: What Does "Just Transition" Actually Mean? Edmonton: CTV News; 2023. Available from: https://edmonton.ctvnews.ca/a-war-of-words-between-alberta-and-ottawa-what-does-just-transition-actually-mean-1.6221188

[20] Healy N, Stephens J, Malin SA. Embodied energy injustices: Unveiling and politicizing the transboundary harms of fossil fuel extractivism and fossil fuel supply chains. ERSS. 2019;**48**:219-234. DOI: 10.1016/j.erss.2018.09.016

[21] Whyte K. Too late for indigenous climate justice: Ecological and relational tipping points. Wiley Interdisciplinary Reviews: Climate Change. 2020;**11**:e603. DOI: 10.1002/wee.603

[22] Bhambra GK, Newell P. More than a metaphor: 'Climate colonialism' in perspective. Global Social Challenges Journal. 2022;**2**:179-187. DOI: 10.1332/EIEM6688

[23] Kumar A, Ferdous R, Luque-Ayala A, McEwan C, Power M, Turner B, et al. Solar energy for all? Understanding the successes and shortfalls through a critical comparative assessment of Bangladesh, Brazil, Mozambique, Sri Lanka and South Africa. ERSS. 2019;**48**:166-167. DOI: 10.1016/j.erss.2018.10.005

[24] Stephens JC. Feminist, antiracist values for climate justice – Moving beyond climate isolationism. In: Agyeman J, Chung-Tiam-Fook T, Engle J, editors. Sacred Civics – Building Severn Generation Cities. London/NY: Routledge; 2022. pp. 177-189

[25] Fattouh B, Poudineh R, West R. The rise of renewables and energy transition: What adaptation strategy exists for oil companies and oil-exporting countries? Energy Transitions. 2019;**3**:45-58. DOI: 10.1007/s41825-019-00013-x

[26] Cajero S. Moving towards transformative climate justice: What needs to be done? Global Development. 2020. Available from: https://globaldev.blog/moving-towards-transformative-climate-justice-what-needs-be-done/

[27] Natural Resources Canada. Energy Fact Book, 2022-2023. 2023. 150 p. Available from: https://natural-resources.canada.ca/sites/nrcan/files/energy/energy_fact/2022-2023/PDF/Energy-factbook-2022-2023_EN.pdf

[28] Ritchie, H, Roser M, Rosado, P. Energy. Our World in Data. 2022. Available from: https://ourworldindata.org/energy/country/canada#per-capita-what-is-the-average-energy-consumption-per-person

[29] Auditor General of Canada. Canadian Net-Zero Emissions Accountability Act – 2030 Emissions Reduction Plan, Reports of the Commissioner of the Environment and Sustainable Development to the Parliament of Canada. Canada; 2023. 43 p. Cat. No. FA1-26/2023-1-6E-PDF. Available from: https://www.oag-bvg.gc.ca/internet/docs/parl_cesd_202311_06_e.pdf

[30] Government of Canada. 2030 Emissions Reduction Plan: Clean Air, Strong Economy/. 2023. Available from: https://www.canada.ca/en/services/environment/weather/climatechange/climate-plan/climate-plan-overview/emissions-reduction-2030.html

[31] Canada Energy Regulator. Canada's Energy Transition: Historical and Future Changes to Energy Systems – Update – An Energy Market Assessment. 2023. Available from: https://www.cer-rec.gc.ca/en/data-analysis/energy-markets/canadas-energy-transition/canadas-energy-transition-historical-future-changes-energy-systems-update-energy-market-assessment-trends-canada.html

[32] Government of Canada. Canada's 2030 Emissions Reduction Plan. 2022. 240 p. Available from: https://www.canada.ca/en/services/environment/weather/climatechange/climate-plan/climate-plan-overview/emissions-reduction-2030/plan.html

[33] Canada Energy Regulator. Canada's energy future 2023. Canada; 2023. 130 p. Available from: https://www.cer-rec.gc.ca/en/data-analysis/canada-energy-future/2023/canada-energy-futures-2023.pdf

[34] Statistics Canada. Population and Demography Statistics. 2023. Available from: https://www.statcan.gc.ca/en/subjects-start/population_and_demography

[35] C.D. Howe Institute. Canada's 2030 Emissions Reduction Plan: what's Realistic? 2022. 21 p. Available from: https://www.cdhowe.org/sites/default/files/2022-07/CR_Reduction_2022_1.pdf

[36] International Energy Agency. Global Energy Transitions Stocktake. 2023. Available from: https://www.iea.org/topics/global-energy-transitions-stocktake

[37] Statistics Canada. Vehicle Registrations. 2022. p. 2023. Available from: https://www150.statcan.gc.ca/n1/daily-quotidien/231102/dq231102b-eng.htm

[38] World Economic Forum. Fostering Effective Energy Transition – Insight Report. 2023. Available from: https://www.weforum.org/publications/fostering-effective-energy-transition-2023/

[39] Canadian Renewable Energy Association. Did you Know? 2023. Available from: https://renewablesassociation.ca/news-release-canada-added-1-8-gw-of-wind-and-solar-in-2022

[40] Statistics Canada. Energy Supply and Demand. 2022. p. 2023. Available from: https://www150.statcan.gc.ca/n1/daily-quotidien/231120/dq231120c-eng.htm#

[41] Ritchie H, Rosado P. Energy Mix. Our World in Data. 2024, first pub 2020 Jul. Available from: https://ourworldindata.org/energy-mix

[42] Stockholm Environment Institute, Climate Analytics, E3G, IISD, and UNEP. The Production Gap Report: Phasing Down or Phasing Up? Top Fossil Fuel Producers Plan Even More Extraction Despite Climate Progress. Sweden: Stockholm Environment Institute, Climate Analytics, E3G, International Institute for Sustainable Development and United Nations Environment Programme; 2023. DOI: 10.51414/sei2023.050

[43] Lovegrove, P. Energy Transition Outlook: Renewables Still Not Replacing Fossil Fuels in the Global Energy Mix. 2023. Available

from: https://w ww.dnv.com/news/energy-transition-outlook-renewables-still-not-replacing-fossil-fuels-in-the-global-energy-mix-247880

[44] Ritchie H, Rosado P, Roser M. Energy Production and Consumption. Our World in Data. 2024, first pub Jul 2020. Available from: https://ourworldindata.org/energy-production-consumption#article-citation

[45] Mortlock L, Hubert A-H. Strategic Partnerships and Innovation Hold the Key to Canada's Energy Transition. EY; 2023. Available from: https://www.ey.com/en_ca/energy-resources/strategic-partnerships-and-innovation-hold-the-key-to-canadas-energy-transition

[46] Evans S. Analysis: Which Countries Are Historically Responsible for Climate Change? Carbon Brief; 2021. Available from: https://www.carbonbrief.org/analysis-which-countries-are-historically-responsible-for-climate-change/

[47] Das RR, Martiskainen M, Bertrand LM, MacArthur JL. A Review and Analysis of Initiatives Addressing Energy Poverty and Vulnerability in Ontario. Canada: University of Sussex; 2022. Available from: https://hdl.handle.net/10779/uos.23489501.v1

[48] Government of Canada. Making Housing More Affordable. 2022. Available from: https://www.budget.canada.ca/2022/report-rapport/chap1-en.html#m5

[49] Statistics Canada. Canadian Income Survey. 2021. p. 2023. Available from: https://www150.statcan.gc.ca/n1/en/daily-quotidien/230502/dq230502a-eng.pdf?st=PSbYdasj

[50] Statistics Canada. Consumer Price Index Portal. 2023. Available from: https://www.statcan.gc.ca/en/subjects-start/prices_and_price_indexes/consumer_price_indexes

[51] Oats LL. Mega Dams Part One: A Tale of Muskrat Fall and Gull Island. Network in Canadian History & Environment; 2022. Available from: https://niche-canada.org/2022/10/07/mega-dams-part-one-a-tale-of-muskrat-falls-and-gull-island/

[52] Gismondi M, Hanson LL. Provincial polyphasia: Community energy generation and the politics of sustainability transition in Alberta, Canada. In: Batel S, Rudolph D, editors. A Critical Approach to the Social Acceptance of Renewable Energy Infrastructures. Cham, CH: Palgrave Macmillan; 2021. pp. 181-198

[53] Hoicka CE, Savic K, Capney A. Reconciliation through renewable energy? A survey of indigenous communities, involvement, and peoples in Canada. ERSS. 2021;**74**:101897

[54] World Bank. The Growing Role of Minerals and Metals for a Low-Carbon Future. 2017. Available from: https://documents1.worldbank.org/curated/en/207371500386458722/pdf/117581-WP-P159838-PUBLIC-ClimateSmartMiningJuly.pdf

[55] Canada Energy Regulator. Market Snapshot: Critical Minerals as Key to Global Energy Transition. 2023. Available from: https://www.cer-rec.gc.ca/en/data-analysis/energy-markets/market-snapshots/2023/market-snapshot-critical-minerals-key-global-energy-transition.html

[56] International Energy Agency. The Role of Critical Minerals in Clean Energy Transitions. 2022. 283 p. Available from: https://iea.blob.core.windows.net/assets/ffd2a83b-8c30-4e9d-980a-52b6d9a86fdc/

TheRoleofCriticalMineralsin CleanEnergyTransitions.pdf

[57] International Renewable Energy Association. Geopolitics of the Energy Transition – Critical Minerals. Abu Dhabi: International Renewable Energy Agency; 2023. 149 p. Available from: https://mc-cd8320d4-36a1-40ac-83cc-3389-cdn-endpoint.azureedge.net/-/media/Files/IRENA/Agency/Publication/2023/Jul/IRENA_Geopolitics_energy_transition_critical_materials_2023.pdf?rev=420aeb58d2e745d79f1b564ea89ef9f8

[58] Government of Canada. Canadian Mining Assets. 2021. Available from: https://natural-resources.canada.ca/maps-tools-and-publications/publications/minerals-mining-publications/canadian-mining-assets/19323

[59] Government of Canada. Mining. 2021. Available from: https://www.tradecommissioner.gc.ca/sectors-secteurs/mining-mines.aspx?lang=eng

[60] Government of Canada. Minerals and the Economy. 2023. Available from: https://natural-resources.canada.ca/our-natural-resources/minerals-mining/mining-data-statistics-and-analysis/minerals-and-the-economy/20529

[61] Keeling A, Sandlos J. Introduction: The complex legacy of mining in northern Canada. In: Keeling A, Sandlos J, editors. Mining and Communities in Northern Canada – History, Politics and Memory. Calgary, AB: University of Calgary Press; 2015. pp. 1-32

[62] World Wildlife Fund for Nature. Extracted Forests – Unearthing the Role of Mining-Related Deforestation as a Driver of Global Deforestation. 2023. Available from: https://wwfint.awsassets.panda.org/downloads/wwf_studie_extracted_forests_1_1.pdf

[63] Keeling A, Sandlos J. Introduction: Critical perspectives on extractive industries in northern Canada. Extractive Industries and Society. 2016;**3**(2):265-268

[64] Raman S. Canada Mining Push Puts Major Carbon Sink and Indigenous Lands in the Crosshairs. Mongabay; 2022. Available from: https://news.mongabay.com/2022/06/canada-mining-push-puts-major-carbon-sink-and-indigenous-lands-in-the-crosshairs/

[65] Canada MW. 77 Global and National Human Rights and Environmental Groups in Six Countries Expose OceanaGold as an Irresponsible Mining Company and Oppose its Expansion. MiningWatch Canada; 2023. Available from: https://miningwatch.ca/news/2023/3/1/77-global-and-national-human-rights-and-environmental-groups-six-countries-expose

[66] Yap J. Extracting Justice: The human rights impact of Canadian mining around the world. Ideas, CBC Radio1; 2023. Available from: https://www.cbc.ca/radio/ideas/canadian-mining-abuses-1.6854852

[67] Gamble J. What's at Stake in Ontario's Ring of Fire. Canadian Geographic. 2022, first published 2017 Aug 24. Available from: https://canadiangeographic.ca/articles/whats-at-stake-in-ontarios-ring-of-fire/

[68] Harris LI, Richardson K, Bona KA, Davidson SJ, Finkelstein SA, Garneau M, et al. The essential carbon services provided by northern peatlands. Frontiers in Ecology and the Environment. 2021;**20**(4):222-230. DOI: 10.1002/fee.2437

[69] McIntosh E. Everything you Need to Know about the Push to Mine Ontario's Ring of Fire. The Narwhal; 2022. Available from: https://thenarwhal.ca/ontario-ring-of-fire-explainer/

[70] Bureau of International Labor Affairs. Child Labor and Forced Labor Reports, Bolivia. United States: U.S. Department of Labor; 2022. Available from: https://www.dol.gov/agencies/ilab/resources/reports/child-labor/bolivia

[71] Human Rights Watch. World Report 2023 – Our Annual Review of Human Rights around the Globe. 2023. Available from: https://www.hrw.org/world-report/202369

[72] Casey L. Inside the Battle over Ontario's Ring of Fire. The Canadian Press; 2023. Available from: https://www.cbc.ca/news/canada/toronto/ont-ring-of-fire-1.6991468#:~:text=electric%20vehicle%20batteries.-,Moonias%2C%20an%20elder%20from%20Neskantaga%20First%20Nation%2C%20and%20many%20in,%2C%20how%20would%20you%20feel%3F%2270

[73] Carver B. Four First Nations Form Land Defence Alliance, Seek to Stop Encroachment on their Territory. Kenora Miner and News; 2023. Available from: https://www.kenoraminerandnews.com/news/local-news/four-first-nations-form-land-defence-alliance-seek-to-stop-encroachment-on-their-territory

[74] Daley DM, Reames TG. Public participation and environmental justice: Access to federal decision making. In: Konisky DM, editor. Failed Promises: Evaluating the Federal government's Response to Environmental Justice. Cambridge, MA: MIT Press; 2015. pp. 143-171

[75] Dunlap A. Does renewable energy exist? Fossil fuel+technologies and the search for renewable energy. In: Batel S, Rudolph D, editors. A Critical Approach to the Social Acceptance of Renewable Energy Infrastructures. Cham, CH: Palgrave Macmillan; 2021. pp. 83-102

[76] Porritt, J. Opinion: Getting Real about Net-Zero. Earth.org. 2021. Available from: https://earth.org/getting-real-about-net-zero-by-jonathon-porritt

Chapter 21

Perspective Chapter: Activism for Social Justice – Moving beyond Counter-Hegemonic Resistance

Lya Sañudo-Guerra and Maria Reyes Hernández-Castilla

Abstract

This study argues that current educational frameworks and public policies fail to meet the needs of marginalized minorities. The study suggests alternative notions of education, knowledge, and educational relationships beyond the current Eurocentric educational model and formal educational institutions. This study is based on the Theory of Resistance and uses a qualitative methodology, counter-narratives method, and a phenomenological perspective to analyze the data. The participants in the study are activists and transformative intellectuals who work toward helping victims of forced disappearances. The key findings reveal that the school system fails to recognize or address internal or contextual violence and how it fails to support the victims or their families, reinforcing the existing hegemonic power structures. This chapter emphasizes the need for institutional educational responses that involve taking a committed stance toward human rights education and implementing strategies that promote peace and social justice. The introduction provides a justification for the topic, background studies, and theoretical perspective. The second section describes the methodology, the method used, and the research subjects. The third section presents the results, and finally, the conclusions and discussion are provided.

Keywords: social justice, school resistance, counter-hegemonic activism, school violence, social solidarity

1. Introduction

The following report is an update on a broader research project. The main objective of the study is to comprehensively analyze the non-hegemonic narratives (counter-narratives) of activists in limited educational contexts to demonstrate forms of resistance against the school tradition that reproduces social injustice. The section consists of two parts. Firstly, it develops the theoretical construction that serves as the basis for the research, including a classification of counter-hegemonic resistance. Secondly, it briefly describes the traditional and most recent authors on this subject.

The topic you mentioned relates to the study of social justice and education, specifically the examination of oppression, discrimination, and resistance within educational institutions. Several authors have contributed to this field and have written extensively on these topics. One prominent author in this area is bell hooks, an American author, feminist, and social activist, who explores the intersection of race, gender, and class in education and highlights the importance of addressing these issues for achieving social justice. Another influential author is Paulo Freire, a Brazilian educator and philosopher, who emphasizes the need for transformative education that empowers marginalized groups and challenges oppressive systems. Giroux's work emphasizes the role of education in shaping social identities and power dynamics. He argues for a transformative and critical approach to education that challenges dominant ideologies and promotes social justice. Giroux's writings often explore the intersections of education, politics, and culture, examining how they influence each other and shape society. These are just a couple of notable authors in the field, and many more scholars and activists have contributed to the study of social justice in education [1, 2].

According to the Education Resources Information Center (ERIC), in the last 5 years, 99 articles have been published on counter-hegemony, education, and social justice. Seventy percent of these articles deal with issues of educational equity, race, politics, and racism, most at the postsecondary and higher education level. Most of the articles were published in Australia, Brazil, Canada, and Sweden. Osterman and Rezende are the authors who present two of the most relevant articles, both from Brazil. In both articles, they use critical theory to understand counter-hegemonic movements in the face of tensions between knowledge and social justice.

The chapter explores the incorporation of curricular theories into science education scholarship. It introduces core concepts from curricular frameworks and investigates their intersection with science education. Focusing on post-critical curricular perspectives, it analyzes the discursive dynamics, shaping meaning in science education. The study delves into hegemonic and counter-hegemonic discourses within Brazilian and international contexts, revealing tensions between traditional and critical viewpoints. Emerging discourses advocating for social justice and democratic equality present a dilemma: if treated prescriptively, they risk standardizing education, undermining its fundamental principles and societal aspirations [3].

The chapter addresses Brazil's sociohistorical context. It advocates for a counter-hegemonic approach rooted in Global South epistemologies, particularly inspired by de Santos' concept of "ecology of knowledge" as essential for global social justice. The chapter proposes a science education framework that prioritizes environmental preservation, critiques capitalist systems, and values knowledge from the Global South [4]. Overall, educational social justice needs further studies that reveal the tensions in education and the contexts in which it develops.

In this framework, the objectives and research questions emerge: (1) the aim is to study the counter-narratives of education activists and how they resist traditional schooling practices. This study aims to provide a peaceful and transformative vision of education from the perspective of these activists; (2) the objective is to find ways to overcome the school's monopoly on education and ensure that everyone has the right to an inclusive and fair education, even outside of traditional educational institutions; (3) lastly, we propose creating spaces for social and educational transformation.

The research questions are (1) What do the counter-narratives of activists say about education and their forms of resistance to the school tradition? (2) How can the monopoly pretension of the school be broken from outside the institutionalized educational processes? (3) What spaces with and from the oppressed and exploited can be proposed?

To conduct research in this field, it is necessary to perform a thorough theoretical review that aids in comprehending the problem and analyzing the findings.

1.1 Critical theory

In this part of the report, the concepts and categories used for the research are developed. The study is grounded in the critical theory of culture and resistance, which draws upon the works of various critical theorists and philosophers. Scholars such as Gramsci, Adorno, Horkheimer, Freire, the Frankfurt School, Gramsci, and Foucault have all contributed to this theoretical framework. Henry Giroux has utilized the theories of these authors to analyze the intersections of education and culture within contemporary society. He has developed a critical theory that aims to foster social transformation.

Critical Theory posits that there exists a profound connection between knowledge, power, and domination. Its purpose is to comprehend how marginalized and oppressed groups confront inequities, and exclusion, and defend themselves against repressive violence. This understanding recognizes defense and resistance as integral to struggles for marginalized voices, human rights, justice, freedom, and peace. However, it is important to note that these groups, striving for social transformation, often face persecution from dominant hegemonic sectors.

The study aims to analyze and question the current power structures and social inequalities within educational institutions and society by engaging with critical theory and its associated thinkers. It strives to promote the voices and experiences of marginalized groups while striving for a more fair and just future.

The theorists believed in studying society systematically with a comprehensive approach, with a conscious commitment toward transformation. They aimed to understand social problems critically and make changes that benefit oppressed groups [5, 6]. Italian neo-Marxism studied the role of culture and education in the reproduction of power relations. They formulated the theory of hegemony, which emphasizes how the ruling class exercises its power not just through force but also through persuasion and achieving consensus from the masses they dominate. This helps sustain the power and continuity of the hegemonic system.

According to these theoretical approaches, a key step is to recognize the oppression and the dominant narratives that sustain it and to confront them with narratives that question them. They use the concept of "transformative resistance" to analyze the ways of denouncing and resisting social oppression and acting for an interest in social justice [7]. "It is about life, not passive and indirect consensus, but active and direct; the participation, therefore, of individuals, even if this produces an appearance of disintegration and tumult [8]."

This theory proposes that those who are marginalized or excluded—the subaltern class—must create their own culture and ideology. By doing so, they can develop an opposing force that can challenge the dominant class's control over society. This would lead to a more equitable and just society. To achieve this change, the dominant class's power structures must be disrupted and replaced with a new emerging block to establish a new hegemony [9].

These processes are shaped by the specific historical culture [2] in which they occur, and this includes collective actions as well. In such situations, practices are influenced by the prevailing social conditions and are also shaped by the unique historical identity of the collective. In this context, the role of organic intellectuals or educational agents is crucial. They are responsible for overseeing the theoretical and ideological construction that legitimizes the dominant group or sometimes leads to the practice of resistance. This is why it is a key concept in research.

The educational practices serve either to maintain a social structure or to transform it through dissent of organic or transformational intellectuals, who are "the educational agents, not as the one who teaches in the school but as a representative of the critical conscience of the society that assumes the role of mediator between the general society and the educational community [9]."

It implies a shift in the individual's mindset, where they acknowledge the need to look beyond the conflicts, inconsistencies, and prevalent ideology, as well as the opposing expressions and efforts of defiance. Education, viewed from this standpoint, is a constantly evolving process that involves confronting challenges and striving to transform oppressive systems. To bring about such transformations, one must comprehend the mechanisms through which power relations are established and upheld within these contexts. Resistance "offers a new theoretical impulse to understand the complex forms under which subordinate groups experience their educational failure and directs their attention toward new ways of understanding and restructuring the modes of critical pedagogy [2]."

Henry Giroux is an American cultural critic and a key contributor to the development of critical pedagogy. He has advanced the Theory of Resistance, which forms a significant foundation of his work. While some of his writings resemble neo-Marxist analyses and critiques, he has developed a unique critical perspective that emphasizes the interconnectedness of education, culture, politics, and society [1, 2, 7]. "His proposal for critical pedagogy is particularly significant [10, 11]." His analysis has inspired many approaches in Epistemologies of the South [12].

He has expressed disapproval of the educational system for playing a role in perpetuating social inequalities and for its reliance on a neoliberal ideology that fosters competition and exclusion. He has also studied the influence of popular culture and the media on the development of identity and subjectivity. As a result, he advocates for a critical pedagogy that encourages reflection, critical thinking, and transformative action.

He believes that education by transformative intellectuals is critical for social transformation, as it provides a space for the construction of a democratic and just society through promoting dialog and reflection. By questioning and transforming reality toward a more equitable society, students are empowered to make a positive impact [2].

The concept of resistance can be applied to the field of education to analyze the culture of schools, including the educational knowledge, values, and interactions that take place within them. This analysis can reveal how certain contents are selected and organized into a particular structure and hierarchy within the dominant curriculum, which can marginalize the knowledge produced by school communities and perpetuate the social effects of colonization and domination [13]. The hidden curriculum can be analyzed critically using a dialectical approach. This approach can reveal how the school's separation from parents, the gender and class dynamics associated with the production and distribution of knowledge, and the distant relationships between school administrators, teachers, and families are all promoted.

The Epistemologies of the South [12] are concerned with democracy and human rights in the Latin American context. They draw inspiration from Critical Culture [7, 8] to examine the prevalence of Eurocentric thought, and to highlight the subaltern groups that resist capitalism, colonialism, and patriarchy.

This perspective values the diversity of knowledge and original wisdom that have been overlooked by the Western model of knowledge. It aims to empower oppressed social groups to represent and transform the world according to their aspirations.

To achieve this, it offers conceptual tools such as the abyssal line, the sociology of absences and emergencies, the ecology of knowledge, the intercultural translation, and the craftsmanship of practices.

Of particular interest for this work are the ecology of knowledge and the intercultural translation. These describe the processes to expand the horizon of possibilities and intelligibilities of the future, and to establish new relationships between diverse types of knowledge. "They focus on the cognitive processes of the meaning, the

Pre-resistance (emerging reactive resistance)	• The aim is to develop personal resilience in the face of oppressive, discriminatory, or exclusionary exercise of power. • These are individual or disorganized practices of resistance that can result in counter-productive or adaptive resistance. • However, there is no evidence of any material, symbolic, or political changes in the dominant hegemonic system.
First parameter (material dimension)	• There are specific practices of collective resistance that challenge the distribution of power and resources. • These practices emerge from the leadership of an organic or transformational intellectual and can either be incorporated or tolerated by the hegemonic model. • The degree or modes of manipulation of the hegemonic system to the collectives are not a closed system. • Resistance practices alleviate the situation of the excluded collectives. • These practices of resistance can serve as a basis for the counter-hegemonic struggle in the future, either alone or articulated with others. • They come to strengthen the material structure of domination.
Second parameter (symbolic dimension)	• These resistance practices encompass patterns of symbolic recognition of values, cultural norms, and preferential recognition. • They aim to challenge and modify the dominant cultural network of meaning. • Additionally, they serve to acknowledge and recognize previously excluded groups. • These practices arise from the leadership of an organic or transformational intellectual. • However, it is important to note that these resistance practices can only challenge the symbolic dimension without questioning the existing patterns of distribution and production of material resources. • Despite this limitation, there may be actions that could alleviate the situation of tension in the resistance.
Third parameter (political practice) material production + symbolic and practical dimension	• Counter-hegemonic resistance, also known as transformative resistance, goes beyond individual or disorganized forms of resistance. • Instead, it involves political articulation of local resistance practices that contribute to redefining the opposition between "us" and "them." • These practices form a new historical block that includes excluded groups and marginalized members of the hegemonic block, referred to as "traditional intellectuals." • Such resistance is led by an organic or transformational intellectual, and it succeeds in improving the living situation of some sectors.

Source: [12, 15, 17].

Table 1.
Parameters to establish the degrees or levels of hegemonic resistance.

justification, and the orientation in the struggle waged by those who resist and rebel against oppression. The question of validity arises from this strong presence [14]." It is possible to bring together knowledge generated in different areas to create languages.

Based on certain ideas, it has been demonstrated that there are various levels of liberation or emancipation for collectives, groups, or countries. To research this topic, a classification system for different forms of resistance has been proposed, [15] which are linked with counter-hegemony. These can be examined based on three criteria: (1) the degree to which certain practices are assimilated or tolerated by the hegemonic model; (2) the degree of articulation between the challenges at the symbolic level and the challenges to the production and distribution of material resources; and (3) if there is a process of articulation between different practices that can generate a collective subject capable of making claims to achieve political power [15, 16]. From these criteria, four parameters have been derived that can be used to organize and interpret the corpus formed through counter-narratives (**Table 1**).

Table 1 is crucial for analyzing and interpreting the activist's narrative as a research subject. It enables establishing the segmentation in the interview corpus and the network of its significance to achieve the following results.

Different models of relations are being sought with more egalitarian power systems, allowing them to organize society according to their conditions, which will eventually be challenged in turn. According to the analysis, counter-hegemonic resistance groups have demanded work ahead. The narratives show that there is an alternative that is not only symbolic but practical. It is a resistance against the hegemonic conditions of oppression and violence. These counter-hegemonic alternatives present internal tensions, contradictions, and reproduction.

2. Methodology and the narrative method

This section explains the research method used and its relationship to the theoretical approach. Additionally, it discusses the epistemological foundations and argues for the reliability of the decisions and results obtained [18].

Research in Social Justice has a long history of exploring and fighting against the invisibility of vulnerable groups. This involves being aware of societal issues and working toward a more equitable and just social order. The research aims to maintain a political and ethical commitment that protects the most disadvantaged and avoids collaborating with injustices.

Unlike the abstract treatment of Social Justice, this research proposes a more open conception that considers improvements in education, society, and personal and collective relationships. It is related to the denunciation of concrete injustices faced by groups such as gender, ethnicity, sexuality, social class, criminality, violence, and forced disappearances. The research strives to use an epistemological and methodological approach that is consistent with education as a means for Social Justice.

This research follows the critical paradigm and is based on the Critical Theory of knowledge. This theory suggests that reflection and social emancipation are necessary responses to hegemonies and forms of domination. It also emphasizes the importance of consciousness in achieving social justice and the common good. The research proposes a concept of science that involves producing social knowledge to generate actions that contribute to the transformation and emancipation of individuals and their communities. This, in turn, enhances their reflection and decision-making abilities concerning the context in which they live.

Research is a space where people can participate, take social responsibility, and commit to meeting the needs and expectations of their communities. Its goal is to promote transformation that frees people from oppression and to make decisions that benefit everyone. The knowledge generated through research is thought about and then put into action by the community, leading to the development of critical thinking. In this way, we can use discussion to reach a shared understanding of the realities of the situation and direct our efforts toward creating a concrete plan for emancipation and decision-making based on self-reflection and critical thinking.

In the field of education, these ideas involve adopting global and dialectical views of the realities that intersect in educational cultures and accepting democratic and participatory approaches when constructing knowledge. This is a quest for knowledge that aims to develop the ability to change the structures of domination—a methodological challenge for researchers.

Critical Research is a transdisciplinary approach to investigating social life, revealing the beliefs of the world and how they operate. Different scientific communities can apply it. "A critical social theory frames its research program and its conceptual framework with an eye to the aims and activities of those oppositional social movements with which it has a partisan, tough critical identification. The questions it asks and the models it designs are informed by that identification and interest [19]."

Dialectics is a method of creating knowledge by analyzing reality and its mutual implications both subjectively and objectively. It acknowledges contradictions as a part of social nature and has various perspectives. In this work, we are interested in critical dialectics, which was developed in the Frankfurt School. It transforms Hegelian dialectics into one based on dialectical materialism [5].

Critical theory uses dialectics as a tool to analyze power relations in society and create a plan for social transformation. This is based on the idea that change comes from the class struggle. The subaltern class must create their own culture and ideology to challenge the hegemony of the dominant class and build a more just society.

In the Frankfurt School, dialectics is used to understand and transform society through criticism. Cultural and social dynamics, such as cultural identity, globalization, resistance, and social change, are analyzed using dialectics. Critical reflection is crucial for people's liberation and social transformation.

Negative dialectics, on the other hand, rejects the "official" representation of reality. It uses principles of negativity, contradiction, and mediation to form critical reflection [5].

The theory of Resistance [7] adopts an orientation called critical dialectics, which aims to enable social transformation. Critical dialectical thinking highlights the potency of human knowledge and activity as a tool for shaping social reality. It is a form of criticism that asserts the existence of a link between knowledge, power, and domination, and emphasizes the importance of social change as the main goal.

Dialectical thinking is used to challenge common sense assumptions and beliefs to demonstrate that internal contradictions can lead to a quality change [20]. Its purpose is to break down self-certainty and self-satisfaction to undermine confidence in the power of facts and language. Dialectics is focused on critiquing power relations and working toward a more just and democratic society.

Critical dialectics is a theory that emphasizes the important link between theory and practice. It asserts that theory is not a separate entity from practice but rather it is founded on the social reality. In the same way that opposites interact, theory and practice also influence each other. The contradictions and social tensions, especially in power relations, can be a catalyst for change and social transformation. Therefore, contemporary dialectics aims to address these contradictions to achieve a deeper comprehension of social reality [2]. Overcoming the crises that arise from the contradictions in the education process is the key to transforming teaching practices and achieving a renewed approach to education. The principles of critical epistemology that underpin this research are evident throughout the entire process, not just in the final reports or written documents.

Qualitative research methods are more suitable for the critical epistemological perspective because they align better with the values of community collaboration and participation, as opposed to quantitative methods which can be exclusive and oppressive. Qualitative methods are also inclusive of diverse cultures and do not discriminate against them [21, 22]. It seeks to study and promote the educational benefits of justice.

This research focuses on the practical experiences of individuals participating in a specific historical and cultural context. It is not only useful but also fair and has the potential to liberate and promote social development. By critically examining collective processes, it creates opportunities for education and empowerment, leading to positive change and progress toward achieving social justice with integrity [10, 19]. In second place, the author explores power imbalances created by social commitments and feelings of isolation.

The counter-narrative is a methodological strategy used by the oppressed to challenge the dominant narratives that silence, marginalize, or dehumanize them [11]. It involves creating and telling their own stories, which is considered a form of resistance and liberation. Through the counter-narrative, the oppressed can recover their voice, regain their sense of identity, and assert their agency.

The counter-narrative is a method of storytelling that involves dialog, awareness, and action. It aims to resist oppressive identities and replace them with ones that inspire respect. This approach is particularly useful for studying the cultural and structural factors in education that contribute to the marginalization of minority groups, both inside and outside the classroom. It can also shed light on the experiences of those who are absent from the educational discourse due to poverty or social and educational exclusion [23, 24].

The individuals involved in this initiative are activists representing various civil society groups that work together to promote education in difficult environments, whether it be through formal or non-formal educational institutions. These groups include social and indigenous movements, women's associations, immigrant communities, and associations of individuals who are at risk of social exclusion or searching for missing persons, among others.

The research technique that will be used is based on narratives derived from the experiences of activists who represent the collective voice of marginalized or absent groups. The data collection process will consist of interviews that focus on narration, complemented by direct observation records that will be reflected in a researcher's diary. The objective is to recover the stories, experiences, and chronicles of the participating individuals who have demonstrated forms of transformative resistance toward the dominant perspectives on education (as shown in **Table 1**). The narratives collected can include experiences of people who have faced various forms of oppression or those who have lived through oppressive situations as part of a vulnerable group.

The analysis uses Resistance Theory [7] and a phenomenological, meta-textual approach, which involves identifying critical incidents and knots of meaning [25]. This type of analysis aims to systematically uncover and describe the internal structures of meaning within language, to understand the lived experiences of activists in their contextual reality. It involves describing the experiential meanings that activists encounter in their daily lives.

The process of analyzing data involves presenting various categories through representative vignettes. The purpose of these vignettes is to offer a comprehensive and subtle comprehension of the ideas presented. The aim is to demonstrate the several types of transformative resistance and produce a narrative that is independent of the school institutions.

3. Findings and discussion

The purpose of this research is to identify strategies that can be used to challenge and overcome oppressive social systems. The pursuit of knowledge is motivated by the goal of acquiring the tools needed to transform these structures of domination. However, achieving this objective poses a significant challenge that needs to be addressed rigorously and methodically.

In this section, we present the findings of the analysis of the first counter-narrative. The interviewee is an activist and university professor from Guanajuato, Mexico. In his presentation, he shares that he has been involved in social movements for the past 25 years, focusing on human rights and HIV/AIDS issues. He assumes that "the educational work in human rights and the university teaching seemed two separate spaces, one more of the community education. But from my point of view, they have always been intertwined. (...) I cannot separate my two practices, my more social practice, let us say as an activist, my practice as a university professor" (Act1.3).[1]

This statement emphasizes his commitment to social awareness as a way of living, "an ethical commitment" (Act.1.9). In this framework, Raymundo establishes a close relationship between the education, the school, and the politics. "The importance of education can vary depending on the context. In situations where there are serious human rights violations or a reduction in the democratic sphere, education can play a crucial role. However, there may be emerging actors who can limit the school's ability to act and even infiltrate spaces that challenge the traditional role of schools and universities. This creates a complex situation and requires careful consideration of the role of education in addressing these challenges" (Act.1.10).

A school model that solely focuses on life inside the school or within its walls fails to acknowledge the important relationship that exists with the external context. Such a model believes that its primary responsibility is to prevent the effects of violence within the institution. However, it is equally important to develop ways of understanding internal violence and support educational agents who may become victims or have someone close to them affected by violence. He talks about containment: "In this sense, universities, social security, professional practices, and psycho-pedagogical work at the community level should all have a perspective of understanding violence" (Act.1. 14).

The main findings consist of six crucial nodes in his counter-narrative, which are organized according to their type. These nodes account for the type and intensity of counter-hegemonic resistance (**Table 1**) in both social and educational spheres and help to understand the limited situation experienced not only by those directly involved, but also by the missing victims, their families, and the school communities to which they belong:

[1] The statements in quotation marks refer to textual evidence expressed in the interviews. The Code identifies the interviewee as Activist 1 and places the vignette in order of appearance in the corpus.

1. The Education for Peace project has implemented a pedagogical response to violence. The PP school, which is run by a religious congregation, has developed a model to address the issue of social exclusion. They aim to create a containment model that not only addresses situations that occur within the school but also considers the commitments of the community and family members. They provide support to those who live in limited situations and ensure their safety within and outside of the school. "The school has tried to contain psychological events with cultural artistic expressions and also generate a demand for an institutional response, especially in cases of femicide, in cases of disappearance and the face of cases of distribution of weapons" (Act.1.21). It is resilient due resilience that is not resistance, as it does not seek to bring about systemic change. "It is resilient due to its reactive nature." He acknowledges that there is a pedagogical approach to modify the reactions of educational agents in extreme cases such as education for peace. Talking about peace in a context of extreme violence should not only be a "perspective of peace awareness or types of peace, but rather develop a much more active perspective, even a reaction to the context strongly devastated by violence, such as human trafficking, femicides, intentional homicide, disappearance and presence of gangs that will not be different from other realities of the region and surely of the country" (Act.1.26).

2. A more fundamental change in basic assumptions is necessary, including identifying and addressing structural violence to achieve comprehensive responses. "The teachers should play a leading role in deciding curricular contents, programs, educational plans, and the purpose of schooling, considering the context they live in" (Act.1.15). The programmatic content reflects an ideology that maintains dominant power structures and is evident in the hidden curriculum.

3. In informative terms, "the goal of education is not just to teach victims, but also those who support and play a key role in the educational process. It is not limited to what happens in the classroom, but is also about building organizational skills, building resilience in the face of violence and other challenging situations" (Act.1.16).

4. The role of a teacher can reveal the challenges of responding to violence and the difficulty of refusing requests from young people who engage in criminal activities. The educational process involves various actors who can respond to such situations. "The emphasis I pointed out in terms of resistance or resilience is at the micro level" (Act.1.32).

5. The accompaniment to the victims of violence "has a strong educational component and emphasizes the processes of organization and capacity development for advocacy, as well as the capacity to recognize oneself as a subject of law" (Act.1.7). Pausing to review the contents of a situation is important, but it is equally important to address the victims. "In this regard, both the school management and administration play a crucial role, as they are responsible for making important decisions. At the same time, the school community must also work together to provide immediate support to those in need" (Act.1.37).

6. A perspective of situated epistemology recognizes the role of educators who are politically committed when faced with violence. It involves a strong political

stance of standing with the victim and demanding justice, access to the truth, and serious violations of human rights to be addressed, such as torture, forced disappearance, and execution.

In terms of institutions, the absence of a response requires a rethinking of established power capacities, possibilities, and forms.

4. Conclusions

This study examines how resistance against dominant forces in education can lead to social justice. It argues that current educational policies and programs do not sufficiently address the needs of oppressed minority groups, which results in perpetuating privileges for dominant groups.

The research report is based on the Theory of Resistance, using qualitative methodology, counter-narratives, and a phenomenological perspective. Activists and intellectuals who promote transformative change in education participated in the study. It also highlights the existence of alternative concepts of education, educational relationships, and knowledge that go beyond the traditional Eurocentric model.

The findings reveal that schools resist acknowledging and addressing internal or contextual violence and fail to support victims or their families, upholding hegemonic dominance. The study emphasizes the pressing need for institutional responses in education that encompass a steadfast commitment to human rights education and strategies to promote peace consciousness.

The study suggests that school curricula should address violence comprehensively. This can be done by incorporating a transversal component that promotes understanding and response to violence along with developing skills and capacities for all individuals involved in the educational process. The author notes that schools may not effectively address daily occurrences of violence due to the evident level of social decay. Therefore, it is necessary to engage in a discussion about how to mitigate the effects and direct consequences of violence. Instead of solely focusing on increasing penalties, a more holistic response to violence should adopt an anti-punitive stance. It should not be limited to institutional measures but also necessitate a critical examination of the prohibitionist perspective. Moreover, exploring transitional justice models could offer valuable insights, such as the restitution and reparation of harm. These models have already been incorporated into some aspects of school coexistence programs, recognizing that violence within families and broader contexts requires a multifaceted response.

In constructing new responses to violence, the role of educators is paramount. I recall the account of a secondary school teacher in Guanajuato, where official schools began receiving projects from productive companies that included the illicit sale of "huachicol"—gasoline obtained through sophisticated illegal pipeline networks. This issue goes beyond mere theft and involves an organized structure. While it may seem unavoidable to include elements of illegality in these projects, it is essential to critically question this decision. However, refusing such projects is not a viable option as it would result in failing students who are part of these networks.

Developing a consciousness of peace and integrating human rights education into the curriculum are crucial aspects that transform human rights from a mere subject matter into a way of life that guides individuals in their actions and decision-making processes.

In addition, schools need to establish strong partnerships with local communities, law enforcement agencies, and social service organizations. By collaborating and sharing resources, these stakeholders can work together to provide comprehensive support systems for students who have experienced or witnessed violence. This support may include access to counseling services, trauma-informed care, and legal or social assistance.

In the future, research projects should focus on topics related to managing extreme situations involving students or teaching staff. Additionally, it is important to explore schools' experiences with resisting hegemonic forces that arise from the evasion of problems presented by their context. Finally, it is essential to study successful cases that have achieved a culture of peace and social justice.

Finally, integrating restorative justice practices into school disciplinary frameworks can be highly advantageous. Instead of focusing solely on punishment, restorative justice emphasizes repairing harm and fostering accountability. By encouraging dialog, understanding, and reconciliation between the parties involved, this approach can help students learn from their mistakes, make amends, and develop healthier relationships within the school community.

Conflict of interests

The authors state that they have no conflict of interest.

Author details

Lya Sañudo-Guerra[1*] and Maria Reyes Hernández-Castilla[2]

1 State Commission for Continuing Education in Jalisco, Guadalajara, Mexico

2 Autonomous University of Madrid, Madrid, Spain

*Address all correspondence to: lya.sanudo@gmail.com

References

[1] Giroux H. La Pedagogía crítica en tiempos oscuros. En Revista Praxis Educativa. 2013;**XVII**(17):13-26. Available from: https://www.redalyc.org/articulo.oa?id=153129924002

[2] Giroux H. Teoría y resistencia en educación. 6° ed. México: Siglo XXI; 2004. p. 331. Available from: https://www.otrasvoceseneducacion.org/wp-content/uploads/2021/07/Teoria-y-Resistencia-Henry-Giroux.pdf

[3] Rezende F, Ostermann F. Hegemonic and counter-hegemonic discourses in science education scholarship from the perspective of post-critical curricular theories. Cultural Studies of Science Education. 2020;**15**(4):1047-1065. DOI: 10.1007/s11422-019-09969-0

[4] Rezende F, Ostermann F, Guerra A. South epistemologies to invent post-pandemic science education. Cultural Studies of Science Education. 2021;**16**:981-993. DOI: 10.1007/s11422-021-10091-3

[5] Adorno TW, Horkheimer M. Dialectic of Enlightenment. Nueva York: Seabury Press; 1972. Available from: https://pdf.yt/d/TJ7HxrAly-MtUP4B

[6] Morrow RA, Brown DD. Critical Theory and Methodology. Vol. 1-3. California, USA: SAGE Publications; 1994. DOI: 10.4135/9781452243696

[7] Giroux H. Teoría y resistencia en educación: Una pedagogía para la oposición. Ciudad de México: Siglo XXI Editores; 1992. 332 p. Available from: https://otrasvoceseneducacion.org/archivos/381814

[8] Gramsci A. Cuaderno De La Cárcel Tomo, 1, 2, 3, 4, 5, 6. Cuadernos 1-29: Ediciones Era/Institución Gramsci.1981. p. 2388. Available from: https://archive.org/details/antonio-gramsci-cuaderno-de-la-carcel

[9] Jarpa CG. Función política de la educación en el pensamiento de Antonio Gramsci. Cinta Moebio. 2015;**53**:124-134. Available from: www.moebio.uchile.cl/53/jarpa.html

[10] Kemmis S, McTaggart R, Nixon R. The Action Research Planner: Doing Critical Participatory Action Research. Singapore: Springer; 1988. p. 200. DOI: 10.1007/978-981-4560-67-2

[11] Freire P. Pedagogy of the Oppressed. USA: The Continuum International Publishing Group, University of California; 2005. p. 181. Available from: https://envs.ucsc.edu/internships/internship-readings/freire-pedagogy-of-the-oppressed.pdf

[12] de Sousa B. Santos: Construyendo las Epistemologías del Sur Para un pensamiento alternativo de alternativas, Volumen I. Buenos Aires, Argentina: CLACSO; 2019. p. 630. DOI: 10.2307/j.ctvt6rmq3

[13] Connel EW, Ashenden DJ, Kessler S, Dowsett GW. Making the Difference. Sydney: George Alien and Unwin; 1982. p. 228. DOI: 10.1086/446422

[14] Meneses MP. Pensando desde el Sur y con el Sur. In: Boaventura de Sousa Santos, editor. Construyendo las Epistemologías del Sur: para un pensamiento alternativo de alternativas. Buenos Aires: CLACSO; 2018. p. 203-2000. Available from: https://estudogeral.uc.pt/bitstream/10316/83438/1/Introduccion%20a%20las%20Epistemologias%20del%20Sur.pdf

[15] Filc D. ¿Es la resistencia siempre contrahegemónica? Journal of Political Ideologies. 2020;**26**(1):23-38. DOI: 10.1080/13569317.2020.1825281

[16] Solorzano B, Delgado-Bernal D. Examining transformational resistance through a critical race and Latcrit theory framework. Chicana and Chicano Students in an Urban Context. 2001;**36**(3):308-342. DOI: 10.1177/0042085901363002

[17] Sañudo L. ¿La resistencia social es siempre contrahegemónica? Llevar la abstracción a la práctica metodológica. In: Actas La Educación Patrimonio Mundial: La Educación en Tiempos de Transformación Digital, Sostenible, Inclusiva y Justa del Congreso EIDE XVII, 14 al 18 de noviembre de 2023, Alcalá de Henares, España: Organización de Estados Iberoamericanos; 2024. p. 87. Available from: https://oei.int/downloads/blobs/eyJfcmFpbHMiOnsibWVzc2FnZSI6IkJBaHBBbkY4Iiwi ZXhwIjpudWxsLCJwdXIiOiJibG9iX2lkIn19--52f9002486e950ae05d7ef603f872a7be88dfd57/ACTAS%2017_%20EIDE_10_05_2024_Rev_Final.pdf

[18] Miranda BS, Ortiz BJ. Los paradigmas de la investigación: un acercamiento teórico para reflexionar desde el campo de la investigación educativa. RIDE. Revista Iberoamericana para la Investigación y el Desarrollo Educativo. 2020;**11**(21):e064. DOI: 10.23913/ride.v11i21.717

[19] Fraser N. Unruly Practices: Power, Discourse, and Gender in Contemporary Social Theory. NED-New ed. Minneapolis, Minnesota, USA: University of Minnesota Press; 1989. Available from: http://www.jstor.org/stable/10.5749/j.cttts7ps

[20] Sañudo L. Epistemología y fundamento lógico para la investigación de las escuelas que resisten (EscuelasQResisten). In: Simposio Investigar las escuelas que resisten. Construyendo el fundamento epistemológico, teórico y metodológico. Memoria Electrónica del XVII Congreso Nacional de Investigación Educativa, 4 al 8 de diciembre de 2023, Tabasco, 2024. Ciudad de México: COMIE; 2024. ISSN 2594-2433

[21] Bell C, Raffe D. Working together? Research, policy, and practice: The experience of the Scottish evaluation of TVEI. In: Walford G, editor. Doing Educational Research. London: Routledge eBook; 1991. ISBN 9780203409480

[22] Somekh B. The contribution of action research to development in social endeavours: A position paper on action research methodology. British Educational Research Journal. 1995;**21**:339-355. DOI: 10.1080/0141192950210307

[23] Goessling KP. Increasing the depth of field: Critical race theory and photovoice as counter storytelling praxis. The Urban Review. 2018;**50**:648-674. DOI: 10.1007/s11256-018-0460-2

[24] Solorzano DG, Yosso TJ. Critical race and LatCrit theory and method: Counterstorytelling. International Journal of Qualitative Studies in Education. 2001;**14**(4):471-495. DOI: 10.1080/09518390110063365

[25] Van Manen M. Investigación educativa y experiencia vivida. Barcelona: Idea Books; 2019. ISBN 84-B236-283-6

Chapter 22

Perspective Chapter: Court Interpreters in the Pursuit of Sustainable Development Goals

Akungah O'Nyangeri, Zaja Omboga and John Habwe

Abstract

The work of court interpreters and translators interconnects with several Sustainable Development Goals (SDGs), particularly those related to access to justice, equality, human rights, and peace. As linguistic intermediaries within the legal system, they make a substantial contribution to fostering profound understanding of the laws and policies that underpin the global commitment to sustainable development. This chapter explores some contributions made by court interpreters and the impediments constraining the realization and achievement of SDGs. In addition, the chapter examines the nuanced challenges faced by interpreters in navigating diverse legal landscapes and cultures and the strategies to empower them to better support the achievement of SDGs.

Keywords: sustainable development goals, legal understanding, court interpretation, access to justice, agenda 2030, global development

1. Introduction

The pursuit toward the realization and achievement of Sustainable Development Goals (SDGs) as outlined by the United Nations [1] necessitates, in several senses, is anchored in a robust understanding of the legal landscape as a fundamental pillar for accessing and achieving justice, equity, and positive and sustainable societal development. The SDGs are a set of 17 goals aimed at addressing global challenges such as poverty, inequality, climate change, and sustainable development [2]. Agenda 2030, a global initiative that all United Nations Member States adopted in 2015, outlines these objectives. Central to this endeavor is the recognition that legal systems are innately implicated in the realization of these goals since they serve as intricate frameworks that govern societies, enable access to justice, and generally underscore the realization and sustainability of virtually all these SDGs. Consequently, in contexts such as these, court interpreters play a crucial role in providing linguistic bridges, dismantling communication barriers posed by language differences, and enabling the realization of common and collective aspirations as informed by diverse legal systems and multicultural contexts. It is critical to note that the diverse strategies deployed for the realization and sustainability of the SDGs in many of the signatory countries are actualized in multilingual and multicultural contexts, which characteristically entail mediated communications. It, therefore, follows that the pursuit of SDGs inevitably obliges the provision of translation and interpretation services in diverse situations, but notably in the communication of the formulation of policies and guidelines geared toward the realization of these goals. Thus, the centrality of translators and interpreters is centrally located in the provision of accurate and culturally sensitive interpretations, the consequence of which means that interpreters not only enable and facilitate communication on issues of SDGs, they empower people—policy formulators, implementers, and beneficiaries, irrespective of their linguistic backgrounds. They enable the comprehension of the legalities of policy, the constitution of implementation frameworks, and most evidently, mediated communications in legal proceedings. Translators and interpreters facilitate the expression of the concerns of the parties to the realization of the SDGs and overall engage meaningfully in the pursuit of justice.

The nexus between legal systems and SDGs is intrinsically implicated in diverse legal frameworks routinely established to provide the bedrock for access to justice, protection of fundamental and human rights, and societal order. The realization of the SDGs as enshrined within the 2030 UN Agenda, as much as they present a holistic roadmap for global development spanning a spectrum of issues from poverty alleviation to environmental sustainability [1], must be predicated on viable and realizable legal frameworks. The achievement of these ambitious goals is, as such, contingent not only on the enactment of legislative and policy measures but also on the development of the capacity of individuals to comprehend and navigate the intricacies of legal processes. Overall, it is important to underscore the fact that mediated communication, as enabled by translation and interpretation, is not an adjunct to the realization of the SDGs but rather a critical component in the overall realization of these goals.

This chapter is thus an exploration of the symbiotic relationship between legal formulation, legal frameworks, and legal understanding and their intricate relations in the pursuit of the SDGs. It encompasses the multifaceted role of court interpreters and translators in the sensitization, implementation, realization, and sustainability of SDGs, highlighting the critical, though usually muted, contributions made by interpreters in enabling the realization and sustainability of SDGs, promoting fairness in the sharing of SDGs' benefits, equality, and social advancement within the intricate fabric of legal systems globally.

2. Sustainability and court interpretation studies

The implementation, realization, and sustainability of SDGs in multilingual and multicultural societies are always enabled through mediated communication, and are principally provided by court interpreters. This is predicated on the assumption that SDGs are not haphazard declarations but rather critical development goals whose realization seeks to guarantee fairness. Yet, in circumstances where such fairness is unrealizable and issues of the distribution and sharing of responsibilities, resources, and benefits of the SDGs require arbitration, interpreters are always critically situated to facilitate access to equity and justice. However, in terms of scholarly engagements that have canvassed diverse implementation and sustainable issues of SDGs, the intersection of legal, particularly courtroom interpretation and translation, informing and underscoring the viability of SDGs is an area of research that is largely unexplored in contemporary scholarship. In view of the foregoing theoretical postulations, this section provides a general overview of the academic research pertaining to sustainability and court interpretation, with a specific emphasis on its potential role in advancing the achievement and sustainability of the SDGs.

Available studies on the genesis, implementation, and sustainability of the SDGs have largely focused on diverse areas of convergence with the SDGs, such as the sustainability of health [3], education [4, 5], and economic goals [6]. However, studies that focus on the legalities of SDGs, especially legal communication as embedded in or as animated by SDGs in general and, specifically manifested in interpreted courtroom communication, are scanty. Yet, studies on interpretation and translation as a language service geared toward enabling the realization and sustainability of the SDGs are severely limited. Interpretation-oriented investigations are critical in highlighting the fundamental role that court interpreters undertake in facilitating communication between participants in legal proceedings pertaining to or in some intricate ways relating to SDGs in overt or covert ways. This is notably so in multilingual and multicultural contexts of the implementation of SDGs, where participants, either in general legal communication or legal arbitrations, do not share a common language spoken in a given country's court system [7]. Whereas access to justice is considered a fundamental human right [8], it is equally a fundamental development right, such that in situations necessitating mediated communication, the availability of the language services of a skilled interpreter is pivotal in ensuring fair and impartial communication and, notably, fair legal proceedings. The need for such studies is normally anchored in the realization that their availability is critical to emphasizing the significance of language access in attaining justice and equality.

The connection between court interpreting and SDGs is a critical study area that has received relatively little scholarly attention. The interplay of justice and sustainable development has, on its part, been investigated and scantly examined [9]. However, there is a significant lack of research that addresses the role of court interpreters in supporting, advancing, and underpinning the SDGs, especially in African multilingual, multiethnic, and multicultural contexts. The Kenyan linguistic, ethnic, and cultural mosaic is illustrative of this diverse and convoluted African reality with regard to the implementation, achievement, and sustainability of the UN's SDGs.

The foregoing discussion above presents a strong basis for rationalizing the recognition of the importance of court interpreters in facilitating equitable access to justice, especially in multilingual and multicultural settings. However, there is a discernible paucity of research that specifically investigates the roles of court interpreters in attaining sustainable development goals. The literature review that has been attempted here, even in its paucity, is an attempt that seeks to emphasize the centrality and significance of court interpreters in enabling access to justice and draws attention to the lack of study on their role in achieving sustainable development goals. This chapter, as such, is an attempt to address the existing scholarly gap in establishing the linguistic and communication nexus between language, law, interpretation, and the achievement and sustainability of the SDGs.

The data analyzed in this chapter was collected from both the field and the library. The overall debate is predicated on data collected from Kisii Law Courts, a regional court within the Kenyan court system. The data was collected from 2020 to 2022, comprising written notes obtained from courtroom observations by the first author and audio recordings of interpreted court proceedings, which were later transcribed and analyzed. These data are complemented by court rulings and judgments that were purposefully selected from the Busia Law Courts, another regional court situated along the Kenya-Uganda border town of Busia. The overarching interest in the examination and analysis of this data is the need to present a general overview of what is entailed in mediated communication generally and courtroom interpretation and translation in particular, and how the same manifests in contributing to the broader understanding of developmental and societal objectives envisaged in the sustainable development agenda. The method of analysis followed a communicative approach to translation and interpretation, largely focusing on the meaning transfer of the ST into the TL, with particular interest in the achievement of associated SDGs.

3. The role of court interpreters in advancing SDGs

We have noted in the foregoing sections that SDGs, as envisioned in the UN development agenda [1], are not insular to legal misinterpretations generally and courtroom arbitrations in particular. Consequently, court interpreters play a crucial role in the realization and achievement of the SDGs in terms of enabling access to and promotion of justice, equity, fairness, the long-term sustainability of development goals, and societal peace. Beyond the confines of language transmission through translation and interpretation, court interpreters stand as unsung heroes who weave communication and linguistic bridges that enable understanding and interaction spanning diverse legal, linguistic, and cultural landscapes. In this section, we explore the diverse roles played by court interpreters in the context of the SDGs. It highlights their crucial participation in promoting access to justice, facilitating conflict resolution and peacebuilding, addressing language inequalities, upholding human rights, supporting immigrants and refugees, promoting education awareness, and ultimately contributing to the creation of more inclusive, just, and robust legal institutions, assuming that these are the basic fundamentals of the realization and sustainability of the SDGs. The interwoven complexities of legal proceedings are explored in terms of underscoring the profound impact interpreters wield in fostering a world where development and the achievement of SDGs are intrinsically intertwined with the tenets of peace, justice, and strong institutions, not merely as aspirations but tangible realities.

3.1 Access to justice, conflict resolution, and peace building

There are several ways in which court interpreters and translators can be shown as playing a fundamental role in enabling and promoting access to justice, and in so doing, their language mediation services overly align with the salient objectives of SDG 16. The challenge of access to justice, especially in contexts of the implementation of SDGs and particularly in ensuring equity in the sharing of development and social benefits, may sometimes compromise social justice, especially for marginalized populations [8, 10]. The significance of the inclusion of minorities in the overall frameworks of the SDGs is linked to the contributions made by minorities in ensuring that individuals, irrespective of their language proficiency, can fully participate in legal proceedings where arbitration becomes necessary, thus underscoring the principles of inclusivity and justice as envisioned in these development goals. SDG 16, for instance, places a great emphasis on peace and justice attainable through building strong institutions. Facilitation of effective communication in multilingual and multicultural legal settings can go a long way in underscoring the realization of this objective. It is in this sense that legal interpreters and translators can be viewed as contributing positively to conflict resolution and the promotion of peace, aligning with the objectives of SDG 16.

The accuracy of translation and interpretation in multilingual and multicultural contexts of SDG implementation is of utmost importance in facilitating the effective functioning of legal and judicial institutions [7], which act as critical components supporting the development of strong and accountable institutions, which are key components of SDG 16. Whereas the following example is not taken from an actual arbitration of an SDG, it nonetheless demonstrates what happens to arbitration where the court system is characteristically unaccountable as an institution. In a case taken from Kisii Law Courts in Kenya, *The Republic v. Amos Moturi Mwagi, Case Number: MCCR/27/2021,* the matter presented for arbitration is theft, in which, because of operational flaws on the part of the court as an enabling institution, the case was repeatedly postponed until the court recognized the need to express its regret to the accused for the delays caused by the court's own operations.

> *Magistrate: The* ***court regrets*** *that this is beyond our capacity, we promise that this is the last time we seek for adjournment.*
>
> *Interpreter: Kesi yako haiwezi kuendelea leo na imeahirishwa hadi tarehe mbili mwezi wa saba. {Your case can't proceed today and has been adjourned until the second of July}.*

In this example, given that the court proceedings are mediated through interpretation from English to Kiswahili, the interpreter's translation accurately transmits the core information of the adjournment but fails to capture the emotional and emphatic components conveyed by the magistrate. Emotions have a bearing on meaning-making in courtroom communication [11]. The magistrate's original statement contains expressions such as regret and the promise by the court that the situation will not reoccur again, which convey a sense of remorse, accountability, and commitment. The fact that a court system resorts to apologizing to a litigant demonstrates the potential dangers weak legal institutions pose to the realization and achievement of sustainable development. The apology, as tendered here, does not only undermine the litigant's confidence in the legal proceedings but also diminishes the overall effectiveness and confidence in the judicial system.

In the context of promoting equal access to justice and aligning with the objectives of SDG 16, whose core principle is the promotion of peaceful and inclusive societies for sustainable development, the provision of access to justice for all and the building of effective, accountable, and inclusive institutions at all levels ideally speak to the necessity of accurate and effective interpretation. For instance, in this example, while the basic information is conveyed, the failure to capture the emotional tone results in a loss of important contextual information. This loss impacts the parties involved in the legal proceedings, as they do not fully understand the court's sentiments regarding the adjournment. Effective communication, including conveying the emotional aspects of the message, is essential for building trust in legal and judicial processes because that is a cardinal principle in the realization of the objectives of SDG 16. The interpreter's failure to convey the emotional nuances frustrates the goal of promoting strong and accountable institutions, as it may lead to misunderstandings or perceptions of insensitivity on the part of the court.

The effectiveness of legal proceedings in the implementation, realization, and achievement of sustainable development presupposes the presence of precise, accurate, and reliable communication, which is critical in upholding principles such as fairness, justice, and the safeguarding of rights. Any shift in meaning, especially in mediated courtroom interactions, portends significant adverse impacts on the trajectory and outcomes of legal proceedings, affecting the presentation of evidence, formulation of arguments, and decision-making processes, ultimately impacting the attainment of SDG objectives of inclusivity. Though the following example is not specifically related to the realization of SDG 16, it demonstrates yet another institutional weakness underlying the realization of fairness and inclusion. In the succession case of *Peter Nyandusi Ombune v. Johnson Mokua Ombune, Case Number: MCSucc/30/2019*, in which a respondent is asked about his opinion on the suggestion proposed by the complainant as a solution to the matter, his response in Ekegusii, a regional language spoken in the locale of Kisii Law courts, may mean he agrees with the suggestion or is totally opposed to it. The court had ordered that the disputing parties reach an out-of-court settlement and report to court in an hour on their resolution. The legal representatives of the parties required translation as the disputants were Ekegusii speakers. After a long argument, the arbitrator asked the complainant for the final word, and he said:

> *Complainant: Bono igo ndatebie momura ominto arore* ***arusie*** *ebituma asimekire kwane rigesa eri rigocha erio antigere ndeme kwane {For now I can tell my brother to see into it that he removes his maize in the forthcoming harvest season to allow me farm in my land}*
>
> *Interpreter: Now I can tell my brother to* ***harvest*** *his maize in the coming harvesting season then pave way for me to farm in my land.*
>
> *Arbiter: What can you say on this?*
>
> *Interpreter: Ninki oratebe ase ayio? {What can you say, on this?}*
>
> *Respondent: Inche ndusie? Ntorore! {I remove [my maize]? We shall see!}*
>
> *Interpreter: I remove maize? I will see to that.*

In the initial statement, the complainant expresses the intention for the brother to harvest his maize to allow him to farm on the land under dispute during the upcoming planting season. But the interpreter's rendition reflects a pragmatic meaning: a shift from "remove" (*arusie*) to "harvest" (*asamore*). This reflects a change in the lexical choice to better convey the intended meaning in the target language. In any case, the Ekegusii term for harvest is *"asamore."* It ensures that the respondent understands that the complainant wants the maize to be harvested, not simply removed. "*Arusie*" conveys the same message but in a more direct and assertive manner, emphasizing the action of harvesting the maize as a prerequisite for the complainant to farm. This shift in pragmatic meaning aims to ensure clarity and effectiveness in communication; thus, in the realization of inclusivity as a critical component of SDG 16, legal institutions are obliged to ensure clarity and effective communication.

The essence of enabling clear communication, as alluded to in this scenario, positions the interpreter as a critical cog in supporting the principles of peace, justice, and strong institutions. The arbiters are accorded the opportunity to follow the interaction between the disputants through an interpreter. The ability to navigate and mediate such disputes contributes to building more inclusive and equitable communities, aligning with broader SDG objectives related to social and economic development. However, it is important to note that in this particular case, the interpreter misrepresents the respondent's stance, which potentially undermines the core tenets envisioned in SDG 16. While the respondent is expressing reluctance and even making a veiled threat by saying, "I remove [my maize]? We shall see!" The interpreter translates this as, "I remove maize? I will see to that." The respondent's reaction is characterized by a challenging tone. This indicates resistance or opposition to the request. In essence, his statement can be interpreted as, "You expect me to remove the maize from the farm? You are damn wrong." He is threatening that he will not do that and is not ready to move out of that farm in the foreseeable future. However, the interpreter's translation, which obviously fails to consider the extralinguistic features of tone and stress, emphasizes the words used by the speaker and renders this as an acceptance of the offer by the complainant, and the respondent has purposed to move out of the land as soon as the harvesting is done. Accordingly, in instances where the translation fails to adequately convey the intense emotional response conveyed by the speaker in the initial statement, such failure undermines the essence of strong legal institutions. The inaccurate portrayal of the respondent's refusal and threat as acceptance could potentially obstruct the achievement of the SDGs, particularly in the areas of peace, justice, and strong institutions. Effective communication is crucial in resolving disputes, and misinterpretation can lead to misunderstandings and conflicts and hinder the establishment of just and inclusive societies.

The critical debate being canvassed here is that misrepresentation of litigants' position in matters brought to court for arbitration contributes to a breakdown in communication and escalates tensions between litigants, thus potentially undermining the facilitation of resolutions. Fair and accurate interpretations are essential for promoting dialogue, understanding, and conflict resolution, all of which are integral to achieving the goals of peace, justice, and strong institutions. The advancement and achievement of the SDGs presuppose that interpreters ordinarily strive to provide unbiased, accurate, and culturally sensitive translations, ensuring that the parties involved fully understand each other's perspectives and consequently foster an environment conducive to conflict resolution and sustainable development.

The examples presented here, highlight the intricate role that interpreters play in legal settings and the broader implications of their work on the administration of justice. Accurate translation and interpretation are not merely linguistic tasks; they are integral linguistic and communication components critical to upholding the principles of fairness, transparency, and accountability within legal and judicial institutions, all of which are central to the objectives of SDG 16. To better align with the goals of SDG 16, interpreters should aim for not only linguistic accuracy but also strive to convey the tone, emotions, and nuances of the original message to ensure fair and effective communication within the legal system.

3.2 Ensuring access to justice for migrants and refugees

Migrants and refugees are in many cases victims of inequality both in their countries or locations of origin and in host countries or locations. It is not uncommon that migrants and refugees do find themselves in need of legal services, especially when they appear in multilingual and multicultural court systems. In essence, therefore, they end up being beneficiaries or victims of mediated language services depending on how interpreters and translators contribute to effective communication in legal proceedings. This is particularly important for achieving SDG 10 by ensuring that vulnerable populations have access to legal services and protection [1]. An apt illustration of this relates to a case where, ten Eritrean nationals appeared before the Senior Principal Magistrate at Busia, a regional court in the Kenya court system, charged with being unlawfully present in Kenya without a valid entry visa or permit. They could neither speak English nor Kiswahili, the languages of the court. The court procured the services of an Arabic interpreter during plea taking, a language they understood and admitted the offense or charge, agreed with the facts of the case, and pleaded for leniency in mitigation. They were eventually convicted, sentenced, and an order made for their repatriation to their country of origin upon completion of the sentence. It can be argued that in this case, the fate of these refugees was decided on account of skewed information gleaned from an inexperienced and non-professional interpreter since the accused persons moved to the High Court and filed an appeal on grounds they never understood the charges. Clearly, therefore, skewed language interpretation was a barrier and an impediment that undermined proper plea-taking process.

The contribution of the interpreter in this instance and other similar scenarios is crucial in facilitating effective communication between the migrants and the legal system, ultimately supporting the achievement of SDG 10. SDG 10 aims to reduce inequalities within and among countries, and ensuring access to justice for vulnerable populations, including migrants and refugees, is a key component of this goal. The interpreter played a vital role in overcoming language barriers during the legal proceedings. By providing translation services in Arabic, a language the accused persons understood, the interpreter enabled the migrants to actively participate in the legal process. This, in turn, ensured that their rights were protected and that they had access to legal services. The interpreter facilitated the migrants' understanding of the charges against them, allowing them to admit to the offense, agree with the case facts, and plead for leniency during the mitigation phase. This underscores the importance of language access in ensuring that individuals, regardless of their linguistic background, can effectively engage with the legal system.

By breaking down language barriers, the interpreter contributes to reducing inequalities in the legal process. The migrants were able to fully participate in their defense, leading to a fair and just legal outcome. This aligns with the overarching goal of SDG 10 to reduce inequalities within the justice system. The legal process, including the conviction, sentencing, and order for repatriation, was carried out with the migrants having a clear understanding of the proceedings. This ensures that the legal actions taken are in compliance with the law, promoting fairness and justice. In summary, the interpreter's role in this case supported SDG 10 by ensuring that migrants had access to legal services, were treated fairly within the legal system, and were able to participate effectively in the legal proceedings despite linguistic differences.

3.3 Educators in the pursuit of sustainable development goals

Interpreters and translators contribute to raising awareness and understanding of legal processes, rights, and responsibilities within diverse linguistic communities [12]. This aligns with various SDGs that emphasize education, awareness, and information dissemination. Quality education, as outlined in the SDGs, specifically in Goal 4, emphasizes the importance of ensuring inclusive and equitable quality education for all [1]. The goal stresses the importance of equitable access to education, aiming to bridge gaps and eliminate disparities among different groups and regions. This involves addressing issues such as gender inequality, economic disparities, and geographic limitations [2]. The emphasis on quality education underscores the need for educational systems that provide effective learning outcomes. Quality education goes beyond merely ensuring access; it also focuses on the relevance, effectiveness, and impact of the education provided.

The goal recognizes the importance of education throughout one's life. It advocates for continuous learning opportunities beyond traditional schooling, acknowledging the dynamic nature of modern societies and economies. The goal promotes the development of skills and knowledge that contribute to global citizenship. This includes fostering an understanding of global issues, cultural diversity, and a sense of responsibility toward sustainable development. In the context of court interpretation and translation, in as much as it is essential to recognize that the primary role of the judiciary is to interpret and apply the law rather than directly providing formal education, the judiciary can be viewed as a para-educational institution and court interpreters as executors of legal education. It is important to note that while the judiciary and court interpreters may indirectly contribute to legal education, their primary functions are centered on administering justice, interpreting laws, and ensuring fair legal processes. The formal responsibility for legal education typically lies with dedicated educational institutions, law schools, and bar associations.

Courts, especially superior courts, set precedents and shape public understanding of the law through their decisions and explanations. These precedents contribute to the development and clarification of legal principles, indirectly serving as educational material for legal professionals and the public. The reasoning provided in judgments can serve as a source of legal education for individuals seeking to understand legal concepts and principles. The critical understanding of this postulation with regard to mediated communication in terms of interpretation and translation is that, court interpreters do not only bridge linguistic gaps but also contribute significantly to upholding the principles of fairness and justice within the legal system, by continuously educating the lay participants on legal processes and procedures. By accurately conveying the spoken words of all parties, they enable a transparent exchange of information, allowing each participant to fully grasp the nuances of the case. This not only aids in preventing misunderstandings but also safeguards the right to a fair trial for individuals who may face language barriers. Moreover, court interpreters must possess not only linguistic expertise but also a deep understanding of legal terminology to accurately convey the legal intricacies involved.

In essence, the multifaceted roles undertaken by court interpreters and translators extends beyond mere language translation to encompass the vital task of preserving the integrity of legal proceedings and ensuring equitable access to justice for all involved parties. Consequently, their multifaceted role extends beyond mere translation, encompassing a crucial function in safeguarding the integrity of the legal process, facilitating the process of legal education by ensuring that all participants fully comprehend the proceedings. In the following example, on the basis of data collected from Kisii Law courts, the education and educating role played by interpreters is illustrated. *The Republic v. Zachariah Onyonka Mogire & Jared Rianga Matibe, Case Number*: *HCCRC/7/2019,* the court interpreter had to educate a witness who was speaking very fast on the manner of presenting her testimony and the rationale behind it.

> *Interpreter: Omongína, okwane ng'ora ngóra* ***ase eng'encho ekoti nkorika ere eki ogokwana*** *erinde eche gosoma eki gwateba ekero ekounenkia konyora obomaene igoro ya nyagosira {mum, speak slowly because the court is writing what you say for purposes of future reference when it will retreat to consider the facts about what led to the death of the late}*

The interpreter provides guidance to the witness by asking her to speak at a slower pace and goes further to explain the significance of the court documenting the testimony for further consultation during the assessment of the evidence. The interpreter's main responsibility is to enable effective communication between individuals who speak different languages. However, in this particular situation, the interpreter seems to be providing additional details regarding courtroom protocols and the importance of the witness's evidence. To the extent that the interpreter is helping the witness understand the necessity of clear and purposeful exchanges within the courtroom, as well as the significance of their testimony in the legal proceedings, it can be concluded that the interpreter is aiding in the litigant's education regarding courtroom protocols. Providing this guidance is crucial to ensuring successful communication and for the witness to fully understand the importance of their role in providing information to the court.

The interpreter's role in educating the witness aligns with SDG 16, which aims to advance peaceful and inclusive societies for sustainable development, guarantee universal access to justice, and establish effective, accountable, and inclusive institutions. The interpreter facilitates the witness's comprehension of courtroom protocols, upholding the desirable principle of equitable and fair access to the legal system. In addition, the interpreter delivers information in a clear and comprehensible manner, aligning with SDG 4, which seeks to guarantee inclusive and fair-quality education and foster lifelong learning opportunities for all. The witness's understanding of courtroom procedures might be seen as a type of education within the context of legal proceedings. The role of the translator is closely linked to SDG 16, which aims to advance peaceful and inclusive societies via the establishment of efficient, responsible, and transparent institutions.

Providing the witness with knowledge about courtroom procedures improves the overall effectiveness and transparency of the legal system. The interpreter's role in enabling efficient communication between the witness and the legal system contributes to minimizing inequalities, as specified in SDG 10. Promoting inclusivity in legal proceedings by accommodating those with linguistic challenges fosters a more equal justice system [11]. However, it is important to note that the interpreter's influence on the attainment of SDGs is contingent upon several factors, such as the accuracy of interpretation, sensitivity to culture, and the overall fairness of the judicial proceedings. Moreover, the SDGs encompass a wide range of interconnected objectives, and the interpreter's role is but one component of an intricate system designed to achieve sustainable development.

The idea of quality education in the SDGs reflects a holistic approach that goes beyond basic literacy and numeracy. Education is perceived as an effective means for individual growth, societal unity, and sustainable development, hence strengthening the overall welfare of individuals and communities [13]. To achieve this goal, we must tackle underlying issues, promote policy reforms, and foster a commitment to education throughout every sector of society. In view of this, we argue that court interpreters and translators improve the quality of education by ensuring that individuals, regardless of their language proficiency, can acquire information about legal processes, rights, and responsibilities. In order to enhance clarity and successfully transmit intended meaning in mediated courtroom communications, thus enabling the ST to be used in the target language. They offer additional information when explaining intricate legal terminologies and concepts, idiomatic expressions, or context-specific information that cannot be directly translated into the target language. By including the necessary additions, translators effectively bridge the linguistic gap between the source and target languages, thus enlightening the target audience

DOI: http://dx.doi.org/10.5772/intechopen.1005081

about legal proceedings. This is demonstrated in the following example drawn from *The Republic v. Victor Nyariki Nyabaro, Criminal Case No. E1285/2021*, a narcotics case, where the accused had pleaded guilty and had been accorded the opportunity to present his mitigation.

*Magistrate: **Mitigation.***

Interpreter: Una malilio yoyote ambayo ungependa korti iyaangalie itakapotoa uamuzi wake? {Do you have any concerns which you will wish the court to consider while making its decision?}

Mitigation, in legal parlance, denotes the deliberate action of reducing the severity or gravity of a certain situation. These are any factors or arguments that can alleviate or reduce the impact of the situation or penalty. The interpreter supplemented the interpretation with a comprehensive clarification to guarantee the accused person's comprehension of the intended message. This modification increases the audience's comprehension of the legal concept under discussion. The interpreter's role involves not only translating but also elaborating on the source text by providing relevant contextual information. This includes ensuring that the "concerns" presented by the accused encompass the factors that the court can take into account when making its decision. It explicitly states that these issues are not ordinary concerns but rather ones that must be considered in the court's decision-making process. Thus, the clarification establishes the specific limits and parameters within which the accused must act while addressing those issues. This additional information serves to ensure that the accused delivers information pertinent to the case, minimizes the risk of miscommunication, and ensures that the accused completely understands the intent of the magistrate's statement.

By removing barriers to language access, increasing knowledge of the legal system, and guaranteeing that people from all walks of life can take part in the judicial process, court interpreters play an essential pedagogical role for lay litigants beyond the scope of conventional education. Their role contributes to the overarching objective of enhancing the accessibility, transparency, and fairness of the legal system for anyone in society. To make sure the accused understood the court's intended purpose and the process itself, the court interpreter gave a rendition in one of the proceedings where a matter was withdrawn without including the legal foundation of the withdrawal. In *The Republic v. Nelson Rogito Ongiri & Benard Kamoi Onkoba, Case Number: MCCR/30/2021*, the translator aims to simplify the language used in order to facilitate understanding and ensure that the magistrate's intended message is effectively communicated. To do this, the translator opts to present a condensed version of the magistrate's statement that excludes the legal foundation of the ruling and instead focuses solely on the ensuing repercussions of this determination.

Magistrate: The case is withdrawn **under section 204 of the CPC**, *the matter is* **closed**

Interpreter: Kesi imeondolewa! {The case has been withdrawn}

The magistrate delivers a comprehensive and legally reasoned verdict in English, outlining the legal grounds for the acquittal of the accused. Conversely, the interpreter provides a concise overview of the ruling to improve understanding, appreciation, and the implications for the lay litigant. This is achieved by omitting intricate legal language. There are two possible explanations for the exclusion of the legal justification for acquittal: either it was intentionally omitted to enhance clarity, or it was considered to have limited analytical significance. The interpreter may have omitted the legal reference to enhance the efficiency of the interpreting process, streamline the interpretation, and enhance the comprehension for the accused. Lay participants in a proceeding face challenges due to the use of legal terms and technical vocabulary; as a result, interpreters may choose to provide a more straightforward and easily understandable rendition by omitting particular legal elements. The interpreter may have also considered the legal reference to be of little significance in relation to the accused's comprehension or overall effect of the message. He intentionally chose to focus on the transmission of the withdrawal of the case and subsequent acquittal as the key factors, rather than placing emphasis on the specific legislative provision involved.

The utilization of summary as a method for interpreting in courtrooms holds immense importance. It can be employed to teach lay people on law-related topics, as it enables the efficient delivery of information. Ideally, this should be done in a methodical manner so that the essence of legal provisions is not compromised just as much as avoiding compromising sustainable growth as envisioned in the SDGs. This occurs when summary as an interpretation method is not effectively deployed, leading to the failure to establish a common understanding among the participants. Misapplied summary usually results in distorted understanding of the messages being transmitted; it is disruptive in several senses. In the following example relating to an acquittal, during the acquittal of the accused in *The Republic v. Justus Ongiri Mainya Gesicho, Case Number: MCCR/15/2021*, the use of summary as an interpretation method, gave rise to significant inconsistencies in the information conveyed by the interpreter when compared to the prior account of the same acquittal.

> *Magistrate: The case is withdrawn* **under section 204 of the CPC,** *the matter is* **closed.**
>
> *Interpreter: Kesi imeisha kortinni, enda nyumbani! {The case ended in court, go home}*

The magistrate's statement establishes the legal grounds for dismissing the case and stresses the finality of the decision; the matter has been withdrawn, therefore suggesting that no more steps need to be taken. However, the interpretation provided in Kiswahili fails to sufficiently incorporate the legal foundation for the termination. The interpreter's goal was to clearly convey that the matter has been withdrawn and no other steps are necessary. The exclusion of this vital information creates perplexity and confusion among the relevant people, most notably the accused, who expected an in-depth clarification or legal context to be provided. A simplified and easily comprehensible explanation of the legal foundation may have convincingly demonstrated to him that the court had actually handled the issue impartially. In this particular instance, it can be debated that the court's ruling, based on constitutional considerations, may have led to the conclusion that the accused did not have a valid defense. However, the interpreter's highly simplified interpretation, delivered with a contemptuous attitude, ultimately portrays the court in an unfavorable light since it fails to adequately communicate the genuine legal rationale behind the withdrawal. The interpretation could downplay the importance of the particular legal action carried out by the magistrate, namely the withdrawal of the case under Section 204 of the Criminal Procedure Code. Consequently, the parties involved may not have full knowledge of the legal consequences of the decision or the exact rationale for the withdrawal of the case. This interpretation does not provide clarity to the lay litigant on the distinction between it and the act of withdrawing a matter under Section 87A of the same code.

This view, in respect of the foregoing arguments, perpetuates the misconception that litigants neither understand nor are interested in legal terminology and that their primary focus is solely on the court's verdict. The interpreter's inadvertent avoidance or oversimplification of legal terminology exacerbates the problem of legal literacy among the litigants. While court interpreters must prioritize accurate and easily understandable communication between parties, they also need to find a balance in their use of legal terminology. As part of the judicial staff, court interpreters have a golden opportunity to educate lay litigants and promote better awareness of legal issues, thus enhancing legal literacy. Interpreters must understand the need for not just facilitating communication but also bridging the divide between legal terminology and the understanding of those participating in court processes. By doing this, they can educate and equip litigants with a deeper understanding of the legal processes they participate in, contributing to a more equitable and knowledgeable court system.

In conclusion, the role of court interpreters in legal proceedings extends far beyond language translation, encompassing a crucial function in safeguarding the integrity of the legal process and facilitating the process of legal education. Their multifaceted responsibilities align with the United Nations Sustainable Development Goals, particularly SDG 4, which focuses on inclusive and fair quality education, and SDG 16, which vouches for peaceful and inclusive societies with universal access to justice. Court interpreters contribute significantly to the transparency, accessibility, and fairness of the legal system by ensuring effective communication and educating lay participants on courtroom protocols. However, the effectiveness of their pedagogical role is contingent upon factors such as accuracy, cultural sensitivity, and the overall fairness of judicial proceedings. As integral members of the judicial staff, court interpreters have a unique opportunity to bridge the gap between legal terminology and lay understanding, thereby promoting legal literacy and contributing to a more equitable and knowledgeable court system. Balancing the use of legal terminology while ensuring clarity is crucial, and by doing so, court interpreters play a vital role in advancing the broader goals of education, justice, and inclusivity outlined in the SDGs.

3.4 Achieving sustainable development through cultural negotiation

Culture is a crucial feature in the establishment, advancement, and realization of the United Nations' SDGs, given that culture, in its multifaceted definitions [14], foundationally and fundamentally influences all aspects of society, including the formation and sustenance of attitudes, beliefs, behaviors, and social norms. The interaction between cultural factors and the SDGs is complex and multifaceted in terms of influencing the way people and societies perceive, embrace, and respond to these global objectives. Understanding and accounting for social and cultural dynamics underscores the eventual, efficient, and sustainable execution and ultimate achievement of the SDGs. Furthermore, cultural diversity can either facilitate collaboration or pose challenges in attaining global goals. Acknowledging and respecting various cultural points of view is crucial for formulating comprehensive approaches that effectively resonate with various cultures in the context of implementing the SDGs. Adapting interpretations to reflect local values, beliefs, customs, and principles increases the chances of being accepted and maintained throughout time.

Cultural practices play a pivotal role in influencing the success or hindrance of specific SDGs. A concrete illustration of this can be found in the cultural practices related to Ekegusii land resource use and conservation, which can significantly impact efforts toward achieving environmental sustainability (SDG 13). Despite the fundamental importance of the concept of cultural equivalents, its practical application in court interpretation and translation poses notable challenges. One major obstacle lies in the careful selection of appropriate equivalents that accurately convey the nuanced cultural and legal aspects of the original text. Given that different languages and cultures possess distinct indicators, the identification of suitable equivalents in the TL demands a profound understanding of both legal and cultural contexts. A pertinent instance of this challenge arises in the courtroom, as evidenced by the decision of a court interpreter to employ the phrase "parental allocation" as the equivalent for the cultural concept "*emonga*" in the land dispute case between *Jesse Ng'ang'a Ombachi and Joyce Kemunto Obara, Rosemary Moraa Obara, and three others, Case Number: MC.ELC/10/2019*. In this case, the petitioner sought the revocation of a title deed and the issuance of a new one, excluding the name of one of the respondents. This example underscores the intricate interplay between cultural intricacies and legal proceedings, emphasizing the need for precise language translation in navigating such complexities.

> *Petitioner: Oboremo obwo nabwo nigo bobwataine* ***n'emonga yomogaka*** *akorema {the parcel is adjacent to the one my father assigned himself to till in his old age}*
>
> *Interpreter: That piece of land is next to the parental allotment, where the parents are farming in the meantime.*

When translating concepts that hold cultural value, choosing phrases that accurately represent the intended meaning is an indispensable strategy in dealing with cultural specifics normally embedded in mediated communication. It enables the consideration and taking into account of the cultural background and subtle distinctions marking one culture from another. Consequently, the interpretation rendition offered in the example above intimates that, though the purported English equivalent term "parental allotment" may suggest the concept of land reserved for a parent's personal use, it does not naturally encompass the precise cultural intricacies connected with "*emonga,*" a piece of land a parent allots to himself to continue farming until the death of both parents, then it is reverted back to the last-born son. The word "parental allotment" is a broad term that encompasses any distribution of land or resources specifically earmarked for a parent's usage. To accurately capture the cultural and familial aspects of "emonga," it would be useful to employ a phrase that expressly encompasses the cultural or traditional setting. As an example, one could consider a phrase such as "customary parental land reserve" to highlight the distinct cultural or familial customs associated with the distribution of land. The phrase appears to be a plausible endeavor to express the notion, as it includes aspects that imply a conventional or habitual practice of setting aside land for parental usage. This suggests that the activity is rooted in tradition or custom, in accordance with the cultural setting of "*emonga*." Additionally, it clarifies that the land reserve is linked to parents, aligning with the concept of a parent allotting land. The expression suggests that the property is deliberately reserved for a particular purpose, resembling the idea of a designated area of land.

The UN's SDG 4 is focused on culture and its place in sustainable development; it defines culture as "who we are and what shapes our identity," assuming that placing culture at the heart of development will ensure human-focused, inclusive, and equitable development [1]. Essentially, this means understanding the place of culture in the achievement of sustainable development that integrates national, regional, and international perspectives of development, thus providing a critical understanding of national developmental experiences, existing and emerging policies, as well as emerging frameworks of innovative and sustainable development. These broad objectives are at the core of what interpreters and translators do, mediating communication and thus subtly fostering cultural understanding and sensitivity, thus contributing significantly to the achievement of various SDGs that aim to build inclusive and diverse societies. Interpreters and translators' work transcends linguistic barriers to create linguistic and communication bridges that allow for effective communication between individuals from different cultural backgrounds. This not only facilitates the exchange of ideas but also promotes a deeper understanding of diverse perspectives informing cultures' involvement in sustainable development. In essence, their work goes beyond language translation; their work serves as a catalyst for building bridges of understanding and appreciation among people from various cultural backgrounds, ultimately fostering the creation of more inclusive and diverse societies.

In the example cited above, the interpretation rendered acts as a medium for conveying the land use practices of the Abagusii people of Kenya to other cultures; it promotes a deeper understanding and appreciation of the community's land use cultural philosophy. The court interpreter, thus, plays a vital role in establishing cross-cultural understanding by effectively articulating the complexities of the Abagusii's land resource management processes, procedures, policies, and practices, as well as the cultural frameworks upon which these practices are enacted. The process entails not only the translation of linguistic expressions but also the communication of the underlying cultural context, values, and traditions related to land use among the Abagusii. This interpretation allows people from different cultures to gain an understanding of the reasoning, importance, and sustainability of the Abagusii land use policy. The interpreter's role in facilitating cultural interchange and mutual understanding is achieved by enabling others to appreciate and comprehend these practices. This intercultural interchange can result in heightened consciousness and admiration for other forms of land utilization, promoting a more comprehensive and cooperative approach to sustainable development.

Court interpreters do not only translate words; their translations are also subtly involved in cultural exchanges that bridge cultural diversities. They help convey the nuances of language and the culture in which language is embedded, ensuring that the court understands the context and background of the individuals involved. This cultural sensitivity is essential for a fair and just legal process. Through cultural accommodation, they ensure that legal language is correctly interpreted and transmitted in an efficient way, thereby incorporating cultural awareness into their job. This optimizes the efficiency of legal education, especially in multicultural and linguistically diverse populations. Court interpretation accommodation entails the interpreter making necessary modifications to culturally loaded utterances so as to ensure that the rendition is in line with the linguistic and cultural backgrounds of the participants [12]. This may involve modifying the vocabulary, tone, register, and cultural references employed in the rendition to better align with the wishes and expectations, values, and cultural norms of the persons concerned. This is an example from the divorce case of *Charles Akora Mochama v. Yunes Kerubo Osoti, Case Number: MC.SEP/1/2019*. The case involves the interpretation of English into Ekegusii. The example demonstrates how the cultural elements of Ekegusii marriage rites influence the process of interpretation.

Magistrate: How are you related?

Interpreter: Naki mokorokana n'oyo? (How do you call this one? [points to the respondent]).

Witness: Oyo ... oyo ... bono ... igo nkomoroka ***baba. Ee, baba****! {This one ... this one ... eeehe ... I call her mother. Yes, mother!}*

Interpreter: She is his mother-in-law.

This example is presented here to enable us to argue that familial relationships are cultural manifestations at the core of the realization of sustainable development. In this case, the court interpreter struggles to capture the meaning of the witness's response while keeping cultural context and relationship-related terminology in mind. The Abagusii culture has unique cultural practices regarding the terminologies employed to mark relationships. The interpreter's decision to translate the witness's response as "*baba*" (mother) instead of "*makobiara*" (mother-in-law) displays a profound understanding of the cultural subtleties and the importance of displaying appropriate sensitivity and respect in the specific situation. In the Abagusii culture, it is normally discouraged for a son-in-law to use the conventional term "*makobiara*" when addressing his mother-in-law because it is not considered pragmatically appropriate, though it is linguistically correct. The phrase "mother-in-law" conventionally denotes the female parent of one's spouse. Nevertheless, it is well recognized that employing this term to refer to one's mother-in-law might be interpreted as impolite, especially when spoken by a son-in-law. Alternatively, it is considered more appropriate for the son-in-law to use the title "mother" as a symbol of respect and in line with prevailing cultural standards. The court interpreter's rendition skillfully includes these cultural subtleties specific to the Abagusii community, resulting in a rendition that can be more accurately described as a paraphrase and summary than a traditional interpretation.

The process of cultural adaptation in court translation and interpretation aims to guarantee equal access to justice for individuals from various cultural backgrounds. It enables efficient communication between legal practitioners and those engaged in legal proceedings. The interpreter demonstrates an in-depth understanding of the cultural context of the Abagusii community through his display of expertise in the specific vocabulary and social norms related to family relations. Interpreters, in performing their trade as such, are obliged to prioritize the recognition and understanding of cultural subtleties, standards, and conventions within the specific community they are interpreting for. This entails ensuring accurate and culturally and pragmatically appropriate renditions that are not offensive. This technique aligns with the aims of the SDGs as it effectively achieves the stated purpose of interpretation within the target culture.

The comprehension of the intricacies of situations involving individuals from different cultures in courtroom contexts requires legal professionals to possess a high degree of cultural knowledge and, subsequently, cultural communication competence. The idea is that such comprehension is critical in ensuring the rendition of interpretations that underscore the realization of equitable legal outcomes, fostering justice, and eliminating inequality within diverse linguistic and cultural communities. Court interpreters, therefore, have a duty to ensure judges and magistrates, who are not fluent in the language of the litigants, are enabled to access a thorough comprehension of the cultural subtleties conveyed by the witnesses. For instance, in the divorce case, *Isaac Nyaberi Nyokayo v. Mary Kwamboka Osuro, Case number: MCDC/2/2020*, it was necessary for the interpreters to educate the magistrate to understand the Ekegusii term "*omosuto*" and its cultural and philosophical framing. During the proceedings, a witness who was opposing the motion stated,

> *Witness: Korwa bakoro, ng'ombe tichiana koirana. Nonya* ***nomosuto*** *twarwete? {Since time immemorial dowry has never been refunded. Wasn't even "omosuto" [a brother's gift to his sister's son as a token to appreciate use of that sister's dowry in his marriage] donated?}*
>
> *Interpreter: From old days, cows have never been returned. They had even given that* **other cow***!*
>
> *Magistrate: Who had given the other cow?*
>
> *Interpreter: The family of the respondent.*

The witness uses a cultural term, "*omosuto*," which refers to an uncle's gift to his nephew as a token of appreciation. The provision of the gift can be attributed to the use of the dowry, which was originally paid for the marriage of the nephew's mother, who happens to be the uncle's sister. This token is usually one cow. The witness states that even "*omosuto*" had been donated, implying that the practice of giving this gift has been done by the brother to the witness. Accordingly, filing for a divorce, in a situation where all requirements had been met was in vain. The inclusion of this example here is critical in speaking to the cultural underpins of property ownership, consequently pointing to familial relationships, property, ownership and development sustainability. Consequently, when the interpreter, in this example, translates the witness's statement and renders it as "They had even given *that other cow!*" This translation fails to adequately convey the specific cultural concept of "*omosuto*" to the court participants yet this is critical for the magistrate to understand given that he will need that understanding in rending a verdict. The rendition provided brings the cultural competence of the interpreter into sharp focus. The attainment of cultural competence is of utmost importance for interpreters in order to effectively and accurately transmit meaning across diverse cultural contexts. In this particular case, it would have been advantageous for the interpreter to acknowledge the significance of the term "*omosuto*" and endeavor to offer a translation that aligns more closely with the cultural context, even if it necessitated a slightly longer explanation to convey its intended meaning. The interpreter should have provided a more comprehensive explanation of the term "*omosuto*" and its underlying significance instead of offering a simplistic translation as "*that other cow*."

The interpreter's failure to effectively communicate the culture-specific connotations of the term "*omosuto*," perplexes the magistrate. The magistrate's inquiry into the party responsible for giving the additional cow demonstrates the necessity of cultural context in legal proceedings. The interpreter, by accurately conveying the familial dynamics surrounding the term "*omosuto*," later on in this interaction enabled the magistrate to comprehend the cultural intricacies of the situation. According to the source statement, the witness affirms that the dissolution of the marriage is deemed impossible, given that the cultural prerequisites have been fulfilled. On the other hand, the interpretation posits that while all cultural prerequisites have been fulfilled, it fails to account for the practical challenges associated with dissolving the marriage contract. This situation highlights the importance of cultural competence and understanding in interpretation and translation given that there are many possibilities of cultural practices projecting into developmental issues and their sustainability.

Cultural competence among legal professionals and interpreters contributes to the effective functioning of legal institutions, particularly in multilingual and multicultural contexts. As in the scenario presented above, accurate interpretation of cultural terminologies, even if by way of explanation, was indispensable to enabling the magistrate to make an informed decision. Misinterpretation during interpretation often leads to an unjust resolution, hindering the goal of establishing a strong and just judiciary. Further, effective communication across cultures requires collaboration and partnerships. In the present instance, the importance of a collaborative effort between legal professionals and interpreters to ensure that cultural nuances are accurately conveyed is evident. The magistrate sought further clarification about the other cow from the court interpreter. Strengthening such partnerships contributes to achieving the broader goal of sustainable development.

The interplay between culture and the implementation and sustainability of SDGs, is intricate and has far-reaching ramifications in shaping attitudes, behaviors and societal norms always subtly embedded in the cultural dynamics of SDGs. The debate canvassed this far, has attempted to underscore the pivotal role culture plays in influencing perspectives on poverty, education, gender equality, and environmental sustainability–all essential elements of the SDGs. The case studies presented, particularly those involving court interpretation in the Abagusii community, highlight the challenges and significance of cultural adaptation in legal proceedings. Interpreters and translators play a vital role in fostering cross-cultural understanding, contributing significantly to the achievement of SDGs that aim to build inclusive and diverse societies. The examples presented this far, also emphasize the need for cultural competence among legal professionals, showcasing instances where accurate interpretation was crucial for just legal outcomes. In fostering collaborative efforts between legal professionals and interpreters, a pathway is paved toward achieving the broader goal of sustainable development, where cultural nuances are respected, understood, and effectively communicated. This debate ultimately reinforces the imperative for a comprehensive and culturally sensitive approach in the pursuit of the SDGs, recognizing that cultural dynamics are integral to the successful execution of global objectives.

3.5 Legal access, inequality reduction and inclusivity

Reduced inequality is one of the UN's SDGs, specifically as outlined by the United Nations in Sustainable Development Goal 10 [1]. The aim of this goal is to address various forms of inequality within and among countries. The goal aims to empower and promote the social, economic, and political inclusion of all, irrespective of age, sex, disability, race, ethnicity, origin, religion, or economic or other status. This involves specific efforts to ensure that marginalized groups have equal opportunities and are not left behind. It also addresses social inequalities, emphasizing the need to promote social inclusion and eliminate discrimination based on factors such as gender, age, disability, race, ethnicity, origin, religion, or economic status. This includes ensuring equal access to education, healthcare, and other essential services.

The recognition of the significance of global collaboration in tackling developmental disparities highlights the interconnected nature of many inequality-related issues, emphasizing that effective solution can only be achieved through collaborative endeavors among states. The court interpreter plays a crucial role in linking various entities that prefer collaboration throughout the entire process, particularly where SDGs demand arbitration. By providing language interpretation services, court interpreters help ensure that individuals with limited proficiency in the official language receive equal protection under the law. This is of utmost importance for marginalized and immigrant communities that may encounter language obstacles during legal proceedings. In the absence of interpreters, persons who have limited language proficiency may encounter challenges when actively engaging in court proceedings, thereby impeding their capacity to present evidence, comprehend charges, or effectively interact with court officials. Court interpreters facilitate meaningful engagement, fostering an equitable legal process.

Reducing inequalities is a multifaceted goal that encompasses the implementation of policies fostering equal opportunities, social protection, and empowerment across diverse segments of society. One critical aspect of achieving this goal involves ensuring that access to legal processes benefits all rather than being concentrated among a privileged few. In this context, court interpreters play a pivotal role in facilitating litigants who do not speak the language of the court, thereby affording them a valuable opportunity to interact with and participate in legal proceedings. They facilitate effective communication within the judicial system by offering interpretation services, thus resolving linguistic hurdles that may hinder full participation. This accessibility is particularly important to safeguarding fairness, as it ensures that all litigants, regardless of their linguistic background, are able to effectively convey their perspectives, present evidence, and understand the proceedings. This is consistent with the overarching goal of reducing inequalities by establishing a more equitable environment within the legal realm.

Furthermore, court interpreters play a crucial role in safeguarding social welfare by guaranteeing that people, irrespective of their linguistic competence, are able to participate in a fair and impartial judicial process. Court interpreters assist litigants by facilitating their ability to exercise their rights and actively engage in judicial proceedings. Empowering individuals is crucial for addressing inequalities within the justice system. Court interpreters play a crucial role in safeguarding this right by ensuring that language barriers do not hinder people of all backgrounds from properly exercising their rights within the legal system, irrespective of their linguistic or cultural background. Language barriers may exacerbate existing inequalities. Interpreters play a crucial role in overcoming language barriers, therefore, guaranteeing that people, irrespective of their level of proficiency in the language, have equitable opportunities to receive legal information and actively participate in the legal system. This is consistent with the objective of SDG 16, which seeks to advance peaceful and inclusive societies by ensuring universal access to justice. This is particularly important when the legal concepts under consideration exhibit a significant level of complexity or specialization.

The intricacy and arduousness of legal terminology pose challenges for comprehension, even among individuals who are fairly proficient in the language. Glossing the message in interpretation can ensure it remains clear and accurate. In *The Republic v. Charles Ayaro Nyakundi, Case Number: MCCR/86/2021*, the accused had appeared in court for an assault matter. During one of the scheduled hearings, the case was scheduled for mention. Nevertheless, the court record did not specify whether the accused person was in custody or had been released on bail. The magistrate sought to verify this information. The court interpreter had to utilize various glossing strategies to accurately convey the magistrate's intention. The numerous intricate legal terminologies obfuscated this intention, but the interpreter's goal was to accurately transmit it.

> *Magistrate: Are you out on* ***bond****?*
>
> *Interpreter: Uko nje kwa bond? (Are you out on bond?)*
>
> *Accused: E-ee? (pardon)*
>
> *Interpreter: Ulitoka ndani kwa bond? (Did you come out [of remand] on bond?)*
>
> *Accused: Ati? (What?)*
>
> *Interpreter: Umetoka nyumbani? (Did you come from your home?)*
>
> *Accused: Ndio. (Yes)*

In this exchange, the interpreter utilizes various glossing tactics to effectively communicate the meaning. The numerous attempts are meant to facilitate the active participation of the accused in the proceedings. He starts with a literal gloss strategy and then switches to rephrasing, ultimately resorting to simplification through adaptation. This demonstrates a commitment to the inclusivity of lay litigants in the judicial system. The accused finally responds in Kiswahili by saying, "*Ndio*," which translates as "Yes." This signifies the successful conveyance and comprehension of the initial aim. In this scenario, the interpreter acknowledges the necessity of modifying the translation to preserve the cultural subtleties and contextual allusions found in the original. To achieve this goal, the terminology is modified to align with the cultural and linguistic conventions of the TL while also preserving the original intent of the magistrate's question. Integrating the localized expression in the TL enables a precise portrayal of the intrinsic nuances seen in the original.

Inequalities can only be reduced through a comprehensive and collaborative approach that addresses disparities both within and between nations. As demonstrated, court interpreters play an important part in this attempt by breaking down language barriers inside the judicial system and allowing equal access to legal processes for people with limited language ability. They empower individuals to actively participate in legal proceedings, promoting a more inclusive and equitable society. Their role extends beyond mere translation, encompassing the intricate adaptation of legal terminology to align with cultural and linguistic norms. They play a crucial role in advancing equality and reducing disparities within the legal system, aligning with the broader global objective of developing a just and inclusive society.

3.6 Peace, justice, strong judiciary and the court interpreter

Peace, justice, and strong institutions are integral components of the SDGs. These goals are primarily encapsulated in Goal 16 [1], which specifically focuses on promoting peaceful and inclusive societies, providing access to justice for all, and building effective, accountable, and inclusive institutions. The targets under SDG 16 include efforts to reduce violence, ensure the rule of law, combat corruption, and strengthen institutions. SDG 16 aims to promote peaceful and inclusive societies by reducing violence, ending abuse, and preventing conflicts. It emphasizes the importance of institutions that can effectively handle disputes and contribute to maintaining peace. It highlights the need to promote the rule of law, ensure equal access to justice, and build effective, accountable, and transparent legal institutions. It seeks to strengthen institutions at all levels, emphasizing the importance of good governance, accountability, and transparency. The goal is to build robust institutions that can effectively respond to the needs of the people and contribute to sustainable development.

Court interpreters play a crucial and multifaceted role in contributing to the realization of these goals related to equal access to justice and upholding the rule of law. Their impact extends across various dimensions of legal proceedings, ensuring that individuals who do not speak the official language of the court can fully participate and have equal access to justice. Interpreters, when seen as linguistic bridges, are critical in breaking down language barriers that might impede effective communication between parties involved in legal proceedings. This inclusive approach facilitates the active and meaningful participation of individuals who may speak different languages, thus promoting equal access to justice. This aligns directly with the goal of reducing inequalities within the legal system. Furthermore, the contribution of court interpreters goes beyond language translation. By enabling effective communication, they play a vital role in upholding the rule of law, and this underscores the fact that understanding and being understood are fundamental to the integrity of legal processes. Interpreters ensure that all parties involved, including litigants, witnesses, and legal professionals, can comprehend the proceedings, present their cases, and engage in the legal discourse.

In multicultural and multilingual societies, the potential for misunderstandings due to language barriers is significant, and these misunderstandings can sometimes escalate into conflicts, thus creating environments in which sustainable development is not feasible. Court interpreters play a crucial role in preventing such conflicts by facilitating accurate communication and understanding among individuals who speak different languages. Language barriers can lead to misinterpretations of statements, confusion about legal processes, and a lack of clarity in conveying one's perspective. In a legal setting, these issues can be particularly sensitive and may contribute to escalating tensions or conflicts. Court interpreters act as essential mediators in such situations, bridging the linguistic gaps and ensuring that all parties involved can effectively communicate and comprehend the proceedings.

By providing accurate and impartial interpretation, court interpreters help prevent miscommunications that could potentially lead to disputes or conflicts. They ensure that individuals have a clear understanding of the information being presented, the questions being asked, and the responses given. This not only promotes a fair and just legal process but also contributes to maintaining a peaceful and respectful environment within the courtroom. In addition to preventing conflicts within the legal setting, the work of court interpreters has broader implications for social harmony in multicultural societies. By fostering clear communication, interpreters contribute to building trust and understanding among individuals from different linguistic and cultural backgrounds. This, in turn, helps to create a more cohesive and inclusive society, mitigating the potential for misunderstandings to escalate into larger conflicts beyond the courtroom. In summary, court interpreters serve as essential peacemakers in multicultural and multilingual societies by preventing conflicts that may arise from language barriers. Their role in ensuring accurate communication and understanding contributes not only to the effectiveness of legal proceedings but also to the broader goal of fostering harmony and unity in diverse communities.

The crucial role of accessible and accountable judicial institutions in the context of achieving sustainable development goals cannot be overstated. Court interpreters and translators play a vital role in shaping perceptions of the court's fairness and justice, facilitating communication in legal settings, promoting understanding of legal processes, and enhancing the overall effectiveness and fairness of judicial institutions. In *The Republic v. Christine Gesare & Another, Case Number: MCCR/1043/2020*, the interpreter, while utilizing the exemplification strategy, ensures that the court is evaluated positively as an institution that is fair and just to all, as opposed to the feeling held by the mother to the deceased, who felt that the court was biased in granting the accused bail. The accused had been granted bond, and the complainant felt that justice was elusive. In this circumstance, it became necessary for the magistrate to inform the complainant of the legal basis of the actions being taken.

> *Magistrate: The accused is presumed* **innocent until proven guilty** *and is accordingly entitled to a* **bond.**

> *Interpreter: Kisheria kila mtu anachukuliwa kuwa hana hatia hadi atapatikana na hatia kwa ushahidi uliotolewa mahakamani. Hii ni kwa sababu kuna uwezekano mtu kufungwa gerezani bila hatia. Kwa sababu hiyo huwa anaruhusiwa kupata bond. {Legally speaking he is presumed innocent until proven guilty based on the evidence presented in court. This is because there is a possibility of a person being jailed wrongfully. On this account he is entitled to a bond}*

In this example, the interpreter has used exemplification to explain the legal principle of "*innocent until proven guilty*" in a way that is relatable to the target audience. By providing an analogy that people can understand—the idea that everyone is considered innocent until there is evidence that they are guilty, just like how someone is considered not to have made a mistake until there is evidence that they did—the interpreter can help to create a sense of trust in the legal system and its institutions. In conclusion, effective communication, facilitated by court interpreters, is integral to creating an environment of trust, fairness, and accountability within judicial institutions—a critical aspect of sustainable development goals related to justice, peace, and strong institutions.

The presence of court interpreters in the judicial system is critically important in instilling confidence in users, in the sense that users are assured of access regardless of their habitual languages of use. The availability of language services in the court system demonstrates the court's commitment as a functional and inclusive institution capable of handling diverse issues of arbitration, including SDG-related issues. Equally, the court system that actively seeks to accommodate language diversity is an inclusive institution that is acquiescent to the principles of sustainable development as espoused in the UN's SDG 16 [1]. Overall, these characteristics of the court as an inclusive institution promote the participation of people from various cultural backgrounds in the legal system without fear of suffering disadvantages as a result of language barriers. Establishing confidence in the judicial system is crucial for promoting greater community engagement. Individuals who have confidence that their language diversity will not be a hindrance to legal procedures are usually more inclined to actively engage in the legal process. This inclusiveness enhances the representation and fairness of the legal system. Furthermore, the overt utilization of court interpreters conveys a strong message regarding the dedication to safeguarding justice, irrespective of language barriers. It indicates that the judicial system acknowledges and values the diversity of individuals it serves. This acknowledgement fosters a feeling of inclusion and motivates people to participate in legal procedures without apprehension of bias or misinterpretation due to linguistic differences.

4. Conclusion

In the preceding sections, we have canvassed the intricate landscape of legal understanding and the instrumental role court interpreters play in supporting SDGs. The symbiotic relationship between the comprehension of legal concepts, principles, and codes of conduct on the one hand, and the implementation and achievement of the SDGs on the other hand, underscores the importance of addressing language barriers within legal systems. As we conclude, several key insights and avenues for future exploration emerge. The interplay between legal understanding and the pursuit of SDGs is undeniable; consequently, the courts, as custodians of justice, are pivotal in shaping societies that align with the principles of development sustainability. Legal systems, when fortified with effective language access, are dependable catalysts for equitable development. It has also emerged that court interpreters are critically situated at the forefront of dismantling linguistic barriers, ensuring that justice is not confined to those fluent in the predominant language of legal practice. The interpreters' role extends beyond mere linguistic translation to embodying cultural exchanges, cultural negotiation, and accommodation. That role also takes into consideration issues of the competency with which cultural issues in interpretation are handled, the ethical considerations of mediated communication, and the muted expectation of interpreters possessing profound understanding of legal intricacies.

The pursuit of sustainable development as envisioned and advocated for in the UN's SDGs recognizes the centrality of legal understanding and language access as imperative. The empowerment of court interpreters is not only a matter of professional development but a fundamental step toward fostering inclusive legal environments that echo the aspirations of the 2030 Agenda. As we look ahead, it is our collective responsibility to champion policies, collaborations, and training initiatives that amplify the impact of court interpreters in realizing the SDGs. By doing so, we contribute to a global legal landscape that transcends linguistic barriers, ensuring justice is truly accessible to all. In essence, the nexus between legal systems and the SDGs highlights the symbiotic relationship between justice, governance, and sustainable development. The realization of the SDGs depends not only on the creation of laws and policies but also on fostering legal literacy and ensuring that individuals can actively participate in and benefit from the legal frameworks that guide societal progress. The symbiotic relationship between legal understanding and sustainable development becomes evident as court interpreters contribute to several dimensions of the SDGs. They facilitate access to justice (SDG 16), promote inclusivity and equality (SDG 10), and support the rule of law (SDG 16). Through their work, court interpreters are critical catalysts for social progress, ensuring that legal processes are not hindered by linguistic barriers and that the benefits of justice are extended to all members of society.

Author details

Akungah O'Nyangeri[1*], Zaja Omboga[2] and John Habwe[2]

1 University of Nairobi, Kenya

2 Kiswahili Department, University of Nairobi, Kenya

*Address all correspondence to: nakungah@gmail.com

References

[1] United Nations. Transforming our world: The 2030 agenda for sustainable development. 2015. Available from: https://sustainabledevelopment.un.org/content/documents/21252030%20Agenda%20for%20Sustainable%20Development%20web.pdf [Accessed: December 17, 2023]

[2] Morton S, Pencheon D, Squires N. Sustainable development goals (SDGs), and their implementation: A national global framework for health, development and equity needs a systems approach at every level. British Medical Bulletin. 2017;**124**(1):1-10. DOI: 10.1093/bmb/ldx031

[3] Morton S, Pencheon D, Bickler G. Public Health. 2019;**174**:65-68

[4] Kleespies MW, Dierkes PW. Thc importance of the sustainable development goals to students of environmental and sustainability studies—A global survey in 41 countries. Humanities and Social Sciences Communications. 2022;**9**:218 [Accessed: January 24, 2023]. DOI: 10.1057/s41599-022-01242-0

[5] Cai Y, Wolff L. Education and sustainable development goals. Sustainability. 2023;**15**(1):643 [Accessed: January 24, 2023]. DOI: 10.3390/su15010643

[6] Singh HP, Ajay S, Fakhre A, Vikas A. Impact of sustainable development goals on economic growth in Saudi Arabia: Role of education and training. Sustainability. 2022;**14**(21):14119. DOI: 10.3390/su142114119

[7] Hale SB. The Discourse of Court Interpreting: Discourse Practices of the Law, the Witness and the Interpreter. Amsterdam: John Benjamins Publishing Co.; 2004. DOI: 10.1075/btl.52

[8] González RD, Vásquez VF, Mikkelson H. Fundamentals of Court Interpretation: Theory, Policy, and Practice. 2nd ed. Durham, North Carolina: Carolina Academic Press; 2012

[9] Muigua K. Sustainable development goals and social justice in Kenya. Journal of CMSD. 2021;**7**(1):23

[10] Blix SB, Wettergren Å. Professional Emotions in Court a Sociological Perspective. London: Routledge; 2018. DOI: 10.4324/9781315306759

[11] Namakula CS. Language and the Right to Fair Hearing in International Criminal Trials. Cham: Springer International Publishing; 2014

[12] Berk-Seligson S. The bilingual courtroom. In: Court Interpreters in the Judicial Process. 2nd ed. Chicago: The University of Chicago Press; 2017

[13] International Institute for Applied Systems Analysis. Education and the sustainable development goals. In: Background Paper Prepared for the 2016 Global Education Monitoring Report, Education for People and Planet: Creating Sustainable Futures for all. Paris: UNESCO; 2016. Available from: https://unesdoc.unesco.org/ark:/48223/pf0000245580

[14] Katan D. Defining culture, defining translation. In: Harding S, Carbonell OC, editors. The Routledge Handbook of Translation and Culture. London: Routledge; 2018. pp. 17-47. DOI: 10.4324/9781315670898

Chapter 23

Perspective Chapter: The International Criminal Court in Service of Sustainable Global Society

Malaika Hassan

Abstract

This chapter aims to analyze the working and understanding of the International Criminal Court, especially in the context of war crimes which specifically affect the preservation and protection of various natural and cultural heritage done so namely in the name of "self-defense." These atrocities that have been committed by convicted criminals by the International Criminal Court allow a better understanding through the punishments which have been granted to prevent the said crimes from occurring again. Looking into the relevant statutes will allow highlight how even though theoretically natural ecosystems, wildlife and various heritages are protected practically many of these are ruined today and the biggest factor of them after a larger ecological footprint is war and destruction which has not only affected the natural environment but as well as its people.

Keywords: ICC, international law, Rome Statute, cultural heritage, natural environment

1. Introduction

The idea of there being a more permanent Court in nature arose in 1998 when various countries throughout the world signed a treaty to legalize at an international level one of the most serious crimes known to mankind which led to the establishment of the International Criminal Court (ICC or the Court) [1]. ICC as we know it today was established in 2002 and is responsible for holding accountable those criminals for committing some of the heinous crimes commonly those fulfilling the elements of what is normally known as war crimes that promotes not only the rule of law but as well as offering justice to the victims of such crimes [2]. This led to the establishment of the Rome Statute of the International Criminal Court (the Rome Statute) that helped mark the milestone in developing international criminal law which was to surpass the *ad hoc* solutions previously provided by the Court before the establishment of ICC and help in building a Court with a much wider jurisdiction which could investigate, try and punish the perpetrators [1]. The reason why ICC should amend existing statutes to ensure that international criminal law has the ability to ensure sustainable development for the global society is that it would be able to meet the needs of resources of the present generation without impacting the ability of the future generations [3]. Through this interconnection, it will allow the environmental objectives amongst others to become the core principle of sustainable development by distinguishing it from other policy adaptation and decision making [3–5].

The protection of the natural environment and cultural heritages from war is necessary because these are considered crimes against future generations as the aim is to penalize such acts which have severe consequences on long-term health, safety and means of survival of individuals. The name 'crime against future generations' however does not mean that these are either future crimes or crimes that will be committed in the future rather they refer to such crimes that take place within the present time which has serious repercussions on the basic survival of those set of individuals. By making this a specific category of an offense in international law it will allow extending the application of international law by virtue of creating and establishing criminal liability for existing prohibitions in terms of both international human rights law and international environmental law. These specific crimes would allow the perpetrator to be held responsible under individual criminal responsibility for causing serious violations against severe harm not only to the environment but as well as against cultural heritages not only during peacetime but during wartime as well. The reason why such incorporations are of uttermost importance is that the current international legal concepts and mechanisms fail to prevent, punish and penalize such behavior as these conducts can only be prosecuted when there is a direct act of physical violence. This concept allows penalizing crimes against future generations of gross violations of not only basic human rights including but not limited to economic, social and cultural rights but as well as massive environmental destruction [6].

In terms of environmental crime, there are many sets of criteria that have yet to be met for proper international development by addressing the global threats the Millennium Development Goals (MDGs) have been able to identify environmental harm as one which affects the development, peace, security as well as human rights of individuals. Compared to the past these issues are slowly starting to show their significance by recognizing the role of how perpetrators are committing war crimes against the environment and are not being held accountable for it. However, slowly but surely environmental crimes are gaining attention and are becoming firmly established yet they have a long way to go to be able to firmly place their mark within international law [7]. In the light of crimes against cultural heritages, the 1954 Hague Convention for the Protection of Cultural Property in the Event of Armed Conflict [8] has laid down that any damage done against cultural heritages is an offense not only to that community but to mankind as well. The reason behind the inclusion of mankind is because the protection of cultural heritage is of such uttermost importance as to be able to preserve history for future generations as well [9]. It was only through the United Nations Educational, Scientific and Cultural Organization (UNESCO) that developed the subject that every present generation not only has the responsibility to protect but also safeguard cultural heritage for future generations to come [10]. It is slowly being recognized that the preservation and integration of cultural heritages is an essential element for sustainable development as it acts as a resource but also for regional development plans as well [8].

The objective of this chapter is not only to highlight the importance of the natural environment and cultural heritages but also how they are drastically affected by war and what the international criminal law specifically the ICC does to protect and penalize it. It also focuses on various inclusions that have yet to be fully recognized dealing with safeguarding these factors but the gaps that exist in ICC to ensure a sustainable global society by failing to incorporate them as an offense in its pre-existing statutes and by initiating new policy decisions to emphasize the devastating nature of damaging the environment and heritages for both the present community and the future generations as well.

1.1 Historical background

The working of the International Criminal Court can be dated back to the time of ancient Greeks as the brutality and nature of war crimes due to their severity can be observed from the fact that they have been prosecuted way before even the International Criminal Court came into existence as it conforms its beliefs from philosophy and religion about the basic fundamental values of the human spirit. However, history shows that when the ICC did not exist such crimes were prosecuted by the national courts and due to the lack of impartiality and balanced nature of such trials, the criminals were still in power and the victims were unfortunately defeated. The Hague Conventions of 1899 and 1907 were the first of their kind that represented the codification of laws of war which were to be applicable on a global level in the form of an international treaty. It included various elements of the protection of civilians, private property as well as but not limited to places of worship, convictions and practice. These conventions however were limited in their nature because they only focused on imposing obligations but failed to create any criminal liability due to the absence of any sanctions being mentioned in the case of breach of such violation [11].

After World War II it was the Allied powers that ended up launching the Tribunal of international war crimes which are popularly known as the Nuremberg Trails to be able to prosecute Nazi officials which eventually led to the idea of the establishment of a permanent court internationally to hold similar perpetrators to account for such heinous crimes. Before the creation of the ICC, the United Nations formulated the *ad hoc* criminal tribunals to prosecute war crimes in the former Yugoslavia and Rwanda due to its inadequacy, but under the request of Trinidad and Tobago the United Nations Commission created a more permanent international court in nature in 1989 [2].

1.1.1 Jurisdiction of the court

The jurisdiction of the Court can be understood as the perpetrators who have committed the crimes mentioned in the Rome Statute as well as those who were indirectly responsible which includes military commanders, political leaders and other superiors. There are four main types of jurisdictions, the firstly is the temporal jurisdiction or *ratione temporis* which states that the Court can only exercise its jurisdiction over crimes which were committed after the enforcement of the Court i.e., all the crimes which occurred in 2002 or the years following. Secondly, the personal jurisdiction or *ratione persona* which states that the Court has jurisdiction over nations who are State Parties of the Statute but also has the authority to prosecute nations of non-State parties if they accept the jurisdiction of the Court on an *ad hoc* basis either by the declaration of the individual being a national to the State party or as per the decision made by the United Nations Security Council [12]. Because of the severity of the crime, the concept of immunity does not exist before the Court and prevents both invoking such immunity or bringing it as a defense as the majority of the perpetrators are those in power of ruling or ensuring administration over the country regardless whether immunity arises from treaty or customary law.

Then comes the territorial jurisdiction or *ratione loci* which means that the Court has jurisdiction over the territory where the crime took place regardless of the fact that whether the individual was of the same nationality or not. ICC exercises jurisdiction when the State accepts its jurisdiction on *ad hoc* basis, conferred by the United Nations Security Council or in accordance with Chapter VII of the Charter of the United Nations. This jurisdiction not only extends towards the land but also includes registered board vessels or aircraft of the State party. Lastly, the subject matter or *ratione materiae* includes the four categories of what is said to be the most heinous international crimes including genocide, war crimes, crimes against humanity and crimes of aggression [13].

1.1.2 Working of the court

The Court functions in a way that it is initiated based on a referral from the United Nations Security Council which is pursuant to Chapter VII of the United Nations Charter [14] or when a State is party to the Rome Statute as mentioned earlier. The Office of the Prosecutor has the authority to initiate investigations independently or when it is subject to the authorization of the Court [15]. When it comes to the working of the Court there is one basic principle which is to be understood that is the principle of complementarity which means that ICC is the last resort when the State is either unable or unwilling to prosecute or conduct a credible investigation over the said offenses [13]. This principle is neither an alternative nor replaces the national Courts but rather complements it when the national Courts fail to properly hold an impartial trial against the perpetrator. Due to the absence of a police force, the Court is heavily dependent on the cooperation of member state authorities to apprehend the suspects. Similarly, the individuals cannot be tried in *absentia* and are only subject to the Courts' warrant when they are in member-state territory [2].

To formally initiate an investigation it is upon the prosecutor to be able to prove after a preliminary examination that the alleged crimes are of sufficient gravity and once it is opened, the Office of the Prosecutor then sends their investigations and other relevant staff to be able to collect the said evidence and all the arrest warrants and/or summons shall be approved by the judiciary based on the information that is provided by the prosecutor after which the judges have to decide whether a case shall be brought forward to trial or not. The defendants have the option to seek outside counsel for their representation and the convictions as well as sentences are subject to at least two out of three judges that make up the trial bench and in case the defendant wants to appeal, the appellate bench of the ICC consists of five judges [2].

2. Understanding the crimes

As discussed earlier, four main types of crimes are considered heinous and fall under the subject matter jurisdiction of the Court as defined under Article 5 of the Rome Statute. Before understanding which crime affects vastly on natural heritage, it is of uttermost importance to know what these different crimes are in their nature as laid down by international criminal law.

2.1 Genocide

In simple terms, genocide under Article 6 of the Rome Statute is the concrete threat which exists for a targeted group of people, where the said conduct leads to the destruction of the said targeted group [16] as it was laid down by the Trial Chamber that genocide is the intention to destroy a considerable group of individuals that are part of the specific group [17]. Genocide throughout history has proven to be one of such nature where the crimes were designed to eradicate, incorporate and exploit members of certain ethnicities many children and women were spared but unfortunately not with good intentions as they were seen as a source of future physical resistance and the latter as sources for future offspring of the dominant group. The term is not exclusive to simple immediate destruction of the group unless it is accomplished by mass killings of members of that nation as the intention to solidify this crime is the coordinated plan to annihilate those specific individuals of a group. The distinguishment is made based on political institution or social institutions such as that of color, race, language, culture, national feelings, religion and economic existence of national groups that leads to the destruction of personal security, liberty, dignity as well as lives of those belonging to the targeted group as it is not the individuals' capacity in which he is targeted rather its membership to that national group [18].

The two phases of genocide include the destruction of the national pattern and national imposition of the pattern of the perpetrator that either allows the oppressed population that remains on that territory to follow the new set of national imposition or after their ethnic cleansing colonization of the oppressors' nations [19]. Through this, it includes enslavement, starvation, mass deportation, degradation and deprivation of basic human rights to be able to cause maximum inhumane suffering and torture similarly physical and mental torture, acts of rape and mutilation are also included within the meaning of genocide [20] and the actions taken to prevent births including but not limited to forced sterilization, separation of sexes, any kinds of sexual traumas and impregnation through rapes and/or gang rapes with the motive behind cleansing the oppressed group and having offspring of the dominant group [18].

2.2 Crimes against humanity

Article 7 of the Rome Statute deals with crimes against humanity which means when an act is committed as a part of a widespread or systematic attack against any civilian population where the knowledge and intent exist of carrying out such an act which includes crimes like murder, extermination, enslavement, deportation, forcible transfer, imprisonment, deprivation of physical liberty, torture, rape, sexual slavery, enforced prostitution, forced pregnancy, enforced sterilization or any kind of sexual violence. The persecution has to take place against such a group which is identified based on their political racial, national, ethnic, cultural, religious or gender affiliation [16].

Medical experimentations, mutilations, severe beating, food deprivation, sterilizations, violations of corpses, forced undressing and any other kinds of crime which severely affects an individual's physical and/or mental health falls under crimes against humanity [20]. By keeping the nature of crimes and the elements of it in mind the name has been given because humanity refers to humankind as individuals as a whole suggesting that the offenses carried out are against a particular party in interest as the amount of damage that is done is not humane because it degrades the humanity of its victim without it implicating interests of entire human race as crimes against humanity is something which is not only carried out against the human kind but injures humanness as well. Because of the severity of the crimes, the perpetrators are referred to as *hostis humani generis* which means that they are an enemy of all mankind. Unlike genocide, these crimes can be committed pre- or during the war by civilians, soldiers, by the government against its people or adversary people as well [21].

By killing members of a group, it has allowed to impose such measures to prevent birth with the intention to destroy and avoid any future generation in terms of the particular group being targeted. The widespread or systematic nature of the offense with the knowledge is sufficient for an offense to be classified under crimes against humanity [22]. The concept of widespread is termed as massive and large-scale action collectively and directed against multiple victims associated with that particular group [23] which is a series of multiple inhumane acts of extraordinary magnitude [24, 25]. Similarly, systematic is termed as repetition of the criminal conduct but as well as its improbable nature at a random occurrence [26]. Similarly, it is necessary to take into account the element of there being a political objective as well as acknowledged policy by using public and private resources with the help of political or military authorities to carry out such criminal activities [26–28]. Lastly, in crimes against humanity the term attack has a different meaning as opposed to genocide, the existence of an armed conflict is not necessary as these crimes can be carried out before, during or after an armed conflict but is not necessary to be a part of the armed conflict in itself as per customary international law [29].

2.3 War crimes

Article 8 of the Rome Statute deals with war crimes are those that are associated with the nature of conflict as well. There are two main types of conflict international armed conflict (IAC) which means when two Member Parties or States are at war with each other and the second type is non-international armed conflict (NIAC) which is when a Member Party or State is internally at war against itself against an armed group. The Court has jurisdiction and the elements of international humanitarian law (IHL) are still applicable regardless of the type of conflict. War crimes can for easy understanding be divided into four major categories which include crimes relating to protected personnel such as civilians, journalists, and healthcare workers, related to protected goods such as private property, unlawful methods of warfare and lastly unlawful means of warfare.

Keeping these factors in mind, the Member States at all times are required to be able to distinguish between civilians and combatants including civilian objects and military objectives. It constitutes a war crime if there are indiscriminate attacks where in the name of eliminating threats, civilians and civilian objects are also targeted without proper distinction between the said two. However, this is where the principle of proportionality comes into play which requires that the balance between collateral damage caused amongst civilians and their objects is in balance and not excessive to the anticipated concrete and direct military advantage to be gained. As principle of proportionality is part of customary international law and applicable in any kind of war which is why it is criminalized if it causes any attack that causes incidental loss to civilians and/or widespread, long-term and severe damage to the natural environment or any damage to civilian objects which is why any military attack which is expected to cause more damage which is excessive to military advantage is prohibited [30].

War crimes because of their extensive nature have also been set with a certain set of requirements the violation must infringe the rule of IHL which must either be set by customary international law or under a treaty, and it should be serious meaning that tits breach causes grave consequences and the said violation should entail individual criminal responsibility. Because of the separate nature of IAC and NIAC, in the former it is referred to as 'wilful killing' and in the latter, it is known as 'murder' but the elements of the crime remain the same. Any kind of torture, inhumane treatment, mutilation, and biological, medical or scientific experiments are also prohibited and the reason behind such prohibition is that it endangers the physical and/or mental health of the person and there is no justification for such procedure to take place medically in the interest of the person or their treatment [31]. Any great suffering or serious injury to the body or health is forbidden because these deliberately cause long-lasting and serious harm even if it does not fulfill the elements of torture [23, 27, 32] these also include all kinds of sexual violence.

Prohibited targets exist to specially protect objects and interests which is why bombarding or attacking any undefended towns, villages, dwellings or buildings which are not military objectives is forbidden [30] going as far as protecting religion, education, art, science or charitable purposes, historic monuments, hospitals etc. [31]. The protection of civilians even includes the use of starvation as a method of war, impeding relief supplies, and using acts of violence to spread terror amongst the population [33]. When it comes to environmental damage it is only considered a crime where the damage is widespread, long term and severe making it disproportionate to the military advantage including but not limited to destruction, appropriation, seizure and pillage in IAC but in NIAC it only includes the prohibition to pillage which is the extensive destruction and appropriation of property, not justified by military necessity and carried out unlawfully and wantonly. Pillage in itself is a war crime because it is the seizure of property for private or personal use as opposed to taking it for any military purposes [31].

2.4 Crimes of aggression

Article 8 of the Rome Statute deals with crimes of aggression which engage with individual criminal responsibility because the conduct of the State is a serious violation of what is prohibited under the UN Charter. However, it is different from the other set of crimes because the jurisdiction cannot be invoked and no complaint can be brought forward unless it is directly related to an act of aggression which has to be determined by the Security Council that indeed a State has committed such a violent attack. It also varies when compared to other crimes because it deals with *ius ad bellum* which is the law governing recourse to conflict as a crime of aggression is a leadership crime and can be committed by those who are leaders and high-level policy makers. This means any individual who is in such a position to be able to effectively exercise control over in terms of directing any political and/or military action of a State. The act of committing the crime of aggression is proved if the perpetrator had the knowledge of the collective intent to be able to initiate and wage an aggressive war and continuously has the ability to participate in such an act and usually this participation is by a State against another State [31].

The term crime of aggression is to be able to ban the violation of Article 2(4) of the UN Charter by eliminating any threat or use of force against the territorial integrity or political independence of any state [34]. The meaning arises from the one set during the Nuremberg Trial which is the planning, perpetration, initiation or execution of such an aggression which is accumulated with evil as a whole. This allows to set an objective that the Court has the authority to exercise jurisdiction over such and to prevent any suffering caused by an armed conflict by preventing state actors from using any aggressive force not laid down by the UN Charter or in excess of the principles laid down under it [35]. It focuses on the actions taken by the state actors which is why states in crime of aggression are seen as weapons in the hands of leaders or state officials which can be used against another state [36].

Because of the threshold of such an aggravated crime, it is necessary that a certain criterion must be met firstly is that the alleged act of aggression in terms of its character, gravity and scale constitutes a manifest violation of the Charter of the UN. Secondly, the potential subjects must range to only a limited group of people who have the authority to effectively exercise control over or to be able to direct any political or military actions of the said State. Lastly, the perpetrator must have an affiliation and personal contribution towards the planning, preparation, initiation or execution which is why it is also known as a leadership crime as lower-ranking State agents do not have the power, authority or influence to be able to exercise the crime. It is also necessary to understand that when a non-state group manages to prepare such a large-scale armed attack against the property and persons of another State it can easily be considered that the host State had substantial involvement in terms of a minimum aiming and aiding towards it which suggests aggressive *mens rea*. From this, it can easily be concluded that a crime of aggression is of such a nature that the use of force must be so unlawful, violent, and devastating that its massive scale cumulates the benchmark of character gravity and scale as laid down in the manifest standard [37].

3. Effect on cultural and natural heritages

The crimes that destroy any cultural and/or natural heritage are of such a nature that they usually fall under war crimes or crimes against humanity as the notion of cultural heritage finds its origins in the Renaissance period and international law it was in the 1907 Hague Regulations concerning the Laws and Customs of War on Land which drafted the first instrument in terms of protecting cultural property especially to protect monuments [38]. It was through UNESCO that established a more concrete level of protection of cultural heritage whenever an armed conflict took place and after 1970 is when both cultural and natural heritage protection became more prominent. Generally, culture means the habits in which a certain community lives and behaves and for it to take that shape it has to meet the standard of settled practice and the practice is constituted as obligatory. The concept behind protecting cultural heritage is that it acts as an interest for all mankind as it acts as a dignity of universal good that further helps in the cooperation amongst the international community promoting mutual understanding. Because cultural property consists of both movable and immovable property ranging from architecture, art or history under Article 1 of the Hague Convention of 1954 it suggests that cultural property is a part of cultural heritage which is why attack on any protected objects and sites is a war crime [39].

Another historical concept that eventually made any cultural property and heritage fall under crimes listed in the Geneva Conventions was that during occupation most of the property consisted of religious, charitable, and educational institutions as well as of those of arts and sciences all of which are protected under private property with no link towards military necessity. This is why any seizure, destruction or wilful damage to any of these institutions is forbidden and the accused is made subject to legal proceedings. To determine the damage that has been done it is tested under the criteria of widespread and deliberate destruction of cultural property which is why all cultural property is referred to within cultural heritage that sheds light on its intergenerational importance as it is protected not only for those present today but the future generations as well. The importance of any cultural heritage is also determined by its extent towards the people as opposed to the location where it is situated [40].

Under Article 8(2)(a)(iv) of the Statute, it criminalizes during IAC any conduct which takes place against property by extensively using such power that aims at destroying property which has been protected under Geneva Conventions 1949 and has not been justified by military necessity [41]. The term destruction includes but is not limited to setting objects on fire, attacking or seriously damaging, appropriation taking, obtaining or withholding property, theft, requisition, plunder, spoliation or pillage [42]. When it comes to cultural and natural heritages an additional level of protection is granted to them under Article 8(2)(b)(ix) of the Statute, mirrored in Article 8(2)(e)(iv) applicable to NIACs. These provisions allow a better approach towards the 1954 Convention for the Protection of Cultural Property in the Event of Armed Conflict and its Protocol II by merging them into a similar level of gravity against offenses conducted against any cultural property. Similarly, the Rome Statute also lays down various degrees on protecting cultural property during armed conflicts depending on the importance of the said cultural property damaged [43] this allows a rather special degree of protection to those heritages by focusing on the principles of IHL during both peacetime and war [44]. The loophole that exists which fails to strengthen the aspect of protection is the necessity to be able to balance the weak level of protection as well as that of military necessity [45] this led the ILC in its Draft Principles lay down certain areas as protected zones regardless of the type of armed conflict [46]. An example of this can be highlighted by the delegation between Ecuador and Peru granting protection to special peace parks [47] and the 1994 peace agreement between Israel and Jordan in terms of water resources [48]. Aside from this, the ILC Draft Principle has also been able to prevent under IHL from disregarding environmental protection by shielding protected zones even if there exists a military objective as well as encouraging parties to ensure the protection of those areas even if there exists a military target [49].

The Office of the Prosecutor has constituted in the past about the importance of cultural heritage as it identifies with the culture and people of the community and its destruction whether it is tangible or not in nature is deemed as a great loss. The Office look for these alleged crimes which affect cultural heritage in terms of preliminary examination, investigation, prosecution and reparations and where evidence allows the Office then has the chance to incorporate any charges which are directed towards the attack against cultural heritage by highlighting all situations that affect it. This shows the commitment of the Office to respecting cultural rights by recognizing its importance towards culture and community by recalling and interpreting the Statute within international law. By incorporating this as a part of universal human rights principles it allows the Office towards the importance of penalizing activities against any cultural heritage [50] in the 2016 Policy Paper on Case Selection and Prioritization it was highlighted by the Office and its commitment towards "paying particular attention to attacks against cultural, religious, historical and other protected objects" as any harm done to them affects every nation in the world [51]. The limitation can be observed as the Office is bound to any harm done to the cultural heritage when it falls within the ambit of the Court's jurisdiction but its reflection can be observed in the rules on the protection of cultural property in IHL and international human rights law. The protection of heritages can be understood from examples such as the right to take part in cultural life, the right of minorities to enjoy their own culture, and the right of indigenous peoples to self-determination and cultural heritage which further co-relates to basic rights like freedom of expression, freedom of thought, conscience and religion, the right to education, economic rights, and the right to development [50].

As attacks against cultural heritage can occur in either of the armed conflicts that are IAC or NIAC it is considered a serious violation of IHL as it has already been recognized that generally cultural heritage is granted the same level of protection as that of civilian property but during an armed conflict these heritages are granted an extra layer of protection [25, 52]. This category of protected cultural heritage enjoying enhanced protection can be observed as far as German Nazis being convicted for plunder as the first internationally enforced protection of cultural property [53]. It was held by the International Military Tribunal (IMT) at Nuremberg that unlawful destruction of any cultural property constituted not only a war crime but a crime against humanity as well [54] and this destruction has the power to entail individual criminal responsibility including applicable to non-state actors as well [55]. Aside from this, the ICTY grants Tribunal jurisdiction over those war crimes which deal with the seizure of institutions dedicated to religion, charity and education, the arts and sciences, historic monuments and works of art and science. The United Nations Transitional Administration in East Timor (UNTAET) Regulation 2000/15102 and the Iraqi Special Tribunal Statute also deal with the same matter but limitations can be observed when it is impossible in practical terms to predict a legitimate military rationale for seizing cultural property though formally it is observed that appropriation of cultural property does not constitute as a war crime if the necessity of war is observed [54]. Through this, it has now set deep roots and has been firmly established that any deliberate act against cultural heritage is a violation of customary international law [56] because of the great importance the sites hold to mankind and history [57].

3.1 Cases protecting cultural heritage

If cultural heritage is perceived from a human rights perspective, it has a great significance over individuals and their collective identity in a community as cultural heritage is perceived as:

> *"Encompassing the resources enabling the cultural identification and development processes of individuals and groups, which they, implicitly or explicitly, wish to transmit to future generations" [58].*

Because of its nature, an attack on cultural property is contingent on to attack on cultural heritage. Throughout the history of the ICC, there has only been one such case where the perpetrator was declared guilty of the destruction of cultural heritage. It occurred in January 2012 when a newly formed Tuareg rebel group known as the National Movement for the Liberation of Azawad (MNLA) attacked military garrisons in Menaka, Aguelhok, and Tessalit and when MNLA declared independence various Islamist groups such as Ansar Dine, Al Qaeda in the Islamic Maghreb (AQIM), and the Movement for the Unicity and Jihad in West Africa (MUJAO) overpowered MNLA. Eventually, Al Mahdi was guilty of a war crime for attacking protected objects under Article 8(2)(e)(iv) of the Rome Statute as he was responsible for attacking ten buildings having a religious and historical character including mausoleums and a mosque all located in Timbuktu [59]. The Chamber also discovered the destruction of such sites that had been classified under UNESCO World Heritage as to be of such gravity that not only directly affected the victims of the crimes but as well as inhabitants of Timbuktu, the people of Mali and the international community [60]. Because of the importance of the destructed sites it was deemed to be breaking the soul of the people of Timbuktu due to their emotional attachment to the said site [59].

This eventually led to a reparation order for the victims who deserve compensation for the criminal activities conducted by Al-Mahdi [60]. Through this case, victims of the destruction of cultural heritage were laid down which includes direct victims including guardians responsible for maintenance of the sites and faithful inhabitants of Timbuktu whereas indirect victims included people of Mali and the international community [61].

The Jokic case also covers the aspect of protection granted to cultural heritage as the perpetrator was sentenced for the shelling of the Old Town of Dubrovnik and by testing the gravity of such acts and the reasons behind the kind of such attacks against a cultural heritage hold such an inherent gravity. The Old Town of Dubrovnik was a protected site under UNESCO World Cultural Heritage as it held at the given time a fundamental value and importance of being a world cultural heritage due to its marvelous architectural ensemble illustrating a significant stage in human history and by attacking such a site the crime was not only against the history and heritage of that particular region but as well as an attack against humankind [62] as the shelling wiped away an important part of history and restoration even though could be made possible however it can never return to its original status and this kind of loss is not only significant but also irreplaceable [63].

3.2 The victims

To understand the victims of cultural heritage aside from the protected sites that are ruined it needs to meet a certain criterion which includes destruction caused to a heritage object which is deemed as unambiguously identifiable, event forming the harmful impact, event not caused or controlled by the victim, event directly linking to the actions attributed by the offender and damage to the heritage which is universally criminalized. The destruction of heritage sites causes psychological and economic harm but not any physical injury which is why these crimes are classified as being non-violent and not falling within the scope of national compensation schemes. Due to which the restitution includes the return of property or payment for harm caused, reimbursement of expenses incurred as a result of the victimization and restoration of rights. Similarly, the United Nations General Assembly Resolution 60/147 of 2005 recognizes that victims of such gross violations benefit more from remedies and reparations which should be proportional to the harm suffered by them ensuring that they are appropriate and effective. These reparations include restitution which is restoring to the original victim as accurately as possible, compensation which is the monetary measure of the loss incurred, rehabilitation which is any medical and psychological care required and lastly guaranteeing non-repetition of the said crime by promoting codes of conduct and ethical norms. It is through these reparations which are considered moral imperative to amend the loss that has occurred by contributing to the aims of rehabilitation and restoration of law [61].

4. Effect on the natural environment

Alongside global warming one of the main other causes that are not discussed enough when it comes to the destruction of the natural environment is a military conflict due to its significance to human populations it has become a part of the conduct of an armed conflict and with the growing technology future warfare include a much wider scope for its destruction [64]. The definition of environmental crime is still not solidified but the most referred one is deemed as a collective term which is used to describe illegal activities that harm the environment but also exploits it causing loss to the community by trading or stealing natural resources [65]. However, despite its growing impact on the natural environment the Rome Statute does not contain any specific crime or element of the crime which deals with crimes against the environment that initiates the jurisdiction of the Court as it fails to lay down any circumstances in which any particular level of harm caused to the natural environment constitutes as a war crime [64]. This has led to many discussions as to whether the elements already somehow constitute obligations to protect the natural environment or need further amendments as there is currently no formal concept under which international criminal law deals with cultural or natural genocide [66]. When it comes to genocide, Article 69(c) in the Elements of Crimes contains a broader range that any acts which poison or affect water wells or destruction of forests do fall within its description. Proving *dolus specialis* is difficult which is why any destruction to the natural environment is rarely applied as the targeted group which is the subject of the natural environment destruction unfortunately fails to constitute an established group that makes it a crime of genocide. Due to its complex nature, it is necessary to prove the existence of *dolus specialis* as well as to show that there existed an intent to target the environment as a victim or that it was used as a weapon to be able to prove to the Court that the destruction was of genocidal. As of now, the deliberate destruction of a habitat as well as access to clean and safe water or food is without a doubt a breach of fundamental human rights of those that are affected as it causes environmental destruction in the longer run against the targeted individuals [64].

To better understand this, it is necessary to recognize that mother nature has a set of rights belonging to itself which can be observed in the 1997 Danube Dam case that land is something which has a spirit and life of its own and is directly linked to the welfare of the community as natural resources are collectively owned by the entire world. To codify these rights was done at the United Nations on Sustainable Development in 2012 to adopt the Universal Declaration of the Rights of Mother Earth (UDRME) by establishing a framework that sees nature as a subject matter having its rights [3]. The Stockholm Conference in 1972 also focused on finding an approach to deal with damage and destruction caused to the natural environment during war which led to the establishment of the United Nations Environment Program (UNEP) [67] and the Stockholm Declaration on the Human Environment (Stockholm Declaration) that introduced 26 non-binding principles of environmental protection in terms of issues regarding sustainability, conservation of wildlife and habitat, toxic substances, pollution of the seas, population growth, nuclear weapons and the ecological balance of the biosphere [68]. Later, the Vienna Convention for the Protection of the Ozone Layer also added aspects like carbon dioxide production, global warming, biological diversity, chemical and biological weapons and ozone depletion [69].

In general, IHL prosecutes crimes which damage the environment first in the aspect that it is necessary to protect civilian objects and prohibits attacks against civilians with the objective that targeted groups are not harmed in any manner which puts their survival at risk and secondly by addressing attacks which are directed specifically at the natural environment that causes widespread, long term, and severe environmental damage [70]. These aspects are protected even when there is no military attack or there does not exist a lack of proportionality between the damage and the military advantage expected to be achieved. According to Professor Mete Erdem the protection of the natural environment is necessary because it is directly linked to the survival of the civilian population and the other is the direct protection of the environment itself [71]. The natural environment is sacred for the growth of the entire population as a whole which is why the adoption of the Environmental Modification Convention (ENMOD) was introduced and then included as part of customary international law. Under Rules 43, 44 and 45 the prohibition of attacks directed against the natural environment was discussed where the latter deals with NIAC and Rules 43 and 44 grant protection via the application of the distinction and precautionary principles. It was these Rules that suggest that during any type of armed conflict directing an attack even if it lacks scientific certainty of environmental harm is prohibited. It was through this it was emphasized that the protection of the natural environment is a separate object instead of its importance to the civil population alone [70].

As the concerns grow regarding the impact of war on the natural environment the term known as 'ecocide' came into existence which dates back to the 1970s which according to Arthur Galston was the wilful destruction of the environment in which individuals can strive and live in their manner of choosing [72]. The Independent Expert Panel for the Legal Definition of Ecocide (IEPLDE') and the Promise Institute for Human Rights started the very much-needed discussion on how the ICC must expand its jurisdiction to deal with crimes related to environmental degradation. Due human-centric fundaments of modern international criminal law limit ICC from exploring newer branches governing environmental law which is why it has been proposed to amend current statutory frameworks in a manner which maximizes the protection of the environment as this will allow a chance to reform statutory provisions from multiple perspectives. The reason why this is becoming a growing necessity is that the environment and heritage are continuously jeopardized during conflicts [73] and are left as a silent casualty of war affecting future generations [41]. The reason why discovering and improving this side of international criminal law is important is because aside from the direct toll on civilian life it has also led to many environmental abuses. After all, prohibiting mass environmental damage is becoming increasingly necessary but ICC has yet failed to incorporate environmental harm into any of its statutes. As environmental law in itself has its legislative gaps because of its unique nature in terms of technicalities, objectives, and narratives it is the job of the ICC to be able to strike the right balance between environmental protection and the integrity of international criminal law [74]. The power that ICC holds has the authority to influence individual States and draw policy attention to recognize these challenges by optimizing the potential of the safety of the environment by drawing policy attention. The lack of its inclusion is because the 1954 Draft Code of Offenses against the Peace and Security of Mankind [41] failed to include any new crimes which were not mentioned in the Nuremberg Principles [75] until later in 1995 when the Draft Code was revised the International Law Commission finally took the initiative in criminalizing any offense committed against the environment [41].

This led to the introduction of a fairly new concept which is known as the "green interpretation" of the Rome Statute refers to the construing provisions in a manner to form a legal basis to be able to prosecute and punish environmentally damaging conduct [76]. The introduction of the green approach was first initiated in 2016 when the Office of the Prosecutor (OP) published a policy paper on Case Selection and Prioritization that allows the investigation and prosecution of such crimes that meet the criteria of threshold in regards to any State that has committed such a crime which has caused the destruction of the environment as well as e illegal exploitation of natural resources or the illegal dispossession of land, however, it is not binding.

As discussed earlier even though the provisions of genocide fail to mention any environmental destruction a green interpretation can allow a perpetrator to still be punished for genocide if the environmental damage is deliberate with the intent to bring about the physical destruction of the group [77] as this allows punishing such crimes which causes deprivation of resources necessary for survival [78]. Article 7(1)(k) of the Rome Statute allows for providing a basis of accountability where environmental damage constitutes inhumane acts leading to great suffering or serious injury to the physical and mental health of a civilian population [77]. Furthermore, the International Criminal Court has adopted a definition that allows to hold accountable any organization which can perform acts that infringe human values [79] further expanding the possibility that the provisions of crimes against humanity can be applicable if environmental damage is caused by a corporation but due to its limited nature it fails to mention any notion which guarantees protection of civilian population from environmental damage directly linked to the protection to the ecosystem on which survival of those individuals is dependent on. The recognition of environmental damage to be linked with war allows an eco-centric orientation to international criminal law [80].

When it comes to the gap between international criminal law and international environmental law the doctrines of *mens rea* and individual criminal responsibility causes complications as Article 30 of the Statute only lays down general subjective element in terms of the standard of crimes where it fails to regulate *mens rea*. The only explicit reference found in the Rome Statute regarding environmental harm is under Article 8(2)(b)(iv) which has to be met by a standard of knowledge of causation. This is the reason why the Office is responsible for proving that the accused knew when committing *actus rea* that it would lead to excessive environmental damage [41]. Another study by Heller and Lawrence highlights how the faulty standard in the Article arises by making it impossible to prove that the accused knew that the attack would be disproportionate [80] as it simply undermines viability in terms of enhancement of environmental protection through the framework of international criminal law as illegal exploitation intensifies during conflicts [41] as the prime example would be extraction of natural resources leading to rise in corruption [81] like allowing a chance to sign and forge import and export certificates, allowing the possibility to transport illicit goods etc. [7]. Not only this but to be able to prove specific intent restricts the burden of proof on ICC Prosecutors as it has yet been unable to prosecute such an offense [41]. The damage caused to the environment can be observed from how in Ongwen, the pillaged objects were limited to foodstuffs, clothing, and households [82] in Katanga, the attackers seized mattresses, tables, chairs and kitchen equipment and took possession of livestock [83] and in Bemba, bicycles, motorcycles, money, 220 and household items such as beds and electrical generators were stolen [84]. Focusing on sustainable management of natural resources will allow the ICC to be able to take a more precautionary measure by not only prohibiting starvation but also illegalizing the destruction of objects indispensable to survival. Prohibition can be understood from examples like the destruction of crops and poisoning water resources that showcase a link between acts of depriving resources by harming the environment that affects the lives of civilians [41].

The nature of this crime is that they are aggravated in nature because of the impact it leaves on the environment that costs future generations to come which can be better understood from examples such as deforestation, dumping of chemicals, illegal fisheries [65], wildlife populations like elephants, rhinos and tigers to pangolins, reptiles, fish and rare birds and plants [85] that impacts the ecosystem and deprives it of clean air and clean water as well as weather mitigation, food security that results in their health being compromised. In terms of transnational organized crime, it includes illegal activities like hazardous waste and chemicals, forest products, pangolins, giant clams, minerals and illegally extracted gold. Within the last decade alone the rise in environmental crime has been estimated to be about 2–3 times of the global economy which shows that resources assigned to protect environmental crimes are not enough to not only contain, protect but also criminalize these offenses. Due to this natural resources are quickly declining which increases the risk of wars and for its early prevention the UN Security Council Resolution S/RES/2195 that aims to widen its scope by allowing State Parties to collect, analyze and exchange information, including law enforcement and intelligence information to be able to prevent terrorism and share relevant data to the Special Representatives of the Secretary-General, the Department of Peacekeeping Operations, the Department of Political Affairs, the Counter-Terrorism Executive Directorate, the UN Office on Drugs and Crime, the Counter-Terrorism Implementation Task Force and the United Nations Development Programme [65].

4.1 Cases protecting natural environment

Though there has not been any particular case held by the International Criminal Tribunal of Yugoslavia (ICTY), the International Criminal Tribunal of Rwanda (ICTR) [86] or the ICC in terms of environmental destruction under the provisions of Article 8(2)(b)(iv) the closest case is the one which was held against Omar Al-Bashir. When any environmental destruction is caused which leads to any harm to the national, religious, racial or ethnic group of people with the intent to destroy the group as a whole it could be charged as genocide. In regards to this, the accusation against Al-Bashir in the Court is of uttermost importance which was for carrying out genocide in Darfur by using such methods that led to the destruction of food, wells and water pumping machines, shelter, crops and livestock, as well as any physical structures capable of sustaining life or commerce [87]. Even though the Trial Chamber had rejected the prosecution's application for a warrant of arrest based on the genocidal charges there existed an intent on behalf of the perpetrator based on reasonable inference based on evidence [88]. As it can be observed that the serious damage to the natural environment is linked to the harm caused to the Darfur population, the prosecution against Al-Bashir could stand as the one of crimes against humanity [87].

4.2 The victims

The "do not see" crime [89] or the "victimless crime" [90] is something which has been commonly referred to as environmental crime as its abstract nature lacks any concrete identifiable victims at first glance as any person or anything which is affected by the destruction of a natural environment destruction is a victim. In the light of violations of criminal and regulatory law, the victims include the general public and the environment but it faces challenges as to be able to understanding the extent to which the damage is accumulated over time as well as the number of acts [91, 92] or the victims which are directly affected may not be able to understand the severity of crime and fail to report the such or the victims which may include wildfire and their habitats is a tremendous task to be able to capture in surveys. These crimes can also be transnational as they often include wildlife trafficking and the illegal trade in ozone-depleting substances, the illegal dumping and transport of hazardous waste, illegal logging and timber trade and illegal, unreported and unregulated fishing as well as pollution of a river that crosses international boundaries. The severity of these crimes can also be reflected in that the harm caused to the natural environment can also be global contributing towards climate change [89].

Understanding who is the actual victim in this scenario of the environment being affected can also lead to the perpetrator also being a victim which is why criminal law must clarify which behavior affecting the natural environment is to be labeled as illegal as it cannot render the whole of society guilty of significant offenses without undermining itself [89]. It was laid down by the studies conducted by the UN Environmental Programme that women are the most vulnerable to environmental harm in terms of pregnant women, nursing women and any other women who plan on bearing children as the harm can lead to an excess mortality rate when it comes to heat waves [93]. Generally, the destruction of the environment can also lead to financial loss in the context of property value, loss of income, consequences to health, emotional distress and loss of quality of their environment. Not only this but it can also impact the health of individuals leading to an increased rate of illness and disease, such as cancer, heart disease, disease of the respiratory system, neurological damage, birth defects and genetic mutations, miscarriages, lowered sperm count, and sterility with exposure to environmental hazards and even causing death. It can also impact the crops because of the damage caused to farming or coastal communities that impacts the living conditions of environmental agents by causing the absence of environmental macro and micronutrients necessary for survival which can lead to malnutrition. Overall, the victims vary greatly when damage or destruction impacts the natural environment as many communities have certain cultures, beliefs and traditions that are scraped away due to environmental degradation [89].

4.3 The remedies

When understanding the remedies that can be provided in a crime against the natural environment the international framework includes certain criteria and standards that guarantee assistance, treatment and remediation to safeguard the human rights of those victims who have suffered the devastating consequences which is provided by the UN Declaration of Basic Principles of Justice for Victims of Crime and Abuse of Power. Strengthening the judicial and administrative mechanisms can give a chance to redress either through formal or informal procedures to obtain remedies. Restitutions can be available via return of property or payment for the harm or loss suffered and when monetary compensation cannot be fully availed from the offender it is the responsibility of the State to ensure financial compensation. Under the 1972 Stockholm Declaration when it comes to rights based on environmental protection it ensures freedom, equality and adequate conditions of life, in an environment that ensures and permits a life of dignity and well-being as well as bears a solemn responsibility to protect and improve the environment for present and future generations. Through this, the protection of an environment by guaranteeing it to be clean, healthy and fit for human survival is deemed to be a human right in itself as the UN Human Rights Council via resolutions has ensured that a healthy natural environment is a right that belongs to every individual [89].

Under the Rome Statute, the remedies have two different aspects firstly it is ensuring that the restitution for those victims that are present before the Court and secondly reparations from a Trust Fund that ensures redressal for a much wider population that is not physically present within the Court as the ICC has the right to provide reparation either individually or on a collective basis or both. The establishment of the Trust Fund for Victims (TFV) by the ICC State Parties has allowed a way to support the victims by implementing court-ordered reparation awards against a perpetrator via voluntary contributions that provide physical, and psychological rehabilitation as well as material and mental support. As it is flexible it allows equitable awards amongst all the victims and in the cases of environmental crimes it provides a chance for reparations as opposed to rigid court procedures [89].

5. Conclusion

From this, we can easily observe that the ICC by setting out various heinous crimes and their elements has allowed a better chance at holding a perpetrator responsible who cannot be tried by the national courts either due to their inability or unwillingness because of the higher status most of these perpetrators hold in their respective States. In terms of natural and cultural heritage, it is deemed to be an already set principle that causing harm to those is indeed a war crime because of not only the loss that is faced by mankind as a whole but especially to those to whom those heritages were of great importance due to its cultural or religious value. By trying Al-Mahdi it has already set down a principle as to the effectiveness that whoever seems to violate and/or destroy such sites shall be held responsible in front of the Court. However, when it comes to the natural environment it is still unfortunately a rather growing concept because of the vast nature of the victims of the crime against the environment and due to the lack of a proper case dealt with by the ICC in terms of crimes against natural environment it is more prone to constant challenges along the way. In short, it can easily be concluded that despite the road bumps faced in these aspects, the importance of cultural and natural heritages as well as the natural environment can be observed not only in the aspect of science but also in the world of national but as well as international law to ensure it applies equally to every country as a whole.

The future possibilities of ensuring more protection to not only cultural heritages but also to the natural environment is to ensure that the various Un Resolutions which have been passed throughout the years not only encourage but incorporate it as a part of their national law, allowing all State Parties to ensure maximum protection to their resources within their territory whether on land or in sea but also allow better accountability to the perpetrators who damage them. This will allow the ICC to charge them of offenses by including such actions as a part of their crimes by allowing them jurisdiction and better access to penalize it for a better future for upcoming generations.

Some policies for the protection of cultural heritage can indulge in amending the Rome Statute by referring to other Conventions dealing with cultural heritage to ensure better protection; by including the destruction of cultural heritage as a part of war crime which will highlight its destruction and/or damage as one with grave consequences; by creating a special tribunal to not only keep a check and balance on those protected sites but also ensuring that those tribunals have the expertise to deal with any kind of illegal activity committed against any cultural artifact; and lastly promoting preventing measures by encouraging State Parties to ratify it for the better protection of these sites during peacetime and especially during any kind of armed conflict. In terms of policies regarding the protection of the natural environment it can include officially recognizing ecocide as a new kind of crime by amending it within the Statute especially in the category of genocide which will allow a steady and protected jurisdiction granting the ICC to try these atrocious crimes; incorporating environmental elements within existing crimes which falls within the ambit of the Court by making it a part of the provisions laid down in the Rome Statute; similarly establishing a specialized environmental chamber allowing ICC to exercise its jurisdiction to handle cases dealing with the destruction of the natural environment by linking it as a direct violation of the basic fundamental human rights granted to humans by making it fall within the ambit of crimes against humanity; and lastly collaborating with environmental organizations by leveraging it with the expertise of the UNEP to investigate and prosecute these cases by granting a special layer of protection and recognizing environmental defenders by linking harm to the environment to those defenders as they act as front line soldiers for the natural environment.

6. Recommendations

The expansion of the Court can play a greater role in creating a sustainable global society for mankind by making it more efficient in a manner that it can strengthen the information collection by analyzing and sharing the peacekeeping missions, Sanctions Committees and across the UN for the protection of natural resources as well as heritages to prevent its destruction during hostility by informing the relevant sub-tribunals created by the Court to ensure sustainable development which can include integrating and supporting inter-government organizations such as INTERPOL liaison officers in peacekeeping missions, United Nations Office on Drugs and Crime (UNODC) and World Customs Organization (WCO). By focusing on mechanisms, the ICC can introduce newer organizations by developing synergy which monitors international liaisons and the Convention's institutions, national enforcement agencies, and existing intergovernmental bodies.

It can also help in creating a rule of law amongst the international community by addressing these crimes as a severe threat to peace and maintaining development by the disruption of overseas tax havens, improving legislation at international and national levels, implementing dissuasive penalties, substantial sanctions and punishments, capacity building and technological support through inter-agency collaborations especially in developing and under-developed states. By creating new institutions and amending the pre-existing Statutes, the Court can focus on gathering the international development community to address these crimes as a severe threat to sustainable development by cooperating with the judicial sector reforms to set a precedent of combating and penalizing these crimes for its prevention.

Author details

Malaika Hassan
Bahria University, Islamabad, Pakistan

*Address all correspondence to: malaika.h.717@gmail.com

References

[1] McTaggart S, Barnett L, Dolin B. The International Criminal Court: History and Role. Ottawa: Canada; 2023:4 p. Available from: https://lop.parl.ca/sites/PublicWebsite/default/en_CA/ResearchPublications/200211E#

[2] Klobucista C, Ferragamo M. The Role of the International Criminal Court [Internet]. 2023. Available from: https://www.cfr.org/backgrounder/role-international-criminal-court

[3] Sarenmalm I. Sustainable Development in International Law and the Protection of the Global Commons [Thesis]. Uppsala, Sweden: Institute for Geovetenskaper, Uppsala Universitet; 2017. 8, 17-21 p

[4] Dernbach C. Achieving sustainable development: The centrality and multiple facets of integrated decision-making. Indiana Journal of Global Legal Studies. 2003;**10**(1):247-285 p

[5] Stoddart H. A Pocket Guide to Sustainable Development Governance. London, UK: Stakeholder Forum; 2011

[6] Jodoin S. Crimes against future generations. World Future Council. 2010;7(2):3, 6-8 p. Available from: https://www.worldfuturecouncil.org/wp-content/uploads/2016/01/WFC_CISDL_2010_Crimes_against_Future_Generations.pdf

[7] Banks D, Davies C, Gosling J, et al. Environmental Crime. UN Document, Environmental Crime Agency; 2008. 2-4, 15 p. Available from: https://www.unodc.org/documents/NGO/EIA_Ecocrime_report_0908_final_draft_low.pdf

[8] O'Keefe R. Cultural heritage and international criminal law. In: O'Keefe R, Segger M, editors. Sustainable Development and International Crimes. Cambridge: Cambridge University Press; 2013. 120-121 p. DOI: 10.1017/CBO9781139507561.010

[9] UNESCO Doc. CUA/120. 3 September 1962. 22

[10] Declaration on the Responsibilities of the Present Generations towards Future Generations, UNESCO Doc. 29C/Res 44, Art. 7, 1997

[11] Schabas W. An Introduction to the International Criminal Court. Cambridge: United Kingdom; 2002:4-8 p. ISBN: 0 521 80457 4

[12] Rules of Procedure and Evidence, ASP/1/3, pp. 10-107, Rule 44

[13] Schabas W. An Introduction to the International Criminal Court. 4th ed. Cambridge: United Kingdom; 2011:69-71, 80-81, 88-89 p. ISBN: 978-0-521-76750-7

[14] Report of the Committee on International Criminal Court Jurisdiction, 1952; UN Doc. A/2135

[15] Report of the Committee on International Criminal Court Jurisdiction, 1954; UN Doc. A/2645

[16] Klamberg M, editor. Commentary on the Law of the International Criminal Court. Brussels: Torkel Opsahl; 2017:18-19, 31, p. Available from: https://www.legal-tools.org/doc/aa0e2b/pdf/

[17] *Prosecutor v. Kayishema* (Case No. ICTR-95-1-T), 21 May 1999

[18] Jones A. Genocide: A Comprehensive Introduction. 1st ed. New York: Routledge; 2006:5, 10-11, 13-14 p. ISBN: 0-415-35385-8

[19] Jones A, editor. Genocide, War Crimes and the West: History and Complicity. London: Zed Books; 2004:217 p. ISBN: 978-1842771914

[20] Ratner S, Abrams J. Accountability for Human Rights Atrocities: Beyond the Nuremberg Legacy. 2nd ed. Oxford: Oxford University Press; 2001:30, 32, 74 p. ISBN: 9780199546671

[21] Luban D. A theory of crimes against humanity. Yale Journal of International Law. 2004;**29**:88-90, 93-94 p. Available from: https://scholarship.law.georgetown.edu/facpub/146/

[22] Makale H, Tekin S. The origins of crimes against humanity: From Nuremberg to Rome. ERÜHFD. 2015;**10**(2):159-160, 167 p. Available from: https://dergipark.org.tr/en/download/article-file/426221

[23] *Prosecutor v. Akayesu,* (Case no. ICTR-96-4-T), 2 September 1998, 502, 580

[24] *Prosecutor v. Kordicet al.,* (Case no. IT-95-14/2-T), 26 February 2001, 179, 245

[25] *Prosecutor v. Kordicet al.,* (Case no. IT-95-14/2-A), 17 December 2004, 89-90, 94

[26] *Prosecutor v. Kunarac et al.,* (Case no. IT-96-23/1-A), 12 June 2002, 94-95

[27] *Prosecutor v. Blaskic* (Case no. IT-95-14-T), 3 March 2000 156, 203

[28] *Prosecutor v. Jelsic* (Case no. IT-95-10-T), 14 December 1999, 53

[29] *Prosecutor v. Naletilic and Martinovic* (Case no. IT-98-34), Judgment, 31 March 2003, 233

[30] Schwarz A. War Crimes. MPEPIL. 2018:39 40 42 46 47. Available from: https://www.researchgate.net/publication/324454876_War_Crimes

[31] Cryer R, Friman H, Robinson D, Wilmshurst E. War crimes. In: An Introduction to International Criminal Law and Procedure. Cambridge: CUP; 2012:226-227, 243-254, 262-267, 271, 274 p. DOI: 10.1017/CBO9780511760808.013

[32] *Prosecutor v. Delalic* (Case no. IT-96-21-T), 16 November 1998, 508-511

[33] *Prosecutor v. Galic* (Case no. IT-98-29-T), 5 December 2003, 87-138

[34] Trahan J. The Rome Statute's Amendment on the Crime of Aggression: Negotiations at the Kampala Review Conference (ICLR'11). Leien, Netherlands: Martinus Nijhoff Publishers; 2011. 57 p

[35] Boas A. The definition of the crime of aggression and its relevance for contemporary armed conflict. ICD Brief. 2013;**6**:7 p. Available from: https://internationalcrimesdatabase.org/upload/documents/20141020T170547-ICD%20Brief%201%20-%20Boas.pdf

[36] Weisbord N. Conceptualizing Aggression. DJCIL. 2009;**20**(1):23 p. Available from: http://scholarship.law.duke.edu/cgi/viewcontent.cgi?article=1053&context=djcil

[37] Sayapin S. The Crime of Aggression in International Criminal Law. Netherlands: The Hague; 2014:257-261 p. DOI: 10.1007/978-90-6704-927-6

[38] Janet B. On defining the cultural heritage. The International and Comparative Law Quarterly. 2000;**49**(1):61 p. Available from: http://www.jstor.org/stable/761578?seq=1#page_scan_tab_contents

[39] Coreiro L. The Protection of Cultural Property under International Criminal Law. Lisboa, Portugal: Univeridade Catolica Portuguesa; 2017. 12-14, 18, 27, 41 p. Available from: https://carlospintodeabreu.com/wp-content/uploads/2018/10/the_protection_of_cultural_property.pdf

[40] Vrdoljak A. The Criminalisation of the Intentional Destruction of Cultural Heritage. Sydney: University of Technology; 2016:3-6 p. Available from: https://www.ohchr.org/sites/default/files/Documents/Issues/CulturalRights/DestructionHeritage/NGOS/A.P.Vrdoljak_text1.pdf

[41] Palarczyk D. Ecocide before the international criminal court: Simplicity is better than an elaborate embellishment. Criminal Law Forum. 2023;**34**:148-152, 160, 164, 178, 182-188 p. DOI: 10.1007/s10609-023-09453-z

[42] Dörmann K. In: Ambos K, editor. The Rome Statute of the International Criminal Court: Article-by-Article Commentary. 4th ed. München, Germany: C H Beck/Nomos/Hart; 2022. 118 p. Article 8

[43] Cottier M, Baumgartner E, Wehrenberg S. In Ambos K, editors. The Rome Statute of the International Criminal Court: Article-by-Article Commentary. 4th ed. München, Germany: C H Beck/Nomos/Hart; 2022. 400-409 p. Article 8

[44] *Prosecutor v. Ahmad Al Faqi Al Mahdi* (Case ICC-01/12-01/15), Trial Chamber, Judgment and Sentence, 27 October 2016, §15

[45] Forrest C. The doctrine of military necessity and the protection of cultural property during armed conflicts California Western international. Law Journal. 2007;**177**:212 p

[46] ILC. Protection of the Environment in Relation to Armed Conflicts: Draft Principles on Protection of the Environment in Relation to Armed Conflicts A/CN4/L968; 20 May 2022. Principle 17

[47] Ecuador and Peru, The Acta de Brasilia. 26 October 1998:Article 3

[48] Treaty of Peace Between the State of Israel and the Hashemite Kingdom of Jordan. Volume 2042, 1-35325: 26 October 1994; Principle 18

[49] Lehto M. UNGA, Third Report on Protection of the Environment in Relation to Armed Conflicts, Special Rapporteur. UN Doc A/CN4/750: 16 March 2022; § 192

[50] Office of The Prosecutor. Policy on cultural heritage. International Criminal Court. 2021:5-6, 8, 11-12, 37-38 p

[51] Office of The Prosecutor. Policy Paper On Case Selection and Prioritisation. International Criminal Court; 2016. 46

[52] Brammertz S. Attacks against cultural heritage as a weapon of war: Prosecutions at the ICTY. Journal of International Criminal Justice. 2016;**14**(5):1152-1153 p

[53] Bassiouni M, Nafziger J. Protection of Cultural Property. International Criminal Law I. 2nd ed. Leien, Netherlands: Brill Nijhoff Publishers; 1999. 954 p

[54] O'Keefe R. Protection of cultural property under international criminal law. Melbourne Journal of International Law. 2010;**11**:341 p

[55] Gerstenblith P. The destruction of cultural heritage: A crime against property or a crime against people? John Marshall Review of Intellectual Property Law. 2016;**15**:390 p

[56] Rep. of the Special Rapporteur in the Field of Cultural Rights, 59, U.N. Doc. A/HRC/31/59, 3 February 2016

[57] Gottlieb Y. Attacks against cultural heritage as a crime against humanity. CWRJIL. 2020;**52**(1):292-293, 311 p. Available from: https://scholarlycommons.law.case.edu/cgi/viewcontent.cgi?article=2574&context=jil

[58] *Prosecutor v. Al Mahdi* (Case no. ICC-01/12-01/15), 17 August 2017, 4

[59] *Prosecutor v. Al Mahdi* (Case no. ICC-01/12-01/15-171), Judgment and Sentence, 27 September 2016, 80

[60] Ba O. Who are the victims of crimes against cultural heritage? Human Rights Quarterly. 2019;**41**(3):580-585 p. DOI: 10.1353/hrq.2019.0044

[61] Kairisa I, Kairiss A. Victims of heritage crimes: Aspects of legal and socio-economic justice. Open Archaeology. 2023;**9**:4-7 p. DOI: 10.1515/opar-2022-0293

[62] *Prosecutor v. Jokic* (Case No. IT-01-42/1-T), 18 March 2004, 51-54

[63] Frulli M. The criminalization of offences against cultural heritage in times of armed conflict: The quest for consistency. EJIL. 2011;**22**(1):208-209 p. DOI: 10.1093/ejil/chr002

[64] Freeland S. Crimes against the environment: What role for the international criminal court? In: Edgardo S, Mead S, Samson B, editors. The Environment through the Lens of International Courts and Tribunals T.M.C. Hague: Asser Press; 2023. 2-5, 7, 12-15 p. DOI: 10.1007/978-94-6265-507-2_6

[65] Nellemann C, Henriksen R, Kreilhuber A, et al. The Rise of Environmental Crime – A Growing Threat to Natural Resources Peace, Development and Security. United Nations Environment Programme and RHIPTO Rapid Response–Norwegian Center for Global Analyses, UNEP-INTERPOL; 2016. 7-9, 12, 17, 26, 28 p

[66] Werle G. Principles of International Criminal Law. Oxford: Oxford University Press; 2005:191 p. DOI: 10.1093/law/9780198826859.001.0001

[67] Institutional and Financial Arrangements for International Environmental Cooperation, GA Res. 2997, 27 UN GAOR Supp. (No. 30) 43 p. 1972. UN Doc. A/8730

[68] Rauxloh R. The role of international criminal law in environmental protection. In: Francis B, editor. Natural Resource Investment and Africa's Development. Cheltenham, UK: Edward Elgar Publishing; 2011

[69] Palmer G. New ways to make international environmental law. American Journal of International Law. 1992;**86**(2):267 p. DOI: 10.2307/2203234

[70] Camilo R, Saavedra E. Protection of the natural environment under international humanitarian law and international criminal law: The case of the special jurisdiction for peace in Colombia. UCLA Journal of International Law & Foreign Affairs. 2020;**25**(1):142-147 p. Available from: https://escholarship.org/content/qt56n1p415/qt56n1p415.pdf?t=qox982

[71] Erdem M. Enforcing conventional humanitarian law for environmental damage during internal armed conflict. Georgetown Environmental Law Review. 2017;**29**(3):440-441, 444-445, 475-476 p

[72] Zierler D. The Invention of Ecocide Agent Orange, Vietnam, and the

Scientists Who Changed the Way we Think about the Environment. Athens, Georgia: University of Georgia Press; 2011. 19 p

[73] Gillett M. Eco-struggles: Using international criminal law to protect the environment during and after non-international armed conflict. In: Stahn C, Iverson J, Easterday J, editors. Environmental Protection and Transitions from Conflict to Peace: Clarifying Norms, Principles, and Practices. Oxford: OUP; 2017. 222 p

[74] Wyatt J. Law-making at the intersection of international environmental, humanitarian and criminal law: The issue of damage to the environment in international armed conflict. International Review of the Red Cross. 2010;**92**(879):645 p

[75] Spier J, Magnus U. Climate Change Remedies: Injunctive Relief and Criminal Law Responses. The Hague, the Netherlands: Sun Press; 2013. 189 p

[76] Anton D. Adding a green focus: The Office of the Prosecutor of the international criminal court highlights the "environment" in case selection and prioritisation. Australian Environment Review. 2016;**31**:346 p

[77] Mwanza R. Enhancing accountability for environmental damage under international law: Ecocide As a legal fulfilment of ecological integrity. Melbourne Journal of International Law. 2018;**19**:11-13 p. Available from: https://law.unimelb.edu.au/__data/assets/pdf_file/0005/2983055/Mwanza-unpaginated.pdf

[78] Report of the Preparatory Commission for the International Criminal Court: Addendum 2, UN Doc PCNICC/2000/1/Add.2; 2 November 2000:7 p

[79] Holvoet M. The State or Organisational Policy Requirement within the Definition of Crimes against Humanity in the Rome Statute: An Appraisal of the Emerging Jurisprudence and the Implementation Practice by ICC States Parties. ICD Brief 2. Washington, DC, USA: International Crimes Database; 2013. 7 p

[80] Lawrence J, Heller K. The first eco-centric environmental war crime: The limits of article 8(2)(b)(iv) of the Rome statute. Georgetown International Environmental Law Review. 2007;**20**(1):61-62 p

[81] Brisman A, South N, White R. Environmental Crime and Social Conflict: Contemporary and Emerging Issues. London: Routledge; 2015. 4 p

[82] *Prosecutor v. Dominic Ongwen* (Case ICC-02/04-01/15-1762-Red), Trial Chamber, Trial Judgement, 04 February 2021, § 2766

[83] *Prosecutor v. Germain Katanga* (Case ICC-01/04-01/07-3436-tENG), Trial Chamber, Judgment Pursuant to Article 74 of the Statute, 07 March 2014, § 883

[84] *Prosecutor v. Jean-Pierre Bemba Gombo* (Case ICC-01/05-01/08), Trial Chamber III, Judgment Pursuant to Article 74 of the Statute, 21 March 2016, § 525

[85] Glaser S, Malsch K. Analysis of the Environmental Impacts of Illegal Trade in Wildlife. 2017. UNEP/EA.2/INF/28

[86] Drumbl M. Waging war against the world: The need to move from war crimes to environmental crimes. Fordham International Law Journal. 1998;**22**(1):145 p

[87] Gillet M. Environmental damage and international criminal law. In: Jodoin S,

Cordonier Segger M, editors. Sustainable Development, International Criminal Justice, and Treaty Implementation. Cambridge: Cambridge University Press; 2013. 90-94 p. DOI: 10.1017/CBO9781139507561.008

[88] *Prosecutor v. Al Bashir* (ICC-02/05-01/09-3), Pre-T.Ch., Decision on the Prosecution's Application for a Warrant of Arrest against Omar Hassan Ahmad Al Bashir, 14 March 2009

[89] Skinnider E. Victims of Environmental Crime – Mapping the Issues. Vancouver, Canada: The International Centre for Criminal Law Reform and Criminal Justice Policy; 2011. 23-28, 30-37, 43-48 p. ISBN: 978-0-9868799-1-3

[90] Korsell L. Big stick, little stick: Strategies for combating environmental crime. Journal of Scandinavian Studies in Criminology and Crime. 2001;**2**(2):132 p

[91] Elliot L. Transnational environmental crime in the Asia-Pacific: Complexity, policy and lessons learned. In: Transnational Crime in the Asia-Pacific: A Workshop Report. Canberra: RSPAS, Australian National University; 2007

[92] Bricknell S. Environmental Crime in Australia. Australian Institute of Criminology Reports, Research and Public Policy Series No. 109. Australian Institute of Criminology: Canberra; 2010:2 p

[93] Buckingham S, Kulcur R. Gendered geographies of environmental justice. Antipode. 2009;**41**(4):659-683 p

Chapter 24

Perspective Chapter: Mining in Colombia from the Perspective of Environmental Justice – Were Judicial Courts an Ideal Setting to Achieve It?

Luis Fernando Sánchez Supelano

Abstract

The Colombian State haspromoted different adjustments to the country's mining model, generally with the intention of increasing foreign investment, increasing production, and increasing State income from the use of mineral resources, but this has led to the emergence of different judicial processes. In which the government's reforms have been questioned. In this context, the article aims to evaluate the jurisprudential developments of the Colombian Constitutional Court and the Council of State from an environmental justice perspective. Thus, this text reconstructs the jurisprudential debates generated by the judicial actions presented by members of Colombian civil society, develops the basic dimensions of an idea of environmental justice, and evaluates the impacts and transformations generated by judicial decisions in terms of mining regulations and public policies. It is based on the jurisprudential analysis of the highest Colombian courts and a critical analysis based on environmental justice. It is found that although jurisprudence has imposed some limits and has served as a counterweight to the actions of the executive, it is still necessary to deepen environmental justice approaches to resolve controversies involving mining.

Keywords: mining, extractive industry, environmental justice, jurisprudence, environmental rights

1. Introduction

In recent years, the extractive industry and especially mining has been at the center of the discussion in Colombia; it is seen as a source of resources to strengthen public finances, the financing of peace agreements, the politics social of the Colombian State or an economic activity that is essential for the economic development, in that perspective, it is argued the need to expand, formalize and deepen mining exploitation in the country. Likewise, for the State, the regulation and control over mining has to do with preventing mining resources from becoming sources of financing for illegal armed groups that participate in the Colombian armed conflict [1–3].

Along these lines, since 2001, the Colombian State has promoted a set of regulatory reforms that have attempted to improve the control, surveillance, and supervision of mining activities to ensure that it is articulated with the State's development policies and strategies [4]. Although it must be recognized that it has also been done so from different approaches that are combined in different ways, in general, in recent years, a vision has prevailed in which mining is an activity carried out by individuals under the surveillance and regulation of the State, exchange of which the State receives some tax and compensatory income [2]. Thus, the idea of the State developing the industry through its own companies has been gradually abandoned [5].

In accordance with the above, in recent times Colombian regulations have attempted to provide an environment that facilitates the private exploration and exploitation of minerals [6] based on at least the following elements: (a) providing a simple regulatory framework and flexible that provides legal security for investors; (b) the generation of fiscal and regulatory incentives that attract investment and generate a competitive environment; and (c) the transformation of the role of the State.

The previous adjustments have at the same time generated multiple tensions, generating environmental and territorial conflicts. Three cores of central conflict can be identified: (i) the environmental one around the discussions about the impacts of mining on ecosystems, the way to control them, and the definition of the areas in which it should be excluded or restricted; (ii) the social in the perspective of the existence of high informality, of families that depend on mining activity to satisfy their basic needs and that do not necessarily have the capital to assume a business exploitation model, and; (iii) the attempt by armed groups financed through the exploitation of these mineral resources [1, 7].

According to Environmental Justice Atlas (EJATLAS), Latin America and the Caribbean have more than 340 conflicts related to the mining sector. The Latin American Mining Conflicts Observatory records more than 289 conflicts involving mining activities, which makes the region the one with the greatest presence of environmental conflicts. Likewise, there is a trend toward progressive growth, given the growing demand for minerals, including those related to the energy transition. For example, the region is a key player in the current and potential supply of critical minerals, especially in the lithium industry, where it has about half of the world's reserves but has participated in around one-third of the production since 2017 [8].

One of the scenarios in which these tensions and challenges have been manifested has been in the judicial courts, in fact different people, collectives, and ethnic groups have filed multiple legal actions before the Colombian courts questioning the norms and public policy issued by the Colombian State. In that sense, this text aims to assess the jurisprudential developments of the Colombian Constitutional Court and the Council of State from a perspective of environmental justice. Hence, this text reconstructs the jurisprudential debates generated by the judicial actions presented by members of Colombian civil society, develops the basic dimensions of an idea of environmental justice, and assesses the impacts and transformations in terms of regulations and mining public policy generated by judicial decisions.

To achieve the above, the text is structured in three large parts: the first presents some methodological elements on the way in which jurisprudential analysis is identified and organized, the second develops some conceptual elements on the idea of environmental justice, the third presents the main lines of jurisprudential debate developed by the Council of State and the Constitutional Court and the main conclusions of the identified judicial decisions. Finally, some partial conclusions are presented.

2. Methodology

Methodologically, the following moments were developed: (1) definition of the universe of rulings on mining; (2) definition of thematic prioritization criteria for the analysis of jurisprudence; (3) the application of prioritization criteria by relevance of the judicial rulings, and; (4) construction of a qualitative analytical matrix.

From the information systems of the Constitutional Court, the Council of State, and the Supreme Court of Justice, the judicial decisions that address the issue of mining are identified using keywords, initially finding a total of 683 judicial rulings.

Taking into account the objectives of the text, the following thematic axes have been determined for the analysis of the judicial decisions (in such a way that all those that do not address at least one of the following thematic axes were discarded): scope of powers mining authorities; environmental criteria for exclusion of mining activities; application of environmental control instruments; citizen participation; environmental, passive and mining damages; environmental conditions for mining exploration and exploitation; mining closure; administration and recovery of minerals. In this way, only those aspects that have a direct and concrete connection with the stated objectives were recorded in the matrix, discarding contents that, although relevant from a legal point of view, do not have this relationship with the objective of the research. In this sense, judicial decisions that were repetitive (since they reiterated previous positions) or judicial rulings that, despite having to do with the issue of mining, did not contribute to the objective of the investigation were excluded (e.g., royalties, denial of mining legalization processes, liability for work accidents, etc.).

In a third moment, a second prioritization of the judicial decisions to be reviewed was carried out, this time establishing a selection by type of judicial decision. Six types of judicial decisions were prioritized: (i) constitutionality judicial decisions, (ii) a unification judicial decision, (iii) popular actions, (iv) judicial decisions considered milestones, (v) "Tutelas," and (vi) legality actions ("nulidad").

In addition, different temporal scenarios were used to analyze the judicial decisions. For constitutionality, "tutela" and popular action judicial decisions issued since 1991 were taken, while for the judicial decisions of nullity, annulment, and contractual controversies, the year 2001 was taken. Once these filtering and prioritization processes of the judicial decisions were carried out, the total universe was reduced to 53 judicial decisions that met the indicated criteria, and that made it possible to identify adjustment criteria or orders of transformation of normative components of the mining rules in the axes of analysis.

Once the previous exercises were completed, a matrix was filled out, which focuses on two major elements: (a) the identification of the central argument (identification of the ratio decidendi of the judicial decisions and central arguments for decision-making); (b) the identification of specific orders of regulatory adjustments required to mining regulations in the thematic axes developed or criteria set forth by the courts.

Finally, a comprehensive analysis of the jurisprudence was carried out to offer three broad readings: (a) the definition of the constitutional standard and rights for the application of the mining code; (b) the evolution of jurisprudence and the twists and transformations it has undergone; (c) the imperatives of adjustment in environmental regulations. In that sense, a mainly dogmatic legal and systematizing study is developed to understand the constitutional axiological limits of mining, from the perspective of environmental justice.

Quantitatively, the judicial decisions analyzed, 41.1% corresponded to constitutionality control rulings and 31% to actions in defense of constitutional rights where no compensation was requested for the economic damages suffered by people. These elements show two initial trends in the litigation against mining: public actions were preferred where the requirements for active legitimacy were lower, and actions that sought the adoption of precise actions to guarantee the enjoyment of rights were preferred over those that sought financial compensation for the plaintiffs. Another important aspect has to do with the preference for actions of abstract control of legality or constitutionality, which together account for almost 50% of the sentences analyzed.

3. Environmental justice

The idea of environmental justice is closely related to the idea of equitable distribution of environmental and natural goods as well as the evils derived from environmental degradation—whether with discriminatory purposes or not [9, 10]. In this way, environmental justice as distributive Justice is defined in terms of four principles. The first is equality, referring to identical or comparable distributions of environmental goods and burdens. The second is proportionality, which implies distributions of goods and burdens under conditions of equity. The third is need, that is, the distribution of goods to satisfy current basic needs. Finally, compensation denotes the need to allocate environmental goods to compensate for undue costs or violations inflicted on a party [11, 12]. Of course, that implies the need to have an intersectional approach.

Going even deeper into the distributive dimensions of environmental justice, it could also be analyzed from its spatial dimensions, which are expressed in the geographical distribution of natural goods and in the calculated distribution and territorial concentration of anthropogenic environmental burdens, and a temporal dimension that indicates the need for a fair distribution of environmental benefits and burdens from the consideration of future generations—intergenerational justice—this must be differentiated from intragenerational environmental justice, which refers to fair distribution in the same generation [13]. In this way, two initial dimensions of environmental justice can be identified in terms of justice with spatial-territorial and temporal approaches.

The next level of discussion about environmental justice came to be around the possibility of participating in decision-making regarding the environment. The incorporation of the so-called procedural justice, within environmental justice would imply the assumption of substantial criteria in the decision-making processes in environmental matters, specifically in terms of four principles. One is fair representation, which refers to the full or balanced representation of the parties in the decision-making process and the interests expected to be affected by the outcome. Another principle refers to the issues of fair treatment and fair play, which refer to the opportunities to participate, to be heard, and to influence the process, as well as the coherence and impartiality in the conduct of the process. A third is the voluntary agreement, that is, accepting without coercion the agreements reached. Finally, the principle of transparency refers to the openness and accessibility of information in decision-making in the negotiation process, as the parties work to reach a negotiated agreement [11]. In this way, the implementation of these principles would allow us to reach better agreements on the form of distribution of burdens and benefits [14–19].

Now, these discussions about participation and procedural justice are framed in the question of whether environmental sustainability can be achieved through democratic forms, which is one of the most important questions around the concepts of governance and sustainability [20, 21]. It is discussed around the criticism of the way in which the market has displaced democracy, and it asks how democracy and sustainability can be revived in such a way that the discussion is articulated around the way of critically theorizing commitments [20, 22, 23].

In this sense, we reflect on the need for a reform of democracy around the idea of environmental sustainability and other values considered fundamental. In this way, concepts such as discursive democracy [24], deliberative democracy [25], substantive democracy [26], cosmopolitan democracy [27], normative democracy [28], ecological democracy [29], sustainable democracy [30], land democracy [31], and environmental democracy [32, 33].

Of course, the previous position also implies that public decision-making exercises must meet material conditions and not just formal ones, as well as articulations between global, regional, and local levels. Conditions such as transparency, accountability, and real participation, the definition of adequate environmental standards for the development of economic activities, and some changes in electoral schemes are desirable [20, 23]. At the same time, a profound change is required in environmental thinking from the perspective of environmental citizenship [20, 33–35].

In this way, it can be summarized that from the perspective of environmental justice, injustices are committed when an individual or group faces (a) disproportionate environmental risks, (b) has unequal access to environmental goods, and (c) has less opportunity to participate in the decision-making process with respect to the environment [36–39].

The conceptualization of environmental justice has revolved around three dimensions: (a) the distribution of burdens and benefits among different subjects, in territories and over time—spatial and distributive dimension temporal, (b) the way in which one participates in decision-making—procedural dimension, and (c) the interests of those who are considered in environmental decisions—recognition dimension [38].

Now, it should be noted that in this last dimension, some have affirmed the need to reflect on the consideration of the non-human. Along these lines, Professor Mesa advances by indicating that a theory of justice should contain the following central elements: incorporate a new vision of rights, highlighting that these are necessarily environmental rights, which can be grouped into two large components (the human rights and the rights of the non-human) and that to contribute to the effective resolution of large and serious conflicts and environmental problems, it is necessary to understand rights from a triple condition: new subjects from an expansion of moral and legal consideration; new temporal vision of them that includes not only the current one but also future generations and; new spatial vision, which transcends the sectoral and partial scope of the narrow limits of the Nation-State and is projected to the future from cosmopolitanism and alternative globalization [40, 41].

References to Pachanama (Mother Land) or Sumak Kawsay (Good Living), of the Kichwa peoples, or Suma Qamaña (Living Well), of the Aymara peoples, belong to these categories. They are founding concepts of indigenous peoples, which mean, in their concrete historical conditions, worldviews, and practices of respect for nature and shared collective life [40]. They can inspire contemporary thought and social organization and restore its strength to the symbol. However, success depends on the necessary adaptations "in such a way that the transformation will have the opportunity to combine the best of ancestral and modern knowledge, with knowledge and technologies synchronized with the functioning of nature" ([42], p. 107).

Finally, an additional dimension could be added to the idea of environmental justice in the sense of indicating that it would also be mediated by the idea of the limits of appropriation—that part of the environment that is valid to appropriate—as well as the idea of respect for nature as an ethical guide for said appropriation. In this sense, the idea of environmental justice encompasses, at least, the following mutually interconnected dimensions: (a) the distribution of burdens and benefits among the different subjects in the territories and over time—spatial and temporal dimension, (b) the how one participants in decision-making—procedural dimension, (c) the interests of those who are taken into account in environmental decisions—dimension of recognition, including the recognition of nature, and (d) the limits of appropriation, which implies that only A sustainable appropriation of nature—sustainable environmental footprint—is legitimate.

4. Colombian jurisprudence, mining, and environmental justice

This section is responsible for presenting a review of the jurisprudence indicating the main points of debate, their scope, and the impacts it has on the development of mining activity, with the purpose of analyzing the transformations in mining activity generated by the decisions of the courts. In this perspective, five main axes of debate are identified as of special relevance for discussions of environmental justice. The first refers to a certain spatial justice in the perspective of identifying the places where mining is permitted, the second to the dimensions of environmental sustainability in the perspective of incorporating environmental variables into the analysis of mining authorizations, the third revolves around the participatory dimensions in decision-making on the development of mining in the dispute between national and local authorities, in the fourth also in the participatory dimensions, this time on the rights of the affected communities, finally in the fifth it turns to the perspective of sustainability and the powers of the State to restrict or revoke authorizations for mining activities that it had previously authorized.

4.1 First line of jurisprudential debate: mining authorization process, versus environmental regulations, which exclude and restrict mining areas

The promulgation of the new Mining Code in 2001 generated a boom in discussions around mining in the country, with the compatibility of mining with some ecosystems being one of the first discussions to emerge. This discussion arose around the concept of excluded and restricted mining areas. At this point, the Constitutional Court, since 2002, indicated that the areas excluded from mining transcended the restrictive list initially established in Law 685 of 2001—which limited them to the System of National Natural Parks, regional natural parks, and areas of forest reserves—for which he indicated that in addition to the exclusion zones provided for in this law, there may be others, already declared previously or that will be declared in the future by the environmental authority, clarifying that the excluded zones must be clearly delimited geographically and this function is assigned to the environmental authority in integration and harmony with the provisions of article 5 of Law 99 of 1993 [43]. This first point forced us to comprehensively consider the different categories of environmental planning and protection regarding compatibility or not with mining, which also had impacts on the development of mining since there were environmental areas that were not adequately delimited or updated.

Now, the mining authority has the duty to collaborate with the environmental authority in the delimitation of protected areas when there are mining interests in those areas, but this duty of collaboration does not limit or condition the exercise of the jurisdiction of the environmental authority, who is the one that can establish exclusion zones [43]. However, the Council of State would then require that the mining authority must actively participate in the preparation of studies that serve as a basis for the environmental authority to finally adopt the respective decision [44, 45].

However, the problems in the granting of mining titles and the deficiency in the protection of the environment would lead the Council of State (rad. 25000234100020130245901) to give several adjustment orders due to the violation of collective rights. Among them, we find orders related to the preservation of areas belonging to Sistema Nacional de Areas Protegidas (SINAP) and *in situ* conservation zones of legal origin and compliance with the commitments adopted in public policy for the National System of Protected Areas (SINAP) with a vision 2021–2030 (CONPES 4050 of 2021).

As temporary actions, the Council of State indicated: (1) The National Mining Agency must require proponents to provide with their titling application a certificate from the competent environmental authorities informing: (a) if their project overlaps or not with any of the ecosystems referred to in subtitles (b) and (c) of chapter II.3.3 of this judicial decision, (d) whether such territory is zoned, and (e) whether mining activities are permitted in the zoning instrument. In the event that doubts arise about the compatibility of the project with the conservation purpose, the National Mining Agency must refrain from substantively resolving the proposal until there is certainty about the aforementioned compatibility, applying the precautionary principle. (2) The National Mining Agency will update the information reported by the Anna Minería platform regarding the protected territories in which there is currently certainty about the exclusion and restriction of the development of mining activities.

This jurisprudential discussion shows us three needs for regulatory articulation: (1) the need to articulate mining authorization processes with the geographical restrictions derived from environmental protection regulations; (2) the articulation of cartographic information systems of the environmental and mining sector; (3) the promotion of compliance with environmental powers regarding the delimitation and zoning of protected areas, as well as the application of the precautionary principle in these matters.

4.2 Second line of jurisprudential debate: the articulation of mining authorization and control processes with environmental variables and regulations

Another aspect of the discussion was developed in the criteria under which mining titles should be granted, so two initial criteria will be developed: (a) the idea of equal treatment between nationals and foreigners in the process of obtaining mining permits and authorizations and (b) the possibility that in certain areas the intervention of state authorities other than environmental and mining authorities may be required, as occurs in the case of those areas of the archeological, historical and cultural importance of the Nation [43].

However, the Court would find that the regulatory structure for granting mining titles suffered from several structural flaws that needed to be corrected. Specifically, the Constitutional Court has urged the Congress of the Republic to issue an adequate regulatory framework to guarantee the inclusion of environmental and social variables in the process of assigning mining areas and in the procedure for granting mining titles. The regulations must be based on differential criteria that respond to the different types and scales in which mining is carried out, respecting the set of constitutional principles. These criteria must include (i) respect for subsistence mining, (ii) standards to adapt small mining to environmental protection and medium and large-scale mining to the highest standards of the industry and the principles of corporate responsibility, and (iii) differentiated mining activity not only by the size of its projects but also by its social, cultural, and legal meaning [46].

In this order of ideas, the Constitutional Court has highlighted that the mining authority must verify minimum labor and environmental suitability before delivering a mining title, taking into account the nature of the concession requested, based on differential criteria between the different types of mining, and extension of the projects, as well as establishing a procedure that ensures citizen participation, without prejudice to the special participation of ethnically differentiated groups [46]. Furthermore, the court has highlighted that it is Congress, with the government initiative, that must establish distribution rules, objectives, purposes, administration, execution, control, efficient use, and destination of the income from mining exploitation [46].

However, the lack of action by the legislature and the executive led the Council of State (rad. 25000234100020130245901) to issue several adjustment orders due to the violation of collective rights. In this sense, the Ministry of Mines and Energy and the National Mining Agency was ordered to, with the intervention of the Ministry of Environment and Sustainable Development, carry out the review and adjustment of the Anna Minería platform with the purpose of implementing a technological module that allows, in real time, to process and monitor the environmental component of mining titles, guaranteeing the authenticity, integrity, and availability of the information. Such review and adjustment must achieve: (1) the usability, accessibility, functionality, and interoperability of the data and technological platforms implemented in both sectors and (2) develop the modules or processes that allow for improving the mining-environmental process of communication, coordination, evaluation, monitoring, and audit.

In the same way, the Ministry of Mines and Energy and the Ministry of Environment and Sustainable Development were ordered to coordinate and prepare a document for the diagnosis and characterization of the problems related to (1) mining projects whose titles overlap with environmentally protected territories, (2) the negative environmental impacts generated by mining projects that do not have environmental licensing in the exploration phase, and (3) mining projects that are not being controlled in the exploration phase. Based on this characterization, both ministerial offices will prepare an inventory of Mining Environmental Liabilities—PAM. Based on the results compiled in the diagnostic document and the inventory of liabilities, the ministries will adopt and implement a preventive and corrective plan that identifies participatory and joint actions aimed at correcting the three problems stated.

In the same perspective, the Council of State decided to urge the Ministries of Mines and Energy and of Environment and Sustainable Development to prepare, review and formulate the draft laws, as well as the draft regulations related to (1) the procedure of evaluation of mining titles, (2) the legal figures for biodiversity conservation, (3) the process of subtraction of some ecosystems for the National System of protected areas, (4) the regulation of environmental liabilities, and (5) the enforceability of environmental licensing from the exploration phase.

In addition, the Ministry of Mines and Energy and the Ministry of Environment and Sustainable Development were ordered to update the mining-environmental guides and the terms of reference with the purpose of adjusting them to the provisions of article 19 of law 1753 of 2015. The documents will include an appendix in which environmental determinants will be defined by type of extraction, according to their specific environmental impacts. In the same way, the Ministry of Mines and Energy and the Ministry of Environment and Sustainable Development were ordered to establish an inter-institutional working group. Said technical table will oversee proposing and executing actions aimed at improving the relationship between the mining and environmental sectors. For this purpose, it will adopt its operating and work regulations.

Likewise, the Ministry of Mines and Energy and the Ministry of Environment and Sustainable Development were ordered to update the National Mining Policy so that it contemplates actions that counteract the problems related to (1) insufficient mining-environmental territorial planning, (2) the lack of coordination between State institutions, and (3) the weaknesses of the control and supervision model of mining titles.

In this order of ideas, this ruling identified the need for several regulatory adjustments in terms of liabilities, procedures, the procedures for the subtraction of protected areas from the national system of protected areas, as well as to strengthen the articulation of mining and environmental procedures, also to facilitate and strengthen the inter-institutional coordination. Giving concrete orders to transform the actions of the State in the perspective of improving the environmental sustainability of mining.

4.3 Third line of jurisprudential debate: the distribution of powers between the national level and the territorial level of government, applying the principles of concurrence and subsidiarity

One of the most controversial points that Colombian jurisprudence has debated has been the distribution and concurrence of jurisdiction between the national level and the territorial level of government to define where mining exploration and exploitation are permitted or not. In this discussion, the Constitutional Court has recognized that mining activity has considerable environmental, social, and economic repercussions, which directly or indirectly affect the people and territories in which said activity is carried out and decisively condition the powers of territorial planning and determination of land uses that correspond to the territorial entities. Reason for which it has determined that no national authority can unilaterally adopt decisions in this regard that exclude the participation of the municipalities, who directly receive the impacts of that activity [47].

In this way, the Constitutional Court has indicated that there is a duty to carry out concerted actions by the national authorities with the territorial entities whenever mining projects are developed, or areas are assigned for mining exploration and exploitation. In relation to article 20 of Law 1753 of 2015—on mining reserve areas—the Court has said that the competent authority to define the mining reserve areas must previously agree with the local authorities of the municipalities where they will be located, to guarantee that their constitutional power to regulate land uses are respected. Also, the National Mining Authority and the Ministry of Mines and Energy must guarantee that the definition and offer of said areas are compatible with the respective territorial planning plans [47].

According to the Constitutional Court, the establishment of a procedure for the National Government and territorial entities to jointly agree on measures to protect the healthy environment, water basins, the economic, social, and cultural development of their communities and the health of the population, against the possible effects that may arise from mining activity, does not ignore the principles of autonomy of territorial entities, since, on the contrary, it is in accordance with the principles of concurrence, coordination and subsidiarity enshrined in article 288 of the Constitution [48].

Recently, the State Council indicated that territorial autonomy appears delimited by the law regarding the protection of the environment and natural resources as a corollary of the principle of subsidiarity provided for in article 311 of the Constitution, by virtue of which, despite the fact that the municipality is the fundamental entity of the political-administrative division of the State, it must sometimes give way to the intervention of other instances. So, in environmental matters in some aspects, it transcends local interests and becomes a matter of national interest and international projection, which as such requires regulation emanating from the central power, since evidently there are environmental aspects that affect the national interest and the global interest [44].

Indeed, to justify the need for concurrence and coordination between the national government and the territorial entities, the Council of State has maintained that policies and definitions of a general environmental nature affect the regulatory power that corresponds to the territorial entities by decision of the Constituent Assembly..., with different intensity, depending on the regulation of a natural resource, whose management affects, to a greater or lesser extent, regional or national ecosystems. In this way, environmental and natural assets require special protection as they are constitutive assets of the national heritage, whose use compromises the present and future of the entire Nation, which merits coordinated and directed action by the State, aimed at preserving and safeguarding them, which prevents the regulatory normative activity that the territorial entities are in charge of from being carried out in an isolated and contradictory manner [44, 49].

For this reason, the Constitutional Court has indicated that for the distribution of powers between the Nation and the territorial entities, the legislator must take into account that the essential content of autonomy focuses on the possibility of managing one's own interests (C.P. art. 287), one of whose most important manifestations is the right to act through one's own bodies in the administration and government of matters of regional or local interest [50].

In congruence with the above, the Constitutional Court has indicated that in the process through which the carrying out of mining exploration and exploitation activities is authorized, the competent authorities at the national level must agree with the territorial authorities concerned, the necessary measures for the protection of the healthy environment, and especially, its water basins, the economic, social, cultural development of its communities and the health of the population, through the application of the principles of coordination, concurrence and subsidiarity [51].

However, the above led the Constitutional Court to conclude that there were two deficits in regulatory matters, one in terms of citizen participation and another in the existence of coordination mechanisms between the territorial level and the national level of government regarding the development of extractive projects. In accordance with what is indicated by jurisprudence and with the purpose of defining constitutional criteria for (1) the institutional and technical strengthening that must be carried out by the central national government, headed by the Ministry of Mines and Energy and (2) the regulation that the Legislator must issue regarding the definition of one or more mechanisms of citizen participation and one or several instruments of coordination and concurrence within the nation and territory, the Court has mentioned constitutional criteria that must be taken into account, both by the central national government and by the Legislator, in the definition and implementation of the same [52]. These criteria are (1) citizen participation and plurality, (2) duty to provide for characteristics of gradualness, availability, access, quality, permanence, and dialog, (3) nation-territory coordination and concurrence, (4) nonexistence of veto power of territorial entities to the exploration and exploitation of the subsoil and non-renewable natural resources, (5) differentially/graduality of the necessary participation mechanisms, (6) territorial approach, (7) legitimacy and representativeness, (8) prior, permanent, transparent, clear, and sufficient information, (9) sustainable development, (10) dialog, communication and trust, (11) respect, protection, and guarantee of human rights, (12) good faith, (13) coordination and strengthening of national and territorial institutional capacity, (14) fiscal sustainability.

4.4 Fourth line of jurisprudential debate: the right to participate of ethnic and non-ethnic peoples in the context of mining processes

One of the important points in the processes of mining activities is related to the protection of the right to participate in the different stages of mining activity. Within this discussion, one of the first issues that the Constitutional Court had to resolve was related to the participation of ethnic peoples in the construction of a Mining Code. On this point, the Constitutional Court expressed that it is not contrary to the Constitution for a government entity to autonomously prepare a bill on an issue that falls within the scope of its powers, even if it is of interest to indigenous peoples, since such activity is part of the exercise of its functions.

However, in that case, the entity must provide the communities, at a time prior to the filing of the project in the Congress of the Republic, the due opportunities so that they not only know the project in depth but, above all, so that they can actively participate and intervene in its modification, if necessary. Notwithstanding the above, the case may arise that the initial composition of the project is the product of the joint and concerted effort of entities and communities, an event in which compliance with indigenous participation would be more clearly evidenced [53]. In accordance with the above, a general principle was developed that the government is obliged to promote effective and reasonable mechanisms for participation in matters that affect indigenous communities. However, if an agreement is not reached with the latter, the legislative process does not have to stop in matters that are also of general interest, as is the case in the mining case [53].

A second issue had to do with the preferential rights of ethnic peoples regarding the use of mining resources and the possibility that the mining authority can grant mining titles when ethnic peoples do not exercise this preferential right. In this case, the Court indicated that the mining authorities must comply with the parameters established regarding prior consultation, that is, giving the ethnic groups the respective opportunities to know, review, debate, and decide on the issue put before their consideration, being able to this end, autonomously decides on the exercise of their right of preference [53]. In this perspective, the Court stated that the indigenous authorities are competent to indicate “within the indigenous mining zone, the places that cannot be the object of mining exploration or exploitation because they have special cultural, social and economic significance for the community or ethnic group," in accordance with their beliefs, uses and customs [53].

Without prejudice to the foregoing, the Court considered that the general interest implicit in mining activity must be harmonized with the interest of the indigenous community in having its own authorities determine the areas that should be understood as restricted to mining, the latter being also of general interest as a manifestation of the ethnic pluralism that the Constitution protects. Thus, to resolve the possible tension between both interests, the Court considered that the indigenous authorities could indicate the aforementioned restricted areas no later than the moment in which the mining authority delimits the indigenous mining zone that, as has been seen, is developed within the framework of a consultation process [53]. In this regard, it is worth remembering that under Law 685 of 2001, the mining authority is competent to indicate and delimit indigenous mining areas in compliance with the paragraph of article 330 of the Constitution and article 15 of Convention 169 of the International Labour Organization (ILO), in terms of sentence C-418 of 2002 [53].

In any case, the Constitutional Court has indicated that the use of natural resources, including miners, found in the territory in which ethnic and Afro-descendant communities are located is a transcendental issue for the definition of identity and various of these towns. Therefore, both legislative and administrative measures that may affect this specific use must have spaces for participation for these communities [50].

The participation of indigenous and Afro-descendant communities regarding measures related to the use of mining resources located in their territories must be compatible with the conditions of their differentiated identity. Therefore, this participation cannot be accredited in any way but must comply with the requirements of prior consultation. This is because such conditions allow the participation of ethnic peoples to be carried out under conditions of good faith, sufficient information for the affected communities, cultural adaptation, and opportunity [50].

The Court has found that the delimitation of strategic areas for the exercise of mining, the granting of mining concession contracts, exploration activities, the grant of environmental licenses for exploitation, the exploitation itself, the adoption of relocation aimed at averting the polluting effects of mining and the transportation and collection of these resources have generated social, cultural, economic and environmental impacts on indigenous and Afro-descendant communities that, as scenarios of potential direct impact, activate the obligation of prior consultation [54, 55].

Without prejudice to the above, the need to guarantee effective participation is not limited to ethnic peoples; on the contrary, the jurisprudence of the Constitutional Court has indicated that the participation of communities affected by mining must be guaranteed, in order to guarantee fair treatment which assumes that "no group of people, including racial, ethnic or socioeconomic groups, should bear disproportionately the burden of negative environmental consequences as a result of industrial, and commercial operations or the implementation of environmental programs and policies at the municipal, regional or national level." Meanwhile, community participation is understood to be meaningful when: (1) potentially affected community residents have an appropriate opportunity to participate in decisions about a proposed activity that will affect their environment and/or health; (2) the public's input and concerns of all participants are effectively taken into account and likely to influence decision-making; and (3) those responsible for deciding promote and facilitate the participation of those potentially affected people and/or groups [46].

Even in the case of Marmato, the Constitutional Court found that there were populations that, due to their social characteristics and their connection to the territory or mining activity, should be protected, guaranteeing them special participation mechanisms. In this sense, the Constitutional Court indicated that the verification of this special interest in participation, considering the specificities of each case—the actors involved in the transfer, the productive vocation of the concessioned territories, the social and economic dynamics of the inhabitants of the area, the presence of ethnic communities, etc., activates the duty to foresee scenarios of effective participation with those potentially affected. Such is the rule established in the jurisprudential precedents that have protected the fundamental right to participate in the adoption of environmental decisions, urban planning, and housing, among other aspects, on the sole assumption that they concerned the people or communities acting, in so much did they compromise their ways of subsistence, the organization of the territories they inhabited or the guarantee of other fundamental rights [55].

Additionally, the Constitutional Court has identified that there is a lack of protection in terms of the right to participation of communities affected by extractive projects because the participation of communities that are in places where they are developed is not specifically legally guaranteed. Activities or operations to explore or exploit non-renewable and subsoil natural resources, and that may be impacted or affected by them, nor an instrument for territorial entities, through their competent authorities, to participate in the definition, execution and monitoring of hydrocarbon and mining activities [52].

Now, the Constitutional Court has also highlighted that for adequate citizen participation, adequate access to information is also required, which is why in mining matters any reservation to public information must satisfy three general conditions, namely: (i) they can only be established by law; (ii) the exceptions are subject to the principles of reasonableness and proportionality, and must also be related to the protection of fundamental rights, such as privacy, or constitutionally protected values, and; (iii) they must be temporary, for which the law establishes a period, after which the documents enter the public domain [56].

4.5 Fifth line of jurisprudential debate: legally consolidated situations and subsequent environmental restrictions

The Constitutional Court has also had several developments on the obligations of people who carry out economic activities such as mining, regarding environmental protection and respect for the fundamental rights of people. Thus, it has indicated that the private interest that economic activity represents is subordinated to the public or social interest that the preservation of the environment demands, in such a way that the individual must carry out their respective activities within the precise frameworks indicated by the Constitution, the environmental law, regulations, and authorizations that must be obtained from the entity responsible for the management of the resource or its conservation. It is an obligation for all individuals, especially those who carry out activities that, in one way or another, may cause environmental impacts, to comply with the duty of protection and conservation of the ecosystem, taking the necessary precautionary measures to avoid the occurrence of harm. Or not to the environment or to reduce or mitigate the consequences generated [57–59].

In particular, the protection of the rights of mining owners has been discussed when collective rights such as a healthy environment and the protection of special ecosystems such as moors are affected. In this sense, the Council of State has indicated that the exercise of the rights arising from the mining concession is subject to obtaining the appropriate environmental permits, which depend on the mining regulations in force at the time of the perfection of the contract, without prejudice to the environmental regulations approved during the development of mining activity and that are immediately applicable [60].

The Council of State has indicated that taking into account that according to the second part of article 58 of the Constitution, even in the event of consolidated legal situations, the State retains the power to configure the exercise of rights when a conflict arises between private interest and public utility or social interest, the foregoing without prejudice to the consequent indemnification duty that may arise in each specific case. In particular, the Council of State has concluded that the mining concession holders if the concession polygon belongs to a high mountain area, do not have a justified expectation of maintaining their economic activity, since, even based on the regulations issued prior to 2009, it is clear that paramo ecosystems are included in situ preservation zones [60].

Along the same lines, the Council of State has maintained that the obligation of the State toward the concessionaire is limited to the delivery of participation in the exploitation of the asset. Therefore, the mere existence of a concession contract does not prevent the State from limiting, conditioning, or prohibiting the activity that is the subject of the concession when this is intended to protect a legal asset of greater constitutional importance. This occurs, for example, when the State acquires new information that a medicine, a food, or the exploitation of certain construction material has harmful consequences for health. In such cases, the State can perfectly prohibit the production, transformation, distribution, and marketing of the respective product. With even more reason, it can then prohibit the extraction of a resource from its property whenever there is a reasonable doubt about the impact on legal assets subject to constitutional protection. Accepting a contrary position, limiting the activity of the Legislator and the government to the absence of concession contracts would be equivalent to subjecting legislative activity to private interests, stagnating the capacity of the system to respond to social changes, and reversing the constitutional rule of prevalence of the general interest [61].

Therefore, the fact that in the first hypothesis, the State has signed a concession contract with the individual does not prevent the Legislator from prohibiting the respective activity, even during the validity of the contract itself. Therefore, the conditions of the concession can be modified by the State, which is the owner of said assets, to protect constitutional assets, such as a healthy environment; and, second, an update and, with it, a periodic modification of the environmental precautionary measures can be expected, it is clear that the holder of a mining concession contract that is developed in an area of national ecological interest is not surprised with the limitation of the rights that the mining concession contract conferred, with the purpose of protecting the appropriate environment [61].

Although jurisprudence has recognized the power to limit the exercise of the rights derived from the mining concession, even after its subscription, multiple doubts persist about the procedure and the appropriate form to develop these actions, which is why it is necessary to go deeper in this subject.

4.6 Some elements for a comparative perspective

Now, the processes of judicialization of mining matters in which communities, ethnic groups and individuals use the judicial system to question the development of mining are found in different countries in Latin America. We can find similar rulings and cases in which the conditions for the development of mining activities were discussed in Guatemala [62], Ecuador [63], and Bolivia [64], there are also cases in Panama, Uruguay, Argentina, and other countries. Although some case studies have been done on these judicial decisions, a more in-depth comparative study is still required to understand the litigation processes against mining at the Latin American level and to understand the similarities and differences in the way in which the courts have addressed the discussions around this economic activity.

In any case, it should be noted that in the case of Latin America, most countries have incorporated rules related to the protection of the environment into their constitutions, and some have even recognized the existence of the rights of nature. It is precisely these regulatory contents that have been used to file judicial actions against mining activity, and others such as so-called climate litigation. Without a doubt, these regulatory contents have contributed to citizen empowerment.

5. Conclusion

Mining as an economic activity is today at an important debate. On the one hand, it is conceived as a necessary activity for the satisfaction of human needs, but on the other, its environmental and social impacts are increasingly being questioned. Which has given rise to the emergence of different environmental conflicts. Now, these conflicts have been increasingly judicialized based on the increasing demand from people and communities for compliance with the postulates of environmental protection that have been incorporated into the different legal systems.

From this perspective, the exercise carried out in this text allows us to highlight some important social demands in terms of environmental justice that involve mining activity.

The first is the need to review the democratic standards of decisions related to the development of mining projects. In fact, the judicial processes initiated in the Colombian case show at least two tensions: first, the level of government that must participate and make the decision (municipal, regional, or national) and second, the degree of involvement of the communities directly affected by extractive projects. In these two aspects, the identified jurisprudence seems to tend toward the construction of spaces for deliberative dialog with the inclusion of substantial rules so that such deliberation is possible. Hence, the requirements for carrying out consultations between local and national authorities on where and how to develop mining projects and the need to have the direct participation of the affected communities can be identified, but which are complemented with the definition of parameters for development of this participation. They involve access to information, deliberative rules of discussion and construction of agreements—which may even be different according to the type of population affected, with a tendency toward more demanding criteria when ethnic communities are affected. Which in any case suggests the need to review the decision-making procedures.

A second need that is identified in the judicial processes studied suggests the need to build increasingly comprehensive and systemic procedures for decision-making regarding the authorization of mining activities. Hence, jurisprudence develops at least two routes of action: the articulation of the mining, labor, environmental and social authorization processes for projects, the need to develop a greater capacity for inter-institutional work, which includes common instruments, the development of models of joint information management, and the existence of concurrent competencies that must be executed with a perspective of collaboration and flexibility.

A third need identified refers to the definition of the limits to mining development, which in the jurisprudence analyzed is specified in two major elements (i) the first related to places, in the sense of identifying the existence of ecosystems that require special protection against mining and in which exploitation can be definitively and even perpetually prohibited. In the Colombian case, this discussion has been especially relevant regarding the paramos due to their importance in ensuring the water supply, in that sense, the question of which places will definitely Mining activities to be developed or their development requires special conditions of restriction or limitation and what they are; (ii) in a second dimension, the discussion arises about the balance between environmental protection and investments, as to the question of whether the State can prohibit or restrict the development of mining projects that it had previously authorized justified by environmental protection, and If you can do it under what conditions to ensure due process and investor rights.

Although jurisprudence does not yet offer definitive answers to the dilemmas identified, it does allow us to identify routes of action that advocate for greater democratization and decentralization in the decision-making processes, greater emphasis on the environmental limits for the development of the activity and the need to develop methodologies for weighing the interests of environmental protection and the development of mining activities. Now, environmental law and the courts are currently beginning to develop criteria and principles that guide these exercises and that will surely be deepened in the coming years. For now, jurisprudence has managed to qualify and limit with a certain degree of success the excesses of the executive and impose limits to improve environmental protection, although the practical effectiveness of these decisions is a point yet to be established.

Author details

Luis Fernando Sánchez Supelano
National University of Colombia, Bogotá, Colombia

*Address all correspondence to: lfsanchezs@unal.edu.co

References

[1] Boza M, Montoya MF. Sector Minero en Colombia. Bogotá: Universidad Externado; 2016

[2] Contraloría General de la República. Minería en Colombia: Institucionalidad y Territorio, Paradojas y Conflictos. Bogotá: CGR; 2013. pp. 1-23

[3] Leguizamo AT, Ruíz JS. Impactos ambientales de la minería de carbón sobre el recurso hídrico en el Departamento de Boyacá. Boletín Semillas Ambientales. 2018;**12**(1):24-63

[4] Asociación Colombiana de Minería—ACM. Minería en Cifras. Bogotá: AMC; 2020

[5] Díaz E, Bastida MJ, Ladeuth Y. La Normatividad Ambiental Colombiana en la Minería Vista Desde Otra Perspectiva. Rioacha: Universidad de La Guajira; 2019

[6] Angarita GN, Gómez LM, Acevedo A. La Minería y su Impacto Económico y Ambiental: Desarrollo de los Pasivos Ambientales Mineros en Colombia. Bogotá: USTA; 2023. pp. 1-23

[7] UPME. Guía para la Incorporación de la Dimensión Minero Energética en los Planes de Ordenamiento Departamental. Bogotá: Unidad de Planeación Minero Energética—UPME; 2019

[8] CEPAL. La Minería en América Latina y el Caribe: Tendencias y Perspectivas. Santiago de Chile: CEPAL; 2023. Available from: https://www.cepal.org/sites/default/files/news/files/presentacioncepal_xiiicamma.pdf

[9] Bellmont Y. El Concepto de Justicia Ambiental, expresión de una idea contemporanea de la justicia Colombia. In: Mesa Cuadros G, editor. Elementos para una Teoría de la Justicia Ambiental y el Estado Ambiental de Derecho. Bogotá: Universidad Nacional de Colombia; 2013. pp. 89-130

[10] Crawford C. Derechos culturales y justicia ambiental. Lecciones del modelo colombiano. In: Maldonado B, editor. Justicia Colectiva, Medio Ambiente y Democracia Participativa. Bogotá: Universidad de los Andes; 2010. pp. 25-68

[11] Albin C, Druckman D. Negotiating effectively: Justice in international environmental negotiations. Group Decision and Negotiation. 2017;**26**(1):93-113

[12] Davies T, Mah A. Tackling environmental injustice in a post-truth age. In: Davies T, Mah A, editors. Toxics Truths. Manchester: Manchester University Press; 2020. pp. 1-27

[13] Klöepfer M. El derecho ambiental en alemania. Revista de Derecho Ambiental. 2013;**4**:15-34

[14] Carre M-N. Stores of metropolitan waste and environmental justice. Ateliers De L'etique-The Ethiscs Forum. 2016;**11**(1):101-122

[15] Dufaux F, Gervais-Lambony P. Justice... spatiale! Annales de Géographie, dossier "Justice spatiale". 2009;**665-666**:3-15

[16] Hollander-Blumoff R. Formation of procedural justice judgments in legal negotiation. Group Decision and Negotation. 2016;**26**:19-30

[17] Pasetto R, Iavarone I. Environmental justice in industrially contaminated sites: From the development of a national surveillance system to the

birth of an international network. In: Davies T, Mah A, editors. Toxic Truths. Manchester: Manchester University Press; 2020. pp. 199-219

[18] Simcock N. Procedural justice and the implementation of community wind energy projects: A case study from South Yorkshire, UK. Land Use Policy. 2016;**59**:467-477

[19] Young IM. Justice and the Politics of Difference. Princeton: Princeton University Press; 1990

[20] Bosselmann K, Engel R, Taylor P. Governance for Sustainability: Issues Challenges, Successes, Environmental Policy and Law Paper No. 70. Switzerland: IUCN; 2008

[21] Engel R. Introducción. In: Engel R, Westra L, Bosselmann K, editors. Democracy, Ecological Integrity and International Law. Cambridge: Cambridge Scholars Publishing; 2010

[22] Allen BL. Making effective participatory environmental health science through collaborative data analysis. In: Davies T, editor. Toxic Truths. Manchester: Manchester University Press; 2020. pp. 59-81

[23] Heinrichs J. A model of value-based democracy as condition of ecological sustainability. In: Engel R, Westra L, Bosselmann K, editors. Democracy, Ecological Integrity and International Law. Cambridge: Scholars Publishing; 2010. pp. 41-49

[24] Habermas J. The Theory of Communicative Action, Volume 2: Lifeworld and System: A Critique of Functionalist Reason. Boston: Beacon Press; 1985

[25] Elster J. Deliberative Democracy. Cambridge: Cambridge University Press; 1998

[26] Bello W. Alternative Security Systems in the Asian Pacific. Bangkok: Conference of Focus Asia; 1997. 27-30 March

[27] Held D. Democracy and the Global Order. Cambridge: Polity Press; 1995

[28] Falk R. Resisting 'globalization-from-above' through 'globalization-from-below'. In: Falk R, editor. Predatory Globalization: A Critique. Cambridge: Polity Press; 1999

[29] Morrison R. Ecological Democracy. Boston: South End Press; 1995

[30] Przeworski A. Sustainable Democracy. Cambridge: University Press; 1995

[31] Shiva VS. Manifiesto para una Democracia de la Tierra. Justicia, Sostenibilidad y Paz. Barcelona: Paidos; 2006

[32] Bruch C, Slobodian L, Nichols SS, Muffett C. Facilitating peace or fueling conflict? In: Bruch C, Muffett C, Nichols SS, editors. Post-Conflict Governance and Natural Resource Management, Governance, Natural Resources, and Post-Conflict Peacebuilding. London: Earthscan; 2016. pp. 953-1040

[33] Mesa Cuadros G. Derechos Ambientales en Perspectiva de Integralidad. Bogotá: Universidad Nacional de Colombia; 2013

[34] Rich E, Negbalee Warner T. Addressing the roots of Liberia's conflict through the extractive industries transparency initiative. In: Lujala P, Rustad SA, editors. High-Value Natural Resources and Peacebuilding. London: Earthscan; 2012. pp. 201-209

[35] López-Navarro MA. Legitimating confrontational discourses by local

environmental groups: The case of air quality monitoring in a Spanish industrial area. In: Thom D, Mah A, editors. Toxic Truths. Manchester: Manchester University Press; 2020. pp. 182-198

[36] Bruch C, Muffett C, Nichols SS. Local institutions and marginalized populations. In: Bruch C, Muffett C, Nichols SS, editors. Governance, Natural Resources, and Post-Conflict Peacebuilding. London: Earthscan; 2016. pp. 1-31

[37] Colleen G, Reed MG. Revealing inadvertent elitism in stakeholder models of environmental governance: Assessing procedural justice in sustainability organizations. Journal of Environmental Planning and Management. 2017;**60**:158-177

[38] Julie Gobert. Éthique environnementale, remédiation écologique et compensations territoriales: Entre antinomie et correspondances. VertigO - la revue électronique en sciences de l'environnement [En ligne], Volume 10 Numéro 1 | avril 2010, mis en ligne le 21 avril 2010, consulté le 18 avril 2024. DOI: 10.4000/vertigo.9535. Available from: http://journals.openedition.org/vertigo/9535

[39] Shrader-Fredchette K. Environmental Justice: Creating Equality, Reclaiming Democracy. Oxford: Oxford University Press; 2002

[40] Houtart F. De los bienes comunes al bien comun de la humanidad. In: Daiber B, Houtart F, editors. Un Paradigma Poscapitalista: El Bien Comun de la Humanidad. Habana: Ruth Casa Editorial; 2012. pp. 15-72

[41] Mesa Cuadros G. Elementos para una Teoría de la Justicia Ambiental Colombia. Bogotá: Universidad Nacional de Colombia; 2011

[42] Quiroga D. Sumak Kawsay hacia un nuevo pacto en armonia con la naturaleza. In: Acosta A, Martinez E, editors. El Buen Vivir: Una via para el Desarrollo. Quito: Abya Yala; 2009. pp. 105-120

[43] Corte Constitucional, Sentencia C-339 de 2002. Available from: https://www.corteconstitucional.gov.co/relatoría/2002/C-339-02.htm

[44] Consejo de Estado. Nulidad Simple Rad. 110010326000200500041 00 (30987). 2010. Available from: https://www.alcaldiabogota.gov.co/sisjur/normas/Norma1.jsp?i=40054

[45] Consejo de Estado. Nulidad Simple Rad. 11001-03-24-000-2005-00084-01(54552). 2018 Available from: https://www.suin-juriscol.gov.co/viewDocument.asp?id=30030557

[46] Corte Constitucional, Sentencia C-389 de 2016. Available from: https://www.corteconstitucional.gov.co/relatoría/2016/C-389-16.htm

[47] Corte Constitucional, Sentencia C-035 de 2016. Available from: https://www.corteconstitucional.gov.co/relatoría/2016/C-035-16.htm

[48] Corte Constitucional, Sentencia C-298 de 2016. Available from: https://www.corteconstitucional.gov.co/relatoría/2016/C-298-16.htm

[49] Consejo de Estado. Tutela 11001-03-15-000-2019-03352-01(AC). 2020. Available from: https://www.consejodeestado.gov.co/wp-content/uploads/2020/05/TutelaMineria.pdf

[50] Corte Constitucional, Sentencia C-395 de 2012. Available from: https://www.corteconstitucional.gov.co/relatoría/2012/C-395-12.htm

[51] Corte Constitucional, Sentencia C-123 de 2014. Available from: https://

www.corteconstitucional.gov.co/relatoría/2014/C-123-14.htm

[52] Corte Constitucional, Sentencia SU-95 de 2018. Available from: https://www.corteconstitucional.gov.co/relatoría/2018/T-095-18.htm

[53] Corte Constitucional, Sentencia C-891 de 2002. Available from: https://www.corteconstitucional.gov.co/relatoría/2002/C-891-02.htm

[54] Corte Constitucional, Sentencia SU-121 de 2022. Available from: https://www.corteconstitucional.gov.co/relatoría/2020/SU-121-22.htm

[55] Corte Constitucional, Sentencia SU-133 de 2017. Available from: https://www.corteconstitucional.gov.co/relatoría/2016/SU-133-17.htm

[56] Corte Constitucional, Sentencia C-221 de 2016. Available from: https://www.corteconstitucional.gov.co/relatoría/2016/C-221-16.htm

[57] Corte Constitucional, Sentencia SU-698 de 2017. Available from: https://www.corteconstitucional.gov.co/relatoría/2017/SU-698-16.htm

[58] Corte Constitucional, Sentencia T-341 de 2016. Available from: https://www.corteconstitucional.gov.co/relatoría/2016/T-341-16.htm

[59] Corte Constitucional, Sentencia T-733 de 2017. Available from: https://www.corteconstitucional.gov.co/relatoría/2017/T-733-17.htm

[60] Consejo de Estado. Acción Popular Rad. 15001233300020140022302. 2018. Available from: https://www.consejodeestado.gov.co/documentos/boletines/235/73001-23-31-000-2011-00611-03.pdf

[61] Consejo de Estado. Nulidad y Restablecimiento del Derecho Rad. 11001-03-26-000-2005-00032-00 (30567). 2018. Available from: https://xperta.legis.co/visor/jurcol/jurcol_a6ba7f69f18d44b0a436f9b7f611c172

[62] Corte Constitucional de Guatemala. Sentencia 452/2019, 7 de Noviembre. 2019. Available from: https://bit.ly/3Bu4Dbc

[63] Corte Constitucional del Ecuador. Sentencia No. 1149-19-JP/21, 2021. 2021. Available from: http://esacc.corteconstitucional.gob.ec/storage/api/v1/10_DWL_FL/e2NhcnBldGE6J3RyYW1pdGUnLCB1dWlkOic2MmE3MmIxNy1hMzE4LTQyZmMtYjJkOS1mYzYzNWE5ZTAwNGYucGRmJ30=

[64] Tribunal Constitucional Sentencia Constitucional Plurinacional 1471/2022. Available from: https://www.noticiasfides.com/docs/news/2023/06/1686937814_SENTENCIA1471_2022-S3.pdf